D1426695

REMOTE SENSING OF COASTAL ENVIRONMENTS

Taylor & Francis Series in Remote Sensing Applications

Series Editor

Qihao Weng

Indiana State University
Terre Haute, Indiana, U.S.A.

Remote Sensing of Coastal Environments, *edited by Yeqiao Wang*

Remote Sensing of Global Croplands for Food Security, *edited by Prasad S. Thenkabail, John G. Lyon, Hugh Turral, and Chandashekhar M. Biradar*

Global Mapping of Human Settlement: Experiences, Data Sets, and Prospects, *edited by Paolo Gamba and Martin Herold*

Hyperspectral Remote Sensing: Principles and Applications, *Marcus Borengasser, William S. Hungate, and Russell Watkins*

Remote Sensing of Impervious Surfaces, *Qihao Weng*

Multispectral Image Analysis Using the Object-Oriented Paradigm, *Kumar Navulur*

REMOTE SENSING OF COASTAL ENVIRONMENTS

Edited by

YEQIAO WANG

CRC Press is an imprint of the
Taylor & Francis Group, an **informa** business

CRC Press
Taylor & Francis Group
6000 Broken Sound Parkway NW, Suite 300
Boca Raton, FL 33487-2742

Printed in the United States of America on acid-free paper
10 9 8 7 6 5 4 3 2 1

International Standard Book Number: 978-1-4200-9441-1 (Hardback)

Library of Congress Cataloging-in-Publication Data

Remote sensing of coastal environments / editor, Yeqiao Wang.
p. cm. -- (Taylor & Francis series in remote sensing applications)
"A CRC title."
Includes bibliographical references and index.
ISBN 978-1-4200-9441-1 (hardcover : alk. paper)
1. Coastal ecology--Remote sensing. 2. Coats--Remote sensing. 3. Coastal zone management--Remote sensing. 4. Environmental monitoring--Remote sensing. I. Wang, Yeqiao. II. Title. III. Series.

QH541.15.R4R45 2010
577.5'10284--dc22 2009035957

Visit the Taylor & Francis Web site at
http://www.taylorandfrancis.com

and the CRC Press Web site at
http://www.crcpress.com

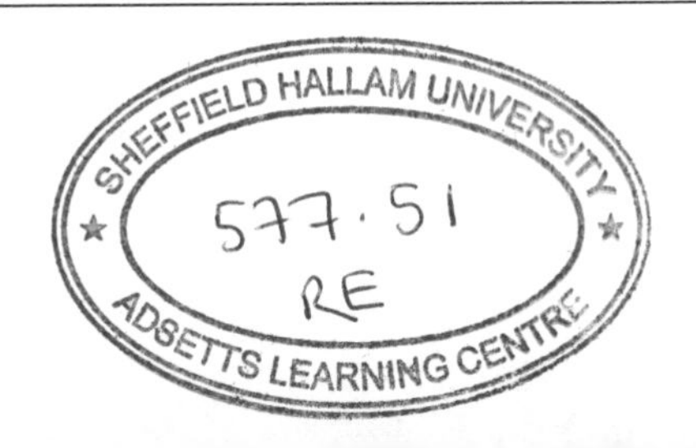

Contents

SECTION V: Effects of Land-Use/Land-Cover Change in Coastal Areas

Series Foreword

Remote sensing refers to the technology of acquiring information about the Earth's surface (land and ocean) and atmosphere using sensors onboard airborne (aircraft and balloons) or spaceborne (satellites and space shuttles) platforms. The technology of remote sensing gradually evolved into a scientific subject after World War II. Its early development was mainly driven by military uses. Later, remotely sensed data became widely applied for civic usages. The range of remote sensing applications includes archeology, agriculture, cartography, civil engineering, meteorology and climatology, coastal studies, emergency response, forestry, geology, geographic information systems, hazards, land use and land cover, natural disasters, oceanography, water resources, and so on. Most recently, with the advent of high spatial-resolution imagery and more capable techniques, commercial applications of remote sensing are rapidly gaining interest in the remote sensing community and beyond.

The Taylor & Francis Series in Remote Sensing Applications is dedicated to recent developments in the theories, methods, and applications of remote sensing. Written by a team of leading authorities, each book is designed to provide up-to-date developments in a chosen subfield of remote sensing applications. Each book may vary in format, but often contains similar components, such as a review of theories and methods, analysis of case studies, and examination of the methods for applying remote sensing techniques to a particular practical area. These books may serve as guides or reference books for professionals, researchers, scientists, and similarly in academics, governments, and industries. College instructors and students may also find them to be excellent sources for textbooks or supplementary to their chosen textbooks.

A coastal environment possesses one of the most dynamic interfaces between human civilization and environmental conservation. In the United States, over half of the human population lives in coastal counties, while worldwide over 38% of the human population lives in the coastal zone. Global climate change has made the coastal zone the most challenging frontier in environmental planning and management, because many of the coastal zones face the danger of being submerged. Remote sensing is one of the most effective technologies for monitoring coastal environments and for assessing their conditions.

Professor Yeqiao Wang is an internationally known expert in the field of remote sensing, especially in applications of this technology to coastal environments. Because of his outstanding achievements, Professor Wang was awarded the prestigious Presidential Early Career Award for Scientists and Engineers (PECASE) by President Clinton in 2000 and a Chang-Jiang endowed professorship by the Ministry of Education of China. In this book, *Remote Sensing of Coastal Environment*, Professor Wang examines three new research frontiers in coastal remote sensing, that is, LiDAR/radar, hyperspectral, and high spatial-resolution sensing. Furthermore, this book analyzes methods for mapping habitats by the integration of remote sensing and *in situ* measurements and investigates the effects of land-use and land-cover change on the coastal environment. Professor Wang has assembled an excellent team of contributors, most of whom are well-respected researchers in the field of remote sensing in universities and government.

I hope that the publication of this book will promote a better use of remote sensing data, science, and technology and will facilitate the monitoring and assessment of global environment and sustaining our common home—the Earth.

Qihao Weng, Ph.D.
NASA, Huntsville, Alabama

Preface

This is an important book that fills a critical niche. We are living in an unprecedented time of change in the condition of our coastal environments. We are also living in a period of exceptionally rapid advancement in technologies to support geospatial data acquisition, imaging, and computing. This volume brings together the world's expert scientists and the best available data to measure and monitor coastal environments. The volume also demonstrates how decision-makers and resource managers are using these data to address complex issues in coastal zone management. A number of over-arching themes establish a context within which the chapters in this volume should be read.

Coastal environments are complex: A large portion of the world's population lives near coasts. Coasts are a vital geographic region in terms of transportation, commerce, and trade. Coasts are dynamic and complex; for example, the quality of our coastal marine ecosystems is largely a result of human activity on land. Point and nonpoint pollution, sedimentation, and changes in freshwater flow all have profound impacts on marine ecosystems. As a number of the chapters in this book attest, land-use changes in coastal watersheds are an important driver of marine ecosystem condition. Thus, our ability to measure existing land-use changes and to model future land-use changes has immediate value to coastal resource managers. The sensors, data, and technologies that allow us to map land-use changes are themselves changing quite rapidly and this book brings together the state-of-the-art in this technically complex arena.

Climate change impacts will be profound in coastal environments: The debate is over; the question is not whether climate change is happening or not, the questions are how much and in what form climate change will manifest itself. Coastal ecosystems will be impacted in many ways. Sea level rise will be an obvious impact, and the inundation modeling that LiDAR data permit are critical for coastal zone managers. These models tell us where we have vulnerable infrastructure, habitats at risk, and potential dispersal corridors for salt marshes and other coastal habitats that will have to migrate inland or drown. There are indications that high-intensity storms will be more frequent under conditions of climate change. Again, accurate terrain data, high-resolution imagery, and detailed land-use data are essential to understand what is at risk under conditions of a 5–7 m storm surge. There is no doubt that coastal submerged, tidal, and terrestrial habitats will see significant changes as near-shore waters warm, freshwater flows become irregular, patterns of coastal erosion and accretion are altered, and coastal waters become more acidic. Within the lifetime of our current students, we anticipate significant negative impacts on the growth and calcification rates of coral reefs, the continuation of a global trend for increasing numbers and areal extent of harmful algal blooms, and other threats to coastal water quality and to coastal marine ecosystems. The chapters in this volume bring together the best experts to discuss how remote sensing data and technology are brought to bear on managing our coastal ecosystems in the context of climate change.

All scales—temporal and spatial—are relevant in coastal ecosystems: Patterns and processes that drive coastal ecosystems occur rapidly at local scales as well as gradually at broad scales. It is important that we be able to map and monitor patterns and processes at the time intervals and spatial resolutions that changes require. Topics addressed in this volume such as salt marsh dieback, breaching of barrier beaches, annual shifts in the extent of submerged aquatic vegetation, and increases in impervious surfaces in coastal watersheds require constant vigilance at fine scales by coastal managers. Remote sensing technology is the only practical way to map and monitor these phenomena. Furthermore, regional changes in sea surface temperature, ocean productivity, and patterns of sediment flux occur over broad scales, and remote sensing data and methods are the most efficient way to track changes over large areas. Our ability to process the massive datasets that are required to monitor large areas of coastal ecosystems or dense datasets where pixel sizes are expressed in centimeters is becoming less and less of an issue.

Remote sensing of the coastal waters is a challenge: The coastal environment is the boundary between the land and the ocean, and remote sensing of the coastal environment straddles the technical boundary of land and ocean satellite remote sensing. Satellite radiometers optimized for remote sensing of terrestrial environments, as well as those for the open ocean, are not well suited for quantitative remote sensing of phytoplankton biomass, sediment, and other constituents of coastal waters. Remote sensing of coastal waters requires high spatial resolution to resolve characteristically small features, as well as daily or higher frequency coverage to understand many of the important coastal processes of these dynamic regions. The ideal satellite radiometer, for measurements of coastal waters, would, for example, have Landsat spatial resolution or, better, revisit times comparable to wide-swath instruments such as MODIS-Aqua, and many narrow spectral bands with high signal-to-noise ratios. The MERIS instrument on ESA's ENVISAT is an important step in this direction, but 300 m pixel resolution is still too coarse to resolve many of the important fine-grain features of coastal waters. The future promises many exciting advances in new and improved sensors for use by researchers and resource managers as well as improved geoprocessing tools and computing technology to study coastal ecosystems. This important volume by Wang and his colleagues establishes the current state-of-the-art in coastal remote sensing.

Peter August
University of Rhode Island Coastal Institute
James Yoder
Woods Hole Oceanographic Institution

Editor

Dr. Yeqiao Wang is a professor at the Department of Natural Resources Science, University of Rhode Island. He received his BS degree from the Northeast Normal University, China, in 1982 and his MS degree in remote sensing and mapping from the Northeast Institute of Geography and Agroecology of the Chinese Academy of Sciences in 1987. He received his MS and PhD degrees in natural resources management and engineering from the University of Connecticut in 1992 and 1995, respectively. From 1995 to 1999, he held the position of assistant professor in geographic information systems (GIS) and remote sensing in the Department of Geography/Department of Anthropology, University of Illinois at Chicago. He has been on the faculty of the University of Rhode Island (URI) since 1999. Besides his tenured position at URI, he held an adjunct research associate position at the Field Museum of Natural History in Chicago between 1998 and 2003. He has been named a Chang-Jiang (*Yangtze River*) scholar lecturing professor at the Northeast Normal University since 2006.

Among his awards and recognitions Dr. Wang was a recipient of the prestigious Presidential Early Career Award for Scientists and Engineers (PECASE) by President William J. Clinton in 2000. The PECASE Award is the highest honor bestowed by the U.S. government on outstanding scientists and engineers beginning their independent careers. He was among the first-place winners of the ESRI Award for Best Scientific Paper in Geographic Information Systems in 2002 by the American Society for Photogrammetry and Remote Sensing. He received recognitions for the PECASE Award at the NASA headquarters in 2000; the Outstanding Contributions to Research by the University of Rhode Island in 2003; and Research Scientist Excellence Award by the College of the Environment and Life Science, University of Rhode Island in 2008.

His specialties are terrestrial remote sensing and modeling in natural resources analysis and mapping. Dr. Wang's particular area of interest is remote sensing of the dynamics of landscape and land-cover/land-use change. He has published over 50 peer-reviewed journal articles and 70 abstracts and conference papers, and contributed over 20 peer-reviewed book chapters. In addition to his professional publications in English, he also has authored several science books in Chinese. His research projects have been funded by NASA, USDA, USDI, USAID, among others, which supported his scientific studies in the Northeast and Midwest of the United States, East Africa, and Northeast China.

Editor

Contributors

Eric R. Akins
Department of Natural Resources Science
University of Rhode Island
Kingston, Rhode Island

Chester L. Arnold
Department of Extension
Center for Land Use Education and Research
University of Connecticut
Storrs, Connecticut

Francisco J. Artigas
Meadowlands Environmental Research Institute
Lyndhurst, New Jersey

Peter V. August
Department of Natural Resources Science
University of Rhode Island
Kingston, Rhode Island

Nels Barrett
Natural Resources Conservation Service
U.S. Department of Agriculture
Tolland, Connecticut

Gregory Bonynge
Department of Natural Resources Science
University of Rhode Island
Kingston, Rhode Island

Cary Chadwick
Center for Land Use Education and Research
University of Connecticut
Storrs, Connecticut

Jingsong Chen
Institute of Space and Earth Information Science
The Chinese University of Hong Kong
Shatin, NT, Hong Kong

Mark Christiano
National Park Service—Gateway National Recreation Area
Fort Wadsworth
Staten Island, New York

Daniel L. Civco
Department of Natural Resources Management and Engineering
University of Connecticut
Storrs, Connecticut

Dongsheng Du
Department of Atmospheric Sciences
Sun Yat-sen University
Guang Zhou, China

Lola Fatoyinbo
Jet Propulsion Laboratory
Pasadena, California

Martha S. Gilmore
Department of Earth and Environmental Sciences
Wesleyan University
Middletown, Connecticut

Arthur J. Gold
Department of Natural Resources Science
University of Rhode Island
Kingston, Rhode Island

Cheryl J. Hapke
U.S. Geological Survey
Woods Hole, Massachusetts

James D. Hurd
Department of Natural Resources Management and Engineering
University of Connecticut
Storrs, Connecticut

Ejaz Hussain
School of Civil Engineering
Purdue University
West Lafayette, Indiana

Darryl J. Keith
Atlantic Ecology Division
National Health and Ecological Effects Research Laboratory
U.S. Environmental Protection Agency
Narragansett, Rhode Island

Oh-Ig Kwoun
Jet Propulsion Laboratory
California Institute of Technology
Pasadena, California

Hui Lin
Institute of Space and Earth Information Science
The Chinese University of Hong Kong
Shatin, NT, Hong Kong

Wenshi Lin
Department of Atmospheric Sciences
Sun Yat-sen University
Guang Zhou, China

Zhong Lu
EROS Center and Cascades Volcano Observatory
U.S. Geological Survey
Vancouver, Washington

Vedast Makota
National Environment Management Council
Dar es Salaam, Tanzania

Amani Ngusaru
National Environment Management Council
Dar es Salaam, Tanzania

Jarunee Nugranad
Department of Natural Resources Science
University of Rhode Island
Kingston, Rhode Island

Naiara Pinto
Jet Propulsion Laboratory
California Institute of Technology
Pasadena, California

Sandy Prisloe
Department of Extension
Center for Land use Education and Research
University of Connecticut
Haddam, Connecticut

Elijah W. Ramsey III
USGS National Wetland Research Center
Lafayette, Louisiana

Amina Rangoonwala
IAP World Services, Inc.
Lafayette, Louisiana

Huiyong Sang
Institute of Space and Earth Information Science
The Chinese University of Hong Kong
Shatin, NT, Hong Kong

Courtney Schupp
Assateague Island National Seashore
Berlin, Maryland

Jie Shan
School of Civil Engineering
Purdue University
West Lafayette, Indiana

Marc Simard
Jet Propulsion Laboratory
California Institute of Technology
Pasadena, California

Sara Stevens
Northeast Coastal and Barrier Network
National Park Service
Kingston, Rhode Island

Tung-Ching Su
Department of Civil Engineering
National Chung Hsing University
Taichung, Taiwan

James Tobey
Coastal Resource Center
University of Rhode Island
Narragansett, Rhode Island

Michael Traber
McDonald Dettwiler and Associates Federal Inc.
Rockville, Maryland

Yeqiao Wang
Department of Natural Resources Science
University of Rhode Island
Kingston, Rhode Island

Emily H. Wilson
Department of Extension
Center for Land use Education and Research
University of Connecticut
Haddam, Connecticut

An-Ming Wu
Division of Systems Engineering
National Space Organization
Hsinchu, Taiwan

Jiansheng Yang
Department of Geography
Ball State University
Muncie, Indiana

Limin Yang
USGS Earth Resources Observation Systems and Science
Sioux Falls, South Dakota

Ming-Der Yang
Department of Civil Engineering
National Chung Hsing University
Taichung, Taiwan

Lu Zhang
Institute of Space and Earth Information Science
The Chinese University of Hong Kong
Shatin, NT, Hong Kong

and

State Key Laboratory for Information Engineering in Surveying, Mapping and Remote Sensing
Wuhan University
Wuhan, China

Guoqing Zhou
Department of Civil Engineering and Technology
Old Dominion University
Norfolk, Virginia

Yuyu Zhou
Department of Earth and Atmospheric Sciences
Purdue University
West Lafayette, Indiana

1 Remote Sensing of Coastal Environments: An Overview

Yeqiao Wang

CONTENTS

1.1 INTRODUCTION

Coastal zone, as defined by the Coastal Institute of the University of Rhode Island, includes areas of continental shelves, islands, or partially enclosed seas, estuaries, bays, lagoons, beaches, and terrestrial and aquatic ecosystems within watersheds that drain into coastal waters. The coastal zone is the most dynamic interface between land and sea and represents the most challenging frontier between human civilization and environmental conservation. Worldwide, over 38% of human population lives in the coastal zones (Crossett et al., 2004). In the United States, about 53% of the human population lives in the coastal counties (Small and Cohen, 2004). An increasing proportion of the global population lives within the coastal zones of all major continents that require increasing attention to agricultural, industrial, and other human-related effects on coastal habitats and water quality and their impacts on ecological dynamics, ecosystem health, and biological diversity.

Coastal environments contain a wide range of natural habitats such as sand dunes, barrier islands, tidal wetlands and marshes, mangrove forests, coral reefs, and submerged aquatic vegetation that provide foods, shelters, and breeding grounds for terrestrial and marine species. Coastal habitats also provide irreplaceable services such as filtering pollutants and retaining nutrients, maintaining water quality, protecting shoreline, and absorbing flood waters. As coastal habitats are facing intensified natural and anthropogenic disturbances by direct impacts such as hurricane, tsunami, harmful algae bloom, and cumulative and secondary impacts such as climate change, sea level rise, oil spill, and urban development, inventory and monitoring of coastal environments become one of the most challenging tasks of the society in resource management and humanity administration. Remote sensing science and technologies that involve space-borne and airborne sensor systems in data acquisition and observation have profoundly changed the practice in monitoring and understanding of the dynamics of coastal environments.

Coarser spatial resolution remote sensing data have been used for broader scale coastal studies. For example, high concentrations of suspended particulate matter in coastal waters directly affect water column and benthic processes such as phytoplankton productivity (Cole and Cloern, 1987; Cloern, 1987), coral growth (Dodge et al., 1974; Miller and Cruise, 1995; Torres and Morelock, 2002; McLaughlin et al., 2003), productivity of submerged aquatic vegetation (Dennison et al., 1993), nutrient dynamics (Mayer et al., 1998), and the transport of pollutants (Martin and Windom, 1991). Although there has been considerable effort toward using remotely sensed images to provide synoptic maps of suspended particulate matter, there are limited routine applications of this technology due in part to the low spatial resolution and long revisit period. Miller and McKee (2004) examined the utility of moderate-resolution imaging spectroradiometer (MODIS) 250 m data, with integration of *in situ* measurements, for analyzing complex coastal waters in the Northern Gulf of Mexico, and mapped the concentration of total suspended matter. The study demonstrates that MODIS near daily coverage of medium-resolution data is useful for examining the transport and fate of materials in coastal environments, particularly smaller bodies of water such as bays and estuaries.

The Sea-viewing Wide Field-of-view Sensor (SeaWiFS) system acquires radiance data from the Earth in eight spectral bands with a maximum spatial resolution of 1 km at nadir. The Global 9 km^2 spatial resolution Level 3 data provide daily, 8-day, monthly, and annual standard data products. Acker et al. (2008) employed monthly chlorophyll *a* data from the 8-year SeaWiFS mission and data from MODIS to analyze the spatial pattern of chlorophyll concentrations and seasonal cycle. The data indicate that large coral reef complexes may be sources of either nutrients or chlorophyll-rich detritus and sediment, enhancing chlorophyll *a* concentration in waters adjacent to the reefs. Lohrenz et al. (2008) reported a retrospective analysis of nutrients and phytoplankton productivity in the Mississippi River plume, in which long-term patterns in riverine nutrient flux in the lower Mississippi River were examined in relationship to spatial and temporal patterns in surface nutrient concentrations, chlorophyll, and primary productivity.

AVHRR data have long been used in the study of coastal waters (Froidefond et al., 1993, 1996). Now the *CoastWatch* Program at the National Oceanic and Atmospheric Administration (NOAA) produces multiple regional daily daytime sea surface temperature images for the United States derived from AVHRR (Ferguson et al., 2006). NASA's remote sensing assets have been used to address coastal issues, such as land-use and land-cover change (LCLUC) detection, harmful algal bloom forecasting, regional sediment management, monitoring coastal forest conditions, and coastal line assessment and synthesis (Peek et al., 2008).

Landsat type of remote sensing data has been used in coastal applications for decades (Munday and Alfoldi, 1979; Bukata et al., 1988; Ritchie et al., 1990). The multispectral capabilities of the data allow observation and measurement of biophysical characteristics of coastal habitats (Colwell, 1983), and the multitemporal capabilities allow tracking of changes in these characteristics over time (Wang and Moskovits, 2001). Landsat and SPOT imageries have been applied in inventory mapping, change detection, and management of mangrove forests through visual interpretation (Gang and Agatsiva, 1992), vegetation index (Blasco et al., 1986; Jensen et al., 1991), classification (Dutrieux et al., 1990; Aschbacher et al., 1995), band ratio (Kay et al., 1991), and to map seagrass coverage (Armstrong, 1993; Ferguson and Korfmacher, 1997; Mumby et al., 1997; Green et al., 2000; Moore et al., 2000; Lathrop et al., 2006), among others (Gao, 1998; Green et al., 1998; Rasolofoharinoro et al., 1998; Pasqualini et al., 1999). As archived Landsat images have been made available at no cost to user communities since early 2009 (Woodcock et al., 2008), coastal applications can take advantage of this type of data.

Remote sensing has been identified as one of the primary data sources to produce land-cover maps that indicate landscape patterns and human development processes (Turner, 1990; Coppin and Baure, 1996; Hansen and Rotella, 2002; Griffith et al., 2003; Rogan et al., 2002; Turner et al., 2003; Wilson and Sader, 2002; Wang et al., 2003). Development of standardized and regional land-cover information is in demand for enabling resource managers to coordinate the planning of shared resources, facilitating an ecosystem approach to coastal environmental issues. The NOAA's Coastal Change Analysis Program (C-CAP), for example, is a nationally standardized database of LCLUC information for the coastal regions of the United States. Developed using Landsat remote sensing imagery, C-CAP products inventory coastal intertidal areas, wetlands, and adjacent uplands with the goal of monitoring those habitats by updating the land-cover maps every 5 years. The C-CAP protocol recommended that shorter time periods may be necessary in regions undergoing rapid economic development or affected by catastrophic events (Dobson et al., 1995).

Although Landsat and SPOT types of multispectral and medium spatial resolution remote sensing data have been broadly applied to address a variety of coastal issues, the complexity of coastal constituents imposes significant challenges in application of remote sensing technologies due to the dynamic spatial and temporal nature of coastal habitats. Recent development in active remote sensing, such as LiDAR (light detection and ranging) and interferometric synthetic aperture radar (InSAR) technologies, and in optical remote sensing, such as hyperspectral and high spatial resolution sensors, bring new types of data and enhanced capacities in the study of coastal environment. LiDAR and InSAR are very effective in coastal studies from three-dimensional (3D) habitat mapping to morphological change analysis. Hyperspectral remote sensing employs hundreds of narrow and contiguous spectral bandwidths in data collection to enhance the capacity in identification of coastal habitats. Space-borne and airborne high spatial resolution remote sensing data, in meter and submeter levels, provide refined spatial details for mapping the coastal landscape and seascape. Remote sensing is well known for its unique role in the study of dynamics of coastal landscape and land-cover/land-use change caused by natural and anthropogenic forces. On the other hand, field survey and *in situ* observations are essential to identify coastal habitats through remote sensing. Almost every remote sensing exercise will require field survey to define habitats, to calibrate remote sensing imagery, and to evaluate the accuracy of remote sensing outputs. With GPS-guided positioning and field spectral measurements becoming routine operation, challenges remain for incorporation of *in situ* measurements with remote sensing observations for quantitative analyses of coastal habitats.

To reflect the most recent development among those mentioned topics, this book is organized into five sections, including applications of active LiDAR and InSAR systems, space-borne and airborne hyperspectral data, high spatial resolution remote sensing data, integration of *in situ* measurements and remote sensing observations, and effects of LCLUC to showcase remote sensing of coastal environments.

1.2 ACTIVE REMOTE SENSING

Part I of this book consists of four chapters under the category of active remote sensing. Active remote sensors are radars and LiDAR systems that transmit electromagnetic pulses with a specific wavelength (λ) and measure the time of return (translated to distance) of the signal reflected from a target. LiDAR systems generally transmit pulses at a wavelength around the visible domain ($\lambda \sim$ nanometers), whereas radar pulses are in the microwave domain ($\lambda \sim$ centimeters) (Simard et al., Chapter 3 of this book). LiDAR and InSAR are among the new developments in active remote sensing that are the topics that have attracted particular interest in coastal-related studies.

Active radar sensors are well known for their all-weather and day-and-night imaging capabilities, which are effective for mapping coastal habitats over cloud-prone tropical and subtropical regions. The SAR backscattering signal is composed of intensity and phase components. The intensity component of the signal is sensitive to terrain slope, surface roughness, and dielectric constant. Studies have demonstrated that SAR intensity images can map and monitor forested and nonforested wetlands occupying a range of coastal and inland settings (e.g., Ramsey III, 1995, 1999; Ramsey III et al., 2006). SAR intensity data have been used to monitor floods and dry conditions, temporal variations in the hydrological conditions of wetlands, including classification of wetland vegetation at various geographic settings (Hess and Melack, 1994; Hess et al., 1995; Kasischke and Bourgeau-Chavez, 1997; Le Toan et al., 1997; Baghdadi et al., 2001; Bourgeau-Chavez et al., 2001, 2005; Costa et al., 2002; Costa, 2004; Simard et al., 2002; Townsend, 2002; Kiage et al., 2005; Grings et al., 2006). When the phase components of two SAR images of the same area acquired from similar vantage points at different times are combined through InSAR processing, an interferogram can be constructed to depict range changes between the radar and the ground, and can be further processed with a digital elevation model (DEM) to produce an image with centimeter to subcentimeter vertical precision under favorable conditions (e.g., Massonnet and Feigl, 1998; Rosen et al., 2000). InSAR has been extensively utilized to study ground surface deformation associated with volcanic, earthquake, landslide, and land subsidence processes. Alsdorf et al. (2000, 2001) found that interferometric analysis of L-band (wavelength = 24 cm), Shuttle Imaging Radar-C (SIR-C) and Japanese Earth Resources Satellite (JERS-1) SAR imagery can yield centimeter-scale measurements of water-level changes throughout inundated floodplain vegetation. Wdowinski et al. (2004) applied L-band JERS-1 images to map water-level changes over the Everglades in Florida.

To showcase InSAR applications in coastal environment, Chapter 2 (Lu and Kwoun) introduces a study that employed multitemporal C-band European Remote Sensing (ERS)-1/-2 satellites and Canadian Radar Satellite (RADARSAT)-1 SAR data over the Louisiana coastal zone to analyze water-level changes of coastal wetlands. The study concluded that radar imagery can provide unique information to characterize coastal wetlands and infer wave-level changes.

LiDAR technology includes small- and large-footprint laser scanners (Hudak et al., 2002; Lefsky et al., 2002). LiDAR has shown promising results for assessing tree heights, forest biomass, and volume (Nelson et al., 1988; Means et al., 2000; Morsdorf et al., 2004), in particular the coastal habitats (Brock et al., 2001, 2002; Gillespie et al., 2001; Nayegandhl et al., 2006). Due to the rapid laser firing, LiDAR pulses can penetrate vegetation cover, which makes LiDAR technology well suited to measure topography in coastal salt marsh areas (Ackermann, 1999; Montane and Torres, 2006). Airborne LiDAR data have been tested for identifying coral colonies on patch reefs (Brock et al., 2006). A recent study compared the capability and accuracy of LiDAR and InSAR for the detection and measurement of individual tree heights and forest plot heights (Huang et al., 2009).

Chapter 3 (Simard et al.) discusses recent advancements in active remote sensing that enable mapping of forest canopy structure. In particular it explores 3D modeling of mangrove forests by Shuttle Radar Topography Mission (SRTM) data. Although the SRTM was designed to produce a global DEM of the Earth's surface, the SRTM InSAR measurement of height z is the sum of the ground elevation and the canopy height contribution. Therefore SRTM DEM can be used to measure vegetation height as well as ground topography (Simard et al., 2006).

Chapter 4 (Hapke) examines two case studies that integrate LiDAR data, digital historical maps, and orthophotos to measure long-term coastal change rates, to map erosion hazard areas, and to understand coastal processes on a variety of timescales from storms to seasons to decades and longer. The two case studies, one in barrier islands along the U.S. Gulf coast and one in rocky coastal environments of the U.S. West coast, represent how similar types of data can be used to map coastal hazards in a variety of geographic settings and at a variety of spatial scales. The first case examined the impacts of hurricanes on shoreline change rates in a barrier island environment. It identified areas within a National Seashore that could be more vulnerable to future storms than indicated by the long-term record. The second case applied the data to a statewide analysis of coastal cliff retreat in California. The approaches and techniques utilized in both the case studies are highly applicable to other coastal regions. Datasets that have commonly been used for coastal zone assessments can be integrated with newer data sources to modernize and update analyses. Such assessments will continue to be critical to coastal management and planning, especially with the currently predicted rates of sea level rise through the twenty-first century (Hapke, Chapter 4 of this book).

Chapter 5 (Zhou) presents a coastal morphological change analysis for the Assateague Island National Seashore using time series LiDAR data acquired in October 1996, September 1997, February and December 1998, and September and November 2000. The result demonstrates that LiDAR sensors can provide an extraordinary data capturing capability for quantitative analysis of coastal topographic morphology.

1.3 HYPERSPECTRAL REMOTE SENSING

Part II of this book consists of three chapters under the category of hyperspectral remote sensing. Hyperspectral technology brings new insights about remote sensing of the coastal environments. Hyperion sensor on board of EO-1 satellite, launched in November 2000, is a representative space-borne hyperspectral system. This grating imaging spectrometer collects data in 30-m ground sample distance over a 7.5-km swath and provides 10-nm (sampling interval) contiguous 220 spectral bands of the solar reflected spectrum from 400 to 2500 nm. Airborne visible infrared imaging spectrometer (AVIRIS) is a representative airborne hyperspectral sensor system. The AVIRIS whiskbroom scanner collects data in the same spectral interval and range as the Hyperion system but with 20 m spatial resolution. Hyperspectral remote sensing has advantages in coastal wetlands characterization due to its large number of narrow, contiguous spectral bands as well as high horizontal resolution from airborne platforms. It has been used to map habitat heterogeneity (Artigas and Yang, 2004), to determine plant cover distribution in salt marshes (Li et al., 2005; Belluco et al., 2006), to map spread of invasive plant species (Rosso et al., 2006), and to map spatial pattern of tree species abundances (Anderson et al., 2007). Hyperspectral images have been used to separate vigor types by detecting slight differences in coloration due to stress factors, infestation, or displacement by invading species (Artigas and Yang, 2005). Expert system approach has been tested for processing hyperspectral remote sensing data for salt marsh mapping (Schmidt et al., 2004).

Chapter 6 by Ramsey III and Rangoonwala introduces a study about mapping the onset and progression of marsh dieback in coastal Louisiana. The authors developed spectral methods to time the onset and monitor the progression of coastal *Spartina alterniflora* marsh dieback by using hyperspectral image data at the plant-leaf, canopy, and satellite levels without a priori information on where, when, or how long the dieback had proceeded. The research plan produced three interlinked products. First, at the plant-leaf level, indicators were developed to identify optical indicators of marsh dieback above natural variability. Indicators were based on blue, green, red, and near-infrared (NIR) spectral bands and NIR/green and NIR/red band transforms derived from leaf spectra obtained from plant samples along dieback transects. Second, these indicators were linked to the marsh canopy reflectance and used to estimate dieback onset and progression at each site. Broadband satellite sensors (e.g., Landsat Thematic Mapper) were simulated with the same spectral bands and band transforms used in the leaf spectral analyses but extracted from the canopy reflectance spectra. High spectral resolution satellite sensors (e.g., EO-1 Hyperion sensor) were simulated by creating and applying characteristic spectra derived from the whole canopy reflectance spectra. Third, dieback indicators developed at the plant-leaf level and applied and tested at the canopy level were used to extract temporal patterns from a suite of Landsat Thematic Mapper images as indicators of dieback occurrence. Application of the satellite sensor level provided a platform more conducive to regional and strategic monitoring. Effectively, broadband sensors like the Landsat Thematic Mapper can map broad divisions of impacted marsh and may provide some discrimination of dieback progression, while high spectral resolution sensors like the EO-1 Hyperion offer an enhanced ability to determine dieback onset and track progression (Ramsey III and Rangoonwala, Chapter 6 of this book).

Increased population and urban development have contributed significantly to environmental pressures along many areas of the U.S. coastal zone. The pressures have resulted in substantial physical changes to beaches, loss of coastal wetlands, declines in ambient water and sediment quality, and the addition of higher volumes of nutrients (primarily nitrogen and phosphorus) from urban, nonpoint source run-off. Algal growth are stimulated when nutrient concentrations from sources such as stream and river discharges, wastewater sewage facilities, and agricultural runoff are increased beyond the natural background levels of estuaries and other coastal receiving waters. These excess nutrients and the associated increased algal growth can also lead to a series of events that can decrease water clarity, cause benthic degradation, and result in low concentrations of dissolved oxygen. Recent research topics that have attracted particular interest include quantification of the effects of sampling design and measurement accuracy, frequency, and resolution on the ability to improve our quantitative knowledge of coastal water quality.

Chapter 7 by Darryl Keith showcases a study that determined ecological conditions of numerous individual embayments and estuaries along the southern New England coast, as well as the adjoining coastal ocean, over an annual cycle using airborne hyperspectral remote sensing data and the criteria for assessing chlorophyll *a* concentrations found in EPA NCA guidelines (USEPA, 2004). The study calculated surface concentrations of chlorophyll *a* using an empirically derived, band ratio model developed from *in situ* hyperspectral data acquired during a multiyear monitoring program in Narragansett Bay, Rhode Island. Then the model was applied to estuaries and bays along the southern New England coastline using aerial survey hyperspectral data to estimate chlorophyll *a* concentrations and assess environmental conditions. Finally, chlorophyll *a* condition assessments are made for individual coastal systems and are aggregated over the survey period to create site-specific annual assessments. The individual site assessments are further used to create regional scale assessments at monthly, seasonal, and annual scales. It concluded that aircraft remote sensing provided near-synoptic, regional views of chlorophyll condition for coastal southern New England. These assessments would have been sample intensive and expensive if conducted over an annual period using traditional field-based monitoring (Keith, Chapter 7 of this book).

In Chapter 8, Yang and Artigas report a study that tested the accuracy of integrated airborne hyperspectral and LiDAR remote sensing data to characterize the patterns and distribution of salt marsh vegetation in the New Jersey Meadowlands. The results indicate that integration of hyperspectral and LiDAR remote sensing is efficient in characterizing the distribution and structure of salt marsh vegetation, particularly for high marsh and tall *Phragmites*.

1.4 HIGH SPATIAL RESOLUTION REMOTE SENSING

High spatial resolution remote sensing imagery data provide much needed spatial details and variations at submeter level for mapping dynamic coastal habitats. For example, besides changing in areas, degradations of mangrove forests due to changing environment and selective harvesting have significant effects on ecosystem integrity and functions. Accurate and effective mapping of mangrove forests is essential for monitoring change in spatial distribution and species composition. Advancement of high spatial resolution multispectral remote sensing data makes such an inventory and monitoring possible (Wulder et al., 2004). Mumby and Edwards (2002) improved thematic mapping accuracy of habitats by incorporation of texture information with high spatial resolution IKONOS satellite data. Wang et al. (2004a) developed an integrated pixel-based and object-based method for improved classification of mangrove forests using IKONOS data. Wang et al. (2004b) compared the performance of QuickBird and IKONOS satellite images for mapping mangrove species and demonstrated the advancement of high spatial space-borne remote sensing in mangrove forest identification. High spatial resolution satellite multispectral and airborne digital remote sensing data have been employed to evaluate benthic habitats (Su et al., 2006; Deepak et al., 2006; Lathrop et al., 2006; Wang et al., 2008). Part III of this book consists of four chapters under the category of high spatial resolution remote sensing.

Chapter 9 (Wang et al.) introduces two case studies that employed QuickBird-2 satellite remote sensing data to map terrestrial vegetation on the Fire Island National Seashore and the salt marsh in the Jamaica Bay of the Gateway National Recreation Area (NRA). Jamaica Bay is one of the three units of the Gateway NRA and one of the largest open space areas in New York City. The Bay played an important role in the development of the city and its surrounding environment. Historically, Jamaica Bay has taken on numerous configurations that eventually evolved into the waters, marshes, and mudflats known today. Salt marshes, one of the most critical habitats in the Jamaica Bay, are rapidly disappearing. Interpretation of historical aerial photographs shows that 51% of salt marshes in the Bay had been lost between 1924 and 1999. Although the salt marshes have been protected since 1972, a recent study shows that 38% of salt marshes in Jamaica Bay have been lost since 1974. Increasingly the losses have occurred within the interior of marsh islands. Salt marsh change in the Jamaica Bay is similar to the trends found in other salt marshes elsewhere in the northeastern United States. Fire Island National Seashore is a member of the Long Island barrier island system. The vegetation communities and spatial patterns are dynamic with the impacts from forces such as sand deposition, storm-driven over wash, salt spray, and surface water. Mapping the vegetation communities and tracking their changes are among important tasks that challenge resource managers. These case studies demonstrate that high spatial resolution satellite remote sensing data provide a successful alternative data source for mapping both terrestrial vegetation in a barrier island setting and salt marsh in a bay setting. The map data and protocols developed provide the references for a long-term monitoring of dynamic coastal ecosystems and aid in management decisions (Wang et al., Chapter 9 of this book).

Chapter 10 (Shan and Hussain) summarizes the use of aerial high spatial resolution color infrared orthoimage, digital elevation model, digital surface model, and road data for the purpose of automated coastal mapping. The images are classified jointly with other geospatial data based on the framework of an object-based fuzzy classification. It demonstrates that integrating high spatial resolution images and multisource spatial data into the object-based fuzzy classification framework can better map the land use and land cover of coastal areas with varying topography, geomorphology, and urban complexities.

In Chapter 11, Zhou and Wang report a study that extracted urban impervious surface area (ISA) from true-color digital orthophotography data for revealing the intensity of urban development surrounding the Narragansett Bay, Rhode Island. ISA is defined as any impenetrable materials that prevent water infiltration into the soil. ISA has been considered a key indicator to explain and predict ecosystem health in relationship to watershed development (Arnold and Gibbons, 1996). Assessment of the quantity of ISA in landscapes has become increasingly important with growing concern about its impact on the environment (Weng, 2001; Civco et al., 2002; Dougherty et al., 2004; Wang and Zhang, 2004). This is particularly true for coastal areas due to the impacts of ISA on aquatic systems and its role in transportation and concentration of pollutants. Urban runoff, mostly over impervious surface, is the leading source of pollution in U.S. estuaries, lakes, and rivers (Booth and Jackson, 1997). A published watershed-planning model predicts that most stream quality indicators decline when watershed ISA exceeds 10%, which could result in altered shape of stream channels, raised water temperatures, and discharge of pollutants into aquatic environments (Schueler, 2003). Precise data of ISA in spatial coverage and distribution patterns in association with landscape characterizations are critical for providing the key baseline information for effective coastal management and science-based decision making. The modeling approach reported in this chapter achieved a reliable result for quantifying spatial patterns of ISA in the populated Southern New England coast.

Remote sensing applications in disaster management have become critically important to support preparation through response to natural and human-induced hazards and events affecting human populations in coastal zones. Increased exposure and density of human settlements in coastal regions amplify the potential loss of life, property, and commodities that are at risk from intense coastal hazards. Remote sensing has a long history of being used to capture towns, harbors, and coastal lines affected by the disasters and has played a key role in recovery efforts following disasters. The history can be traced back from the earthquake and tsunami that struck Alaska on March 1964, to the recent five hurricanes of Dennis, Katrina, Ophelia, Rita, and Wilma that directly impacted the U.S. coastal regions in 2005 (White and Aslaksen, 2006), among others. On December 26, 2004, the Sumatra earthquake struck South Asia and triggered monstrous waves that turned into a tsunami hitting the ocean regions and caused the most severe damages. To provide real-time information for rescue and rehabilitation plans, remote sensing images and geographic information systems (GIS) were applied to monitor and evaluate the damage over several devastated spots (Kelmelis et al., 2006). Yang et al. introduce in Chapter 12 an application of the FORMOSAT-2 satellite to assess damages over the tsunami-devastated areas. FORMOSAT-2 acquired post-tsunami images of the affected areas in Thailand and Indonesia on December 28. The high spatial resolution images with timely coverage were proved to be efficient and useful to make rescue and recovery plans.

1.5 INTEGRATION OF REMOTE SENSING AND *IN SITU* DATA

Part IV of this book consists of four chapters under the category of integration of remote sensing and *in situ* measurements for habitat mapping. Integration of multispectral, multitemporal, multisensor airborne and space-borne remote sensing data with GPS-guided *in situ* observations is becoming necessary for effective and timely inventory and monitoring of the coastal environments. For example, a significant amount of the coastal wetlands along the Long Island Sound in the northeastern United States has been lost over the past century due to urban development, filling and dredging, or damaged due to human disturbance and modification. Beyond the physical loss of marshes, the species composition of marsh communities is changing. With the mounting pressures on coastal wetland areas, it is becoming increasingly important to identify and inventory the current extent and condition of coastal marshes located on the Long Island Sound estuary, implement a cost-effective way by which to track changes in wetlands over time, and monitor the effects of habitat restoration and management. The identification of the distribution and health of individual marsh plant species like *Phragmites australis* using remote sensing is challenging, because vegetation spectra are generally similar to one another throughout the visible to near-infrared (VNIR) spectrum and the reflectance of a single species may vary throughout the growing season due to variations in the amount and ratios of plant pigments, leaf moisture content, plant height, canopy effects, leaf angle distribution, and other structural characteristics.

Gilmore et al. report a study in Chapter 13 that addresses the use of multitemporal field spectral data, satellite imagery, and LiDAR top of canopy data to classify and map common salt marsh plant communities. VNIR reflectance spectra were measured in the field to assess the phenological variability of the dominant species of *Spartina patens*, *P. australis*, and *Typha* spp. The field spectra and single-date LiDAR canopy height data were used to define an object-oriented classification methodology for the plant communities in multitemporal QuickBird imagery. The classification was validated using an extensive field inventory of marsh species.

The wetland in Poyang Lake at the coastal Southeast China is one of the first national natural reserves listed in Ramsar Convention in 1992. The Lake also plays an important role in flood control of the Yangtze River watershed. Overexploitation of wetlands in the Poyang Lake has altered the ecosystem and reduced the biodiversity. Recognizing the urgency, a series of programs have been initiated by the local and provincial agencies since the late 1980s aiming at reversing the trends of wetland loss, including monitoring status and trends of the wetlands using remote sensing augmented by ground surveying. Yang et al. report a study in Chapter 14 that evaluated the feasibility of quantifying biophysical conditions and seasonal dynamics of wetland vegetation in the Poyang Lake region using multitemporal, multipolarization

SAR imagery and *in situ* measurements. The study quantified biophysical conditions and seasonal dynamics of two herbaceous wetland species (*Phragmites communis* Trin., and *Carex* spp.) through field survey and ENVISAT ASAR imagery. The results revealed both opportunity and challenge in using ENVISAT ASAR data for monitoring wetland biophysical conditions.

As submerged aquatic vegetation (SAV) habitats are mostly in shallow coastal waters, short wavelengths in the light blue portion of the spectrum possess a certain capacity for penetration of water column and therefore could provide the information for improved SAV mapping. Chapter 15 (Akins et al.) reports a study that employed EO-1 Advance Land Imager (ALI) data, which include short wavelengths of the light blue spectrum, for identification and mapping of temperate coastal SAV. In particular, the study was augmented by intensive *in situ* data such as benthic conditions and underwater videography data that helped SAV habitat identification and classification.

The advantages and potentials of remote sensing have not only been the topics that have attracted particular interest of the scientific community, but are also recognized by resource managers and the user groups. For example, managers of the National Park Service (NPS) across the country are confronted with increasingly complex and challenging issues that require a broad-based understanding of the status and trends of each park's natural resources as a basis for making decisions, working with other agencies, and communicating with the public to protect park natural systems and native species. The Northeast Coastal and Barrier Network (NCBN) of the NPS Inventory & Monitoring Program is made up of eight parks found along the Northeastern Atlantic seaboard. The parks are within the coastal zone and consist of critical coastal habitat. Chapter 16 of this book (Stevens and Schupp), from management perspective, reports how remote sensing data are being employed at the NCBN for inventory and monitoring of the natural resources.

1.6 EFFECTS OF LCLUC

Part V of this book consists of four chapters under the category of effects of LCLUC on coastal environments. LCLUC is an interdisciplinary scientific theme that includes to perform repeated inventories of landscape change from space; to develop scientific understanding and models necessary to simulate the processes taking place; to evaluate consequences of observed and predicted changes; and to further understand consequences on environmental goods and services and management of natural resources. The study of the LCLUC is critical to improve our understanding of human interaction with the environment, and provides a scientific foundation for sustainability, vulnerability, and resilience of land systems and their use. This is even true in coastal zones where land and water are constantly affected by natural and anthropogenic forces. An emerging area of increased emphasis in physical oceanography studies is the research on the coastal ocean. As identified by NASA's Research Opportunities in Space and Earth Sciences in 2008 that many of the practical problems with respect to human interaction with the ocean lie within the coastal seas. Deforestation and loss of topsoil can affect the amount of sediment deposited in rivers and streams, which empty into coastal regions, and eventually to the open ocean. Treatment of agricultural crops and changes in land-cover and land-use practices can also lead to chemical constituents being added to waters of rivers, lakes, and oceans.

Change detection has often been discussed in the literature (Mouat et al., 1993; Lambin and Strahler, 1993; Roberts et al., 1998; Mas, 1999; Hayes and Sader, 2001; Rogan et al., 2002; Woodcock and Ozdogan, 2004; Rhemtulla et al., 2007; Wilkinson, et al., 2008).

Hurd et al. summarizes a study in Chapter 17 about land-cover change analysis of the Connecticut coast. The coastal area of Connecticut is diverse, but is under significant pressure from both anthropogenic and natural forces. For example, Connecticut's two largest populated cities, Bridgeport and New Haven, are located within this region. Once thriving industrial centers, these cities represent some of the most intensively developed areas of the state. Several towns along the western coast comprise the "Gold Coast" of Connecticut, wealthy suburbs of New York City, which provide their own unique form of housing, commercial development, and open space patterns. Also, the two largest casinos in the United States, Foxwoods and Mohegan Sun, in addition to Pfizer Pharmaceutical's new Global Research and Development headquarters are located in the eastern region, all recently contributing to significant housing growth and related impacts on the coastal landscape of Connecticut. This chapter describes the development of the Connecticut's Changing Landscape (CCL) land-cover dataset, and how these data were used in two projects, the Coastal Area land Cover Analysis Project (CALCAP) and the Coastal Riparian Buffers Analysis Project to assess the land-cover condition of coastal Connecticut.

Along this line, Chapter 18 (Zhou et al.) reports a study about the effects of increasing urban ISA on the hydrology of coastal watersheds. The study developed a distributed object-based rainfall–runoff simulation (DORS) model with incorporation of ISA derived from high spatial resolution remote sensing data. This modeling approach simulated hydrologic processes of precipitation interception, infiltration, runoff, evaporation and evapotranspiration, soil moisture and change of water table depth, runoff routing, ground water routing, and channel routing. The study investigated the relationship between watershed characteristics and hydrology response. The results demonstrate that ISA plays an important role in watershed hydrology. The hydrologic model, regression methods, and relationships between watershed characteristics and hydrology pattern provided important tools and information for decision makers to evaluate the effect of different scenarios in land management.

The developed remote sensing data capability and modeling approaches help answer science questions such as what are the consequences of LCLUC by increased human activities for coastal regions. Chapter 19 (Yang et al.) introduces a study about quantifying major LCLUC at the coastal Pearl River Delta (PRD) region in southeastern China and the impact of changes on regional climate. The PRD region is one of the fastest growing regions in the world resulted from rapid urbanization since the 1980s. About 60 million people reside in the region, including 7 millions in Hong Kong, a half million in Macao, and the rest in Mainland China. The region has undergone an unprecedented rapid economic development that has fundamentally changed the region's landscape (Yeh and Li, 1997, 1998; Seto et al., 2002; Weng, 2002). The changes could have affected the local and regional climate over time (Shepherd et al., 2002; Shepherd, 2005; Kaufmann et al., 2007; Lo et al., 2007).

Through numerical simulation by mesoscale climate model with land-cover data, the results suggest that the distinct heat islands formed in expanded urban areas as a result of increased monthly mean air temperature in both summer and autumn. Other impacts due to urban expansion include decreasing mixing ratio and relative humidity, and changing of sensible latent heat flux.

Finally, Chapter 20 (Wang et al.) reports a study about using remote sensing-derived land-cover change data for sustainable development considerations in coastal East Africa. Balancing natural resource management and the needs for sustainable development is a challenging task that affects virtually all humanity. Geographic information and technologies are central to the transition from traditional environmental management to sustainable development. The coastal region of Kenya and Tanzania in East Africa has been undergoing notable changes. Primary coastal issues important to sustainable development include intensification of agriculture and mariculture, declining resource base, destruction of critical habitats such as mangrove forest, rapid expansion of coastal cities, increasing population, and overfishing. This case study developed land-cover maps of the Tanzania coastal regions and analyzed the change. The study helped identify priority locations for coastal conservation, aquaculture, and land-use zoning. It also helped build capacity and provide a boost to a long-term sustained effort in the use of remote sensing and geospatial science and technology for coastal management.

1.7 REMARKS AND ACKNOWLEDGMENTS

As dynamic as the coastal environments, remote sensing is among the fascinating frontiers of science and technology that are constantly changing. It is impossible to work out a book that would cover all hot topics and latest developments under the general term of remote sensing of coastal environments. My hope is that this book will provide a snapshot about remote sensing applications in coastal issues so that it will inspire a broader scope interests on remote sensing applications in scientific research and resource management of the coastal environments.

I am very fortunate to have this opportunity to work on such an important book. So many people have helped me tremendously during the process and deserve acknowledgment. First of all I thank all the contributors of this book. Their dedications and hard work make this book the best possible. I appreciate gratefully the contributions from the reviewers. Their expertise, insights, and professional services help improve the quality of this book significantly. The reviewers include Eric Akins, Mike Bradley, Mark Christiano, Roland Duhaime, James Hurd, Wei Ji, Sandy Prisloe, Karen Seto, Joel Stocker, Yang Shen, Le Wang, Jiansheng Yang, Limin Yang, Xiaojun Yang, Guoqing Zhou, and Yuyu Zhou. Dr. Qihao Weng, the Series Editor of this Taylor & Francis Series in Remote Sensing Applications, offered encouragement and guidance in preparation of this book.

I appreciate the Preface by Dr. Peter August, the Director of the Coastal Institute and a professor of the Department of Natural Resources Science, University of Rhode Island, and Dr. James Yoder, the Vice President for Academic Programs and Dean of the Woods Hole Oceanographic Institution and a former professor and the Associate Dean of the Graduate School of Oceanography, University of Rhode Island. Their insightful and visionary understanding of the coastal environments represents the authority on the issues of coastal management and the advancement of geospatial science and technology in coastal research and applications.

My research experience in coastal remote sensing and inspiration of working on such a book came partially from my faculty career at the University of Rhode Island, where I have the opportunity of working with different agencies for funded scientific research and working with many top notch scientists, scholars, staffs, and enthusiastic students who are interested in the management and conservation of coastal environments. In particular, the staff members and graduate students at the Laboratory for Terrestrial Remote Sensing and the Environmental Data Center, within the Department of Natural Resources Science, University of Rhode Island, made valuable contributions and offered constructive comments during the preparation of this book.

Special appreciation is due to my wife and daughters for their patience, understanding, and encouragement.

REFERENCES

Acker, J., Leptoukh, J., Shen, S., Zhu T., and Kempler, S., 2008, Remotely-sensed chlorophyll a observations of the northern Red Sea indicate seasonal variability and influence of coastal reefs, *Journal of Marine Systems*, 69: 191–204.

Ackermann, F., 1999, Airborne laser scanning-present status and future expectations, *ISPRS Journal of Photogrammetry and Remote Sensing*, 54: 64–67.

Alsdorf, D., Birkett, C., Dunne, T., Melack, J., and Hess, L., 2001, Water level changes in a large Amazon lake measured with spaceborne radar interferometry and altimetry, *Geophysical Research Letters*, 28: 2671–2674.

Alsdorf, D., Melack, J., Dunne, T., Mertes, L., Hess, L., and Smith, L., 2000, Interferometric radar measurements of water level changes on the Amazon floodplain, *Nature*, 404: 174–177.

Anderson, J.E., Plourde, L.C., Martin, M.E., Braswell, B.H., Smith, M.L., Dubayah, R.O., Hofton, M.A., and Blair, J.B., 2008, Integrating waveform lidar with hyperspectral imagery for inventory of a northern temperate forest, *Remote Sensing of Environment*, 112(4): 1856–1870.

Armstrong, R.A., 1993, Remote sensing of submerged vegetation canopies for biomass estimation, *International Journal of Remote Sensing*, 14(3): 621–627.

Arnold, C.A., Jr. and Gibbons, C.J., 1996, Impervious surface coverage: The emergence of a key urban environmental indicator, *Journal of the American Planning Association*, 62: 243.

Artigas, F.J. and Yang, J., 2004, Hyperspectral remote sensing of habitat heterogeneity between tide-restricted and tide-open areas in New Jersey Meadowlands, *Urban Habitat*, 2: 1.

Artigas, F.J. and Yang, J., 2005, Hyperspectral remote sensing of marsh species and plant vigor gradient in the New Jersey Meadowland, *International Journal of Remote Sensing*, 26: 5209.

Aschbacher, J., Ofren, R.S., Delsol, J.P., Suselo, T.B., Vibusresh, S., and Charrupat, T., 1995, An integrated comparative approach to mangrove vegetation mapping using remote sensing and GIS technologies, preliminary results, *Hydrologia*, 295: 285–294.

Baghdadi, N., Bernier, M., Gauthier, R., and Neeson, I., 2001, Evaluation of C-band SAR data for wetlands mapping, *International Journal of Remote Sensing*, 22(1): 71–88.

Belluco, E., Camuffo, M., Ferrari, S., Modenese, L., Silvestri, S., Marani, A., and Marani, M., 2006, Mapping salt marsh vegetation by multispectral and hyperspectral remote sensing, *Remote Sensing of Environment*, 105: 54–67

Blasco, F., Lavenu, F., and Baraza, J., 1986, Remote sensing data applied to mangroves of Kenya coast, *Proceedings of the 20th International Symposium on Remote Sensing of the Environment*, 3: 1465–1480.

Booth, D.B. and Jackson, C.R., 1997, Urbanization of aquatic systems: Degradation thresholds, stormwater detection, and the limits of mitigation. *Journal of American Water Resources Association*, 35: 1077–1090.

Bourgeau-Chavez, L.L., Kasischke, E.S., Brunzell, S.M., Mudd, J.P., Smith, K.B., and Frick, A.L., 2001. Analysis of space-borne SAR data for wetland mapping in Virginia riparian ecosystems, *International Journal of Remote Sensing*, 22(18): 3665–3687.

Bourgeau-Chavez, L.L., Smith, K.B., Brunzell, S.M., Kasischke, E.S., Romanowicz, E.A., and Richardson, C.J., 2005, Remote monitoring of regional inundation patterns and hydroperiod in the greater Everglades using synthetic aperture radar, *Wetlands*, 25(1): 176–191.

Brock, J., Wright, C.W., Hernandez, R., and Thompson, P., 2006, Airborne Lidar sensing of massive stony coral colonies on patch reefs in the northern Florida reef tract, *Remote Sensing of Environment*, 104: 31–42.

Brock, J.C., Sallenger, A.H., Krabill, W.B., Swift, R.N., and Wright, C.W., 2001, Recognition of fiducial surfaces in Lidar surveys of coastal topography, *Photogrammetric Engineering and Remote Sensing*, 67(11): 1245–1258.

Brock, J.C., Wright, C.W., Sallenger, A.H., Krabill, W.B., and Swift, R.N., 2002, Basis and methods of NASA Airborne Topographic Mapper Lidar surveys for coastal studies, *Journal of Coastal Research*, 18(1): 1–13.

Bukata, R.P., Jerome, J.H., and Bruton, J.E., 1988, Particulate concentrations in Lake St. Clair as recorded by a shipborne multispectral optical monitoring system, *Remote Sensing of Environment*, 25: 201–229.

Civco, D.L., Hurd, J.D., Wilson, E.H., Arnold, C.L., and Prisloe, S., 2002, Quantifying and describing urbanizing landscapes in the Northeast United States, *Photogrammetric Engineering and Remote Sensing*, 68: 1083–1090.

Cloern, J.E., 1987, Turbidity as a control on phytoplankton biomass and productivity in estuaries, *Continental Shelf Research*, 7(11): 1367–1381.

Cole, B.E. and Cloern, J.E., 1987. An empirical model for estimating phytoplankton productivity in estuaries. *Marine Ecology. Progress Series*, 36: 299–305.

Colwell, R.N., 1983, *Manual of Remote Sensing*, 2nd edition. American Society of Photogrammetry, Falls Church, VA.

Coppin, P.R. and Bauer, M.E., 1996, Digital change detection in forest ecosystem with remotely sensed imagery, *Remote Sensing Review*, 13: 207–234.

Costa, M.P.F., 2004. Use of SAR satellites for mapping zonation of vegetation communities in the Amazon floodplain, *International Journal of Remote Sensing*, 25(10): 1817–1835.

Costa, M.P.F., Niemann, O., Novo, E., and Ahern, F., 2002, Biophysical properties and mapping of aquatic vegetation during the hydrological cycle of the Amazon floodplain using JERS-1 and RADARSAT, *International Journal of Remote Sensing*, 23(7): 1401–1426.

Crossett, K.M., Culliton, T.J., Wiley, P.C., and Goodspeed, T.R., 2004, Population trends along the coastal United States: 1980–2008. Coastal Trends Report Series. National Oceanic and Atmospheric Administration, National Ocean Service Management and Budget Office Silver Spring, MD, 54pp.

Deepak, M., Narumalani, S., Rundquist, D., and Lawson, M., 2006, Benthic Habitat Mapping in Tropical Marine Environments Using QuickBird Multispectral Data, *Photogrammetric Engineering and Remote Sensing*, 72(9): 1037–1048.

Dennison, W.C., Orth, R.J., Moore, K.A., Stevenson, J.C., Carter, V., and Kollar, S., 1993, Assessing water quality with submersed aquatic vegetation, *Bioscience*, 43: 86–94.

Dobson, J.E., Bright, E.A., Haddad, K.D., Iredale, H., III, Jensen, J.R., Klemas, V.V., Orth, R.J., and Thomas, J.P., 1995, NOAA Coastal Change Analysis Program (C-CAP): Guidance for Regional Implementation. NOAA Technical Report NMFS 123. Department of Commerce, 140 pp (http://www.csc.noaa.gov/crs/lca/pdf/protocol.pdf).

Dodge, R.E., Aller, R., and Thompson, J., 1974, Coral growth related to resuspension of bottom sediments, *Nature*, 247: 574–577.

Dougherty, M., Randel, L.D., Scott, J.G., Claire, A.J., and Normand, G., 2004, Evaluation of impervious surface estimates in a rapidly urbanizing watershed. *Photogrammetric Engineering and Remote Sensing*, 70: 1275–1284.

Dutrieux, E., Denis, J., and Populus, J., 1990, Application of SPOT data to a base-line ecological study the Mahakam Delta mangroves East Kalimantan, Indonesia, *Oceanologica Acta*, 13: 317–326.

Ferguson, R.L. and Korfmacher, K., 1997, Remote sensing and GIS analysis of seagrass meadows in North Carolina, USA, *Aquatic Botany*, 58: 241–258.

Ferguson, R.L., Krouse, C., Patterson, M., and Hare, J.A., 2006, Automated Thematic Registration of NOAA, Coast Watch, and AVHRR Images, *Photogrammetric Engineering and Remote Sensing*, 72(6): 677–685.

Froidefond, J.M., Castaing, P., and Jouanneau, J.M., 1996, Distribution of suspended matter in a coastal upwelling area. Satellite data and *in situ* measurements, *Journal of Marine Systems*, 8: 91–105.

Froidefond, J.M., Castaing, P., Jouanneau, J.M., Prud'homme, R., and Dinet, A., 1993, Method for the quantification of suspended sediments from AVHRR NOAA-11 satellite data, *International Journal of Remote Sensing*, 4(5): 885–894.

Gang, P.O. and Agatsiva, J.L., 1992, The current status of mangroves along the Kenyan coast, a case study of Mida Creek mangroves based on remote sensing, *Hydrobiologia*, 247: 29–36.

Gao, J., 1998, A hybrid method toward accurate mapping of mangroves in a marginal habitat from SPOT multispectral data, *International Journal of Remote Sensing*, 19: 1887–1899.

Gillespie, T.W., Brock, J., and Wright, W., 2001, Prospects for quantifying structure, floristic composition, and species richness of tropical forests, *International Journal of Remote Sensing*, 24: 1–9.

Green, E.P., Mumby, P.J., Edwards, A.J., and Clarke, C.D., 2000. *Remote Sensing Handbook for Tropical Coastal Management*, Coastal Management Sourcebooks 3, UNESCO, Paris, x+316.

Green, E.P., Mumby, P.J., Edwards, A.J., Clark, C.D., and Ellis, A.C., 1998, The assessment of mangrove areas using high resolution multispectral airborne imagery, *Journal of Coastal Research*, 14: 433–443.

Griffith, J.A., Stehman, S.V., Sohl, T.L., and Loveland, T.R., 2003, Detecting trends in landscape pattern metrics over a 20-year period using a sampling-based monitoring programme, *International Journal of Remote Sensing*, 24: 175–181.

Grings, F.M., Ferrazzoli, P., Jacobo-Berlles, J.C., Karszenbaum, H., Tiffenberg, J., Pratolongo, P., and Kandus, P., 2006, Monitoring flood condition in marshes using EM models and ENVISAT ASAR observations, *IEEE Transactions on Geoscience and Remote Sensing*, 44(4): 936–942.

Hansen, A.J. and Rotella, J.J., 2002, Biophysical factors, land use, and species viability in and around nature reserves, *Conservation Biology*, 16: 1112–1122.

Hayes, D.J. and Sader, S.A., 2001, Comparison of change-detection techniques for monitoring tropical forest clearing and vegetation regrowth in a time series, *Photogrammetric Engineering and Remote Sensing*, 67(9): 1067–1075.

Hess, L.L. and Melack, J.M., 1994, Mapping wetland hydrology and vegetation with synthetic aperture radar, *International Journal of Ecology and Environmental Sciences*, 20: 197–205.

Hess, L.L., Melack, J.M., Filoso, S., and Wang, Y., 1995, Delineation of inundated area and vegetation along the amazon floodplain with the SIR-C synthetic aperture radar, *IEEE Transactions on Geoscience and Remote Sensing*, 33(4): 896–904.

Huang, S., Hager, S.A., Halligan, K.Q., Fairweather, I.S., Swanson, A.K., and Crabtree, R., 2009, A comparison of individual tree and forest plot height derived from LiDAR and InSAR, *Photogrammetric Engineering and Remote Sensing*, 75(2): 159–167.

Hudak, A.T., Lefsky, M.A., and Cohen, W.B., 2002, Integration of LiDAR and Landsat ETM+ data for estimating and mapping forest canopy height, *Remote Sensing of Environment*, 82: 397–416.

Jensen, J.R., Ramset, E., Davis, B.A., and Thoemke, C.W., 1991, The measurement of mangrove characteristics in south-west Florida using SPOT multispectral data, *Geocarto International*, 2: 13–21.

Kasischke, E.S. and Bourgeau-Chavez, L.L., 1997, Monitoring south Florida wetlands using ERS-1 SAR imagery, *Photogrammetric Engineering and Remote Sensing*, 63: 281–291.

Kaufmann, R.K., Seto, K.C., Schneider, A., Liu, Z., Zhou, L., and Wang, W., 2007, Climate response to rapid urban growth: Evidence of a human-induced precipitation deficit, *Journal of Climate*, 20(10): 2299.

Kay, R.J., Hick, P.T., and Houghton, H.J., 1991, Remote sensing of Kimberley rainforest, In: N.I. McKenzie, R.B. Johnston, and P.G. Kendrick (Eds), *Kimberley Rainforests*, pp. 41–51, Survey Beatty & Sons, Chipping Norton.

Kelmelis, J.A., Schwartz, L., Christian, C., Crawford, M., and King, D., 2006, Use of geographic information in response to the Sumatra-Andaman Earthquake and Indian Ocean Tsunami of December 26, 2004, *Photogrammetric Engineering and Remote Sensing*, 72(8): 862–876.

Kiage, L.M., Walker, N.D., Balasubramanian, S., Babin, A., and Barras, J., 2005, Applications of RADARSAT-1 synthetic aperture radar imagery to assess hurricane-related flooding of coastal Louisiana, *International Journal of Remote Sensing*, 26(24): 5359–5380.

Lambin, E.F. and Strahler, A.H., 1993, Change-vector analysis in multitemporal space: A tool to detect and categorize land-cover change processes using high-temporal resolution satellite data, *Remote Sensing of Environment*, 48: 231–244.

Lathrop, R.G., Montesano, P., and Haag, S., 2006, A multi-scale segmentation approach to mapping seagrass habitats using airborne digital camera imagery, *Photogrammetric Engineering and Remote Sensing*, 72(6): 665–675.

Lefsky, M.A., Cohen, W.B., Parker, G.G., and Harding, D.J., 2002, LiDAR remote sensing for ecosystem studies, *Bioscience*, 52: 19–30.

Le Toan, T., Ribbes, F., Wang, L.-F., Floury, N., Ding, K.-H., Kong, J.A., Fujita, M., and Kurosu, T., 1997, Rice crop mapping and monitoring using ERS-1 data based on experiment and modeling results, *IEEE Transactions on Geoscience and Remote Sensing*, 35(1): 41–56.

Li, L., Ustin, S.L., and Lay, M., 2005, Application of multiple endmember spectral mixture analysis (MESMA) to AVIRIS imagery for coastal salt marsh mapping, a case study in China Camp, CA, USA, *International Journal of Remote Sensing*, 26: 5193.

Lo, C.F., Lau, A.K.H., Chen, F., Fung, J.C.H., and Leung, K.K.M., 2007. Urban modification in a mesoscale model and the effects on the local circulation in the Pearl River Delta region, *Journal of Applied Meteorology and Climatology*, 46: 457–476.

Lohrenz, S.E., Redalje, D.G., Cai, W-J., Acker, J. and Dagg, M., 2008, A retrospective analysis of nutrients and phytoplankton productivity in the Mississippi River plume, *Continental Shelf Research*, 28: 1466–1475.

Martin, J.M. and Windom, H.L., 1991, Present and future roles of ocean margins in regulating marine biogeochemical cycles of trace elements, In: R.F.C. Mantoura (Ed.), *Ocean Margin Processes in Global Change*. Report, Dahlem workshop, Wiley, Berlin, pp. 45–67, 1990.

Mas, J.F., 1999, Monitoring land-cover changes: A comparison of change detection techniques, *International Journal of Remote Sensing*, 20(1): 139–152.

Massonnet, D. and Feigl, K., 1998, Radar interferometry and its application to changes in the Earth's surface, *Reviews of Geophysics*, 36: 441–500.

Mayer, L.M., Keil, R.G., Macko, S.A., Joye, S.B., Ruttenberg, K.C., and Aller, R.C., 1998, The importance of suspended particulates in riverine delivery of bioavailable nitrogen to coastal zones, *Global Biogeochemical Cycles*, 12: 573–579.

McLaughlin, C.J., Smith, C.A., Buddemeier, R.W., Bartley, J.D., and Maxwell, B.A., 2003, Rivers, runoff and reefs, *Global and Planetary Change*, 39(1–2): 191–199.

Means, J.E., Acker, S.A., Brandon, J.F., Renslow, M., Emerson, L., and Hendrix, C.J., 2000, Predicting forest stand characteristics with airborne scanning LiDAR, *Photogrammetric Engineering and Remote Sensing*, 66: 1367–1371.

Miller, R.L., and Cruise, J.F., 1995, Effects of suspended sediments on coral growth: Evidence from remote sensing and hydrologic modeling, *Remote Sensing of Environment*, 53: 177–187.

Miller, R.L. and McKee, B.A., 2004, Using MODIS Terra 250 m imagery to map concentrations of total suspended matter in coastal waters, *Remote Sensing of Environment*, 93(1–2): 259–266.

Montane, J.M. and Torres, R., 2006, Accuracy assessment of LiDAR saltmarsh topographic data using RTK GPS, *Photogrammetric Engineering and Remote Sensing*, 72(8): 961–967.

Moore, K.A., Wilcox, D.J., and Orth, R.J., 2000, Analysis of the abundance of submerged aquatic vegetation species in the Chesapeake Bay, *Estuaries*, 21(1): 115–127.

Mouat, D.A., Mahin, G.G., and Lancaster, J., 1993, Remote sensing techniques in the analysis of change detection, *Geocarto International*, 2: 39–50.

Mumby P., and Edwards, A., 2002, Mapping marine environments with IKONOS imagery: enhanced spatial resolution does deliver greater thematic accuracy, *Remote Sensing of Environment*, 82: 248–257.

Mumby, P.J., Green, E.P., Edwards, A.J., and Clark, C.D., 1997, Measurement of seagrass standing crop using satellite and digital airborne remote sensing, *Marine Ecology Progess Series*, 159: 51–60.

Munday, J.C., Jr. and Alfoldi, T.T., 1979, Landsat test of diffuse reflectance models for aquatic suspended solids measurements, *Remote Sensing of Environment*, 8: 169–183.

Nayegandhl, A., Brock, J.C., Wright, C.W., and O'Connell, J., 2006, Evaluating a small footprint, waveform-resolving LiDAR over coastal vegetation communities, *Photogrammetric Engineering and Remote Sensing*, 72(12): 1407–1417.

Nelson, R., Krabill, W., and Tonelli, J., 1988, Estimating forest biomass and volume using airborne laser data, *Remote Sensing of Environment*, 24: 247–267.

Pasqualini, V., Iltis, J., Dessay, N., Lointier, M., Gurlorget, O., and Polidori, C., 1999, Mangrove mapping in North-Western Madagascar using SPOT-XS and SIR-C radar data, *Hydrobiologia*, 413: 127–133.

Peek, A.H., Griffith, B.O., Underwood, L.W., Hall, C.M., Armstrong, C.D., Spruce, J.P., and Ross, K.W., 2008, Stennis Space Center Uses NASA Remote Sensing Assets to Address Coastal Gulf of Mexico Issues, *Photogrammetric Engineering and Remote Sensing*, 74(12): 1449–1453.

Ramsey, E., III and Rangoonwala, A., 2006, Canopy Reflectance related to marsh dieback onset and progression in coastal Louisiana, *Photogrammetric Engineering and Remote Sensing*, 72(6): 641–652.

Ramsey, E.W., III, 1995. Monitoring flooding in coastal wetlands by using radar imagery and ground-based measurements, *International Journal of Remote Sensing*, 16(13): 2495–2502.

Ramsey, E.W., III, 1999. Radar remote sensing of wetlands, In: R.S. Lunetta and C.D. Elvidge (Eds), *Remote Sensing Change Detection*, pp. 211–243, Ann Arbor Press, Chelsea, MI.

Rasolofoharinoro, M., Blasco, F., Bellan, M.F., Aizpuru, M., Gauquelin, T., and Denis, J., 1998, A remote sensing based methodology for mangrove studies in Madagascar, *International Journal of Remote Sensing*, 19: 1873–1886.

Rhemtulla, J.M., Mladenoff, D.J., and Clayton, M.K., 2007, Regional land-cover conversion in the U.S. upper Midwest: Magnitude of change and limited recovery (1850–1935–1993), *Landscape Ecology*, 22(Supplement 1/December, 2007): 57–75.

Ritchie, J.C., Cooper, C.M., and Shiebe, F.R., 1990, The relationship of MSS and TM digital data with suspended sediments, chlorophyll and temperature in Moon Lake, Mississipi, *Remote Sensing of Environment*, 33: 137–148.

Roberts, D.A., Batista, G.T., Pereira, J.L.G., Waller, E.K., and Nelson, B.W., 1998, Change identification using multitemporal spectral mixture analysis: Application in Eastern Amazonia, In: Lunetta and Elvidge (Eds), *Remote Sensing Change Detection: Environmental Monitoring Methods and Applications*, pp. 137–161, Ann Arbor Press, Chelsea, MI.

Rogan, J., Franklin, J., and Roberts, D.A., 2002, A comparison of methods for monitoring multitemporal vegetation change using Thematic Mapper imagery, *Remote Sensing of Environment*, 80: 143–156.

Rosen, P., Hensley, S., Joughin, I.R., Li, F.K., Madsen, S.N., Rodriguez, E., and Goldstein, R.M., 2000, Synthetic aperture radar interferometry, *Proceedings of the IEEE*, 88: 333–380.

Rosso, P.H., Ustin, S.L., and Hastings, A., 2006, Use of Lidar to study changes associate with Spartina invasion in San Francisco Bay marshes, *Remote Sensing of Environment*, 100: 295.

Schmidt, K.S., Skidmore, A.K., Kloosterman, E.H., van Oosten, H., Kumar, L., and Janssen, J.A.M., 2004, Mapping coastal vegetation using an expert system and hyperspectral imagery, *Photogrammetric Engineering and Remote Sensing*, 70(6): 703–715.

Schueler, T., 2003, *Impacts of Impervious Cover on Aquatic Systems*, 142pp, Center for Watershed Protection (CWP), Ellicott City, MD.

Seto, K.C., Woodcock, C.E., Song, C., Huang, X., Lu, J., and Kaufmann, R.K., 2002, Monitoring land-use change in the Pearl River Delta using Landsat TM, *International Journal of Remote Sensing*, 23: 1985–2004.

Shepherd, J.M., 2005, A review of current investigations of urban-induced rainfall and recommendations for the future, *Earth Interactions*, 9: 1–27.

Shepherd, J.M., Pierce, H., and Negri, A.J., 2002, Rainfall modification by major urban areas: Observations from spaceborne rain radar on the TRMM satellite. *Journal of Applied Meteorology*, 41: 689–701.

Simard, M., De Grandi, G., Saatchi, S., and Mayaux, P., 2002, Mapping tropical coastal vegetation using JERS-1 and ERS-1 radar data with a decision tree classifier, *International Journal of Remote Sensing*, 23(7): 1461–1474.

Simard, M., Zhang, K., Rivera-Monroy, V.H., Ross, M.S., Ruiz, P.L., Castañeda-Moya, E., Twilley, R.R., and Rodriguez, E., 2006, Mapping height and biomass of mangrove forests in Everglades National Park with SRTM elevation data, *Photogrammetric Engineering and Remote Sensing*, 72(3): 299–311.

Small, C. and Cohen, J.E., 2004, Continental physiography, climate, and the global distribution of human population, *Current Anthropology*, 45(2): 269–277.

Su, H., Karna, D., Fraim, E., Fitzgerald, M., Dominguez, R., Myers, J.S., Coffland, B., Handley, L.R., and Mace, T., 2006, Evaluation of eelgrass beds mapping using a high-resolution airborne multispectral scanner, *Photogrammetric Engineering and Remote Sensing*, 72(7): 789–797.

Torres, J.L. and Morelock, J., 2002, Effect of terrigenous sediment influx on coral cover and linear extension rates of three Caribbean massive coral species. *Caribbean Journal of Science*, 38(3–4): 222–229.

Townsend, P.A., 2002, Estimating forest structure in wetlands using multitemporal SAR, *Remote Sensing of Environment*, 79: 288–304.

Turner, M.G., 1990, Spatial and temporal analysis of landscape patterns, *Landscape Ecology*, 4: 21–30.

Turner, W., Spector, S., Gardiner, N., Fladeland, M., Sterling, E., and Steininger, M., 2003, Remote sensing for biodiversity science and conservation, *Trends in Ecology and Evolution*, 18(3): 306–314.

USEPA, 2004, National Coastal Condition Report II, Office of Research and Development/Office of Water, EPA-620/R-03/002, Washington, DC.

Wang, L., Silvan-Cardenas, J.L., and Sousa, W.P., 2008, Neural network classification of mangrove species from multi-seasonal IKONOS imagery, *Photogrammetric Engineering and Remote Sensing*, 74(7): 921–927.

Wang, L., Sousa, W.P., and Gong, P., 2004a, Integration of object-based and pixel-based classification for mangrove mapping with IKONOS imagery, *International Journal of Remote Sensing*, 25(24): 5655–5668.

Wang, L., Sousa, W.P., Gong, P., and Biging, G.S., 2004b, Comparison of IKONOS and QuickBird images for mapping mangrove species on the Caribbean coast of Panama, *Remote Sensing of Environment*, 91: 432–440.

Wang, Y., Bonynge, G., Nugranad, J., Traber, M., Ngusaru, A., Tobey, J., Hale, L., Bowen, R., and Makota, V., 2003, Remote sensing of mangrove change along the Tanzania Coast, *Marine Geodesy*, 26(1–2): 35–48.

Wang, Y. and Moskovits, D.K., 2001, Tracking fragmentation of natural communities and changes in land cover: Applications of Landsat data for conservation in an urban landscape (Chicago Wilderness), *Conservation Biology*, 15(4): 835–843.

Wang, Y., Traber, M., Milstead, B., and Stevens, S., 2006, Terrestrial and submerged aquatic vegetation mapping in Fire Island National Seashore using high spatial resolution remote sensing data, *Marine Geodesy*, 30(1): 77–95.

Wang, Y. and Zhang, X., 2004, A SPLIT model for extraction of subpixel impervious surface information, *Photogrammetric Engineering and Remote Sensing*, 70: 821–828.

Wdowinski, S., Amelung, F., Miralles-Wilhelm, F., Dixon, T., and Carande, R., 2004, Space-based measurements of sheet-flow characteristics in the Everglades wetland, Florida, *Geophysical Research Letters*, 31, L15503, doi: 10.1029/2004GL020383.

Weng, Q., 2002, Land use change analysis in the Zhujiang Delta of China using satellite remote sensing, GIS and stochastic modeling, *Journal of Environmental Management*, 64: 273–284.

Weng, Q.H., 2001, Modeling urban growth effects on surface runoff with the integration of Remote Sensing and GIS, *Environmental Management*, 28: 737–748.

Wilson, E.H. and Sader, S.A., 2002, Detection of forest type using multiple dates of Landsat TM imagery, *Remote Sensing of Environment*, 80: 385–396.

Wilkinson, D.W., Parker, R.C., and Evans, D.L., 2008, Change detection techniques for use in a statewide forest inventory program, *Photogrammetric Engineering and Remote Sensing*, 74(7): 893–901.

White, S., and Aslaksen, M., 2006, NOAA's Use of direct georeferencing to support emergency response, *Photogrammetric Engineering and Remote Sensing*, 72(6): 623–627.

Woodcock, C.E., Allen, R., Anderson, M., Belward, A., Bindschadler, R., Cohen, W., Gao, F., Goward, S.N., Helder, D., Helmer, E., Nemani, R., Oreopoulos, L., Schott, J., Thenkabail, P.S., Vermote, E.F., Vogelmann, J., Wulder, M.A., and Wynne, R., 2008, Free access to Landsat imagery, *Science*, 320: 1011.

Woodcock, C.E. and Ozdogan, M., 2004, *Trends in Land Cover Mapping and Monitoring, Land Change Science* (Gutman, Ed.), pp. 367–377, Springer, New York.
Wulder, M.A., Hall, R.J., Coops, N.C., and Franklin, S.E., 2004, High spatial resolution remotely sensed data for ecosystem characterization, *Bioscience*, 6: 511–521.
Yeh, A. and Li, X., 1997, An integrated remote sensing and GIS approach in the monitoring and evaluation of rapid urban growth for sustainable development in the Pearl River Delta, China, *International Planning Studies*, 2: 193–210.
Yeh, A. and Li, X., 1998, Sustainable land development model for rapid growth areas using GIS, *International Journal of Geographical Information Science*, 2: 169–189.

Section I

LiDAR/Radar Remote Sensing

2 Interferometric Synthetic Aperture Radar (InSAR) Study of Coastal Wetlands over Southeastern Louisiana

Zhong Lu and Oh-Ig Kwoun

CONTENTS

2.1 INTRODUCTION

Coastal wetlands constitute important ecosystems in terms of flood control, water and nutrient storage, habitat for fish and wildlife reproduction and nursery activities, and overall support of the food chain [1]. Louisiana has one of the largest expanses of coastal wetlands in the conterminous United States, and these wetlands contain an extraordinary diversity of habitats. The unique habitats along the Gulf of Mexico, complex hydrological connections, and migratory routes of birds, fish, and other species place Louisiana's coastal wetlands among the nation's most productive and important natural assets [2].

The balance of Louisiana's coastal systems has been upset by a combination of natural processes and human activities. Massive coastal erosion probably started around 1890, and about 20% of the coastal lowlands (mostly wetlands) have eroded in the past 100 years [3]. For example, the loss rate due to erosion for Louisiana's coastal wetlands was as high as 12,202 and 6194 ha/year in the 1970s and 1990s, respectively [4]. Marked environmental changes have had significant impacts on Louisiana's coastal ecosystems, including effects from frequent natural disasters such as the hurricanes Katrina and Rita in 2005 and Ike in 2008. Therefore, an effective method of mapping and monitoring coastal wetlands is essential to understand the current status of these ecosystems and the influence of environmental changes and human activities on them. In addition, it has been demonstrated that measurement of changes in water level in wetlands and, consequently, of changes in water storage capacity provides a governing parameter in hydrologic models and is required for comprehensive assessment of flood hazards (e.g., [5]). Inaccurate knowledge of floodplain storage capacity in wetlands can lead to significant errors in hydrologic simulation and modeling [5]. *In situ* measurement of water levels over wetlands is cost-prohibitive, and insufficient coverage of stage recording instruments results in poorly constrained estimates of the water storage capacity of wetlands [6]. With frequent coverage over wide areas, satellite sensors may provide a cost-effective tool to accurately measure water storage.

A unique characteristic of synthetic aperture radar (SAR) in monitoring wetlands over cloud-prone subtropical regions is the all-weather and day-and-night imaging capability. The SAR backscattering signal is composed of intensity and phase components. The intensity component of the signal is sensitive to terrain slope, surface roughness, and the dielectric constant of the target being imaged. Many studies have demonstrated that SAR intensity images can be used to map and monitor forested and nonforested wetlands occupying a range of coastal and inland settings (e.g., [7–9]). Most of those studies relied on the fact that, when standing water is present beneath the vegetation canopies, the radar backscattering signal intensity changes with water-level changes, depending on vegetation type and structure. As such, SAR intensity data have been used to monitor flooded and dry conditions, temporal variations in the hydrological conditions of wetlands, and classification of wetland vegetation at various geographic settings [7,10–23].

The phase component of the signal is related to the apparent distance from the satellite to ground resolution elements as well as the interaction between radar waves and scatterers within a resolution element of the imaged area. Interferometric SAR (InSAR) processing can then produce an interferogram using the phase components of two SAR images of the same area acquired from similar vantage points at different times. An interferogram depicts range changes between the radar and the ground and can be further processed with a digital elevation model (DEM) to produce an image of ground deformation at a horizontal resolution of tens of meters over large areas and centimeter to subcentimeter vertical precision under favorable conditions (e.g., [24,25]). InSAR has been extensively utilized to study ground surface deformation associated with volcanic, earthquake, landslide, and land subsidence processes [24,26].

Alsdorf et al. [27,28] found that interferometric analysis of L-band (wavelength = 24 cm) Shuttle Imaging Radar-C (SIR-C) and Japanese Earth Resources Satellite (JERS-1) SAR imagery can yield centimeter-scale measurements of water-level changes throughout inundated floodplain vegetation. Their work confirmed that scattering elements for L-band radar consist primarily of the water surface and vegetation trunks, which allows double-bounce backscattering returns as illustrated in Section 2.3.2 of this chapter. Later, Wdowinski et al. [29] applied L-band JERS-1 images to map water-level changes over the Everglades in Florida. All these studies rely on this common understanding: flooded forests permit double-bounce returns of L-band radar pulses, which allow maintaining InSAR coherence—a parameter quantifying the degree of changes in backscattering characteristics (see Sections 2.3.2 and 2.6 for details). Loss of coherence renders an InSAR image useless to retrieve meaningful information about surface movement. However, it is commonly recognized that the shorter wavelength radar, such as C-band (wavelength = 5.7 cm), backscatters from the upper canopy of swamp forests rather than the underlying water surface, and that a double-bounce backscattering can only occur over inundated macrophytes and small shrubs [14,30–32]. As a consequence, C-band radar images were not exploited to study water-level changes beneath swamp forests until 2005, when Lu et al. [33] found that C-band InSAR images could maintain coherence over wetlands to allow estimates of water-level change.

The primary objectives of this study are to utilize multitemporal C-band SAR images from different sensors to differentiate vegetation types over coastal wetlands and explore the potential utility of C-band InSAR imagery for mapping water-level changes. SAR data acquired from two sensors during several consecutive years are used to address these objectives.

The rest of the chapter is composed of seven sections. The study site is introduced in Section 2.2. Section 2.3 describes the fundamental background of the SAR backscattering mechanism and the relationship between InSAR phase measurements and water-level changes. Section 2.4 describes SAR data, calibration, processing, and InSAR processing. Section 2.5 describes temporal variations of radar backscattering signal over different vegetation classes and their usefulness to infer vegetation structures. Section 2.5 also evaluates the relationship between radar backscattering coefficients and normalized difference vegetation index values derived from optical images to provide additional information to differentiate wetland classes. In Section 2.6, interferometric coherence measurements are systematically analyzed for different vegetation types, seasonality, and time separation. In Section 2.7, we present the InSAR-derived water-level changes over swamp forests and discuss the associated potentials and challenges of our approach. Section 2.8 provides discussions and conclusion. Although this chapter largely explores C-band radar images for wetland mapping and water dynamics, a few L-band images are introduced in Section 2.8 to highlight the potential of integrating C-band and L-band images for improved vegetation and water mapping.

2.2 STUDY SITE

The study area is over southeastern Louisiana (Figure 2.1) and includes the western part of New Orleans and the area between Baton Rouge and Lafayette. The area primarily consists of eight land-cover types: urban, agriculture, bottomland forest, swamp forest, freshwater marsh, intermediate marsh, brackish marsh, and saline marsh. Agriculture and urban land cover are found in higher elevation areas and along the levees. Bottomland forests exist in less frequently flooded, lower elevation areas and along the lower perimeter of the levee system, while swamp forests are in the lowest elevation areas. Bottomland forests are dry during most of the year, while swamp forests are inundated. Both types of forests are composed largely of American elm, sweetgum, sugarberry, swamp red maple, and bald cypress [23,34].

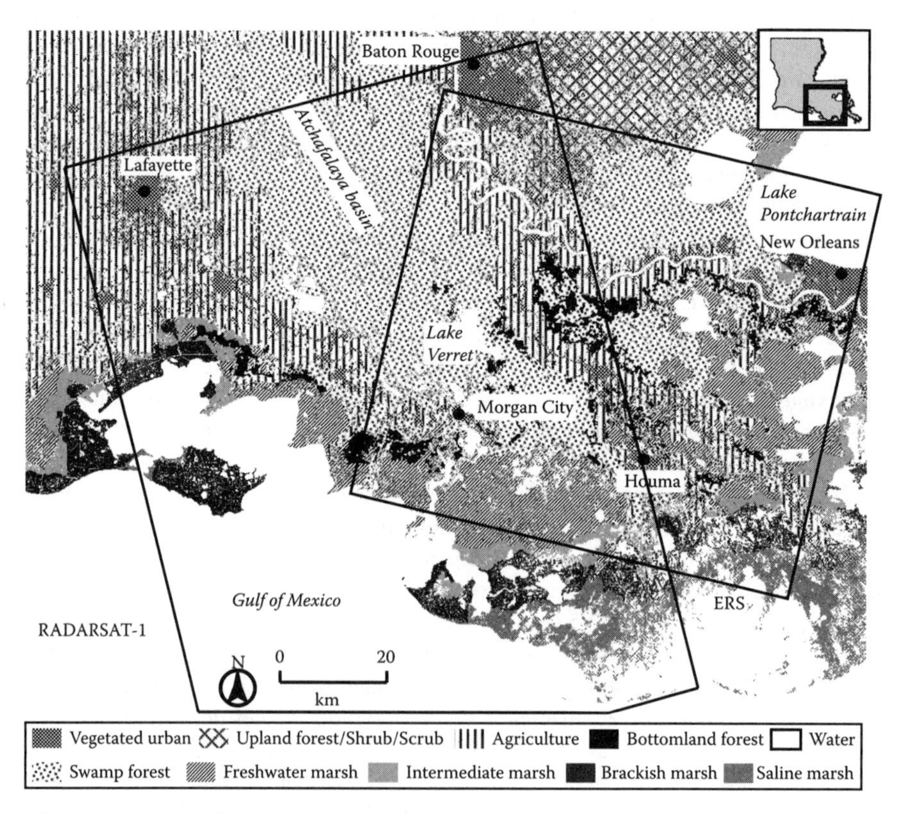

FIGURE 2.1 (See color insert following page 206.) Thematic map. Modified from GAP and 1990 USGS National Wetland Research Center classification results, showing major land-cover classes of the study area. Polygons represent extents of SAR images shown in Figure 2.3 for the ERS-1/ERS-2 and RADARSAT-1 tracks.

The freshwater marshes (Figure 2.1) are composed largely of floating vegetation. Maidencane, spikerush, and bulltongue are the dominant species. Freshwater marshes have the greatest plant diversity and the highest soil organic matter content of any marsh type throughout the study area [35]. However, plant diversity varies with location, and many areas of monotypic marshes are found in the Louisiana coastal zone. The salinity of freshwater marshes ranges from 0 to 3 ppt [3,34].

With a salinity level of 2–5 ppt, intermediate marshes (Figure 2.1) represent a zone of mild salt content that results in fewer plant species than freshwater marshes have [35]. The intermediate marsh is characterized by plant species common to freshwater marshes but with higher salt-tolerant versions of them toward the sea. Intermediate marshes are largely composed of bulltongue and saltmeadow cordgrass [3]. The latter, also called wire grass, is not found in freshwater marshes [36].

Brackish marshes (Figure 2.1) are characterized by a salinity range of 4–15 ppt and are irregularly flooded by tides; they are largely composed of wire grass and three-square bullrush. This marsh community virtually contains all wire grass—clusters of three-foot-long grass-like leaves with little variation in plant species [36].

Saline marshes (Figure 2.1) have the highest salinity concentrations (12 ppt and higher) [3]. With the least diversity of vegetation species, saline marshes are largely composed of smooth cordgrass, oyster grass, and saltgrass.

The hydrology of marsh areas produces freshwater marshes in relatively low-energy environments, which are potentially subject to tidal changes but not ebb and flow. They change slowly and have thick sequences of organic soils or floating grass root mats. Saline and brackish marshes are found in high-energy areas and are subject to the ebb and flow of the tides [3].

2.3 RADAR MAPPING OF WETLANDS

2.3.1 Possible Radar Backscattering Mechanisms Over Wetlands

Over vegetated terrain, the incoming radar wave interacts with various elements of the vegetation as well as the ground surface. Part of the energy is attenuated, and the rest is scattered back to the antenna. The amount of radar energy returned to the antenna (backscattering signal) depends on the size, density, shape, and dielectric constant of the target, as well as SAR system characteristics, such as incidence angle, polarization, and wavelength. The dielectric constant, or permittivity, describes how a surface attenuates or transmits the incoming radar wave. Live vegetation with high water content has a higher dielectric constant than drier vegetation, implying that a stronger backscattering signal is expected from wet vegetation than drier vegetation. The transmission of the radar signal through the canopy is also directly related to the characteristics of the radar as well as the canopy structure. Therefore, comparing the backscattering values (σ°) from various vegetation canopies in diverse environments can provide insight into the dominant canopy structure.

Radar signal backscattering mechanisms over wetlands are simplified into four major categories: *surface backscattering*, *volume backscattering*, *double-bounced backscattering*, and *specular scattering*. Figure 2.2 illustrates how different structural layers of vegetation affect the way a radar signal returns. Forested wetlands

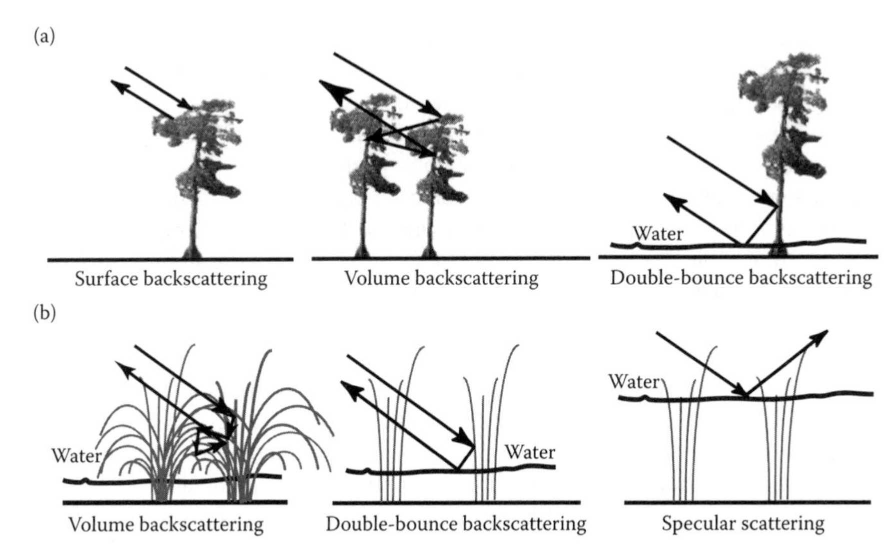

FIGURE 2.2 Schematic figures showing the contributions of radar backscattering over (a) forests and (b) marshes due to canopy surface backscattering, canopy volume backscattering, specular scattering, and double-bounce backscattering.

often develop into distinct layers, such as an overstory of dominant tree species, an understory of companion trees and shrubs, and a ground layer of herbaceous plants [37]. Therefore, over a dense forest, the illuminating radar signal scatters from the canopy surface, and a fraction of the energy is returned to the antenna. This phenomenon is called *surface backscattering*. The remaining radar wave penetrates into and interacts with the vegetation volume, and a portion of the energy is returned to the antenna. This results in *volume backscattering*. Volume backscattering can also dominate moderately dense forests with dense understory (Figure 2.2a). In a moderately dense forested canopy, some microwave energy penetrates through the overstory and interacts with tree trunks and the ground layer. If the ground is flooded, a large portion of the microwave energy is forward-scattered off the tree trunks, bounced off the smooth water surface, and then back to the radar antenna. This phenomenon is called *double-bounced backscattering* (Figure 2.2a). Because double-bounced backscattering returns more microwave energy back to the antenna than other types of backscattering, the SAR image should have an enhanced intensity compared to other types of vegetation canopy where volumetric backscattering dominates.

Over herbaceous canopies, SAR can often penetrate through the vegetation to reach the ground surface depending on the vegetation density. If the soil is dry, multiple backscatterings between vegetation and the ground surface can attenuate the incoming radar signal, reducing the energy returned to the radar. If the soil is wet, the higher dielectric constant of the soil reduces the transmission of the radar wave and enhances the backscattering return. If the ground is flooded, and the above-water stems are large enough and properly oriented to allow double bounce between the water surface and stems, the backscattering signal is significantly enhanced (i.e., double-bounced backscattering) (Figure 2.2b). If the ground is completely flooded, and the vegetation canopy is almost submerged, there is little chance for the radar signal to interact with the canopy stems and the water surface. Instead, most of the radar energy is scattered away from the antenna (i.e., *specular scattering*) (Figure 2.2b) and little energy is bounced back to the radar. Floating aquatic vegetation and short vegetation in flooded areas may exhibit similar backscattering returns and are therefore indistinguishable from SAR backscattering values. In general, the overall bulk density of these vegetation classes may determine the total amount of SAR signal potentially backscattered to the sensor.

2.3.2 Mapping Water-Level Changes Using InSAR

Interactions of C-band radar waves with water surface are relatively simple [38]. As SAR transmits radar pulses at an off-nadir look-angle, if the weather is calm, a smooth open-water surface causes most of the radar energy to reflect away from the radar sensor, resulting in little energy being returned to the SAR receiver. When the open-water surface is rough and turbulent, part of the radar energy can be scattered back to the sensor. However, SAR signals over open water are not coherent if two radar images are acquired at different times. Thus, it has been generally accepted that InSAR is an inappropriate tool to use in studying changes in the water level of open water. As described in the previous section, the radar backscattering over flooded wetlands consists of contributions from the interactions of radar waves with the canopy surface, canopy volume, and water surface. Neglecting specular scattering, the total radar backscattering over wetlands can be approximated as the incoherent summation of contributions from (a) canopy surface backscattering, (b) canopy volume backscattering that includes backscattering from multiple path interactions of canopy water, and (c) double-bounce trunk-water backscattering (Figure 2.2a). The relative contributions from those three backscattering components are controlled primarily by vegetation type, vegetation structure (and canopy closure), seasonal conditions, and other environmental factors. Over marsh wetlands, the primary backscattering mechanism is volume backscattering with possible contributions from stalk-water double-bounce backscattering, or specular scattering if the aboveground vegetation is short and the majority of the imaged surface is water (Figure 2.2b).

Ignoring the atmospheric delay in SAR data acquired at two different times, and assuming that topographic effect is removed, the repeat-pass interferometric phase (ϕ) is approximately the incoherent summation of differences in surface backscattering phase (ϕ_s), volume backscattering phase (ϕ_v), and double-bounce backscattering phase (ϕ_d):

$$\phi = (\phi_{s2} - \phi_{s1}) + (\phi_{v2} - \phi_{v1}) + (\phi_{d2} - \phi_{d1}), \quad (2.1)$$

where ϕ_{s1}, ϕ_{v1}, and ϕ_{d1} are the surface, volume, and double-bounce backscattering phase values from the SAR image acquired at an early date, and ϕ_{s2}, ϕ_{v2}, and ϕ_{d2} are the corresponding phase values from the SAR image acquired at a later date.

As the two SAR images are acquired at different times, the loss of interferometric coherence requires evaluation. Only when coherence is maintained are interferometric phase values useful to map water-level changes. Loss of InSAR coherence is often referred to as decorrelation. Besides the thermal decorrelation caused by the presence of uncorrelated noise sources in radar instruments, there are three primary sources of decorrelation over wetlands (e.g., [9,39,40]): (a) geometric decorrelation resulting from imaging a target from different look angles, (b) volume decorrelation caused by volume backscattering effects, and (c) temporal decorrelation due to environmental changes over time.

Geometric decorrelation increases as the baseline—the distance between satellites—increases, until a critical length is reached when coherence is lost (e.g., [41,42]). For surface backscattering, most of the effect of baseline geometry on the measurement of interferometric coherence can be removed by common spectral band filtering [43]. Volume backscattering describes multiple scattering of the radar pulse occurring within a distributed volume of vegetation; therefore, InSAR baseline geometry configuration can significantly affect volume decorrelation. Volume decorrelation is most often coupled with geometric decorrelation and is a complex function of vegetation canopy structure that is difficult to simulate. As a result, volume decorrelation cannot be removed. Generally, the contribution of volume backscattering is controlled by the proportion of transmitted signal that penetrates the surface and the relative two-way attenuation from the surface to the volume element and back to the sensor [9]. Canopy closure may significantly impact volume backscattering; the volume decorrelation should generally be disproportional to canopy closure. Both surface backscattering and volume backscattering consume and attenuate the transmitted radar signal; hence, they reduce the proportion of radar signal available to produce double-bounce backscattering that is utilized to measure water-level changes [44].

Temporal decorrelation describes any event that changes the physical orientation, composition, or scattering characteristics and spatial distribution of scatterers within an imaged volume. Temporal decorrelation is the net effect of changes in radar backscattering and therefore depends on the stability of the scatterers, the canopy penetration depth of the transmitted pulse, and the response to changing conditions with respect to the wavelength. Over wetlands, these decorrelations are primarily caused by wind changing leaf orientation, moisture condensation, rain, and seasonal phenology changing the dielectric constant of the vegetation, flooding changing the dielectric constant and roughness of the canopy background, as well as anthropogenic activities such as cultivation and timber harvesting [9,44].

The above discussion has clarified how the geometric, volume, and temporal decorrelation are interleaved with each other and collectively affect InSAR coherence over wetlands. The combined decorrelation estimated using InSAR images (quantitatively assessed in Section 6.2) determines the ability to detect water-level changes through radar double-bounce backscattering. When double-bounce backscattering dominates the returning radar signal, a repeat-pass InSAR image is potentially coherent enough to allow the measurement of water-level changes from the interferometric phase values. The interferometric phase (ϕ) is related to the water-level change (Δh) by the following relationship [44]:

$$\Delta h = -\frac{\lambda\phi}{4\pi\cos\theta} + n, \quad (2.2)$$

where ϕ is the interferogram phase value, λ is the SAR wavelength (5.66 cm for C-band ERS-1, ERS-2, and RADARSAT-1), θ is the SAR local incidence angle, and n is the noise caused primarily by the aforementioned decorrelation effects.

2.4 DATA AND PROCESSING

2.4.1 SAR Data

SAR data used in the study consist of 33 scenes of European Remote Sensing (ERS)-1 and ERS-2 images and 19 scenes of RADARSAT-1 images (Table 2.1). The ERS-1/ERS-2 scenes, spanning 1992–1998, are from the descending track 083 with a radar incidence angle of about 20°–26°. The ERS-1/ERS-2 data are vertical-transmit and vertical-receive (VV) polarized. The RADARSAT-1 scenes, spanning 2002–2004, are from an ascending track with a radar incidence angle of about 25°–31°. Unlike ERS-1/ERS-2, RADARSAT-1 images are horizontal-transmit and horizontal-receive (HH) polarized. SAR raw data are processed into single-look complex (SLC) images with antenna pattern compensation. The intensity of the SLC image was converted into the backscattering coefficient, $\sigma°$, according to Wegmüller and Werner [45]. Southern Louisiana's topography is almost flat; therefore, additional adjustment of $\sigma°$ for local terrain slope effect is not necessary.

ERS-1 and ERS-2 SLC images are coregistered to a common reference image using a two-dimensional *sinc* function [45]. The coregistered ERS SLC images are multilooked using a 2×10 window to represent a ground-projected pixel size of about 40×40 m^2. The same procedure is used to process RADARSAT-1 data

TABLE 2.1
SAR Sensor Characteristics: Sensor, Band, Orbit Direction, and Incidence Angle

Satellite	Image Acquisition Dates (year: mm/dd)
ERS-1 (C-band, VV)	1992: 06/11, 07/16, 08/20, 09/24, 10/29
Orbit pass: Descending	1993: 01/07, 04/22, 09/09
Incidence angle: 23.3°	1995: 11/11
	1996: 01/20, 05/04 (11 scenes)
ERS-2 (C-VV)	1995: 11/12, 12/17
Orbit pass: Descending	1996: 01/21, 05/05, 06/09, 07/14, 08/18, 09/22, 10/27, 12/01
Incidence angle: 23.3°	1997: 01/05, 03/16, 05/25, 09/07, 10/12, 11/16
	1998: 01/25, 03/01, 04/05, 07/19, 08/23, 09/27 (22 scenes)
RADARSAT-1 (C-HH)	2002: 05/03, 05/27, 06/20, 07/14, 08/07, 08/31, 11/11
Orbit pass: Ascending	2003: 02/15, 05/22, 06/15, 07/09, 08/02, 10/12, 12/23
Incidence angle: 27.7°	2004: 02/09, 03/28, 04/21, 07/02, 09/12 (19 scenes)
PALSAR (L-HH)	2007: 02/27, 04/14 (2 scenes)
Orbit pass: Ascending	
Incidence angle: 38.7°	

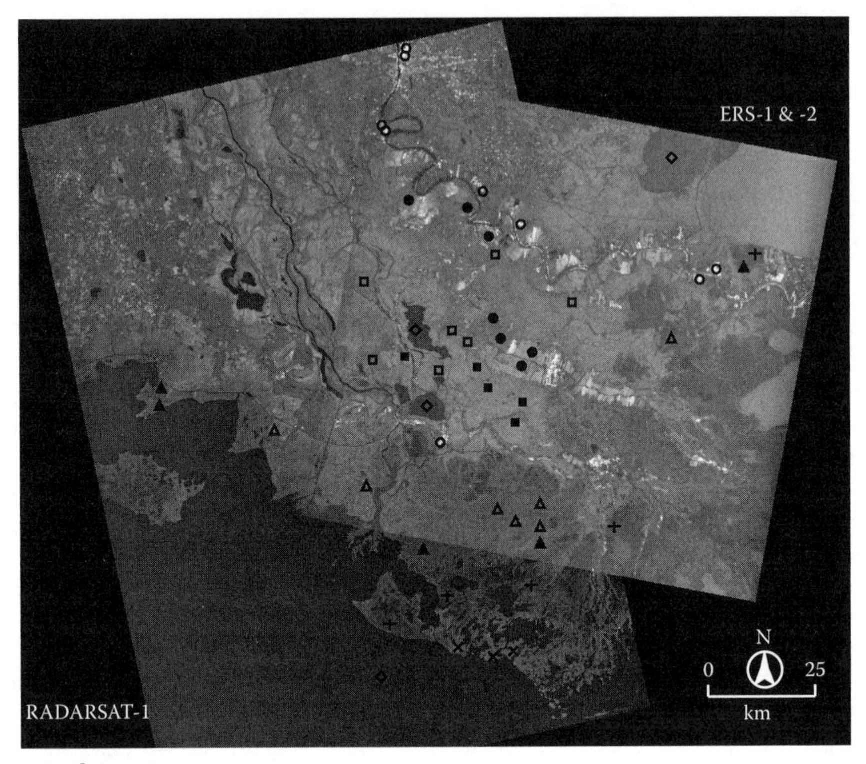

FIGURE 2.3 Averaged ERS-1/ERS-2 and RADARSAT-1 intensity images showing locations where quantitative coherence analyses are conducted.

(Table 2.1). All coregistered RADARSAT-1 images are multilooked with a 3×10 window to represent a ground-projected pixel size of about 50×50 m^2. Speckle noise in the images is suppressed using the Frost adaptive despeckle filter [46] with a 3×3 window size on the coregistered and multilooked images. Finally, SAR images are georeferenced and coregistered with the modified GAP land-cover map (Figure 2.1) [23]. A SAR image mosaic composed of both the ERS-1/ERS-2 and RADARSAT-1 images is shown in Figure 2.3.

2.4.2 SAR Data Calibration

Many data samples across the study area were selected to examine seasonal variations of $\sigma°$ for different vegetation types. Locations of data samples are shown in Figure 2.3. For each of the nine land-cover classes, between three and nine locations distributed across the study area were chosen for backscattering analysis. The 2004

Digital Orthophoto Quarter Quadrangle (DOQQ) imagery for Louisiana [47] is used to verify the land-cover type over the sampling sites. The size of the sampling boxes varies between 3 × 3 and 41 × 41 pixels so that each box covers only a single land-cover type. The DOQQ imagery is also used to ensure the homogeneity of land cover at each site.

The results of average $\sigma°$ for each class are shown in Figure 2.4. The overall difference in the average $\sigma°$ between Figure 2.4a and b is due to differences in sensors and environmental change. The $\sigma°_{ERS}$ shows a generally downward trend (Figure 2.4a). This long-term declination is present for all land-cover classes, suggesting that ERS-2 has a temporal decrease of antenna power at a rate of about 0.5 dB per year, similar to the report by Meadows et al. [48]. Therefore, this long-term declination of $\sigma°$ has been compensated before further analysis.

Unlike ERS, $\sigma°_{RADARSAT}$ exhibits strong temporal variation for all land-cover types (Figure 2.4b), which is particularly evidenced by the observation that $\sigma°_{RADARSAT}$ over water mimics the variation of $\sigma°$ over other land classes. This strongly suggests that the temporal variation in RADARSAT-1 is caused not only by changes in environmental conditions but also by some systematic changes that are not well understood. Such variations warrant removal prior to any further analysis of $\sigma°$.

Extensive homogeneous surfaces with known backscattering characteristics, such as the Amazon forests, or carefully designed corner reflectors, are ideal for radiometric calibration, but no such locations exist in our study site. However, many artificial structures and objects in cities, such as buildings, roads, and industrial facilities, may be considered corner-reflector units and behave like permanent scatterers whose backscattering characteristics do not change with time despite environmental variation [49]. Under ideal conditions, backscattering coefficients from urban areas should remain almost constant over time and therefore usable as an alternative to calibrate time-varying radar backscattering characteristics. This led Kwoun and Lu [23] to propose a relative calibration of radar backscattering coefficients for vegetation classes using $\sigma°$ over urban areas: for each SAR scene, the averaged $\sigma°$ value of urban areas from the corresponding image is used as the reference and subtracted from $\sigma°$ values of other land-cover classes in that individual image. The "calibrated" $\sigma°$s are then used to study backscattering characteristics of different land-cover types and their seasonal changes (Figure 2.5). Figure 2.6 shows seasonally averaged $\sigma°$ of each land-cover type for leaf-on and leaf-off seasons after relative calibration.

2.4.3 InSAR Processing

A total of 47 ERS-1/ERS-2 interferograms with perpendicular baselines less than 300 m (Figure 2.7a) and 31 RADARSAT-1 interferograms with perpendicular baselines less than 400 m (Figure 2.7b) were produced. The common spectral band filtering is applied to maximize interferometric coherence [43]. Interferometric coherence was calculated using 15 × 15 pixels on ERS-1/ERS-2 interferograms that were generated with a multilook factor of 2 × 10 from the SLC images, and 11 × 11 pixels for RADARSAT-1 interferograms with a multilook factor of 3 × 11. Therefore, the coherence measurements were made over a spatial scale of about 600 × 600 m^2. As significant fringes were observed over swamp forest areas, we "detrended" the

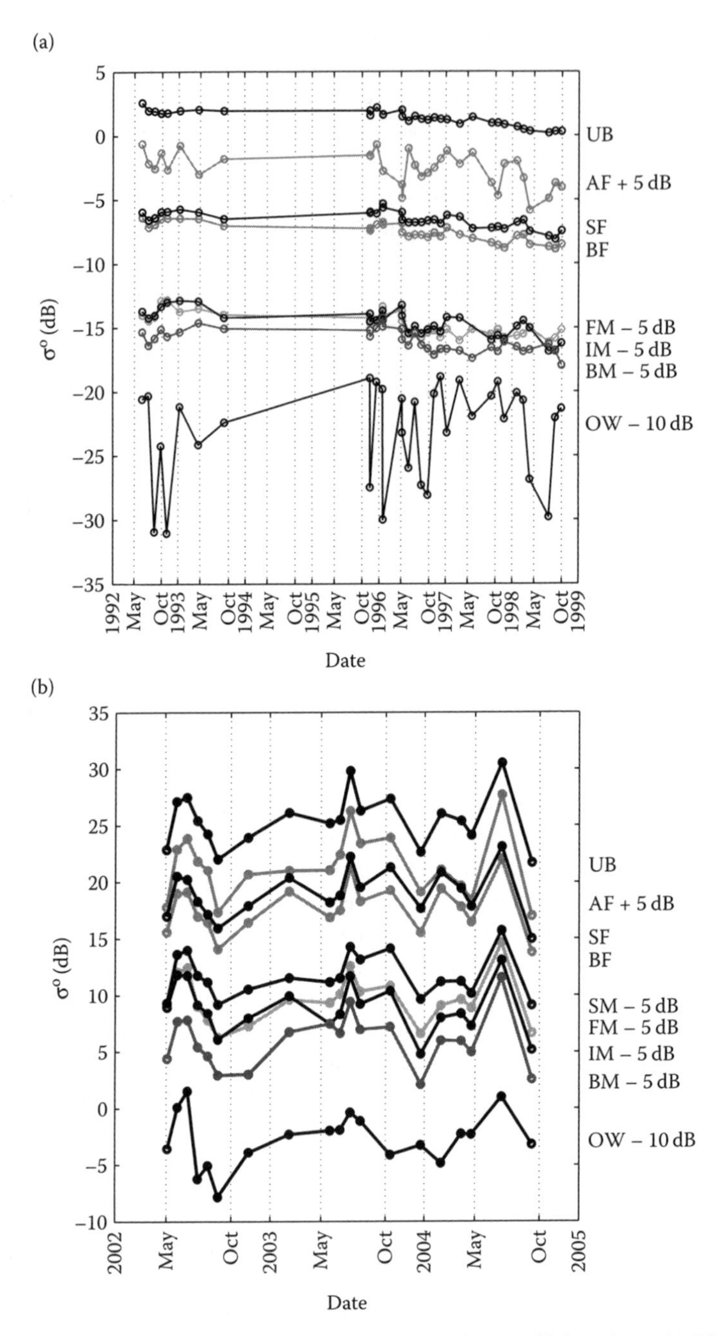

FIGURE 2.4 Temporal variations of radar backscattering coefficient from (a) ERS-1 and ERS-2 and (b) RADARSAT-1. UB: urban, AF: agricultural fields, SF: swamp forests, BF: bottomland forests, FM: freshwater marshes, IM: intermediate marshes, BM: brackish marshes, SM: saline marshes, OW: open water.

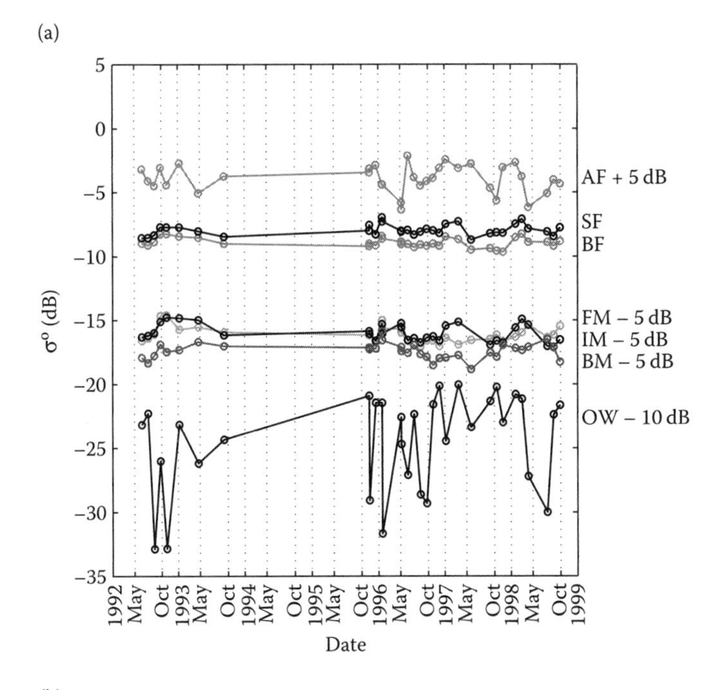

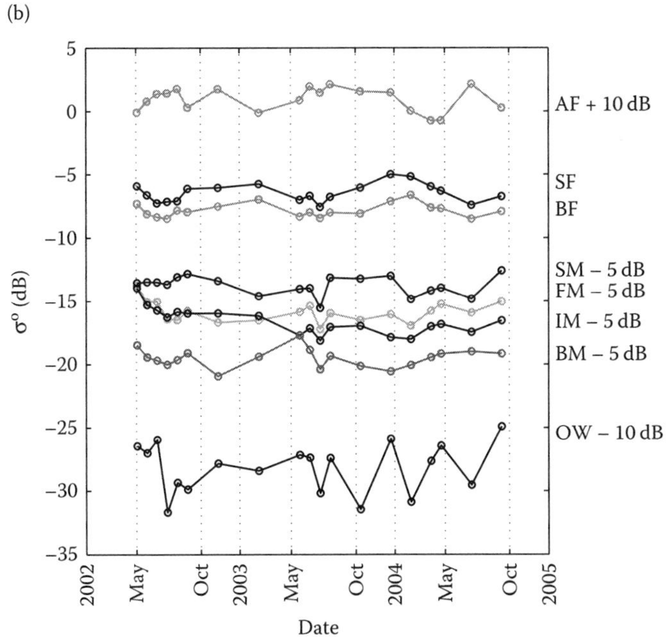

FIGURE 2.5 Calibrated radar backscattering coefficient (σ°) of each land-cover type from (a) ERS-1 and ERS-2 and (b) RADARSAT-1 images. The relative calibration is achieved with the averaged σ° of urban areas. AF: agricultural fields, SF: swamp forests, BF: bottomland forests, FM: freshwater marshes, IM: intermediate marshes, BM: brackish marshes, SM: saline marshes, OW: open water.

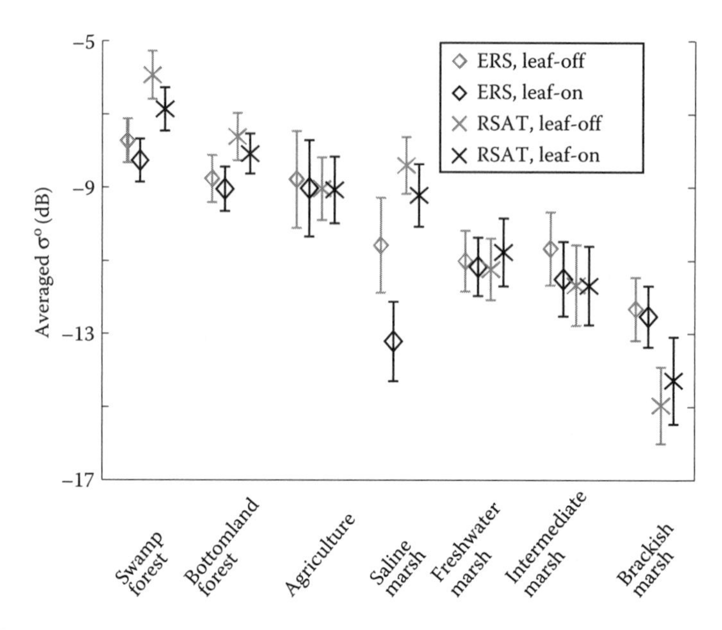

FIGURE 2.6 Multiyear seasonally averaged $\sigma°$ of each land-cover type. The error bars represent 1 standard deviation.

fringes to calculate the coherence based on Lu and Kwoun's [44] procedure to reduce artifacts caused by dense fringes on the coherence estimation.

2.5 SAR BACKSCATTERING ANALYSIS

2.5.1 Radar Backscattering Over Different Land-Cover Classes

For the purpose of analyzing seasonal backscattering changes, a typical year is split into two seasons. As summarized in Section 2.5.2, the Normalized Difference Vegetation Index (NDVI) is used to identify the peaks of "green-up," which occur around early May and early October. For convenience, the period between May and October is referred to as the "leaf-on" season, and the rest of the year as the "leaf-off" season. However, our definition of "leaf-off" does not necessarily mean that the vegetation has no leaves, as one would expect of deciduous trees in high latitude regions. Over our study area, some marsh types exhibit little, if any, seasonal variation (e.g., black needlerush); others change in their green biomass percentage; and others completely overturn. The "calibrated" $\sigma°$s within a season are averaged to study backscattering characteristics of different land covers and their seasonal changes (Figure 2.6).

The agricultural fields in the study area do not follow the natural cycle of vegetation. Multiple harvests and plowing drastically change surface roughness and moisture conditions, which significantly alter radar backscattering values. Therefore, agricultural fields are excluded from further analysis.

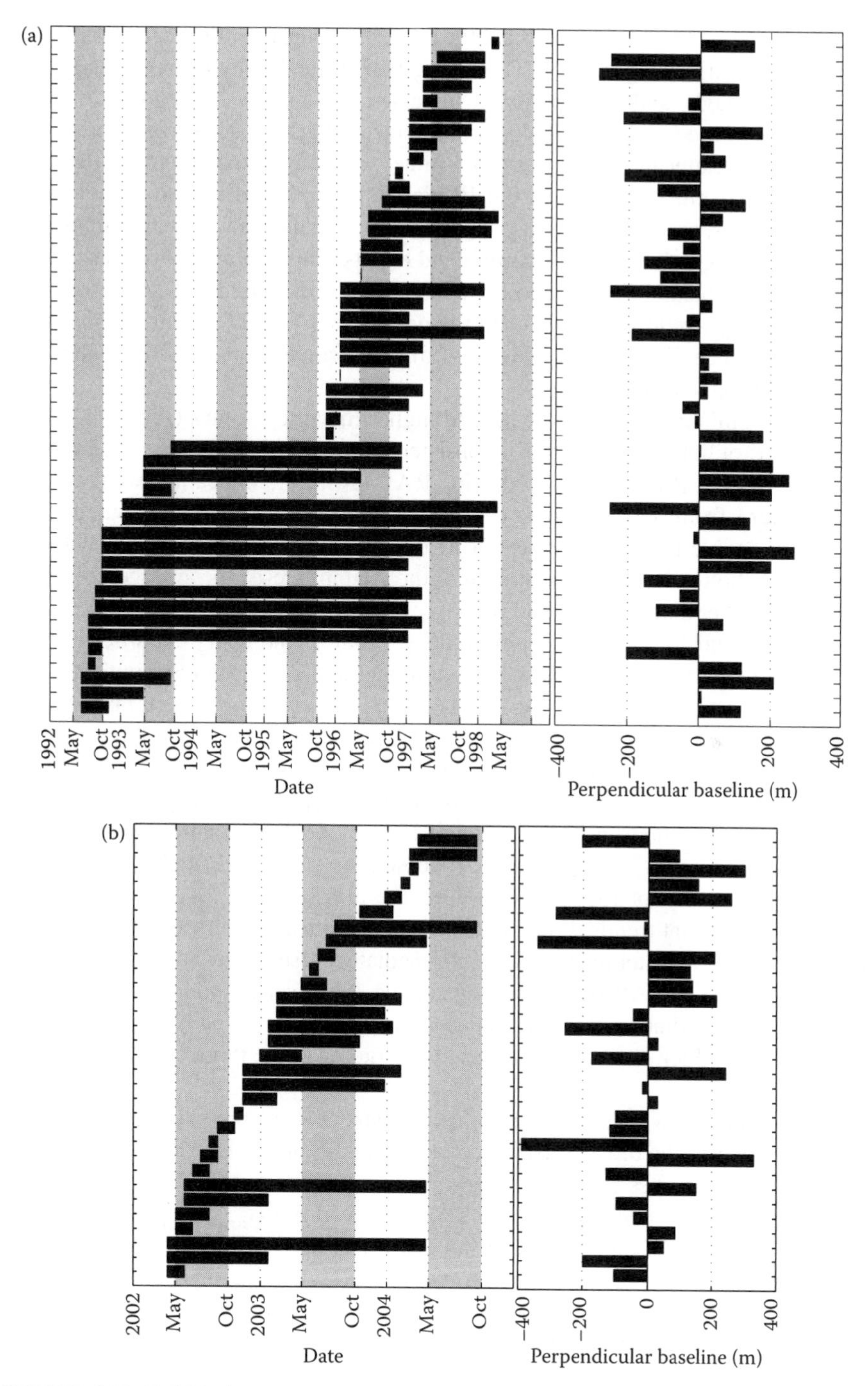

FIGURE 2.7 InSAR image pair characteristics, including image acquisition times and their corresponding baselines for both ERS-1/ERS-2 and RADARSAT-1 data used in this study.

The σ° values of swamp forests are the highest among all of the vegetation classes under investigation (Figure 2.6). This suggests that the density of trees is moderate or sparse enough, and the density of understory, if any, is low enough to allow penetration of the C-band SAR signal to interact with the water surface for double-bounce backscattering. The mean σ° values of swamp forests from ERS and RADARSAT-1 are about 0.5 and 0.9 dB higher during leaf-off seasons than during leaf-on seasons, respectively. Seasonal backscatter changes over swamp forests are consistently larger than those of bottomland forests. This is probably because during the leaf-on season, radar attenuation at the overstory is increased and double-bounced backscattering is reduced, which results in decreases in both σ° values and interferometric coherence. The $\sigma^\circ_{RADARSAT}$ during leaf-on season is around 0.9 dB higher than the σ°_{ERS} during leaf-off season.

The bottomland forest has the second highest mean σ° values (Figure 2.6). The averaged σ° of bottomland forests is consistently lower than that of swamp forests by 0.5–1.3 dB for ERS and 0.8–1.7 dB for RADARSAT-1, indicating weaker radar signal return from bottomland forest than swamp forest. This is attributed to the decreased double-bounced backscattering due to dense understory canopy, which is abundant in bottomland forests. Similar to the swamp forests, the averaged $\sigma^\circ_{LEAF\_OFF}$ is slightly higher than the averaged $\sigma^\circ_{LEAF\_ON}$; however, the difference is much smaller over bottomland forests than swamp forests. From a land-cover classification perspective, comparison of σ° between bottomland and swamp forests indicates that averaged intensity of RADARSAT-1 data during any single year contains sufficient information to differentiate the two classes (Figure 2.6).

Freshwater and intermediate marshes show relatively similar σ° for both ERS and RADARSAT-1. The seasonally averaged values of σ°_{ERS} and $\sigma^\circ_{RADARSAT}$ for freshwater marshes do not show any distinct trends (Figure 2.6). As for intermediate marshes, the mean values of σ°_{ERS} during leaf-off season are 0.9–1.5 dB higher than during leaf-on season, except for 1996; however, the averaged $\sigma^\circ_{RADARSAT}$ values do not show any consistent trends. From a land-cover classification perspective, Figure 2.6 indicates that freshwater marshes and intermediate marshes may not be easily distinguishable based on SAR backscattering signals. Also worth noting is that although fresh and intermediate marshes are outside the direct inundation of most tides, they could be flooded for extended time periods. In our analysis, those conditions are not included.

The seasonally averaged σ°_{ERS} of brackish marshes have the lowest mean σ° values (Figure 2.6). The drastic difference between ERS and RADARSAT is probably because sampling sites for the two sensors are not colocated due to a limitation in the image coverage (Figures 2.3 and 2.6). The averaged $\sigma^\circ_{RADARSAT}$ during leaf-on seasons is 0.8–0.9 dB higher than during leaf-off seasons, while σ°_{ERS} does not show any significant difference (Figure 2.6). From a land-cover classification perspective, Figure 2.6 indicates that single-year SAR data, particularly RADARSAT-1, are potentially sufficient to distinguish brackish marshes from other vegetation communities.

As in the case of brackish marshes, the sampling sites for RADARSAT and ERS data for saline marshes cannot be colocated (Figures 2.3 and 2.6). The mean $\sigma^\circ_{ERS\_LEAF\_ON}$ of saline marshes is comparable to that of brackish marshes, and the mean $\sigma^{\circ}_{ERS\_LEAF\_OFF}$ shows considerably dynamic interseasonal change and is in the range of freshwater and intermediate marshes (Figure 2.6). The averaged $\sigma^{\circ}_{RADARSAT}$ is comparable to that of bottomland forests (Figure 2.6). Both ERS and RADARSAT data show that the averaged $\sigma^{\circ}_{LEAF\_OFF}$ is higher than the averaged $\sigma^{\circ}_{LEAF\_ON}$, as is the case with forests. The saline marsh community is inundated daily with salt water tides and is subjected to the ebb and flow of the tides [3]. Therefore, it provides a favorable condition for double-bounced scattering between stems and the water surface underneath. From the image classification perspective, RADARSAT data are probably sufficient to distinguish saline marshes from other marsh classes. The mean value of $\sigma^{\circ}_{ERS\_LEAF\_ON}$ of saline marshes is so distinct that some level of ambiguity in $\sigma^{\circ}_{RADARSAT}$ between bottomland forests and saline marshes can be resolved. In addition, the proximity to salt water is another potential indicator that separates these two communities.

In summary, to classify wetland classes over the study area, the seasonal σ° values averaged over multiple years are useful to distinguish among bottomland forests, swamp forests, saline marshes, brackish marshes, and freshwater and intermediate marshes. Forests versus marshes are identifiable because the $\sigma^{\circ}_{ERS\_LEAF\_ON}$ of marshes is significantly lower than that of forests. Swamp forests are marked with the highest σ° values from both ERS and RADARSAT-1.

Among the marshes, brackish marshes are characterized by the consistently lowest σ° of RADARSAT-1. A saline marsh may be identified by its highest averaged $\sigma^{\circ}_{RADARSAT}$ among marsh classes. Freshwater and intermediate marshes have very similar σ°. However, the averaged $\sigma^{\circ}_{ERS\_LEAF\_OFF}$ for intermediate marshes is marginally higher than $\sigma^{\circ}_{ERS\_LEAF\_ON}$. The seasonally averaged σ°s of both saline and brackish marshes behave quite distinctly compared to those of freshwater and intermediate marshes, which may help map changes in salinity in coastal wetlands.

2.5.2 SAR Backscattering versus Vegetation Index

The previous section has shown that seasonal variation of radar backscattering signals responds to changes in structural elements of vegetation classes. The seasonal changes of vegetation cover are also detectable by optical sensors. NDVI is a numerical indicator that utilizes the ratio between spectral reflectance measurements acquired in the red and near-infrared regions to assess whether the target being observed contains live green vegetation (e.g., [50]). We now show how the radar signal can be related to land-cover information derived from optical sensors by comparing σ° to NDVI derived from Advanced Very High Resolution Radiometer (AVHRR) imagery. The multiyear NDVI curves (Figure 2.8) are averaged into a single year to determine typical leaf-on and leaf-off seasons [23]. The peaks of averaged NDVI are found in the intervals of April 23–May 6 in the spring and September 24–October 7 in the fall; therefore, the time window from around May 1 until about the end of September is chosen as the "leaf-on" season. The leaf-on season is meant to represent the time when leaves maintain fully developed conditions and is characterized by peaks in the NDVI curves in the spring and fall (Figure 2.8). The rest of the year is defined as the "leaf-off" season.

Because the leaf-on season is the time period when leaves are fully developed, significant changes in radar backscattering are not expected. Regressions between

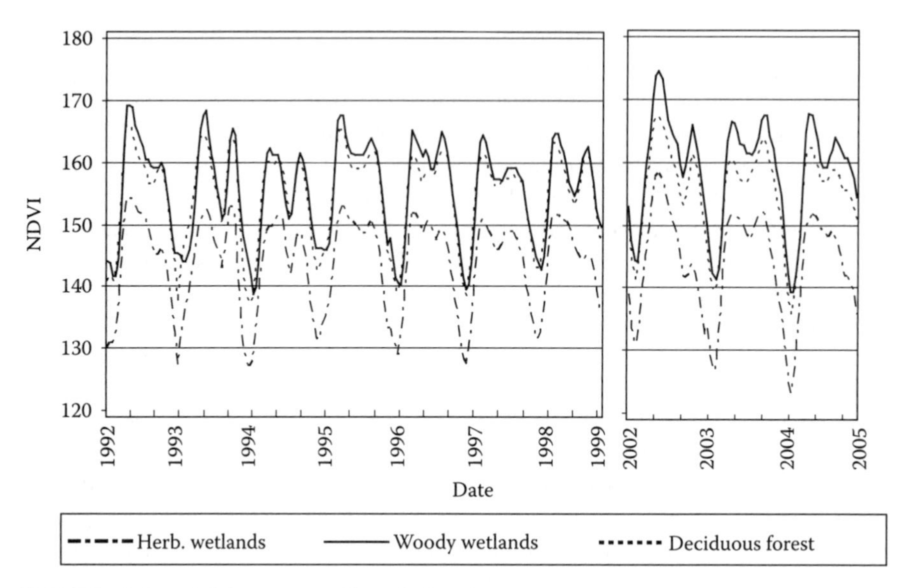

FIGURE 2.8 NDVI values adjusted for long-term trends during 1992–1998 and 2002–2004.

$\sigma°$ and NDVI for all vegetation types do not show any significant correlation during leaf-on season because the dynamic range of the variation of NDVI is too narrow compared to radar backscatter changes. During the leaf-off season, both bottomland and swamp forests show moderate to strong negative correlations with NDVI (Figure 2.9a through d). The negative correlation during the leaf-off season is likely associated with the attenuation of radar backscatter due to the growth of leaves, which reduces the amount of radar signal available for double-bounce and volume scattering. As a result, radar backscatter decreases with the increase in NDVI for swamp and bottomland forests. Therefore, negative correlation between NDVI and $\sigma°$ is anticipated.

For marshes, $\sigma^{\circ}_{ERS\_LEAF\_OFF}$ does not show any correlation with NDVI. However, $\sigma^{\circ}_{RADARSAT\_LEAF\_OFF}$ shows impressive positive correlation with NDVI (Figure 2.9e through h). This positive correlation implies that the radar backscattering is enhanced with an increase in NDVI, suggesting surface or volume scattering of the radar signal. For freshwater and intermediate marshes, positive correlation is consistent with our interpretation of $\sigma°$ in Section 5.1. Brackish marshes show marginally positive correlation, which implies that brackish marshes are not as dense as the other marshes to enhance $\sigma°$ sufficiently with the growth of vegetation. Saline marshes show moderate positive correlation. This may sound contradictory to our previous interpretation in Section 2.5.1. The increase in NDVI is probably associated with the thickening of saline marsh stems, which may be translated into an increase in the double-bounced radar backscattering signal. By combining NDVI and radar backscatter signal, forests versus wetland marshes are classifiable. NDVI maps derived from higher spatial resolution images and more detailed classes than those defined by NLCD may improve our current results.

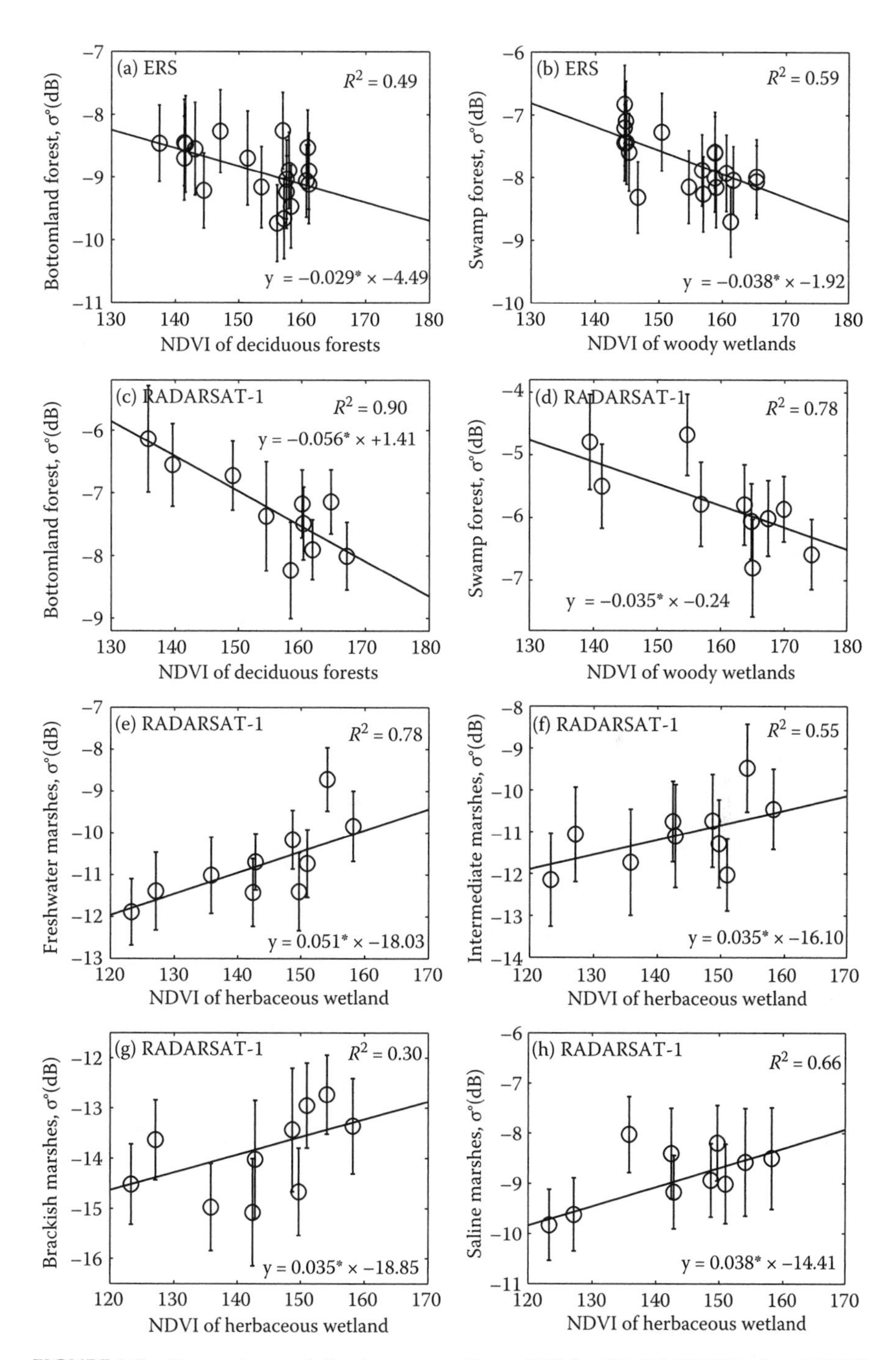

FIGURE 2.9 Regression modeling between calibrated ERS or RADARSAT-1 σ° and NDVI for the leaf-off season. ERS σ°s for marshes do not show any correlation with NDVIs and are therefore not presented in this figure.

2.6 InSAR COHERENCE ANALYSIS

2.6.1 Observed InSAR Images

A few examples of ERS-1/ERS-2 and RADARSAT-1 interferograms for portions of the study area are shown in Figure 2.10. Figure 2.10a through d shows ERS-1/ERS-2 interferograms acquired during leaf-off seasons, with a time separation of 1 day (Figure 2.10a), 35 days (Figure 2.10b), 70 days (Figure 2.10c), and 5 years (Figure 2.10d). The 1-day interferogram (Figure 2.10a) during the leaf-off season is coherent for almost every land-cover class except open water. In the 1-day interferogram, a few localized areas exhibit interferometric phase changes, which are most likely a result of water-level changes over the swamp forests. The large-scale phase changes over the southeastern part of the interferogram are likely caused by atmospheric delay anomalies. Most of the land-cover classes (Figure 2.1), except open water, bottomland forests, and some of the freshwater and intermediate marshes, are coherent in the 35-day interferogram (Figure 2.10b). The interferogram clearly shows the

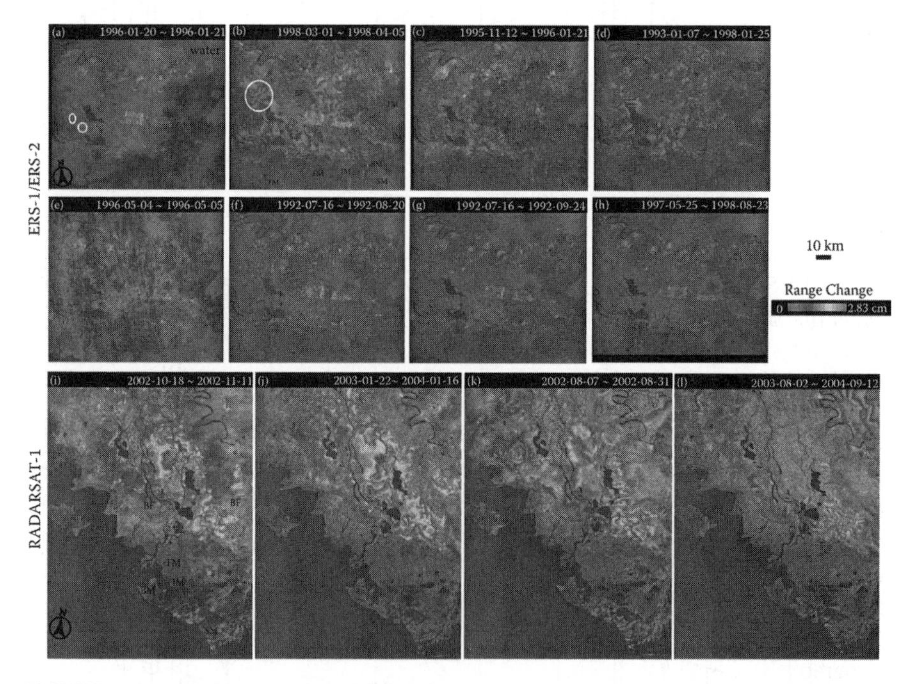

FIGURE 2.10 **(See color insert following page 206.)** (a–d) ERS-1/ERS-2 InSAR images with different time separations during leaf-off seasons. (e–h) ERS-1/ERS-2 InSAR images with different time separations during leaf-on seasons. (i, j) RADARSAT-1 InSAR images during leaf-off seasons. (k, l) ERS-1/ERS-2 InSAR images during leaf-on seasons. Each fringe (full color cycle) represents 2.83 cm of range change between the ground and the satellite. The transition of colors from purple, red, yellow and green to blue indicates that the water level moved away from the satellite by an increasing amount in that direction. Random colors represent loss of InSAR coherence, where no meaningful range change information can be obtained from the InSAR phase values. AG: agricultural field, SF: swamp forest, BF: bottomland forest, FM: freshwater marsh, IM: intermediate marsh, BM: brackish marsh, SM: saline marsh, OW: open water.

water-level changes over both swamp forests and marshes (Figure 2.10b). The overall coherence for the 70-day interferogram (Figure 2.10c) is generally lower than the 35-day interferogram (Figure 2.10b). In 70 days (Figure 2.10c), bottomland forests, freshwater marshes, and intermediate marshes completely lose coherence, although some saline marshes and brackish marshes can maintain coherence. Over 5 years, some swamp forests and urban areas can maintain coherence (Figure 2.10d). Coherence can be maintained for swamp forests for over 5 years.

Figure 2.10e through h shows interferograms from ERS-1/ERS-2 SAR images acquired during leaf-on seasons, with a time separation of 1 day (Figure 2.10e), 35 days (Figure 2.10f), 70 days (Figure 2.10g), and 1 year (Figure 2.10h). Compared with the corresponding interferograms acquired during leaf-off seasons with similar time intervals (Figure 2.10a through d), the leaf-on interferograms generally exhibit much lower coherence. All the land-cover classes (Figure 2.1) maintain coherence in 1 day (Figure 2.10e). For most land-cover classes, except urban, agriculture, and portions of swamp forests, interferometric coherence cannot be maintained after 35 days (Figure 2.10f through h). With a time interval of 70 days, only urban and some agricultural fields have coherence (Figure 2.10g). Over 1 year, only urban areas maintain some degree of coherence (Figure 2.10h). The overall reduction in interferometric coherence for swamp forests during leaf-on seasons is because the dominant backscattering mechanism is not double-bounce backscattering but a combination of surface and volume backscattering.

Figure 2.10i and j shows two RADARSAT-1 interferograms acquired during leaf-off seasons. Interferometric coherence for the 24-day HH-polarization RADARSAT-1 interferogram (Figure 2.10i) is generally higher than the 35-day VV-polarization ERS-1/ERS-2 interferogram (Figure 2.10b). In 24 days, only water and some freshwater marshes do not have good coherence (Figure 2.10i). Bottomland forests (Figure 2.10i and j) can maintain good coherence for 24 days. From Figure 2.10j, it is obvious that some swamp forests and urban areas maintain coherence for more than 1 year.

Figure 2.10k and l shows RADARSAT-1 interferograms acquired during leaf-on seasons. Again, the 24-day HH-polarization RADARSAT-1 interferogram maintains higher coherence than the 35-day VV-polarization ERS-1/ERS-2 images (Figure 2.10f) for most land-cover types. The 1-year RADARSAT interferogram surprisingly is able to maintain relatively high coherence over parts of swamp forests and saline marshes in leaf-on seasons. In general, even though RADARSAT-1 coherence is reduced during leaf-on seasons, the reduction in coherence for RADARSAT-1 is much less than that for ERS-1/ERS-2. Over a similar time interval, HH-polarized RADARSAT-1 interferograms have higher coherence than VV-polarized ERS-1/ERS-2 interferograms.

During a very dry season or a period of extremely low water, even swamp forests are potentially exposed to dry ground. Among the interferograms in our study area, patches of swamp forests can lose coherence. This is probably because there was no water beneath the swamp forest. Therefore, the double-bounce backscattering mechanism is diminished, and the dominant backscattering mechanism for "dried" swamp forests become very similar to that for bottomland forests. Alternatively, bottomland forests can be flooded occasionally. The presence of water on bottomland forests produces double-bounce backscattering and, accordingly, makes the radar backscattering return from a bottomland forest similar to that of a swamp forest.

2.6.2 Interferometric Coherence Measurement and Analysis

Only coherent InSAR images enable detection of water-level changes beneath wetlands; hence interferometric coherence variations over the study area are quantitatively assessed. Figure 2.11 shows the InSAR coherence measurements for different land-cover types. Thresholds of complete decorrelation for ERS-1/ERS-2 and RADARSAT-1 interferograms are determined by calculating interferometric coherence values over open water. In Figure 2.11, the coherence measurements from both leaf-on and leaf-off seasons are combined for all classes except for swamp and bottomland forests because seasonality is a critical factor that controls the interferometric coherence of forests. The dependence of the interferometric coherence on a spatial baseline was also explored. For the interferograms used in this study, no dependence between the interferometric coherence and the perpendicular baseline is observed. This is because more than 70% of the interferograms have perpendicular baselines of less than 200 m, and the common spectral band filtering [43] was applied during the interferogram generation.

The coherence over open water is shown in Figure 2.11a. Because open water completely loses coherence for repeat-pass interferometric observations, its coherence value can be regarded as the threshold of complete decorrelation (loss of coherence). The coherence values for both ERS-1/ERS-2 and RADARSAT-1 InSAR images are about 0.06 ± 0.013. Therefore, coherence values smaller than 0.1 are deemed as complete decorrelation.

Several sites (Figure 2.3) were chosen to show coherence over urban areas: six of them are located along the Mississippi River, and one is in Morgan City, which is more vegetated than the other urban sites. Figure 2.11b shows interferometric coherence measurements over urban sites from both ERS-1/ERS-2 and RADARSAT-1 interferograms. Overall coherence measurements from both ERS-1/ERS-2 and RADARSAT-1 images are similar, and they are higher than any other land-cover class. However, they vary in the range of about 0.2–0.7. The variations are most likely a result of decorrelation caused by vegetation over the urban areas. Vegetation in urban areas can alter radar backscattering coefficients by more than 8 dB [23]. Urban areas with lower radar backscattering intensities are usually associated with lower interferometric coherence values, suggesting that the vegetation in urban areas causes lower radar backscattering coefficients as well as reduced coherence measurements.

Coherence measurements over agricultural fields are shown in Figure 2.11c. Frequent farming activity with multiple harvests leads to a complete decorrelation in about 100 days. This implies that the vegetation condition of these fields changes completely in about 100 days.

Coherence measurements from swamp forests are shown in Figure 2.11d. The comparison of coherence measurements from ERS-1/ERS-2 and RADARSAT-1 images produces the following inferences:

1. Coherence is higher during leaf-off seasons than during leaf-on seasons for both ERS-1/ERS-2 and RADARSAT-1 images.
2. The coherence from HH-polarization RADARSAT-1 images is generally higher than that from VV-polarization ERS-1/ERS-2 images.

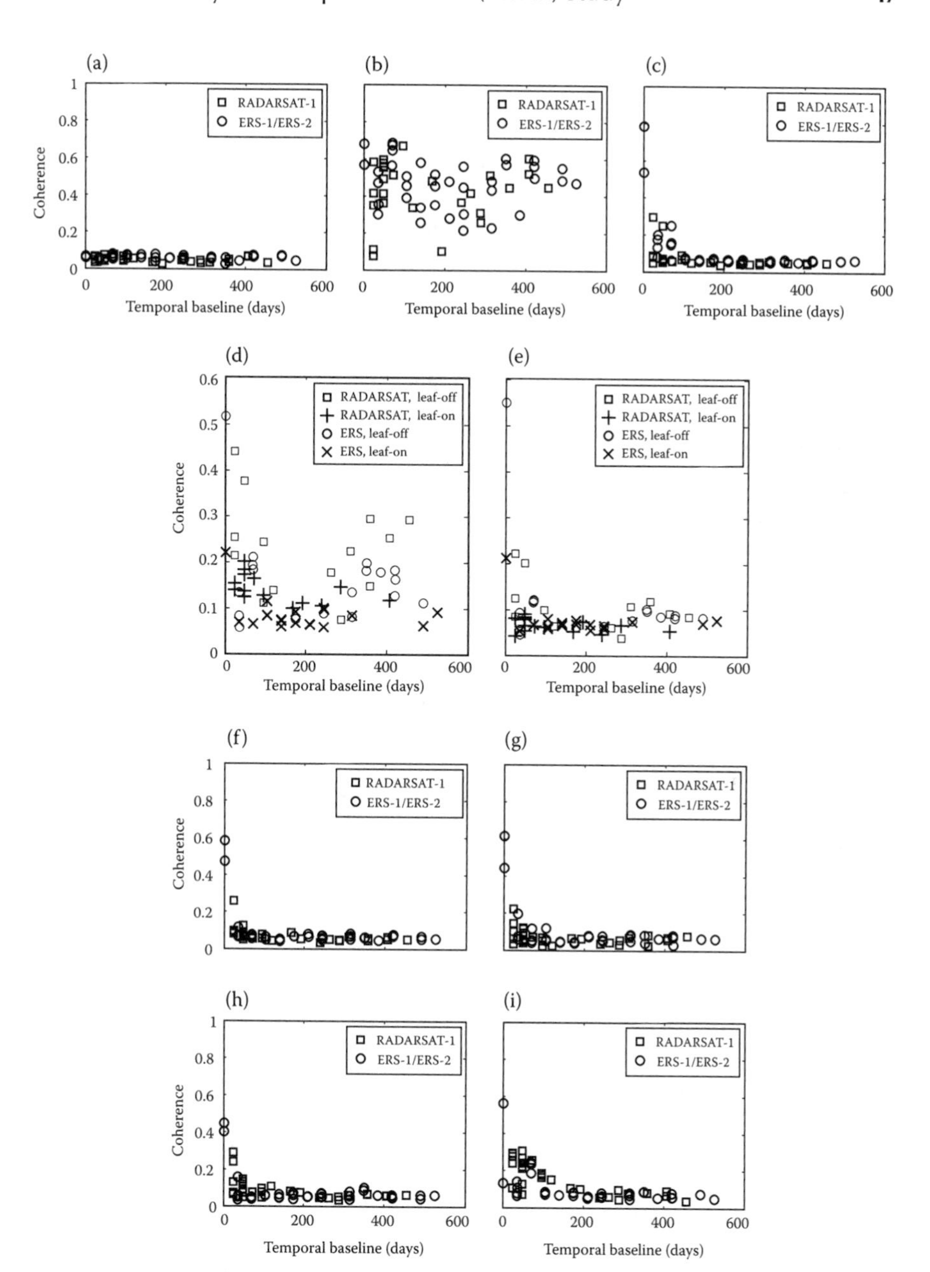

FIGURE 2.11 InSAR coherence as a function of time separation for seven major land-cover classes including (a) open water, (b) urban, (c) agriculture, (d) swamp forest, (e) bottomland forest, (f) freshwater marsh, (g) intermediate marsh, (h) brackish marsh, and (i) saline marsh for both ERS-1/ERS-2 and RADARSAT-1 interferograms acquired during leaf-off and leaf-on seasons. The scale for (d) and (e) is different from others to illustrate that seasonality is one of the factors controlling coherence for forests.

3. The coherence from both RADARSAT-1 images and ERS-1/ERS-2 images during leaf-off season can last over 2 years (Figure 2.11d).

If the scattering elements came primarily from the top of the forest canopy, it is unlikely that the SAR signals are coherent over a period of about 1 month or longer (e.g., [39,40]) because leaves and small branches that make up the forest canopy change due to weather conditions. Based on interferometric coherence (Figure 2.11d) and backscattering coefficient values (Figure 2.11c) during leaf-off and leaf-on seasons, we conclude that the dominant radar backscattering mechanism over swamp forests during the leaf-off seasons is double-bounce backscattering. As a result, RADARSAT-1 and ERS-1/ERS-2 images during leaf-off seasons are capable of imaging water-level changes over swamp forests. During leaf-on seasons, HH-polarization RADARSAT-1 images can maintain coherence for a few months, reaching up to about 0.2. If HH-polarized C-band radar images were acquired for shorter time intervals during leaf-on seasons, they could also be used for measuring water-level changes.

Coherence measurements for bottomland forests are shown in Figure 2.11e. The coherence is higher during leaf-off than leaf-on seasons. The HH-polarization RADARSAT-1 images tend to have higher coherence than the VV-polarized ERS-1/ERS-2 images for short temporal separations (less than about 2 months). The coherence from both ERS-1/ERS-2 and RADARSAT-1 images decreases exponentially with time. ERS-1/ERS-2 images become decorrelated in about 1 month, but RADARSAT-1 can maintain coherence for up to about 2 months (Figure 2.11e). Two major factors affect the difference in radar backscattering and coherence between swamp and bottomland forests. First, the double-bounce backscattering is enhanced over swamp forests. The water beneath trees enhances the double-bounce backscattering for swamp forests, producing high InSAR coherence as well as a high backscattering coefficient. For bottomland forests, forest understory attenuates radar signal returns and the double-bounce backscattering is retarded, resulting in relatively lower coherence as well as smaller backscattering values than swamp forests. Second, there are structural differences between the two forest types. The bottomland forests have broad leaves and deterrent structures where the lateral branches form a wide and bell-shaped crown, which enhances surface and volume backscattering. The above coherence analysis suggests that SAR images, preferably HH-polarized, can maintain good coherence over both swamp and bottomland forests for about 1 month. Accordingly, shorter temporal separations (a few days) will significantly improve the utility of InSAR coherence for the detection of water-level changes.

Coherence measurements over marshes are shown in Figure 2.11f–i. Coherence measurements are generally higher from HH-polarized RADARSAT-1 images than from VV-polarized ERS-1/ERS-2 images; ERS-1/ERS-2 can barely maintain coherence for about 1 month, whereas RADARSAT-1 maintains coherence up to about 3 months. The coherence values for intermediate, freshwater, and brackish marshes are similar, and they are lower than those for saline marshes. Saline marshes have nearly vertical stalks. Freshwater marshes have broadleaf plants that form a mostly vertical canopy and the plants die in the winter but retain the canopy structure until spring turnover and green-up [9]. Overall, saline marshes have the highest coherence

(Figure 2.11i) as well as the highest backscattering value (Figure 2.6) among marsh classes, suggesting that the saline marshes tend to develop more dominant vertical structure than other marshes to allow double-bounce backscattering of C-band radar waves. As marshes can only maintain coherence in less than 24 days, acquiring repeat-pass SAR images over short time intervals (a few days) would help robustly detect water-level changes.

Based on the findings in Sections 2.5 and 2.6, a decision-tree vegetation classification approach is proposed as shown in Figure 2.12. In the classifier, $\langle\rho\rangle$ is the

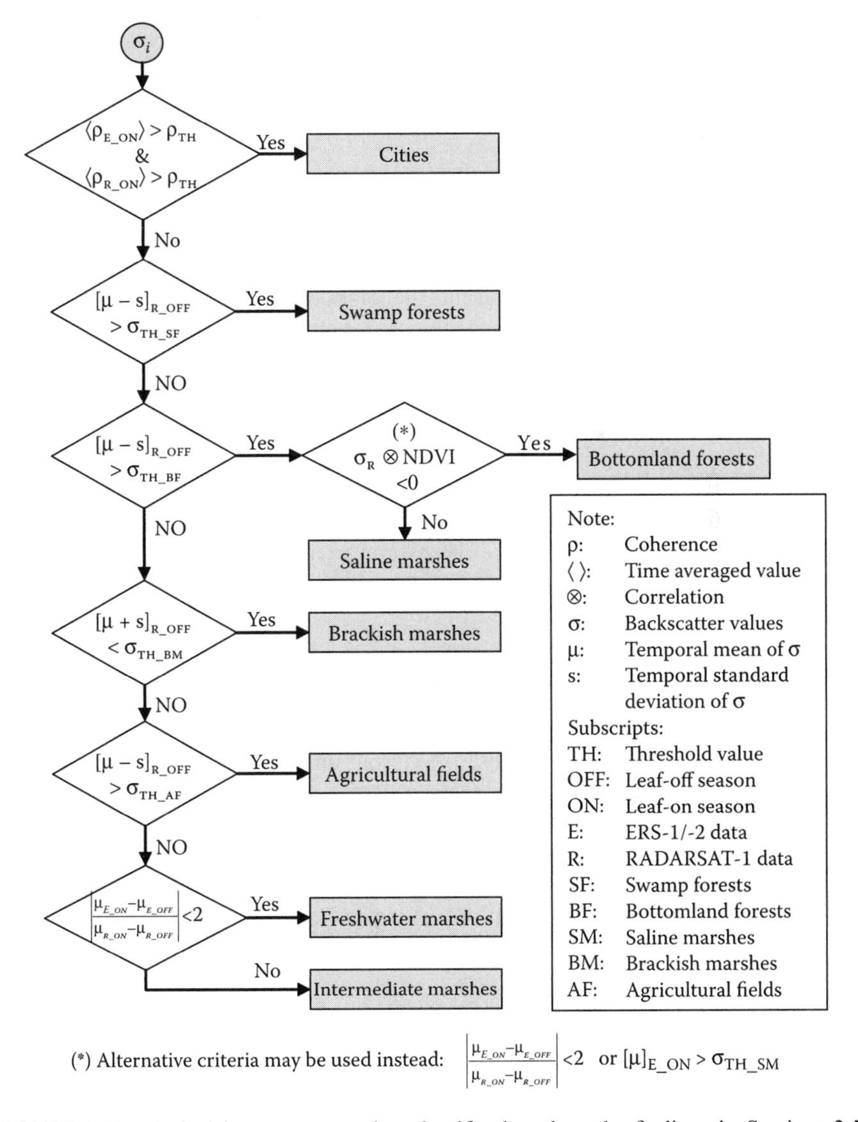

FIGURE 2.12 A decision-tree vegetation classifier based on the findings in Sections 2.5 and 2.6.

TABLE 2.2
Threshold Values for the Decision-Tree Classifier Based on the Data Used in this Study

Parameters	Threshold Values
ρ_{TH}	0.4
σ_{TH_SF}	−6.8 dB
σ_{TH_BF}	−9.5 dB
σ_{TH_SM}	−11.0 dB
σ_{TH_BM}	−12.5 dB
σ_{TH_AF}	−10.5 dB

average coherence value from image pairs with temporal baselines ranging from 1 day through over a year. "$\sigma \otimes$ NDVI" means the computation of correlation coefficient between temporal radar backscatter data and the NDVI data of woody wetlands or deciduous forests. "μ" represents the temporal mean value of "σ," and "s" is the standard deviation of "σ" of the sample under consideration for classification.

The assumptions for this classifier are as follows: (1) multiyear time series ERS-1/ERS-2 and RADARSAT-1 data are available, and the backscatter values are calibrated, (2) season-averaged NDVI values are at least available for woody wetlands and deciduous forests. Given a sample location to classify, the time series of backscatter data should be extracted for both ERS and RADARSAT-1 data. The threshold values noted by the subscript "TH" in Figure 2.12 can be derived based on the discussion in Sections 2.5 and 2.6. For the dataset used in this study, those threshold values are suggested in Table 2.2.

2.7 InSAR-DERIVED WATER-LEVEL CHANGES

Examples of coherent RADARSAT-1 interferograms with short temporal separations are shown in Figure 2.13. These interferograms are unwrapped to remove the intrinsic ambiguity of 2π in phase measurements (e.g., [25]), and interferometric phase values are used to study changes in the water level of swamp forests. Each interferogram shows the relative changes in water level between dates when the two images were acquired. Each fringe represents a range (distance from the satellite to ground) change of about 3.20 cm water-level change for RADARSAT-1 images. From the interferograms in Figure 2.13, the following inferences are made:

2.7.1 Water-Level Changes are Dynamic

Water-level changes reach as much as 50 cm over a distance of about 40 km (Figure 2.13b). The direction and the density of fringes within the Atchafalaya Basin vary spatially. Such changes in water level reflect local differences in topographic

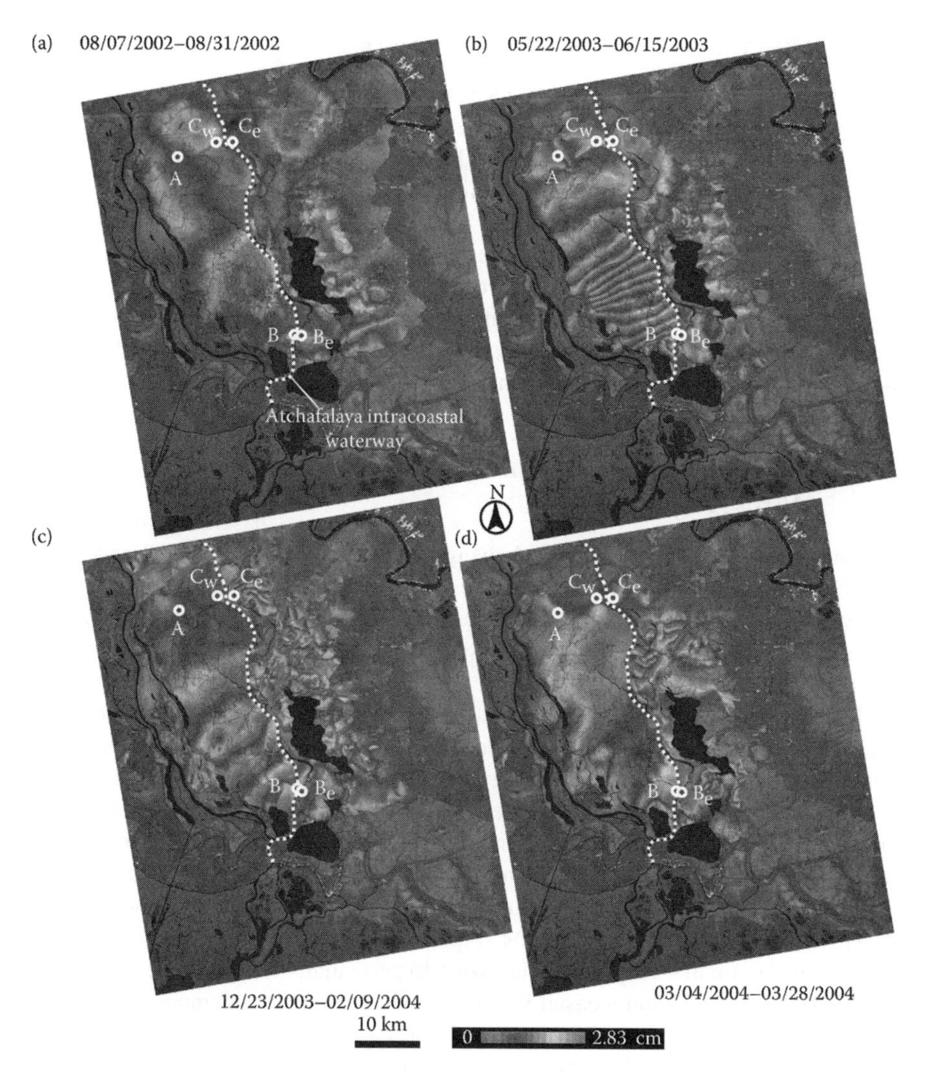

FIGURE 2.13 (See color insert following page 206.) Unwrapped RADARSAT-1 images of the Atchatalaya Basin are used to quantify water-level changes over Atchafalaya Basin Floodway. InSAR-derived water-level changes at the selected locations are compared with gage readings (see Table 2.3 for details).

constrictions and vegetation resistance to the surface flow. Flooding throughout this area is primarily by sheet flow after the rivers and bayous leave their banks. Under ideal circumstances, water should flow placidly and smoothly over a symmetrically smooth surface devoid of obstructions. Thus, the sheet flow should not be symmetric throughout the study area, that is, it should not be a smooth, even surface of constant elevation from one edge of the swamp to the other. Instead, a water surface with bulges and depressions reflecting the topographic constrictions and vegetation resistance in sheet flow is ideal.

2.7.2 Water-Level Changes are Heterogeneous

First, the observed fringes exhibit evidence of control by structures such as levees, canals, bayous, and roads, resulting in abrupt changes in interferometric phase value. The heterogeneity in water-level change is due primarily to these man-made structures and artificial boundaries. Second, within the Atchafalaya Basin Floodway (ABF), the observed interferometric fringes are bent. This suggests that local variations in vegetation cover resist water flow variably. Heterogeneous water-level changes such as these make it impossible to accurately characterize water storage based on measurements from a few sparsely distributed gauge measurements. This demonstrates the unique capability of InSAR to map water-level changes in unprecedented spatial detail. This is the most promising aspect of mapping water-level changes with InSAR.

2.7.3 Interferograms Reveal Both Localized and Relatively Large-Scale Water-Level Changes

On one hand, for example, localized changes in water flow are evident in the 24-h interferogram (outlined in white in Figure 2.10a) and the 35-day interferogram during March–April 1998 (outlined in white in Figure 2.10b). On the other, relatively large-scale changes in water level are observed across much of the water basin (e.g., Figure 2.13b and c).

The interferometric fringes are dissected by rivers, canals, levees, roads, and other structures; therefore, the interferometric phase measurements are perhaps disconnected at these boundaries. In other words, interferometric phase measurements at two nearby pixels separated by these boundaries are discontinuous. This adds enormous complexity to understanding water-level changes inferred from InSAR measurements. Furthermore, calculating water-level changes along two different paths that are separated by these boundaries may lead to different estimates.

The RADARSAT-1 interferograms (Figure 2.13) are used to illustrate this. The interferograms are first unwrapped piecewise. In particular, the regions to the west and east of the Atchafalaya Intracoastal Waterway (AICWW) were unwrapped separately. The interferometric coherence along the AICWW is often lost. To investigate water-level changes quantitatively, several locations including two gauge locations (Cross Bayou station at A and Sorrel station at B) were selected. Both A and B lie within the swamp forests west of the AICWW, and the phase measurements at the exact locations of A and B can be extracted. To the east of B and across the AICWW, a location B_e (Figure 2.13) over the swamp forest east of the AICWW, where InSAR coherence is maintained, is chosen. Finally, a location C in the upper portion of the AICWW (Figure 2.13) is selected. The interferometric coherence is not maintained at C; therefore, two locations immediately adjacent to C are chosen: one is over the swamp forest to the west of the AICWW (C_w in Figure 2.13) and the other, over the swamp forest to the east of AICWW (C_e in Figure 2.13). Interferometric coherence is maintained at C_w and C_e, and consequently InSAR phase values at these two points are extractable.

First, the water-level changes measured by InSAR are compared with those recorded at gages at A and B to validate the reliability of the InSAR-based measurements of water-level changes. Table 2.3 summarizes the results of water-level changes

TABLE 2.3
Comparison of Water-Level Change Measurements between InSAR and Gage Stations

Date	InSAR Measurements A–B (cm)	Gage Readings A–B (cm)	InSAR Measurements	
			C_w–B (cm)	C_e–B_e (cm)
08/07/2002–08/31/2002	−0.01	−0.60	−1.38	−2.01
05/22/2003–06/15/2003	36.10	N/A	31.34	6.61
12/23/2003–02/09/2004	13.63	11.44	13.99	1.34
03/04/2004–03/28/2004	3.99	5.19	5.68	−0.96

Notes: A and B are two gage stations within the swamp forest west of the AICWW (Figure 2.13). C is a point over the AICWW, and InSAR images are not coherent at this location. C_w and C_e, two coherent points near C, are located over the swamp forests west and east of the AICWW, respectively. B_e is a point adjacent to B. Both C_w and B are within the swamp forest west of the AICWW, and C_e and B_e are over the swamp forest east of the AICWW (Figure 2.13).

from gages and from the interferograms in Figure 2.13. The InSAR-derived water-level changes at A and B are in good agreement with gage readings and within about 2 cm overall discrepancy. This indicates that water-level change measurements by InSAR are probably as good as those of the gages. We can infer that the InSAR-detected water-level changes in non-gage areas are trustworthy. If this is the case, then the InSAR technique provides a unique way to map dynamic and heterogeneous water-level changes at accuracies comparable to gages and at a spatial resolution unattainable by gages. The gage data at Cross Bayou (A in Figure 2.13) on May 22, 2003, do not exist, so one could not confirm water-level changes of about 36 cm detected by InSAR (Figure 2.13b). If the perceived correspondence of about 2 cm between InSAR and gage measurements is extended, the gage reading at A can be estimated to be about 437 cm on May 22, 2003. This demonstrates the utility of InSAR-based water-level change measurement to augment the missing gage data. InSAR measures relative elevation changes between image acquisition dates; hence, it requires calibration with respect to absolute water-level measurements. For the Achafalaya Basin, the gage station over the swamp forest is used for the absolute water-level change calibration. Combining the InSAR image (Figure 2.13c) and the gage station reading, one can, therefore, derive volumetric water storage change during 12/23/2003 and 2/9/2004 (Figure 2.14).

Next, the water-level changes measured along two different paths (C_w–B and C_e–B_e) within two swamp forest bodies separated by the AICWW are compared. Please note that locations C_w and B are within the swamp forest west of the AICWW and locations C_e and B_e are within the swamp forest east of the AICWW. Integrating interferometric phase measurements along the western path (C_w–B) and the eastern path (C_e–B_e) gives different water-level changes that depend on the path (Table 2.3). This is interpreted as the result of structures that obstruct smooth and rapid water flow, primarily within the swamp forests west of the AICWW. The change in fringe

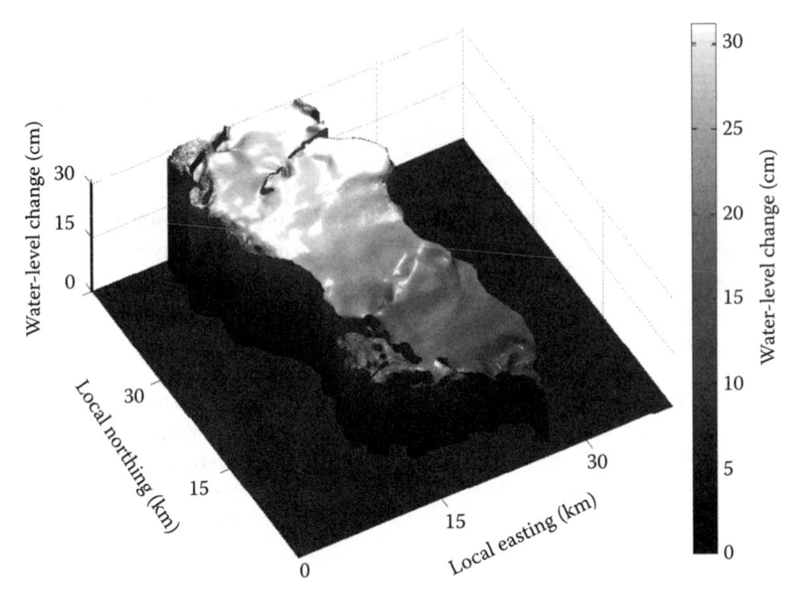

FIGURE 2.14 Volumetric rendering of absolute water-level changes over the portion of swamp forests west of the AICWW during December 23, 2003 and February 9, 2004, derived from the InSAR image in Figure 2.13c.

pattern across the AICWW suggests that parts of the AICWW also act as barriers to continuous water flow in this area. Water-level changes in swamp forests over the study area are heterogeneous and disconnected by structures and other barriers and therefore not represented adequately by sparsely distributed gage stations. This finding is useful for hydrologists to enhance surface water flow models by correctly defining the spatial extent of homogeneous continuum and for emergency planners to simulate the dynamics of flood waters in the region with enhanced accuracy.

2.8 DISCUSSIONS AND CONCLUSION

Multitemporal RADARSAT-1 and ERS SAR images over southern Louisiana are used to study characteristics of the radar backscattering coefficient over vegetation classes. Calibrated radar backscattering coefficients over six land-cover classes—bottomland forest, swamp forest, freshwater marsh, intermediate marsh, brackish marsh, and saline marsh—provide insights about the relationship between seasonal variation of $\sigma°$ and vegetation canopy structure. Double-bounced backscattering is the dominant scattering mechanism for swamp forests and saline marshes. Volume backscattering dominates freshwater and intermediate marshes and bottomland forests. Brackish marshes are likely dominated by volume backscattering and specular scattering. RADARSAT-1 backscattering coefficients offer better separability among different wetland land-cover types than ERS data, suggesting that C-band HH polarization is more sensitive to structural differences than C-band VV polarization.

Radar backscattering coefficients during leaf-off seasons have strong correlations with NDVI. Swamp and bottomland forests show negative correlations between NDVI and SAR data, while marshes exhibit positive correlations with RADARSAT data only. The correlation between σ° and NDVI is useful in differentiating between forests and coastal marshes and in refining our understanding of vegetation structure.

Wherever InSAR coherence is maintained, InSAR's utility to map water-level changes at high spatial resolution makes it an attractive tool for studying many hydrological processes. Adequately characterizing the heterogeneous water-level changes over a complex wetland system requires many ground-based measurements, which are cost-prohibitive. In this chapter, C-band InSAR images have proved useful for mapping water-level changes of coastal wetlands in Louisiana. Particularly, HH-polarized C-band InSAR can maintain good coherence for mapping water-level changes over coastal wetlands if the SAR images are acquired in a few days. The L-band InSAR images generally maintain much higher coherence than C-band InSAR; therefore, it is expected that L-band InSAR can be used to map water-level changes over denser forests.

Figure 2.15a and b shows two interferograms over the study area. Figure 2.15a is a 46-day L-band ALOS PALSAR interferogram acquired from HH-polarized images acquired on February 27 and April 14, 2007, and Figure 2.15b is a 24-day C-band RADARSAT-1 interferogram acquired from HH-polarized SAR images acquired on March 4 and 28, 2004. This study shows that the 46-day L-band interferogram is generally more coherent than the 24-day C-band interferogram, which can maintain relatively higher coherence than VV-polarized C-band ERS/ENVISAT images as demonstrated in the previous sections. Particularly, the L-band PALSAR interferogram can maintain coherence over bottomland forests and marshes where C-band coherence is often lost (Figure 2.15b). Therefore, L-band interferograms allow robust monitoring of water-level changes over coastal wetlands due to the higher coherence. Ultimately, combining C-band and L-band InSAR images can significantly improve temporal sampling of water-level measurements.

However, there are at least two shortfalls regarding water-level measurements from InSAR images. First, InSAR requires the presence of emergent vegetation [27–29,33,44] or structures in water [51] to allow radar signals to be scattered back to the antenna to measure water-level change. Over open-water bodies, InSAR is useless for detecting water-level change. Second, a repeat-pass InSAR image measures the relative spatial gradient of water-level change between two time periods. In other words, from interferometric phase measurements alone, the absolute volumetric change of water storage within a wetland is not derivable without additional constraints. In a simple case, for example, let us assume the water level over a wetland moves up or down by a constant height. The volumetric change of the wetland can be calculated by the area of the wetland and the constant water-level change. However, an InSAR image can only exhibit a constant phase shift with an ambiguity of multiples of 2π. This can be mistakenly interpreted as being no water-level changes. To estimate the volumetric change of water storage, the absolute water-level change at a single location with a wetland body is required. The situation can become even more complicated if the wetland system consists of many wetland bodies, which are

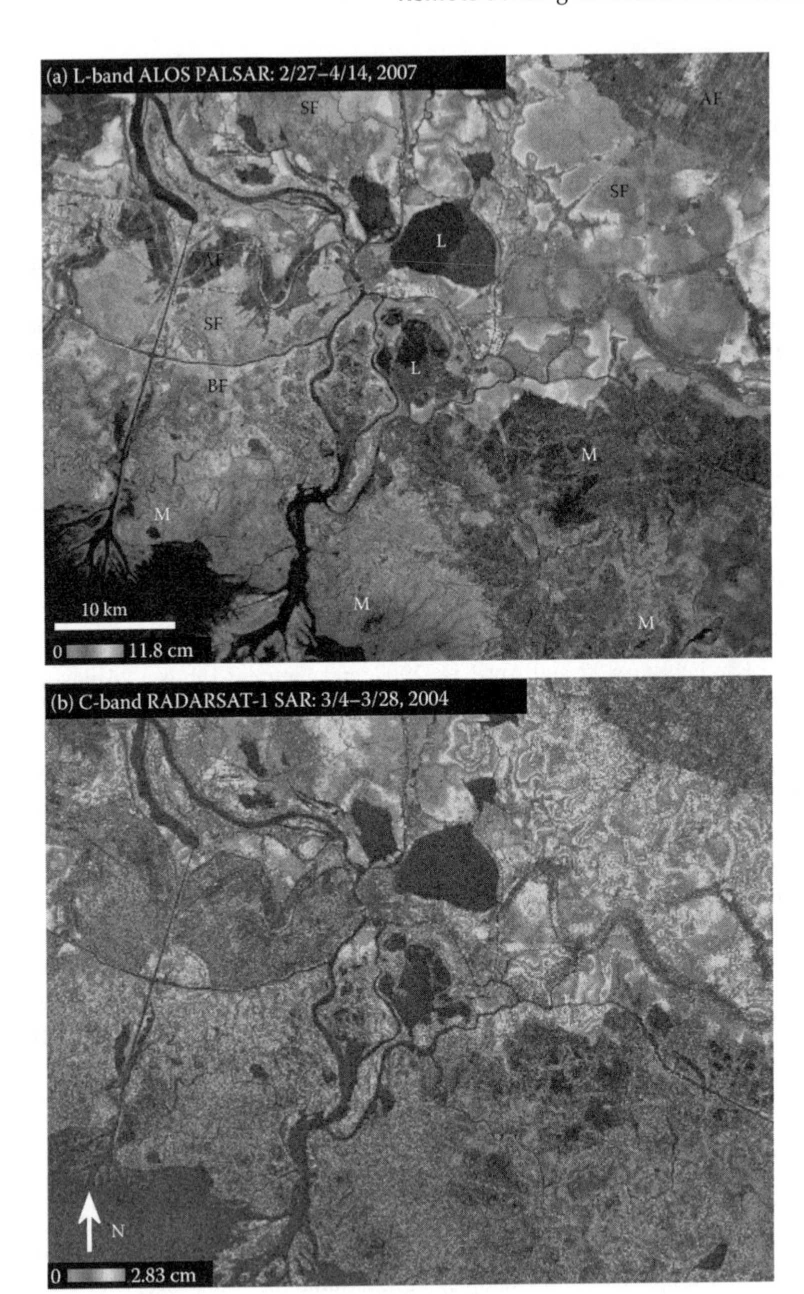

FIGURE 2.15 **(See color insert following page 206.)** (a) L-band ALOS and (b) C-band RADARSAT-1 InSAR images showing water-level changes in coastal wetlands over southeastern Louisiana. Each fringe (full color cycle) represents a line-of-sight range change of 11.8 and 2.83 cm for ALOS and RADARSAT-1 interferograms, respectively. Interferogram phase values are unfiltered for coherence comparison and are draped over the SAR intensity image of the early date. Areas of loss of coherence are indicated by random colors. M: marshes (freshwater, intermediate, brackish, and saline marshes); L: lake, SF: swamp forest; BF: bottomland forest; AF: agricultural field.

bounded by anthropogenic structures such as levees and canals, which can disconnect the InSAR phase changes. In this case, it may be impossible to estimate volumetric storage change of the whole wetland system without knowing the absolute water-level change in each wetland body.

A more feasible solution is to include water-level measurements from a radar altimeter with adequate spatial resolution and vertical accuracy, such as the Water And Terrestrial Elevation Recovery (WATER) satellite mission [52]. Similar to the Shuttle Radar Topography Mission (SRTM), WATER is a dual-antenna radar interferometer that can achieve centimeter-level height accuracy at tens of meters spatial resolution. A similar system will not only provide temporal and spatial variations of water-level height but also provide measurements to facilitate the use of InSAR measurements over wetlands from other satellite radar imagery. Optimized radar images with short repeat-pass acquisitions from multiple satellite sensors, combined with available ground-based gage readings, will improve the characterization of surface water hydraulics, hydrological modeling predictions, and the assessment of future flood events over wetlands.

ACKNOWLEDGMENTS

ERS-1/ERS-2 SAR images are copyright ©1992–1998 European Space Agency (ESA) and were provided by ESA and Eurimage. RADARSAT-1 images are copyright ©2002–2004 Canadian Space Agency and were provided by Alaska Satellite Facility (ASF). PALSAR data are copyright JAXA/METI 2007 and were provided by ASF. This research was supported by funding from the U.S. Geological Survey (USGS) Director Venture Capital Fund, USGS Eastern Region Venture Capital Fund, USGS Land Remote Sensing Program, NASA Solid Earth and Natural Hazards Program. We thank ASF and ESA for their support in programming SAR data acquisitions and delivering data timely. We thank Christopher Swarzenski (USGS) for providing us the water-level readings at gauge stations used in this study, and Kurtis Nelson and Bhaskar Ramachandran for technical reviews and comments.

REFERENCES

1. Karszenbaum, H., Kandus, P., Martinez, J.M., Le Toan, T., Tiffenberg, J., and Parmuchi, M.G., ERS-2, RADARSAT SAR backscattering characteristics of the Parana River Delta Wetland, Argentina, Special Publication SP-461, 2000.
2. USACE, *Louisiana Ecosystem Restoration Study: Louisiana Coastal Area (LCA)*, http://www.lca.gov/final_report.aspx, 2004.
3. LCWCRTF/WCRA, *Coast 2050: Toward a Sustainable Coastal Louisiana*, Louisiana Department of Natural Resources, Baton Rouge, LA, 1998, p. 161.
4. Barras, J., Beville, S., Britsch, D., Hartley, S., Hawes, S., Johnston, J., Kemp, P., et al., Historical and projected coastal Louisiana land changes: 1978–2050, USGS Open File Report 03-334, 2003, p. 39.
5. Coe, M., A linked global model of terrestrial hydrologic processes: Simulation of the modern rivers, lakes, and wetlands, *J. Geophys. Res.*, 103, 8885, 1998.
6. Alsdorf, D., Lettenmaier, D., and Vörösmarty, C., The need for global, satellite-based observations of terrestrial surface waters, *EOS Trans.*, 84, 269, 2003.

7. Ramsey, E.W., III, Monitoring flooding in coastal wetlands by using radar imagery and ground-based measurements, *Int. J. Remote Sens.*, 16, 2495, 1995.
8. Ramsey, E.W., III, Radar remote sensing of wetlands, In: R.S. Lunetta and C.D. Elvidge (Eds), *Remote Sensing Change Detection*, Ann Arbor Press, Chelsea, MI, 1999, p. 211.
9. Ramsey, E.W., III, Lu, Z., Rangoonwala, A., and Rykhus, R., Multiple baseline radar interferometry applied to coastal land cover classification and change analyses, *GISci. Remote Sens.*, 43 (4), 283–309, 2006.
10. Grings, F.M., Ferrazzoli, P., Jacobo-Berlles, J.C., Karszenbaum, H., Tiffenberg, J., Pratolongo, P., and Kandus, P., Monitoring flood condition in marshes using EM models and ENVISAT ASAR observations, *IEEE Trans. Geosci. Remote Sens.*, 44, 936, 2006.
11. Costa, M.P.F., Use of SAR satellites for mapping zonation of vegetation communities in the Amazon floodplain, *Int. J. Remote Sens.*, 25, 1817, 2004.
12. Costa, M.P.F., Niemann, O., Novo, E., and Ahern, F., Biophysical properties and mapping of aquatic vegetation during the hydrological cycle of the Amazon floodplain using JERS-1 and RADARSAT, *Int. J. Remote Sens.*, 23, 1401, 2002.
13. Hess, L.L. and Melack, J.M., Mapping wetland hydrology and vegetation with synthetic aperture radar, *Int. J. Ecol. Environ. Sci.*, 20, 197, 1994.
14. Hess, L.L., Melack, J.M., Filoso, S., and Wang, Y., Delineation of inundated area and vegetation along the Amazon floodplain with the SIR-C synthetic aperture radar, *IEEE Trans. Geosci. Remote Sens.*, 33, 896, 1995.
15. Kasischke, E.S. and Bourgeau-Chavez, L.L., Monitoring south Florida wetlands using ERS-1 SAR imagery, *Photogrammetric Eng. Remote Sens.*, 63, 281, 1997.
16. Le Toan, T., Ribbes, F., Wang, L.-F., Floury, N., Ding, K.-H., Kong, J.A., Fujita, M., and Kurosu, T., Rice crop mapping and monitoring using ERS-1 data based on experiment and modeling results, *IEEE Trans. Geosci. Remote Sens.*, 35, 41, 1997.
17. Baghdadi, N., Bernier, M., Gauthier, R., and Neeson, I., Evaluation of C-band SAR data for wetlands mapping, *Int. J. Remote Sens.*, 22, 71, 2001.
18. Bourgeau-Chavez, L.L., Kasischke, E.S., Brunzell, S.M., Mudd, J.P., Smith, K.B., and Frick, A.L., Analysis of space-borne SAR data for wetland mapping in Virginia riparian ecosystems, *Int. J. Remote Sens.*, 22, 3665, 2001.
19. Simard, M., Grandi, G.D., Saatchi, S., and Mayaux, P., Mapping tropical coastal vegetation using JERS-1 and ERS-1 radar data with a decision tree classifier, *Int. J. Remote Sens.*, 23, 1461, 2002.
20. Townsend, P.A., Estimating forest structure in wetlands using multitemporal SAR, *Remote Sens. Environ.*, 79, 288, 2002.
21. Bourgeau-Chavez, L.L., Smith, K.B., Brunzell, S.M., Kasischke, E.S., Romanowicz, E.A., and Richardson, C.J., Remote monitoring of regional inundation patterns and hydroperiod in the greater Everglades using synthetic aperture radar, *Wetlands*, 25, 176, 2005.
22. Kiage, L.M., Walker, N.D., Balasubramanian, S., Babin, A., and Barras, J., Applications of RADARSAT-1 synthetic aperture radar imagery to assess hurricane-related flooding of coastal Louisiana, *Int. J. Remote Sens.*, 26, 5359, 2005.
23. Kwoun, O. and Lu, Z., Multi-temporal RADARSAT-1 and ERS Backscattering Signatures of Coastal Wetlands at Southeastern Louisiana, *Photogrammetric Eng. Remote Sens.*, 75 (5), 607–617, 2009.
24. Massonnet, D. and Feigl, K., Radar interferometry and its application to changes in the Earth's surface, *Rev. Geophys.*, 36, 441, 1998.
25. Rosen, P., Hensley, S., Joughin, I.R., Li, F.K., Madsen, S.N., Rodriguez, E., and Goldstein, R.M., Synthetic aperture radar interferometry, *Proc. IEEE*, 88, 333, 2000.
26. Lu, Z., Kwoun, O., and Rykhus, R., Interferometric synthetic aperture radar (InSAR): Its past, present and future, *Photogrammetric Eng. Remote Sens.*, 73, 217, 2007.

27. Alsdorf, D., Melack, J., Dunne, T., Mertes, L., Hess, L., and Smith, L., Interferometric radar measurements of water level changes on the Amazon floodplain, *Nature*, 404, 174, 2000.
28. Alsdorf, D., Birkett, C., Dunne, T., Melack, J., and Hess, L., Water level changes in a large Amazon lake measured with spaceborne radar interferometry and altimetry, *Geophys. Res. Lett.*, 28, 2671, 2001.
29. Wdowinski, S., Amelung, F., Miralles-Wilhelm, F., Dixon, T., and Carande, R., Space-based measurements of sheet-flow characteristics in the Everglades wetland, Florida, *Geophys. Res. Lett.*, 31, L15503, doi: 10.1029/2004GL020383, 2004.
30. Richards, J., Woodgate, P., and Skidmore, A., An explanation of enhanced radar back-scattering from flooded forests, *Int. J. Remote Sens.*, 8, 1093–1100, 1987.
31. Beaudoin, A., Le Toan, T., Goze, S., Nezry, E., Lopes, A., Mougin, E., Hsu, C.C., Han, H.C., Kong, J.A., and Shin, R.T., Retrieval of forest biomass from SAR data, *Int. J. Remote Sens.*, 15, 2777, 1994.
32. Wang, Y., Hess, L., Filoso, S., and Melack, J., Understanding the radar backscattering from flooded and nonflooded Amazonian forests: Results from canopy backscatter modeling, *Remote Sens. Environ.*, 54, 324, 1995.
33. Lu, Z., Crane, M., Kwoun, O., Wells, C., Swarzenski, C., and Rykhus, R., C-band radar observes water-level change in swamp forests, *EOS Trans. AGU*, 86, 141, 2005.
34. Nelson, S.A.C., Soranno, P.A., and Qi, J., Land-cover change in upper Barataria Basin Estuary, Louisiana, 1972–1992: Increase in wetland area, *Environ. Manag.*, 29, 716, 2002.
35. Chabreck, R.H., Vegetation, water, and soil characteristics of the Louisiana coastal region, *LSU Agric. Exp. Stn. Bull.*, 664, 1, 1972.
36. Barras, J., Bolles, J., Carriere, R., Daigle, D., Demcheck, D., Dittmann, D., Etheridge, et al., *Types of Wetlands*, http://www.americaswetlandresources.com/wildlife_ecology/plants_animals_ecology/wetlands/TypesofWetlands.html, 2006.
37. IFAS, *Forest Ecosystems*, http://www.sfrc.ufl.edu/4h/Ecosystems/ecosystems.html, 2006.
38. Ulaby, F.T. and Dobson, C.M., *Handbook of Radar Scattering Statistics for Terrain*, Artech House, Inc., Norwood, MA, 1989, p. 357.
39. Hagberg, J., Ulander, L., and Askne, J., Repeat-pass SAR interferometry over forested terrain, *IEEE Trans. Geosci. Remote Sens.*, 33, 331, 1995.
40. Wegmüller, U. and Werner, C., Retrieval of vegetation parameters with SAR interferometry, *IEEE Trans. Geosci. Remote Sens.*, 35, 18, 1997.
41. Zebker, H.A. and Villasenor, J., Decorrelation in interferometric radar echoes, *IEEE Trans. Geosci. Remote Sens.*, 30, 950, 1992.
42. Lu, Z. and Freymueller, J., Synthetic aperture radar interferometry coherence analysis over Katmai volcano group, Alaska, *J. Geophys. Res.*, 103, 29887, 1998.
43. Gatelli, F., Guarnieri, A.M., Parizzi, F., Pasquali, P., Prati, C., and Rocca, F., The wave-number shift in SAR interferometry, *IEEE Trans. Geosci. Remote Sens.*, 32, 855, 1994.
44. Lu, Z. and Kwoun, O., RADARSAT-1 and ERS interferometric coherence analysis over southeastern coastal Louisiana: Implication for mapping water-level changes beneath swamp forests, *IEEE Trans. Geosci. Remote Sens.*, 46, 2167, 2008.
45. Gamma, Gamma MSP Reference Manual, http://www.gamma-rs.ch/software, 2003.
46. Frost, V.S., Stiles, J.A., Shanmugan, K.S., and Holtzman, J.C., A model for radar images and its application to adaptive digital filtering of multiplicative noise, *IEEE Trans. Pattern Anal. Machine Intell.*, PAMI-4, 157, 1982.
47. USGS-NWRC, *2004 Digital Orthophoto Quarter Quadrangles for Louisiana*, http://www.lacoast.gov/maps/2004doqq/index.htm, 2005.

48. Meadows, P.J., Rosich, B., and Santella, C., The ERS-2 SAR performance: The first 9 years, in *Proceedings of the ENVISAT and ERS Symposium*, Salzburg, Austria, September 6–10, 2004.
49. Ferretti, A., Prati, C., and Rocca, F., Permanent scatterers in SAR interferometry, *IEEE Trans. Geosci. Remote Sens.*, 39, 8, 2001.
50. Sellers, P.J., Canopy reflectance, photosynthesis, and transpiration, *Int. J. Remote Sens.*, 6, 1335, 1985.
51. Kim, S.-W., Hong, S.-H., and Won, J.-S., An application of L-band synthetic aperture radar to tide height measurement, *IEEE Trans. Geosci. Remote Sens.*, 43, 1472, 2005.
52. Alsdorf, D., Rodríguez, E., and Lettenmaier, D., Measuring surface water from space, *Rev. Geophys.*, 45, RG2002, doi: 10.1029/2006RG000197, 2007.

3 Mangrove Canopy 3D Structure and Ecosystem Productivity Using Active Remote Sensing*

Marc Simard, Lola E. Fatoyinbo, and Naiara Pinto

CONTENTS

3.1 INTRODUCTION

Mangrove forests are the dominant coastal ecosystem in tropical and subtropical regions. They form an important link between aquatic and terrestrial ecosystems and are one of the most productive ecosystems with a primary productivity of 2.5 g carbon/m^2 per day [1]. Mangrove litter provides large amounts of carbon (C) to coastal and offshore marine ecosystems and contributes over 10% of the dissolved organic C (DOC) to ocean sediments worldwide [2]. Mangroves also protect coastal areas from wave action, erosion, storms, and tidal waves. Despite their known benefits as coastal buffers and biodiversity harbors, mangrove areas are being altered and destroyed by anthropogenic impacts and their cover has decreased by 35% in the past 20 years [3]. Mangrove forests are ecologically and economically important. They have been estimated at 200,000–900,000 USD/km^2 per year by the United Nations Environment Programme World Conservation Monitoring Center [4].

Remotely sensed images provide an efficient and cost-effective way to gain insight into mangrove areas that are often difficult to access and survey [5–11]. Both optical and active remote sensing techniques have been commonly used to study mangrove forests, and in the past years the combination of radar (RAdio Detection And Ranging) and LiDAR (Light Detection And Ranging) has yielded interesting results that reach further than determining mangrove cover alone. Generally, optical remote sensing instruments such as Landsat, MODIS, and SPOT measure the color of forests and are often considered more intuitive. To this day, mangrove forests have also been studied using polarimetric and interferometric radar and airborne and space-borne LiDAR systems. Several studies were carried out using radar data for mangrove mapping and monitoring [12]. In addition, three-dimensional (3D) modeling of mangrove forests was made possible by the Shuttle Radar Topography Mission (SRTM) data. The 3D rendition was validated with airborne and space-borne LiDAR and field data to provide large-scale canopy height and biomass estimates of mangrove forests [9–11].

In this chapter, we discuss recent advancements in remote sensing technology that enable mapping of forest canopy structure in three dimensions. In the first section, we introduce active remote sensing technologies (i.e., radar and LiDAR) and a methodology to link field and remote sensing data through a compromise between parameter estimation accuracy and efficiency. Then, we define vegetation 3D structure in terms of both horizontal and vertical distribution of its physical components (i.e., wood and leaves) before we use ecological models to estimate productivity, which is another parameter of interest as it is related to vegetation health and carbon cycle.

3.2 ACTIVE REMOTE SENSING

There are two main categories of remote sensing instruments: *passive* sensors that require sunlight or emission from the target (e.g., infrared and microwave) and *active* sensors transmitting their own signal and measuring the energy reflected by targets on the surface. Examples of passive sensors useful in vegetation studies include those carried by Landsat, SPOT, QuickBird, and MODIS by Terra and Aqua satellites. Active sensors are radars and LiDAR systems that measure the time of return (translated to distance) of the signal reflected from a target, which is directly related to the physical structure of the target. To help the reader, we list a few remote sensing terms in Table 3.1. Active sensors transmit electromagnetic pulses with a specific wavelength (λ). LiDAR systems generally transmit pulses at wavelengths around the visible domain (λ around a thousand nanometers), whereas radar pulses are in the microwave domain (λ on the scale of centimeters). Pulse wavelength influences the contribution of the different vegetation components to the measured energy. As a rule of thumb, the energy pulse is reflected by scatterers that are larger than its wavelength. This means that LiDAR pulses can be reflected by all vegetation components but mainly by the upper canopy: the LiDAR pulse is reflected by upper canopy layers and loses energy as it travels toward the ground. On the other hand, the radar pulse can travel through smaller components (twigs, leaves, etc.) and can be reflected by trunks, large branches, and the ground surface.

TABLE 3.1
Common Vocabulary in Active Remote Sensing

Term	Definition
Sensor	Instrument used to measure the remote sensing signal
Backscatter	Energy reflected by the target directly toward the sensor
Target	Object or surface that reflects the signal (vegetation, buildings, ground, etc.)
Scatterer	Generally refers to a physical component of a target. For example, a target may be a forest that is composed of scatterers: branches and leaves
Range	Distance between the sensor and the target
Path (track)	Projection of the sensor's trajectory on the ground
Nadir (*Z*)	Direction below the sensor that is perpendicular to the ground
Swath	Cross-track (i.e., perpendicular to instrument path as it moves) ground distance illuminated by the signal
Ellipsoid	Representation of the Earth's shape as a 3D ellipse defined by 3 axes
Geoid	Deviations, from an ellipsoid, of the Earth's surface as defined by a gravitational equipotential (i.e., elevation at which gravitation is equal). Mean sea level follows the Geoid
Look angle	Angle at which the sensor is aimed with respect to nadir
Footprint	Area on the ground illuminated by the active sensor

LiDAR systems generally transmit pulses toward nadir. The LiDAR pulse is spatially confined to concentrate energy and get accurate range estimates. There are several LiDAR technologies that can be generally classified as profiling LiDARs and imaging LiDARs. The first, and simplest, collects a profile of nearly equidistant points along the sensor's path. The second technology collects range samples within a swath by transmitting the LiDAR pulses away from nadir, effectively collecting a "cloud" of data points. If the spatial sampling is dense enough, measurements can be used to form an elevation image (i.e., digital elevation model, DEM). Currently, only airborne LiDARs achieve swath mapping and, depending on the technology and altitude, can cover swaths of several hundreds of meters, producing 3D renditions of the landscape. Some LiDAR systems record full waveforms by fast sampling (on the order of nanoseconds) the returned pulse as a function of time. Since the measurement is collected around nadir, the waveform provides a direct measurement of height distribution of scatterers within the LiDAR footprint (Figure 3.1).

Radar systems transmit electromagnetic pulses sideways in a direction perpendicular to the sensor's path (i.e., crosstrack) (Figure 3.2). Compared to LiDAR systems, radar covers very large areas typically on the order of tens of kilometers. The technology is also very different and produces images of backscattered energy and signal phase. The crosstrack range resolution is determined by the signal bandwidth (Δf). The along-track spatial resolution (i.e., azimuth resolution) uses advanced signal-processing techniques that can follow the apparent trajectory of targets as seen by the sensor, using their Doppler signature (frequency shift) to separate their respective signals. This technique, called synthetic aperture radar (SAR), is used to simulate a large antenna and produce fine spatial resolution images (on the scale of meters). Using at least two radar antennas that observe the same terrain with slightly

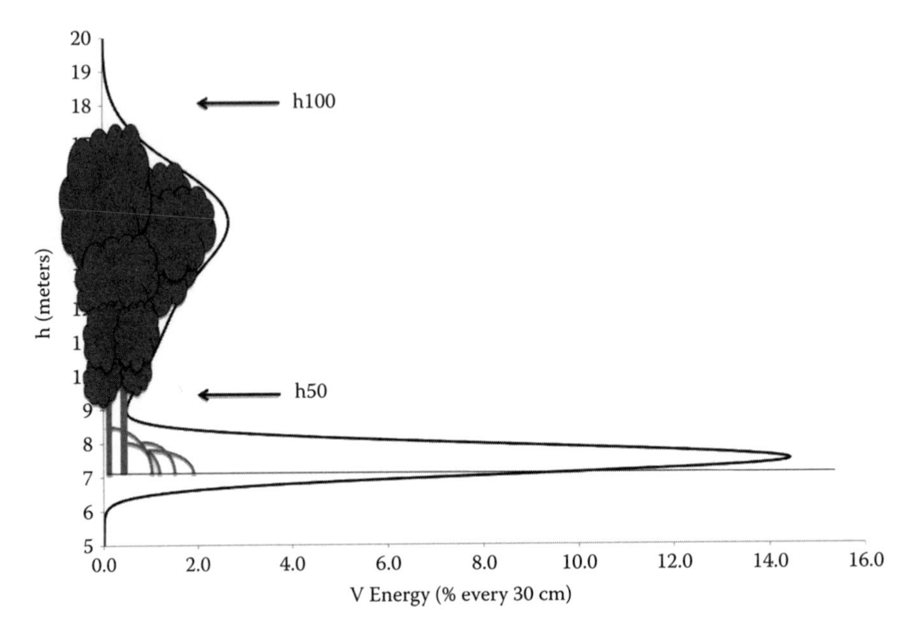

FIGURE 3.1 LiDAR waveform reflected by a mangrove forest with the bottom peak near 7 m, which is due to the ground, and the wider upper peak near 17 m due to mangrove structure. The canopy height can be defined as waveform percentile energy. The points h100 and h50 are the 100th and 50th percentile, which is where 100% and 50% of the pulse energy is contained between the top and ground pulse.

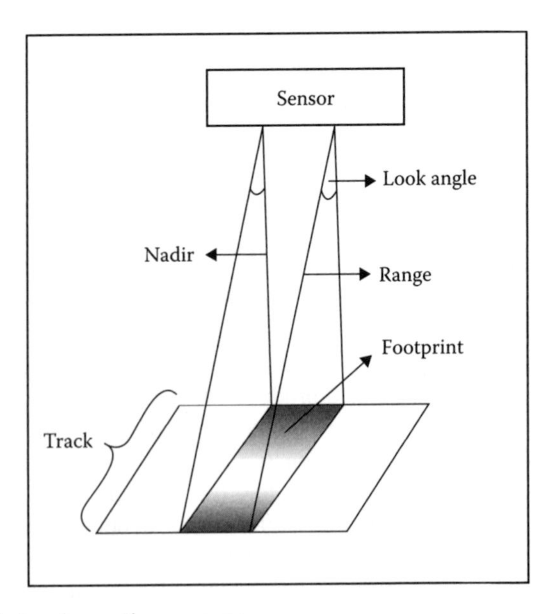

FIGURE 3.2 Radar observation geometry.

different viewing geometries, we can obtain a 3D rendition of the landscape. We introduce this technique, called SAR interferometry (InSAR), later in this chapter.

In general, the application of LiDAR and radar images to vegetation studies involves a trade-off between vertical accuracy and spatial coverage. InSAR data cover several kilometers with a height measurement for every image pixel capturing only dominant features of the vegetation [13]. However, the large swaths make it possible to estimate canopy height and biomass over very large areas. On the other hand, LiDAR systems produce height measurements with higher vertical accuracy [14], but sparse spatial sampling or small swath width.

3.3 MEASUREMENT OF VEGETATION STRUCTURE

The term "vegetation structure" describes the size and distribution of vegetation material within a landscape. We generally characterize individual trees in terms of their species, trunk size (i.e., diameter at breast height, DBH), and height. They can also be described by other parameters such as crown size, both horizontally and vertically. Some of these parameters can be related through allometric equations derived from labor-intensive fieldwork. Productivity is estimated by monitoring wood growth, measuring tree DBH and height over several years, and collecting debris (wood, leaves, and twigs).

There is a significant disconnection between ecological investigations performed through field measurements and remote sensing due to the different parameters being measured and the differences in scales at which these are measured. The amount of work required for field measurements limits its availability to small and sparse plots. On the other hand, remote sensing produces a large-scale view of a forest with relatively low spatial resolution. In addition, remote sensing cannot measure field parameters such as DBH and the amount of debris, and it is difficult to clearly identify species and see vertical vegetation layers in the canopy. In the next sections, we describe a methodology to relate remote sensing and field data to derive ecological parameters such as height, biomass, and productivity of mangrove forests.

3.4 FIELD DATA

A major issue in large-scale studies is to extrapolate field measurements to landscape, regional, continental, and global scales. Field data are required for calibration and validation of the remotely sensed parameters, but are generally collected at the plot level and at scales of meters. To be useful to large-scale remote sensing applications, field sampling methods must capture the large-scale structures of forests that follow variations of environmental factors, as well as the structure at the scale of tree competition, in addition to being systematic and objective. Most remote sensing systems used in large-scale science studies (e.g., Landsat, SPOT, ERS-1, ERS-2, JERS-1, ALOS, Radarsat, MODIS, etc.) have spatial resolution on the order of tens of meters, often around 30 m. For calibration and validation of remotely sensed parameters, the field data should be collected at least at that scale. Collecting vegetation parameters such as tree height, DBH, species, crown size/cover, and tree biomass for all trees within plots encompassing at least one pixel (say 1 ha) is ideal to calibrate remote sensing data. Such data can be mapped within a single or several image pixels. An image pixel size is generally on the same order of the sensor's spatial resolution. However, this amounts to an enormous amount of work in the field resulting in a few plots available to calibrate the remote sensing data. Therefore, one may consider a more field efficient technique such as the variable plot method. This method allows for inclusion of large trees that are furthest from the center of the plot. Due to competition for resources (i.e., water, nutrients, and light), large trees are located further from one another than are smaller trees. The variable plot method is based on the angle subtended by the trees as seen from the center of the plot. A fixed angle is selected for a plot and all trees that subtend an angular DBH that is larger than the fixed angle are tallied (also said "in" trees). This requires an angle gauge (e.g., slot, cruiser or basal area prism) located at a fixed distance from the user's eye standing at the center of the plot. Since larger trees subtend a larger diameter they are tallied further from the center of the plot, whereas small trees are tallied only if close to the center. This means that the probability of tallying a tree increases with basal area, which improves accounting of dominant trees without ignoring the smaller trees. The radius "r_v" of the imaginary plot depends on the DBH of a tree following the formula

$$r_v = \frac{50 * \text{DBH}}{\sqrt{(\text{BAF})}}, \tag{3.1}$$

where BAF is the basal area factor. Another advantage of this method is that counting "in" trees amounts to measuring the basal area of the plot. The BAF is associated with the gauge using the following formula:

$$\text{BAF} = 2500 * (\tan \theta_g)^2, \tag{3.2}$$

where θ_g is the fixed angle in radians. Thus, if the fixed angle is 0.02 radians, each "in" tree adds a basal area of 1 m^2/ha to the plot. In a forest of 50 m^2/ha, a BAF of 1 m^2/ha implies that on average 50 trees will be tallied. The BAF should be selected so that large trees located near "r_v" can be clearly seen through the understory from the center of the plot. Otherwise, the measurement may be biased.

Canopy height is an important parameter as it can be measured both in the field and through remote sensing. It can then be related to other canopy parameters through allometric equations. To estimate canopy height from field data, we compute a crown weighted mean of tree heights. This is closer to the canopy height measured by remote sensing [10]. The weight is based on the tree crown area that we compute from the following empirical relation between tree DBH and mangrove crown radius r_t of tree t [15]:

$$r_t = 0.222 \text{ DBH}^{0.654}. \tag{3.3}$$

As expected, this equation indicates that larger trees have larger crowns. Similarly to Lorey's mean height [16], we define the crown weighted mean canopy height H as

$$H = \frac{\sum_t \pi r_t^2 h_t}{\sum_t \pi r_t^2}, \tag{3.4}$$

where h_t and r_t are the height and crown radius of an individual tree t. The summation is over all trees within a fixed plot size, say 1 ha. Thus, the number of "in-trees" selected with the variable plot method must be extrapolated based on Equation 3.1 to reflect their density within the entire 1-ha plot. For closed mangrove canopies, the denominator of Equation 3.4 should be very close to the total area of the plot.

3.5 COMBINING RADAR AND LiDAR TO ESTIMATE CANOPY HEIGHT AND BIOMASS

A few mangrove studies [5] use radar backscatter to characterize mangroves in terms of species composition and growth stage. The mechanisms governing the scattering of the pulse energy by the target are absorption, reflection, and transmission (Figure 3.3). The backscatter is the portion of the energy that is reflected directly back to the sensor and thus can be measured by it. The remaining energy is absorbed by the target, transmitted or reflected away from the sensor. The contribution of each scattering mechanism depends on sensor parameters such as wavelength and polarization as well as target characteristics. Generally, longer radar wavelengths penetrate deeper in the canopy. Target reflectivity at λ, volume distribution, size, and orientation of the scatterers (e.g., leaves, branches, trunk, debris, etc.) determine where and how much of the pulse energy is reflected. In mangroves, smooth surfaces such as soil and water can act like mirrors reflecting the pulse away from the sensor, whereas prop roots increase surface roughness and produce reflections at various

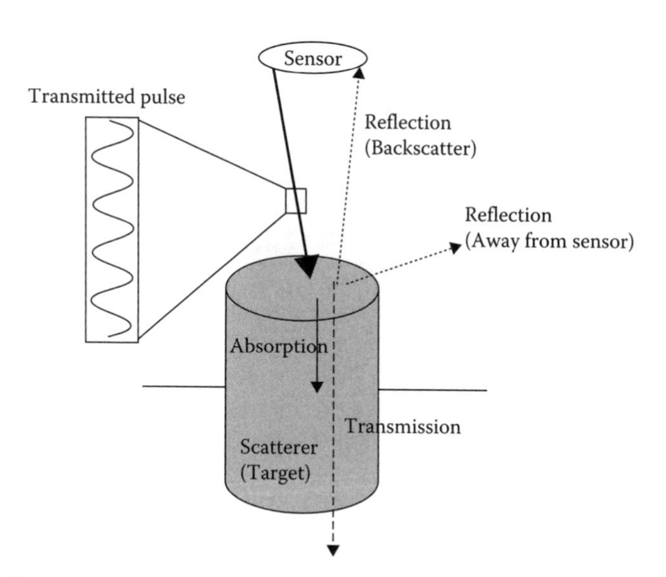

FIGURE 3.3 Scattering mechanisms. The sensor transmits a pulse with a given amplitude (or energy). Upon reaching a scatterer, the pulse can be absorbed, transmitted, or reflected. The sensor captures only the pulse that is reflected back to the sensor. The received pulse is called backscatter and contains only a proportion of the transmitted power.

angles [12]. The contribution of various scattering mechanisms and their geometry can be analyzed by polarimetric SAR (PolSAR) [17].

A method to estimate canopy height employs a radar technique called InSAR [18]. In this technique, scenes are acquired over the study area from two slightly different angles. This can be accomplished by using a sensor carrying two antennas (i.e., single-pass interferometry, Figure 3.4), or by flying over the study area twice with one antenna (repeat-pass interferometry). Platforms carrying radar sensors include the Shuttle Imaging Radar (SIR) [19], the Canadian Radarsat [20], the Japanese Earth Resource Satellite (JERS) [21], the European Remote Sensing (ERS) [22], and more recently the Advanced Land Observing Satellite (ALOS) [23]. The geometry of radar interferometry is shown in Figure 3.4. The measured quantity is the height z. Recall that the received signals are waves with characteristic amplitude and phase. Difference in travel time between the two antennas and the target results in received signal with a distinct phase. The phase difference between signals S1 and S2 is the interferometric phase φ (not shown). Given φ, we can obtain an accurate value of δr, the difference in target range between S1 and S2. Next, using trigonometric relations we derive z as a function of δr. For more details, refer to Franceschetti and Lanari [24]. The accuracy of the measurement is influenced by decorrelation factors that complicate the estimation of z [25]. Decorrelation is a measure of dissimilarity between the two scenes due to differences in the integrated contributions of target components within the canopy volume (i.e., volume decorrelation), system noise (e.g., electronics), and temporal decorrelation. The latter arises in the case of repeat-pass interferometry, due to changes in the target characteristics between the two image acquisitions. This is especially relevant for vegetation studies given the

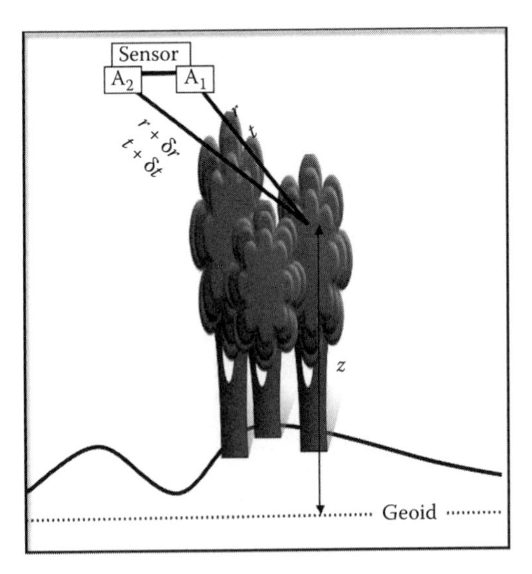

FIGURE 3.4 Simplified view of the geometry of InSAR. The figure illustrates a single-pass approach whereby the canopy is surveyed simultaneously by two antennas A_1 and A_2. We wish to estimate the target height z. The range to the target is denoted r for signal S_1 and $r + \delta r$ for signal S_2. The travel time to the target is denoted t for signal S_1 and $t + \delta t$ for signal S_2.

expected temporal variation in structure from winds, rains, seasonality, growth, and mortality. However, temporal decorrelation is not an issue with single-pass InSAR such as SRTM since both scenes are acquired simultaneously.

Radar microwaves penetrate deep inside the canopy and each InSAR height measurement results from the sum of contributions of all canopy scatterers at the pixel location. As a result, the point z (Figure 3.4) is located somewhere between the ground and the top of the canopy. It is also important to note that the estimated height z is the sum of both vegetation height and topography. If ground elevation is known from ancillary data, it can be subtracted from z to obtain an estimate of canopy height. Thus the application of InSAR to vegetation studies depends on the availability of a DEM for the ground. However, since mangroves are located near sea level and within the tidal zone, an assumption can be made that the ground topography is negligible at least within the accuracy of the remote sensing measurement. In upland forests and in the absence of ground information, more advanced methods based on radar scattering models, such as polarimetric InSAR (PolInSAR), must be used [26].

In February 2000, the SRTM collected InSAR data over 80% of the Earth's land surfaces [27]. Although the mission was designed to produce a global DEM of the Earth's surface, the InSAR measurement of height z is, as mentioned earlier, the sum of the ground elevation and the canopy height contribution. The SRTM DEM can therefore be used to measure vegetation height when ground topography is known (Figure 3.5a). SRTM data are freely available through various resources such as the Global Land Cover Facility (http://glcf.umiacs.umd.edu).

Since the InSAR estimates of canopy height are located below the top of the canopy, they need to be calibrated with field data or LiDAR-derived estimates of canopy height. Assuming that the LiDAR estimates of height are more accurate than radar estimates, calibration is performed through linear regression between the InSAR and LiDAR estimates. Airborne LiDAR data provide the largest amount of

FIGURE 3.5a SRTM DEM over Everglades National Park. The gray level indicates the elevation from 0 m (black) to 12 m (white).

data points; however, they are rarely available. The current space-borne system ICESat (Ice, Cloud, and Land Elevation Satellite) [28] provides the best alternative for global canopy height calibration and is available for free from the National Snow and Ice Data Center (http://nsidc.org/data/icesat). The GLAS (Geoscience Laser Altimeter System) onboard ICESat is a profiler that records the full waveform (Figure 3.1). It has been collecting data since 2003 within a varying footprint size of about 65 m in diameter, sampled every 170 m (Figure 3.5b). The profiles are separated by several kilometers at the equator and do not sample all mangrove forests. Therefore, it may be necessary to use generic SRTM calibration equations that are available for similar mangrove forest settings.

Since mangrove forests are located at sea level, the SRTM elevation dataset enables the systematic canopy height measurement of mangrove forests globally. Mangrove forests thereby provide an ideal ecosystem for SRTM vegetation height estimations. SRTM data were used to estimate mangrove canopy height with accuracies of about 2 m over several mangrove forests such as Everglades National Park [9], the 1280 km^2 Cienaga Grande de Santa Marta wetland in Colombia [10], and the coastline of Mozambique [11]. The methodology to produce these estimates using SRTM and LiDAR data is described in detail in the above-mentioned publications and is summarized below.

First, the mangrove areas must be identified from available land-cover maps or can be produced from Landsat images, as SRTM elevation data will only provide an estimate of canopy height and does not work well to determine forest types.

FIGURE 3.5b Map of SRTM-derived mangrove canopy height overlaid on Google Map in the Everglades National Park, Florida. The markers indicate the location of ICEsat/GLAS LiDAR footprints along the orbit tracks (i.e., profiles).

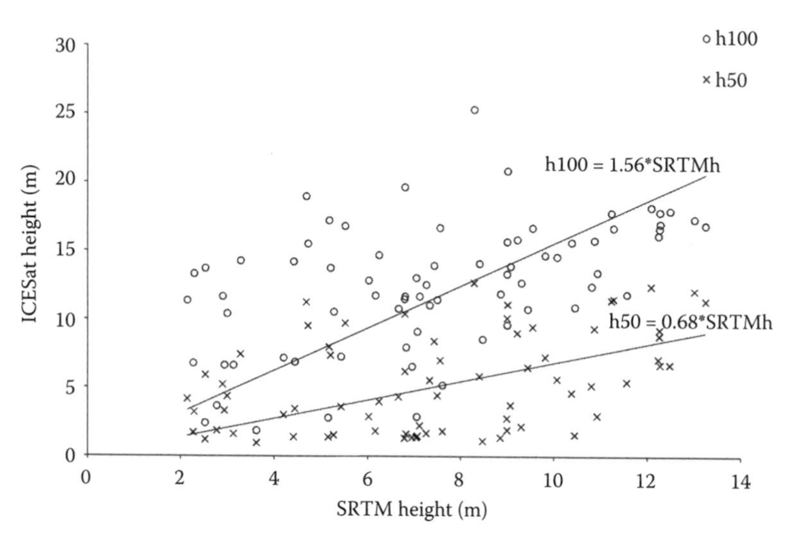

FIGURE 3.6 Calibration/validation between SRTM and ICEsat/GLAS estimates of canopy height. The data can be noisy due to forest structure variability and geolocation errors.

The resulting land-cover map is used to produce a mask or subset of mangrove areas in the SRTM image. Then, the GLAS points that fall within these mangrove areas are compared with the SRTM height data and a linear regression between the two height measurements is computed (Figure 3.6). In addition to the LiDAR measurements, field data should be used, when possible, to increase accuracy and validate the height estimates. The regression curve is then applied to the SRTM DEM to obtain a calibrated estimate of canopy height at the landscape scale. In the above studies, field data from variable plots were also used to derive allometric relations between mean canopy height and biomass at the field plot scale. Although more elaborate models could be designed, we used these equations as first-order estimates to convert mangrove canopy height maps as in Figure 3.5b into regional estimates of aboveground biomass.

3.6 USING LiDAR TO ESTIMATE TREE HEIGHT CLASS DISTRIBUTION

Tree height class distribution is a physical characteristic of forest canopies that is useful to estimate above ground biomass, aerodynamic roughness, and habitat suitability. Both radar and LiDAR can be used to quantify height class distribution, but at different scales. Because InSAR depicts elevation over large areas, the canopy height distribution is obtained by collecting pixels within a window to form a histogram distribution of canopy height. Similarly, for high spatial resolution LiDAR systems with mapping capability, single crowns can be mapped and the tree height distribution is obtained by collecting the points within a window to form the histogram distribution. In the case of a LiDAR system that records full waveforms as illustrated in Figure 3.7, it can be used to characterize the height distribution of the canopy components locally, that is, within the LiDAR footprint. However, in the case of a large footprint LiDAR,

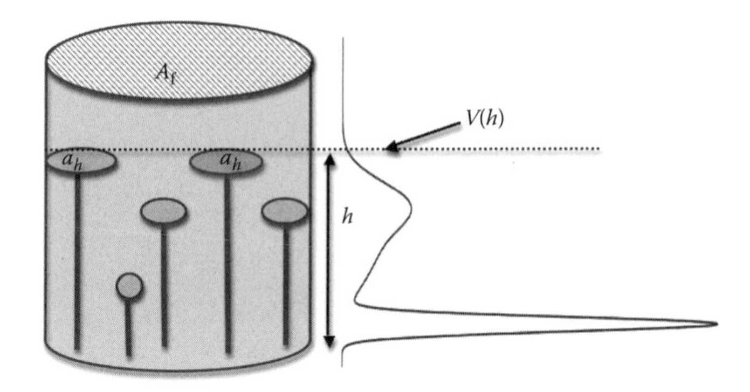

FIGURE 3.7 Using a LiDAR waveform to estimate the number of trees n_h of a given height h. A_f is the LiDAR footprint size and $V(h)$ is the normalized energy associated with height h. The visible crown area of a tree (a_h) is a function of tree height (Equation 3.3). $V(h) \sim (a_h * n_h)/A_f$.

the position of individual trees remains unknown, and the waveform describes the canopy's intercepted surface height distribution within the footprint. From a waveform we can extract $V(h)$, the proportion of energy measured at a given height h. Assuming all crowns have the same reflectivity, the energy backscattered by a single tree of height h is directly proportional to its crown area a_h. Thus $V(h)$ can be modeled as the proportion of the footprint area A_f covered by visible crowns (Figure 3.7). Furthermore, since there exists a relationship between crown area a_h and tree height h (Equation 3.3), we can infer the number n_h of trees that are h meters tall required to reflect energy $V(h)$. Here we specify visible crowns meaning that the LiDAR pulse is unobstructed by upper canopy layers. After normalizing the recorded waveform V such that the total energy equals 1 (Figure 3.7):

$$n_h = \frac{V(h) * A_f}{a_h}. \tag{3.5}$$

We use Equation 3.3 to relate a_h and h, and field data collected in the Everglades National Park in Florida, where $h \sim 77 * \text{DBH}$ (Figure 3.8), to obtain the number of trees of height h within a footprint:

$$n_h = \frac{V(h) * A_f}{[\pi * (0.222 * (h/77)^{0.654})^2]}. \tag{3.6}$$

By computing n_h for each height h, we obtain the tree size class distribution within the footprint. As an example, we applied this analysis to ICESat/GLAS waveforms of Everglades National Park and found that the average number of trees n_h from the LiDAR waveforms was comparable to the average number of trees found in the field. However, the accuracy strongly depends on the knowledge of the LiDAR footprint size A_f, which varies significantly for ICESat/GLAS, and the accuracy of the allometric equations. In addition, the geolocation error of both LiDAR and field data has a major impact on the calibration validation of methods presented in this chapter. In the case of GLAS, geolocation can be on the order of ten meters, while the field data can generally be located to within a similar accuracy with a handheld global positioning system (GPS), that is, if one receives any GPS signal under the thick canopy. Therefore,

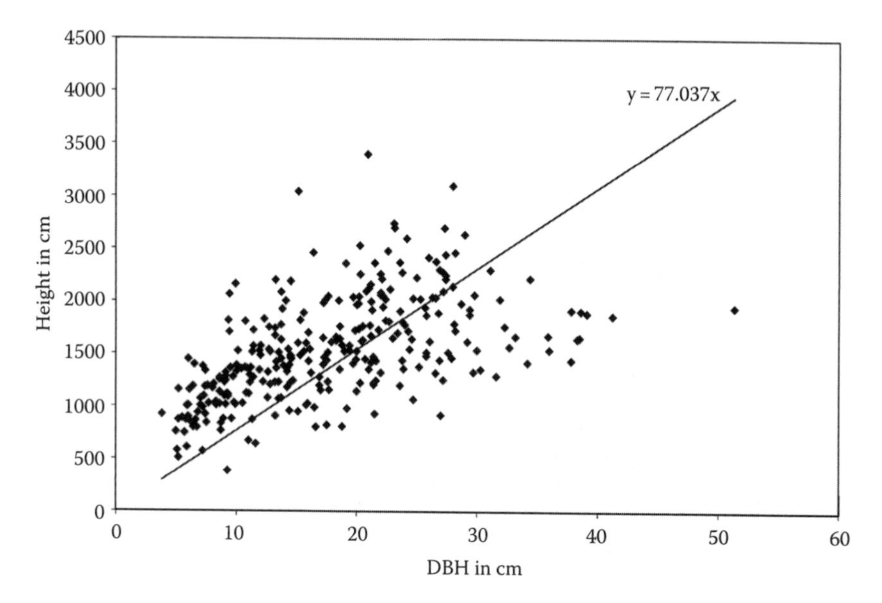

FIGURE 3.8 Allometric DBH–height relationship for the mangroves of Everglades National Park calculated from field data.

it is very difficult to geographically match exactly the space-borne LiDAR data with field data. In Figures 3.9a and b, we show an example of height class distribution derived from the field and LiDAR data. Qualitatively they have a similar shape, exhibiting many trees of medium height with fewer very large and small trees.

3.7 ESTIMATING ECOSYSTEM PRODUCTIVITY

Chen and Twilley [29] have developed an individual-based ecological model with the goal of predicting growth rates of mangrove species:

$$\frac{\mathrm{d}D}{\mathrm{d}t} = \frac{[G * \mathrm{DBH}(1 - (\mathrm{DBH} * h/\mathrm{DBH}_{\mathrm{max}} * h_{\mathrm{max}}))]}{[274 + (3 * b2 * \mathrm{DBH}) - (4 * b3 * \mathrm{DBH}^2)]} * S * N * T * L. \tag{3.7}$$

In this model, the individual annual diameter increase $\mathrm{d}D/\mathrm{d}t$ is a measure of productivity and is a function of height (h) and DBH. Salinity (S), temperature (T), and light availability (L) are parameters that range between 0 and 1 representing environmental resources and stress factors. G, $b2$, and $b3$ are species-specific parameters.

As mentioned in the previous section, the LiDAR waveform can be used to estimate tree size class distribution for a plot (Equation 3.6). The following steps are needed to calculate the growth rate $\mathrm{d}D/\mathrm{d}t$ for a plot:

a. Estimate the relative proportion of each species $p_s = [p_1, p_2, p_3, \ldots, p_{ns}]$ from field data.
b. Calculate the number of trees per size class h: for a given height h, if there are n_h individuals, the number of individuals from species s is assumed to be $n_h * p_s$.

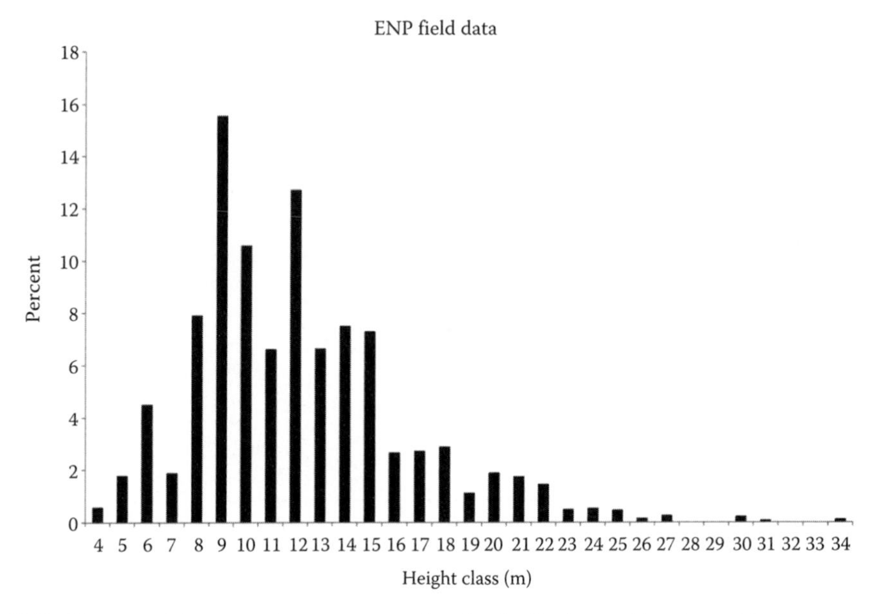

FIGURE 3.9a Histogram of mangrove tree height distributions in Everglades National Park as calculated from field data for a forest with an average height of 12 m.

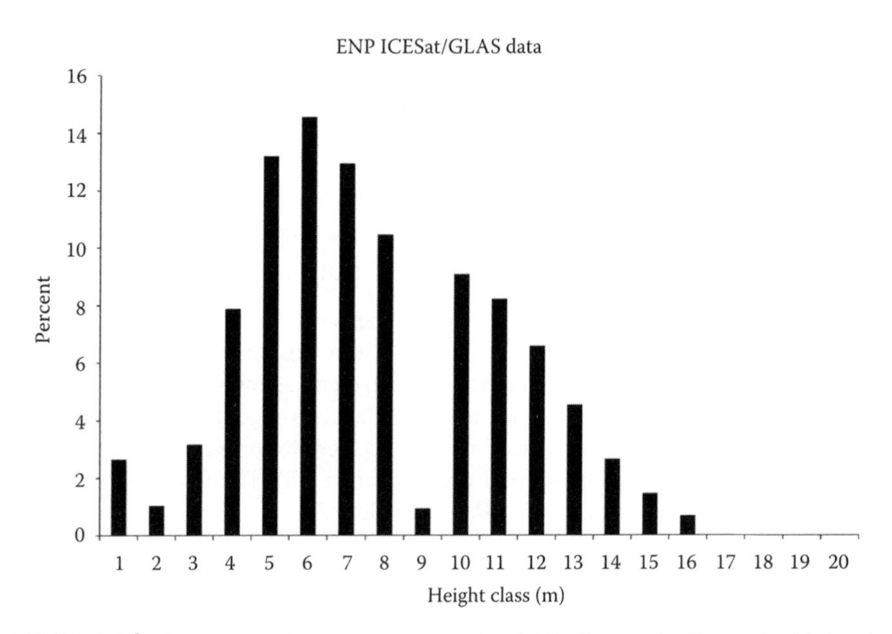

FIGURE 3.9b Histogram of mangrove tree height distributions in the Everglades National Park as calculated from a single GLAS footprint for a forest with a crown weighted mean canopy height of 10 m.

c. Apply allometric equations to derive DBH from tree height.
d. Use Equation 3.7 to calculate d*D*/d*t* for each individual tree, taking into account its species and size class *h*.
e. Finally, results are summed over all individuals in the plot or footprint.

An interactive software applying this algorithm to mangroves can be found at http://www.radar.jpl.nasa.gov/coastal/. As an example, we calculated the productivity of a mangrove plot located in the Florida Everglades (Figure 3.10). A GLAS waveform was used to estimate the size class distribution within a 65×65 m^2 footprint. The proportion of mangrove species calculated from field data was 0.2 for *Avicennia germinans*, 0.4 for *Laguncularia racemosa*, and 0.4 for *Rhizophora mangle* [9]. DBH was derived from tree height using the relationship DBH = $h/77$. The parameters S, T, L, G, $b2$, and $b3$ are from Table 1 of Chen and Twilley [29]. Results were multiplied by 2.37 to obtain estimates for a 1-ha plot (i.e., 1 ha/A_f).

Using Equation 3.7, the increase in basal area (cm^2/year/ha) for this plot was estimated at 1162 for *A. germinans*, 2498 for *L. racemosa*, and 1498 for *R. mangle* (total ~ 5156 cm^2/year). As a comparison, Chen and Twilley [29] estimated the productivity for one site (S4) at 965 for *A. germinans*, 191 for *L. racemosa*, and 4938 for *R. mangle* (total = 6094 cm^2/year). Two other sites [28] produced estimates of 11670 and 11873 cm^2/year/ha. This variation in productivity is expected given the range of vegetation structures and species composition found in the Everglades. The same procedure can be applied to all GLAS waveforms to obtain a landscape scale representation showing spatial patterns of ecosystem productivity. Validation of the results with field data remains difficult, because of the geolocation errors. Furthermore, the accuracies of footprint size knowledge and allometric equations, and the validity of the model parameters must be considered in the validation process and landscape analysis.

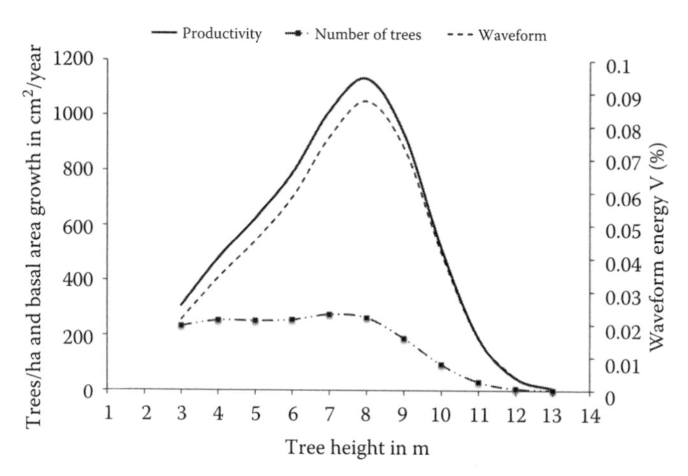

FIGURE 3.10 Calculation of annual productivity for a mangrove plot in Everglades National Park. The size class distribution was estimated from a LiDAR waveform. (Productivity values were calculated from Chen, R. and Twilley, R.R., *Journal of Ecology*, 86, 27, 1998.)

3.8 CONCLUSIONS

Accurately quantifying the structure and biomass of mangrove ecosystems on a large scale is crucial for studies of carbon storage, deforestation, biodiversity, forest quality, and habitat suitability. These measurements can be used in natural resource management and conservation such as the establishment of priority areas of conservation or the implementation of strategies geared at reducing emissions from deforestation. The use of remote sensing technology is a method that addresses a lack of field data and can produce accurate large-scale maps of forest extent, structure, composition, and biomass. In particular, the combination of radar and LiDAR technologies permits the derivation of three-dimensional maps of tree heights, biomass, and carbon storage.

We described the use of active remote sensing instruments such as LiDAR and radar to characterize mangrove canopy structure in three dimensions. The LiDAR waveforms provide an explicit representation of mangrove canopy 3D structure within small areas, and the radar enables mapping of the canopy height at the landscape scale. The SRTM elevation dataset provides a systematic tool to map mangrove canopy height globally. However, this technique is uniquely applicable to mangrove forests since they are located at the sea level and within the tidal zone (i.e., low topography). Indeed, the radar height measurement is the sum of the canopy height and ground elevation, which can be neglected in this case. In this methodology, first, a land-cover map is necessary to extract areas of mangrove forests from the SRTM dataset, and then the LiDAR data (e.g., ICESat/GLAS) is used to calibrate SRTM and obtain a landscape scale map of canopy height. In addition, we showed how to use the LiDAR waveform to derive the size class distribution of mangrove trees to finally estimate productivity assuming simple inversion models of the LiDAR waveforms, allometric equations, and individual-based ecological models.

ACKNOWLEDGMENTS

This study was conducted by the Jet Propulsion Laboratory, California Institute of Technology, under contract with the National Aeronautics and Space Administration (NASA), and funded by the NASA Interdisciplinary Science (IDS), ICESat science team and Terrestrial Ecology (TE) programs. The authors would like to thank Victor H. Rivera-Monroy, Edward Castañeda-Moya, and Robert R. Twilley from Louisiana State University and Michael Ross and Keqi Zhang from Florida International University for their fruitful collaboration.

REFERENCES

1. Jennerjahn, T.C. and Ittekot, V., Relevance of mangroves for the production and deposition of organic matter along tropical continental margins, *Naturwissenschaften*, 89, 23, 2002.
2. Dittmar, T., Hertkorn, N., Kattner, G., and Lara, R.J., Mangroves, a major source of dissolved organic carbon to the oceans, *Global Biogeochemical Cycles*, 20, 1012, 2006.
3. Valiela, I., Bowen, J.L., and York, J.K., Mangrove forests: One of the world's threatened major tropical environments, *BioScience*, 51, 807, 2001.

4. UNEP-WCMC, *In the Front Line: Shoreline Protection and Other Ecosystem Services from Mangroves and Coral Reefs*, p. 33, UNEP-WCMC, Cambridge, UK. 2006.
5. Held, A., Ticehurst, C., Lymburner, L., and Williams, N., High resolution mapping of tropical mangrove ecosystems using hyperspectral and radar remote sensing, *International Journal of Remote Sensing*, 24, 2739–2759, 2003.
6. Dadouh-Guebas, F., Qualitative distinction of congeneric and introgressive mangrove species in mixed patchy forest assemblages using high spatial resolution remotely sensed imagery (Ikonos), *Systematics and Biodiversity*, 2(2), 113, 2004.
7. Krause, G., Bock, M., Weiers, S., and Braun, G., Mapping land-cover and mangrove structures with remote sensing techniques: A contribution to a synoptic GIS in support of coastal management in north Brazil, *Environmental Management*, 34, 429–440, 2004.
8. Wang, L., Sousa, W.P., Gong, P., and Biging, G.S., Comparison of IKONOS and Quickbird images for mapping mangrove species on the Caribbean coast of Panama, *Remote Sensing of Environment*, 91, 432–440, 2004.
9. Simard, M., Zhang, K., Rivera-Monroy, V.H., Ross, M.S., Ruiz, P.S., Castaneda-Moya, E., Twilley, R.R., and Rodriguez, E., Mapping height and biomass of mangrove forests in Everglades National Park with SRTM elevation data, *Photogrammetric Engineering and Remote Sensing*, 72, 299–311, 2006.
10. Simard, M., Rivera-Monroy, V.H., Mancera-Pineda, J.E., Castañeda-Moya, E., and Twilley, R.R., A systematic method for 3D mapping of mangrove forests based on Shuttle Radar Topography Mission elevation data, ICEsat/GLAS waveforms and field data: Application to Ciénaga Grande de Santa Marta, Colombia, *Remote Sensing of Environment*, 112, 2131–2144, 2008.
11. Fatoyinbo, T.E., Simard, M., Washington-Allen, R.A., and Shugart, H.H., Landscape-scale extent, height, biomass, and carbon estimation of Mozambique's mangrove forests with Landsat ETM+ and Shuttle Radar Topography Mission elevation data, *J. Geophys. Res.*, 113, G02S06, 2008, doi:10.1029/2007JG000551.
12. Lucas, R.M., Mitchell, A.L., Rosenqvist, A., Proisy, C., Melius, A., and Ticehurst, C., The potential of L-band SAR for quantifying mangrove characteristics and change: Case studies from the tropics. *Aquatic Conservation: Marine and Freshwater Ecosystems*, 17, 245–264, 2007.
13. Treuhaft, R.N., Law, B.E., and Asner, P.G., Forest leaf area density profiles from the quantitative fusion of radar and hyperspectral data, *Journal of Geophysical Research*, 107, 4568, 561–571, 2002.
14. Breidenbach, J., Koch, B., Kandler, G., and Kleusberg, A., Quantifying the influence of slope, aspect, crown shape and stem density on the estimation of tree height at a plot level using Lidar and InSAR data, *International Journal of Remote Sensing*, 29, 1511, 2008.
15. Cintron, G. and Shaeffer Novelli, Y., Methods for studying mangrove structure, In: S.C. Snedaker and J.G. Snedaker (Eds), *The Mangrove Ecosystem: Research Methods*, p. 3, UNESCO, Paris, 1984.
16. Husch, B., Miller, C.I., and Beers, T.W., *Forest Mensuration*, 3rd edition, p. 402, Wiley, New York, 1982.
17. Mougin, E., Proisy, C., Marty, G., Fromard, F., Puig, H., Betoulle, J.L., and Rudant, J.P., Multifrequency and multipolarisation radar backscattering from mangrove forests, *IEEE Transactions on Geoscience and Remote Sensing*, 34, 94–102, 1999.
18. Graham, L.C., Synthetic interferometer radar for topographic mapping, *Proceedings of the IEEE*, 62, 763, 1974.
19. Elachi, C., Brown W.E., Cimino, J.B., Dixon, T., Evans, D.L., Ford, J.P., Saunders, R.S., et al., Shuttle imaging radar experiment, *Science*, 218, 996–1003, 1982.
20. CASI RADARSAT2-Special Issue, *Canadian Journal of Remote Sensing,* 30(3), 365, 2004.

21. Yoneyama, K., Koizumi, T., Suzuki, T., Kuramasu, R., Araki, T., Ishida, C., Kobayashi, M., and Kakuichi, O., JERS-1 development status, *Acta Astronautica*, 21(11), 783–794, 1990.
22. Attema, E.P.W., The active microwave instrument on-board the ERS-1 satellite, *Proceedings of the IEEE*, 79, 791, 1991.
23. Igarashi, T., ALOS mission requirement and sensor specifications, *Advances in Space Research*, 28, 127, 2001.
24. Franceschetti, G. and Lanari, R., *Synthetic Aperture Radar Processing*, Chapter 4, CRC Press, New York, 1999.
25. Zebker, H.A. and Villasenor, J., Decorrelation in interferometric radar echoes, *IEEE Transactions on Geoscience and Remote Sensing*, 30, 950, 1992.
26. Papathanassiou, K.P. and Cloude, S.R., Single baseline polarimetric SAR interferometry, *IEEE Transactions on Geoscience and Remote Sensing*, 39, 2352, 2001.
27. Farr, T.G. and Kobrick, M., Shuttle Radar Topography Mission produces a wealth of data, *American Geophysical Union EOS*, 81, 583, 2000.
28. Zwally, H.J., Schutz, B., Abdalati, W., Abshirea, J., Bentley, C., Brenner, A., Bufton, J., et al., ICESat's laser measurements of polar ice, atmosphere, ocean, and land, *Journal of Geodynamics*, 34, 405–445, 2002.
29. Chen, R. and Twilley, R.R., A gap dynamic model of mangrove forest development along gradients of soil salinity and nutrient resources, *Journal of Ecology*, 86, 27, 1998.

4 Integration of LiDAR and Historical Maps to Measure Coastal Change on a Variety of Time and Spatial Scales

Cheryl J. Hapke

CONTENTS

4.1 INTRODUCTION

This chapter examines two case studies that integrate modern Light Detection And Ranging (LiDAR) data and historical maps to calculate long-term coastal change rates, to map and identify erosion hazard areas, and to understand coastal processes on a variety of timescales, from storms to seasons to decades and longer. The case studies presented represent how similar types of data can be used to map coastal hazards in a variety of geographic settings and at a variety of spatial scales: in this case, barrier islands along the U.S. Gulf coast and rocky coastal environments of the U.S. West coast.

The first study examines the impacts of hurricanes on shoreline change rates in a barrier island environment, and identifies areas within a National Seashore that may be more vulnerable to future storms than is indicated by the long-term record. The second study is a regional analysis, using the same data sources and types as those used in the first study, but applying them to a statewide analysis of coastal cliff retreat in California.

4.2 LONG-TERM AND STORM-RELATED SHORELINE CHANGE TRENDS IN THE FLORIDA GULF ISLANDS NATIONAL SEASHORE

The Gulf Islands National Seashore (GUIS), in the northern Gulf of Mexico, is composed of a series of barrier islands that lie along the Florida Panhandle and Mississippi coasts. GUIS consists of seven barrier islands that extend along a total of 240 km of coast, with two in Florida (Perdido Key and Santa Rosa Island; Figure 4.1) and five in Mississippi. This case study focuses on the portions of Santa Rosa Island that fall within the boundaries of GUIS, and are described by the naming convention used by the National Park Service: the Santa Rosa Unit (SRU) to the east and the Fort Pickens Unit (FPU) to the west (Figure 4.1).

In 1995, Santa Rosa Island was almost entirely overwashed by Hurricane Opal, which made landfall just east of Pensacola Beach. Opal caused extensive erosion and removed dunes as high as 5 m [1], leaving the island and park resources more susceptible to future storms and erosion. In 2004 and 2005, a series of large and catastrophic hurricanes and tropical storms again directly impacted GUIS, causing extensive damage to the coastal barrier islands and to coastal infrastructure.

During such catastrophic events and/or storm seasons, dynamic but fragile barrier islands can become completely geomorphologically reconfigured, which directly affects habitat availability and functionality. Management of the highly impacted coastal resources in GUIS requires a working knowledge of how the system has changed. In order to understand the signal of severe storm shoreline change in the long-term record, the spatial patterns were assessed to determine whether long-term erosion rates can indicate where future storm-related hazards are greatest.

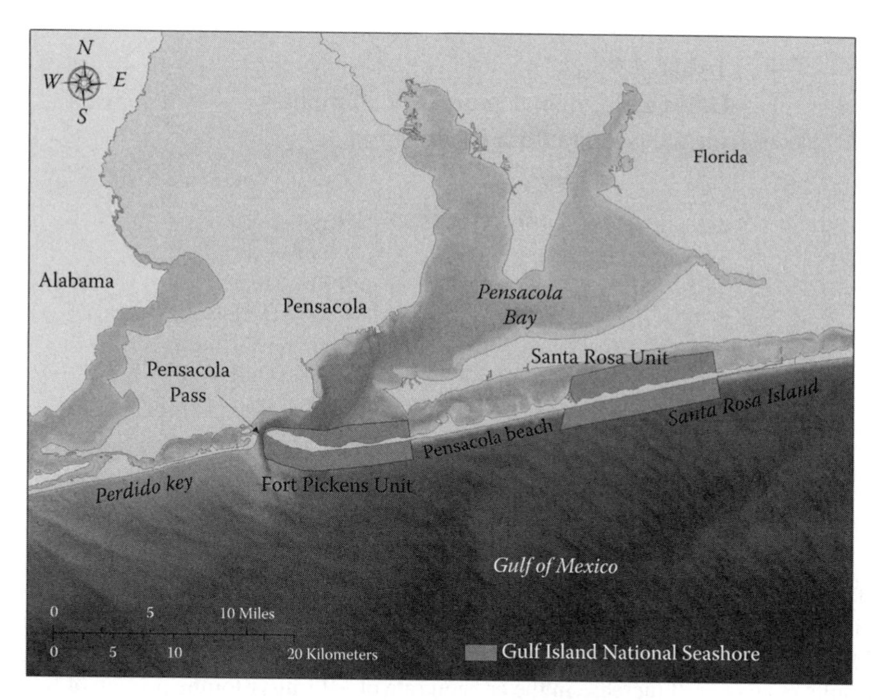

FIGURE 4.1 Location map for the GUIS and surrounding features.

The objective of this research is to build on existing long-term data that are critical to the understanding of short-term storm response in the Gulf Islands in Florida. The intention of this study is to determine, based on long-term trends (>100 years) and short-term (storm or seasonal) changes, where high rates of shoreline erosion may occur during future severe storms and thus provide useful information on the placement and maintenance of park infrastructure.

4.2.1 Methods

The shoreline change rates used for this analysis are derived from both historical maps (the 1800s, 1930s, and 1970s) and recent (2001, 2004, and 2005) LiDAR data. Table 4.1 outlines the years and the sources of the data used for the shoreline change analysis. The historical shorelines were digitized from georeferenced National Ocean Service (NOS) T-sheets. The historical shorelines represent the field interpretation of a high water line (HWL) on the beach [2]. The dates of the oldest map-derived shorelines are 1856, 1858, and 1859 for the SRU and 1871 for the FPU. For the 1930s- and 1970s-era T-sheets, the dates of the map-derived shorelines are 1934 and 1978 for both units. The modern shorelines are mean high water (MHW) shorelines derived from LiDAR data. For the long-term (~150 years) shoreline change the most recent shoreline is from May 2004 LiDAR data.

The use of two different shoreline proxies (HWL and MHW) may introduce a bias in the data. An HWL shoreline is higher on the beach than the MHW shoreline, which does not take total water level (wave run-up and tide height) into account [3,4].

TABLE 4.1
Dates and Original Sources of Shorelines for GUIS Santa Rosa Island

Date	Original Data Source
1856, 1858, 1859, or 1871	NOS T-sheets
1934	NOS T-sheets
1978	NOS T-sheets
2001	USGS ATM[a] LiDAR
May 2004 (2004a)	USGS EAARL[b] LiDAR
September 2004 (2004b)	USGS EAARL[b] LiDAR
September 2005	USGS EAARL[b] LiDAR

[a] ATM, Airborne Topographic Mapper.
[b] EAARL, Experimental Advanced Airborne Research LiDAR.

If a correction is not applied, the shoreline change rates will have an accretionary bias. A proxy bias was calculated using methods developed by Ruggiero et al. [3] and Moore et al. [4], and ranged from 5.0 to 10.2 m. The application of the proxy bias results in an overall increase in the erosion rate of –0.1 m/yr for the portion of Santa Rosa Island extending from Pensacola Pass to Navarre Beach.

In order to assess storm-driven shoreline change, pre-Hurricane Ivan (May 2004), post-Hurricane Ivan (Sept. 2004), and post-Hurricane Katrina (Sept. 2005) LiDAR data were acquired for Santa Rosa Island, from Pensacola Pass to Navarre Beach. The same 2300 transects were used to calculate short-term (May 2004 to September 2005) storm-driven shoreline change rates and the long-term, prestorms analysis. However, due to numerous gaps in the poststorm shoreline from island breaching and washover, only about 60% of the original transects crossed all three LiDAR shorelines. The data presented are in Universal Transverse Mercator (UTM) projection with the North American Datum of 1983 (NAD83).

Rates of shoreline change were generated in a Geographical Information System (GIS) with the Digital Shoreline Analysis System (DSAS), an ArcGIS© tool developed by the U.S. Geological Survey (USGS) in cooperation with TPMC Environmental Services [5]. This tool contains three main components that define a baseline, generate transects perpendicular to the baseline that intersect the shorelines at a user-defined separation along the coast, and calculate rates of change. The baseline was constructed seaward of, and roughly parallel to, the general trend of the coastline. Using DSAS, transects were spaced at 20 m intervals. Rates of shoreline change were calculated at each transect using a linear regression rate where there were four or more shorelines and an end-point rate for periods when there were three or less shorelines.

4.2.2 Uncertainties and Errors

Trends and calculated rates of shoreline change are only as reliable as the measurement errors that determine the accuracy of each shoreline position and statistical

TABLE 4.2
Maximum Estimated Rate Uncertainties for GUIS Shorelines

Shoreline Change Rate Uncertainties (m/yr)	Time Period			
	1800s–2004a	2004a–2004b	2004a–2005	1800s–2005
Linear regression rate	<0.1	—	—	<0.1
End-point rate	—	2.1	2.1	—

errors associated with compiling and comparing shoreline positions. A number of authors have provided general estimates of the typical measurement errors associated with mapping methods and materials for historical shorelines, referencing shoreline position relative to geographic coordinates, and shoreline digitizing [6–9].

The largest errors in this analysis were positioning errors of ±10 m, which were attributed to scales and inaccuracies in the original T-sheet surveys. However, the influence of large position errors on long-term rates of change can be reduced if the rate is calculated over a sufficiently long period of time. Additional data source errors implicit in this analysis result from Global Positioning System (GPS) positioning errors (±1 m), which Stockdon et al. [10] associated with the LiDAR data. Estimates of the maximum measurement errors for this study are provided in Table 4.2 to show how each error contributes to inaccuracy in the shoreline position.

End-point rates were calculated for those time periods with three or fewer shorelines, and include the 2004a–2004b (Ivan) time period and the 2004b–2005 (Katrina, Dennis, and Rita) time period. The total shoreline position error for the end-point retreat rate is the quadrature sum of the LiDAR shoreline position uncertainty for each LiDAR shoreline divided by the time period over which the rate was determined. For this analysis, all end-point rates included only LiDAR shorelines, and each of these has an estimated total positional error of 1.5 m. The shoreline change rate uncertainties were determined to be 2.1 m/yr for the end-point rates (Table 4.2).

Linear regression is a commonly applied statistical technique for determining rates of shoreline change [7] where there are a statistically valid number of samples. The long-term rates of shoreline change (1800s–2004a and 1800s–2005) were determined at each transect by taking the slope of the regression line applied to all shoreline positions. The resulting rates are reported in units of m/yr (Table 4.3). The uncertainty term reported for the long-term shoreline change rates is the 90% confidence interval of the linear regression shoreline change rate for each transect. For this study, the error associated with the linear regression rates was determined to be <0.1 m/yr.

4.2.3 Discussion

The shoreline change analysis covers a total of approximately 25 km of Santa Rosa Island. The long-term average rate of shoreline change prior to the 2004–2005 storm seasons was −0.3 ± 0.1 m/yr. However, the rates of shoreline change vary dramatically within the two park units. In both the long-term and storm-driven change

TABLE 4.3

Average and Maximum Long-Term and Storm-Associated Erosion Rates for the Florida GUIS Park Units: FPU and SRU

Unit	Long-Term (Prestorms) (1800s–May 2004) Erosion Rate (m/yr)		Hurricane Ivan (May 2004–May 2005) Erosion Rate (m/yr)		2004–2005 Hurricanes (Sep. 2004–Sep. 2005) Erosion Rate (m/yr)		Long-Term (Poststorms) (1800s–2005) Erosion Rate (m/yr)	
	Average	Maximum	Average	Maximum	Average	Maximum	Average	Maximum
FPU	−0.7 ± 0.1	−1.3 ± 0.1	−22.6 ± 2.1	−43.6 ± 2.1	−38.5 ± 2.1	−70.2 ± 2.1	−0.9 ± 0.1	−1.5 ± 0.1
SRU	−0.1 ± 0.1	−0.3 ± 0.1	−24.6 ± 2.1	−59.3 ± 2.1	−35.2 ± 2.1	−60.3 ± 2.1	−0.4 ± 0.1	−0.6 ± 0.1

analyses, rates of change within the FPU were much more variable than those within the SRU. During the hurricanes of 2004 and 2005, the shoreline change rates increased by two orders of magnitude (Table 4.3). Historically, the maximum rate of change, –1.3 ± 0.1 m/yr, was measured within the FPU. The maximum rate of change driven by the storms was –70.2 ± 2.1 m/yr.

Overall, the impacts of the 2004–2005 storms on the long-term trends of shoreline change and the mean rates of change in both the FPU and the SRU were variable, and the SRU was more greatly impacted, as discussed below. A Student's *t*-test run on the datasets indicate that there is a statistically significant difference between the 1800s–May 2004 (pre-Ivan) and the 1800s–September 2005 (poststorms) rates ($p < 0.0001$).

4.2.3.1 Fort Pickens Unit

The average rate of long-term, prestorm (1858–2004a) shoreline erosion (the average of transects that had negative change) in the FPU was –0.7 ± 0.1 m/yr (Table 4.3). The maximum erosion rate during this period, –1.3 ± 0.1 m/yr, occurred near the eastern portion of the unit where the island is narrow (Figure 4.2). The long-term, prestorm shoreline change rates reverse from an erosional trend to an accretional trend at the western end of the FPU where the island width increases to greater than 370 m. Indicative of the documented east-to-west sediment transport direction, the western end of Santa Rosa Island has been prograding over the ~140 years of the study. However, this accretional trend was interrupted during the 2004–2005 hurricanes. As a result of storm surge related to Hurricane Ivan, much of this previously stable (or accretional) portion of the FPU became erosional.

Large stretches of the FPU were overwashed or completely breached during the 2004–2005 hurricanes and, as a result, there are sizable gaps in the shoreline datasets (Figure 4.2b–d). The overwash/breach areas correspond spatially to the narrowest portions of the island within the FPU.

Average rates of shoreline erosion that resulted from the storms were two orders of magnitude greater than the long-term trends. During the Hurricane Ivan time period, the average rate of erosion increased from the long-term average of –1.3 ± 0.1 m/yr to –22.6 ± 2.1 m/yr and the shoreline erosion associated with both Ivan and the 2005 storms was –38.5 ± 2.1 m/yr. Although the proximity of the hurricane eye and northeast quadrant to Santa Rosa Island were more favorable for causing high rates of beach erosion during Ivan than those produced during the 2005 storms (Katrina, Rita, and Dennis), these storm likely resulted in more damage because the island was decimated by Ivan and was still in a state of recovery when the storms of 2005 occurred.

The combined impact of the 2004–2005 storms on the long-term shoreline change trend was measurable, although not as dramatic as expected, especially in the FPU. The average long-term erosion rate increased by –0.2 m/yr to –0.9 ± 0.1 m/yr, and the maximum rate increased by the same amount to –1.5 ± 0.1 m/yr. The overall patterns of long-term shoreline change within the FPU remained similar (Figure 4.2d). However, the point of reversal from erosion to accretion near the western end of the park shifted 1 km to the west in the long-term dataset that include the storms, resulting in a shorter extent of shoreline that is accretional.

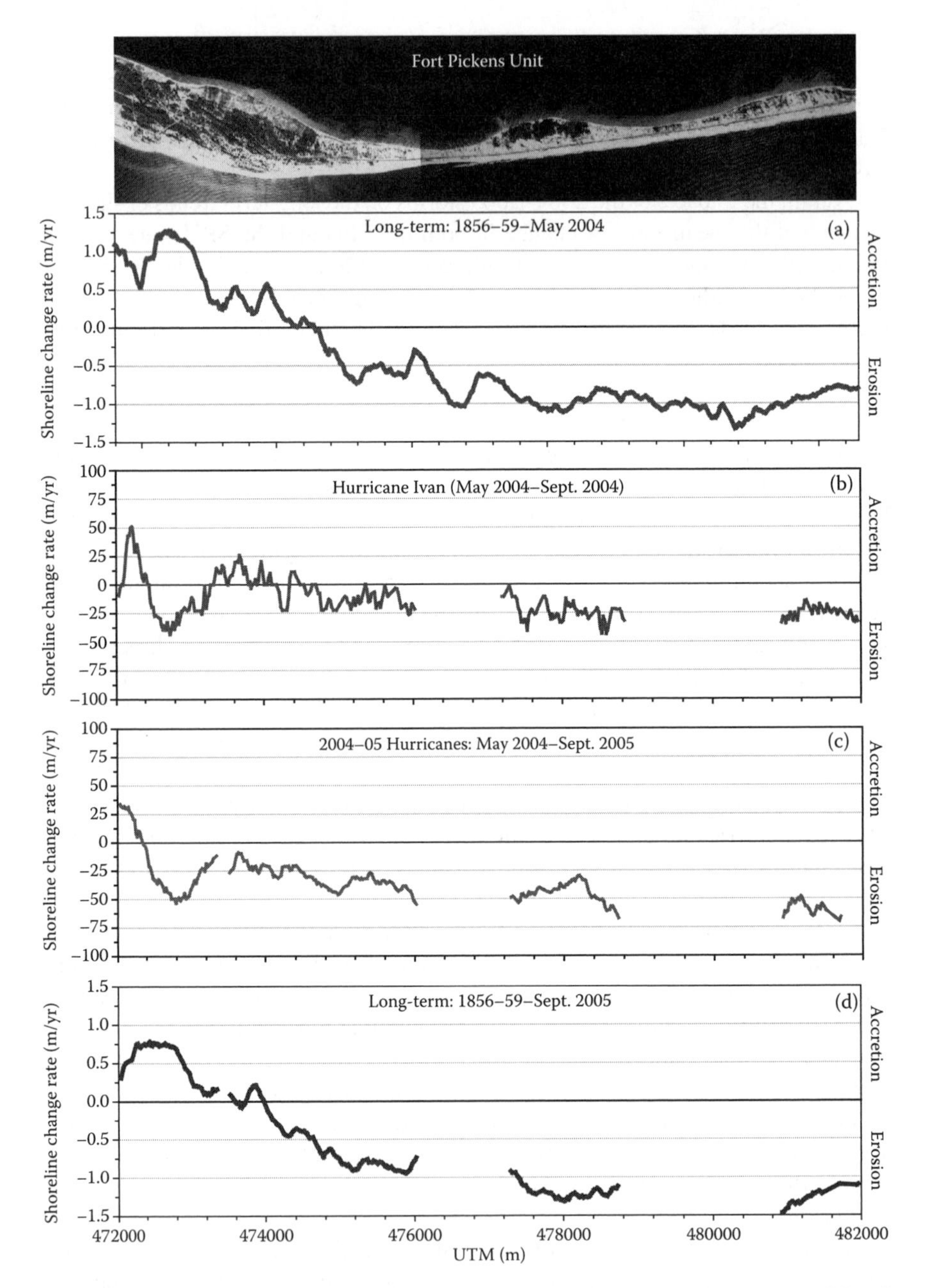

FIGURE 4.2 Orthophotomosaic from 1993 and shoreline change plots for the FPU of GUIS.

4.2.3.2 Santa Rosa Unit

The shoreline change trends in the SRU are much less variable along-shore than the trends in the FPU. Rates of long-term, prestorm erosion averaged −0.1 ± 0.1 m/yr, a nonsignificant signal given the uncertainties in the data. The maximum long-term (prestorm) erosion rate was −0.3 ± 0.1 m/yr, and was measured near the western end of the SRU, at one of the narrowest portions of the island within the SRU (Figure 4.3a). As shown in Figure 4.3a, rates are systematically low throughout this unit, with a slight trend toward erosion at both the east and west ends of the unit, and a slightly more accretional (or less erosional) trend in the center portion of the unit. Spatially, the pattern of shoreline change tends to mimic the shape of the island, with low erosion rates or accretion occurring where the island is wider—in general, at the locations of the bayside cuspate shoreforms.

Similar to the FPU, rates of shoreline change increased by two orders of magnitude, both from Hurricane Ivan and from the combined impacts of all the 2004–2005 storms (Table 4.3). The shoreline erosion rate associated with Hurricane Ivan averaged −24.6 ± 2.1 m/yr, with a maximum erosion rate of −59.3 ± 2.1 m/yr located on the eastern end of the SRU, immediately adjacent to the largest breach site, where there is a gap in the shoreline data (Figure 4.3b). The storms of 2005 resulted in additional areas of overwash and breaching (additional data gaps in Figure 4.3c). The maximum erosion rate during this period, −60.3 ± 2.1 m/yr, was again located on the eastern end of the SRU, adjacent to a locale that breached during Hurricane Ivan.

Unlike the FPU, the storms of 2004 and 2005 resulted in an overall regime shift in the long-term shoreline change pattern in the SRU (Figure 4.3d). Whereas the historical prestorm shoreline change data indicated that this portion of the island was relatively stable, the long-term, poststorm data show a completely erosional regime with erosion rates averaging 0.4 ± 0.1 m/yr, and an increase of −0.3 m/yr.

4.2.4 Summary/Conclusions

Long-term (~140-year) shoreline change rates were generated for the FPU and SRU of GUIS in Florida to assess the historical trends prior to the 2004 and 2005 hurricanes. Results indicate that the rates of shoreline change were higher and more variable in the FPU and that the shoreline was relatively stable within the SRU. In the FPU, the western 1.5 km of the island had a long-term progradational trend, but the average rate of shoreline erosion for the rest of the unit was −0.7 ± 0.1 m/yr.

Shoreline change rates along the same stretches of coast were calculated to assess the impacts of Hurricane Ivan in 2004 and for the total impacts of the 2005-05 hurricane season, which included Hurricanes Ivan, Cindy, Dennis, Katrina, and Rita. During these time periods, the rates of shoreline change increased by two orders of magnitude from the long-term trends. Numerous areas of the island were breached and overwashed, leading to significant gaps in the data, especially within the FPU. In general, these areas tended to correspond to the narrowest portion of the island.

Overall, the average and maximum rates of shoreline erosion were greater in the SRU during the Hurricane Ivan period, which was historically the more stable part of GUIS, and higher in the FPU during the combined periods that include both Ivan and the storms of 2005. This, in part, is likely related to the large data gaps in the FPU, where storm shoreline change rates could not be measured due to breaching or severe island elevation deflation.

Long-term rates of shoreline change were measurably influenced by the 2004 and 2005 hurricanes, although the impacts are greater in the SRU, which has direct park

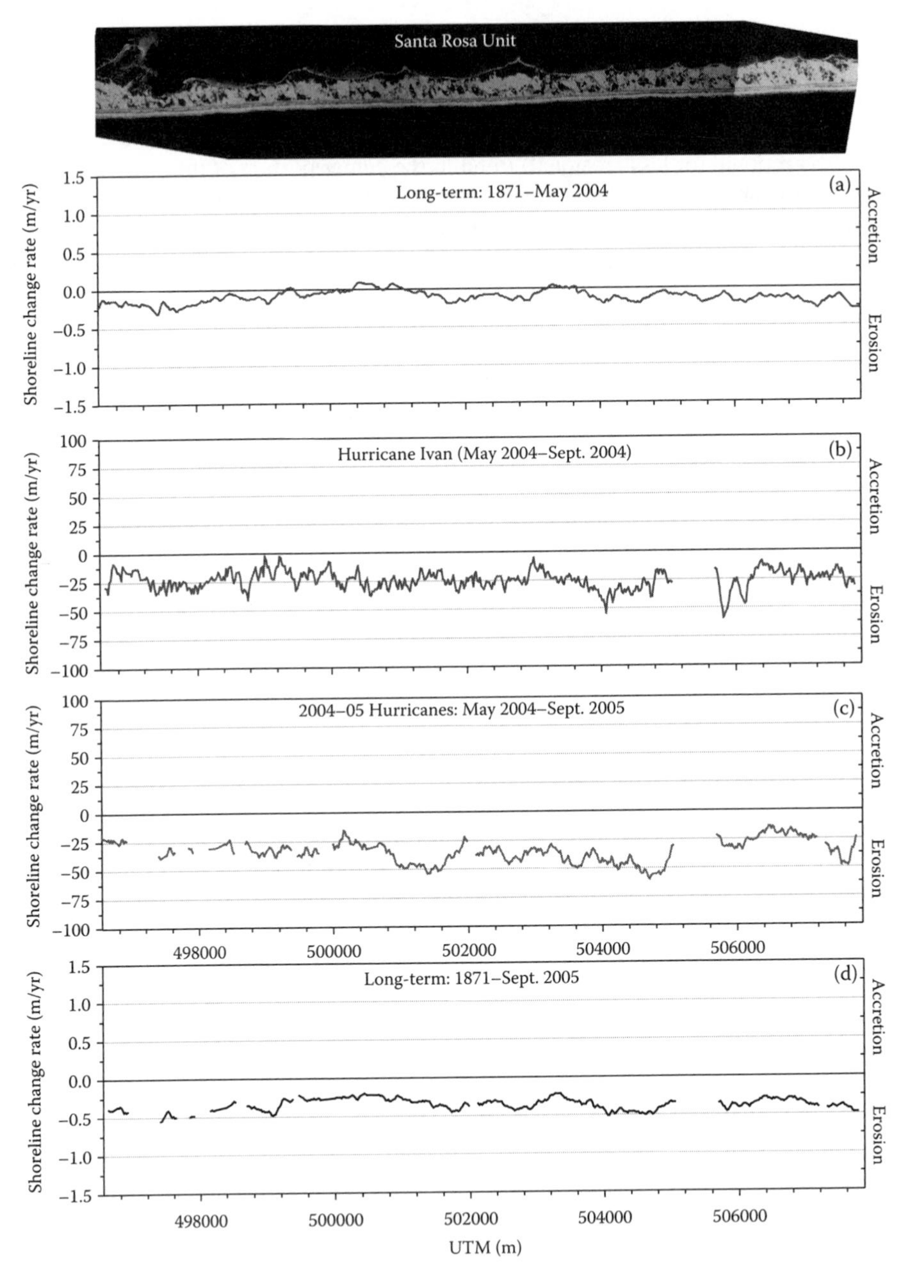

FIGURE 4.3 Orthophotomosaic from 1993 and shoreline change plots for the SRU of GUIS.

management implications. In the FPU, areas that had high rates of historical erosion prior to the storms were also those areas that experienced high rates of erosion during the storms, and thus the spatial patterns and distributions of shoreline change did not change significantly. One exception is an ~1 km section of coast in the western half of the FPU where the long-term shoreline change signal reversed from historically prograding to eroding at rates as high as –0.7 m/yr. In the SRU, the shoreline shifted from a historically stable system to one that has consistently significant rates of shoreline erosion along its entire length. The shoreline will likely experience rapid recovery in lieu of additional extreme storm events, and an important component of the long-term analysis will be the incorporation of such data as they become available.

4.3 COASTAL CLIFF RETREAT FOR THE CALIFORNIA COAST

4.3.1 Introduction

According to Griggs and Patsch [11], 72%, or 1300 km, of the California coast has eroding coastal cliffs. The retreat of these cliffs results in land loss and damage to private and community properties. Besides being popular tourist and recreation locales, the coastal cliff environments of California constitute some of the most valuable real estate in the country. Accurately documenting historical change rates is crucial to understanding the potential future behavior of the cliffs and the hazards they present.

Prior to this study, no consistent, systematic methodology existed to address regional coastal cliff retreat. This is partially due to the lack of available datasets to provide sufficient regional historical coverage. Traditionally, aerial photography has provided the most extensive datasets and has been used to derive two or more dates of cliff edges for retreat analyses. However, it is generally difficult to interpret an edge feature from 2D orthophotographs or georeferenced imagery. Visualization in 3D using photogrammetric techniques is therefore favorable for identifying the true break in slope at the cliff crest, but this technique can be costly and time-consuming. Additionally, historical aerial photographs are not always widely available and can be difficult to acquire and process for regional scale assessments. Alternatively, aerial-based LiDAR datasets are now becoming readily available and provide a consistent topographic model from which a cliff edge may be derived. However, because it is a relatively new technology as applied to coastal mapping, there is no historical data with which to make comparisons. This study uses a cliff edge digitized from historical maps as a base for comparison to the modern LiDAR dataset. The maps provide the oldest, regional dataset along the California coast for which a cliff edge can be derived. The specifics of the historical dataset are described in more detail below.

4.3.2 Background

There are few studies of regional coastal cliff erosion for California. Numerous analyses have been conducted for specific sites by private contractors, cities, and counties where erosion rates have been required for regulatory or management purposes. Some of these analyses were incorporated into Dolan et al. [12] and Griggs and Savoy [13], where rates of change were presented on maps and the long-term trends of erosion were summarized in an accompanying text. The Griggs and Savoy [13] compilation has recently been updated [14], and most of the erosion hazards addressed therein pertain to coastal cliff erosion, with the exception of Southern California. These compilations rely on existing data and erosion rates calculated using different methods, and therefore the results from one section of the coast to the next cannot be validly compared. The most regionally comprehensive modern cliff retreat analysis is presented by Moore et al. [15], where retreat rates were determined for two counties in California: Santa Cruz and San Diego. This analysis used digital photogrammetric techniques wherein the cliff edge is digitized while viewing in 3D. Even more recently, airborne and ground-based LiDAR have been used to map coastal cliffs at high resolution and measure short-term cliff retreat in California [16–18]. Whereas these methods can be accurate and precise, the analyses lack a historical component, simply due to the youth of the technology.

4.3.3 General Characteristics of the California Coast

For this analysis, the California coast is broadly divided into three sections: Northern, Central, and Southern California. In addition, and for the purposes of assessment and interpretation of trends, the sections are further divided into 15 analysis regions, as shown in Figure 4.4. The coast of Northern California can be characterized as a rugged landscape with low population. The coast from the Oregon Border to Point Arena is dominated by steep coastal cliffs that are dissected by numerous streams. Franciscan Complex rocks are common and the more resistant units often result in a coast with steep cliffs, small offshore islands, and sea stacks. Marine terraces and wave-cut bluffs are common between the areas dominated by the steep cliffs.

Central California is the most diverse coastal region of the state, having characteristics of both the north and south regions. Marine terraces and coastal bluffs are well developed south of Point Reyes, in the Monterey Bay region and along parts of the Big Sur coast (Figure 4.4). High-relief coastal slopes occur just north and south of San Francisco, respectively, and along most of the Big Sur coast.

The coast of Southern California, extending from Point Conception to the Mexican border (Figure 4.4), is markedly different from the rest of the state. Point Conception marks a dramatic change in coastal orientation due to tectonic movement along the Transverse Ranges that has resulted in an east–west trending coast. Farther south, the coast gradually returns to the northwest–southeast trend. Coastal cliffs and marine terraces are widespread and are typically fronted by narrow beaches. This portion of the coast is the most urbanized stretch of coast in California.

4.3.4 Methods of Analyzing Coastal Cliff Retreat

4.3.4.1 Compilation of Historical Cliff Edges

Coastal scientists in U.S. universities and government agencies have been quantifying rates of coastal cliff retreat and studying coastal change for decades. Before GPS and LiDAR technologies were developed, the most commonly used sources of historical cliff edges were aerial photographs.

This study incorporates cliff top edge positions from two time periods and two unique data sources. The two time periods and data sources are georeferenced 1920s–1930s NOS Topographic maps (T-sheets) and 1998 or 2002 LiDAR data, depending on availability. The historical cliff edges from the 1920s to the 1930s, which were clearly delineated on most of the T-sheets, were digitized from historical

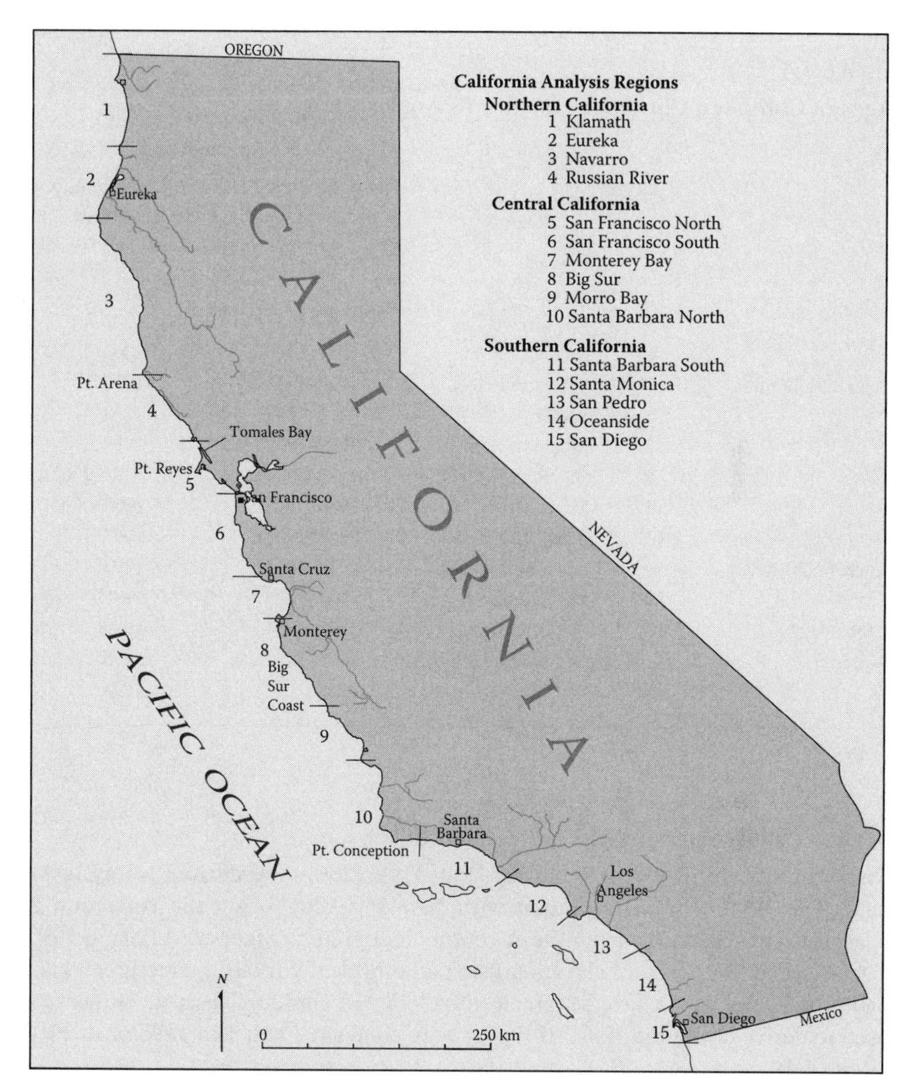

FIGURE 4.4 Location map of California showing the 15 analysis regions and specific locations of geographic places discussed in the text.

T-sheets, georeferenced as part of this effort. Table 4.4 lists the range of years for cliff edges compiled for each period by region.

For this analysis, the seaward edge of the coastal cliff, a commonly used indicator of coastal cliff retreat, is used to measure change. The cliff top is used instead of the base for several reasons, including the following: (a) the base is sometimes obscured by shadowing in our data sources, (b) the cliff base is irregularly interpreted on the historical maps used in the study, and (c) emplacement of seawalls and revetment, some of which may not be identifiable on the LiDAR data, can result in apparent accretion of a cliff base.

TABLE 4.4
Ages of Compiled Cliff Edges for the 15 Analysis Regions

			Selected Periods	
		Region	1920s–1930s NOS T-sheets	1998/2002 LiDAR
Northern California	1	Klamath	1926–1929	2002
	2	Eureka	1929	2002
	3	Navarro	1929–1935	2002
	4	Russian River	1929–1930	2002
Central California	5	San Francisco N	1929–1931	1998
	6	San Francisco S	1929–1932	1998
	7	Monterey Bay	1932–1933	1998
	8	Big Sur	1933–1934	1998–2002
	9	Morro Bay	1934	1998–2002
	10	Santa Barbara N	1933–1934	1998
Southern California	11	Santa Barbara S	1932–1934	1998
	12	Santa Monica	1933	1998
	13	San Pedro	1920–1934	1998
	14	Oceanside	1933–1934	1998
	15	San Diego	1933	1998

4.3.4.2 Derivation of a LiDAR-Derived Cliff Edge

The most recent cliff edge used in this study (1998/2002) was derived from LiDAR data. The USGS, in collaboration with NASA [19,20], begin the collection of coastal LiDAR datasets using the Airborne Topographic Mapper (ATM) in 1997. The ATM surveys ground elevation using an elliptically rotating blue-green laser and ground elevations have accuracies of about ±15 cm [20]. The LiDAR surveys used to derive cliff edges for this report were conducted either in 1998 or in 2002 (Table 4.4).

To compare with historical cliff edges, a methodology was developed to digitize cliff edges from the LiDAR surveys. The LiDAR point cloud data were gridded using a natural neighbors algorithm, at a 1-m cell size. A hillshade (or shaded relief map) was created from each grid, and the resulting 3D rendering was used to hand-digitize the cliff edge using the visual break in slope (Figure 4.5). This visual rendering approach has advantages over slope or second-derivative (gradient) methods of edge enhancement in that objects such as buildings or vegetation that are near the cliff edge are easier to identify and omit from the dataset. Use of slope and gradient approaches was explored but found to be noisy, especially in areas where the top of the cliff is developed or vegetated close to the edge. For the slope method, data were gridded at a 1-m cell size and a slope map generated. More gently sloping areas were shaded darker and steeply sloping areas were shaded lighter. The cliff edge was interpreted as the change in shade where there is an abrupt change in slope

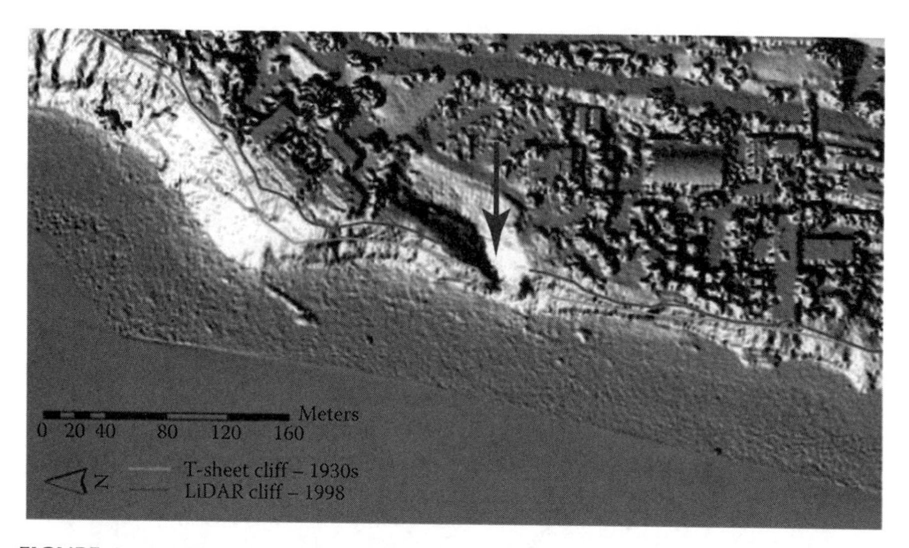

FIGURE 4.5 Hillshade of LiDAR data from a location in Southern California. Features such as gullies (arrow), trees, and buildings can easily be seen in the rendering.

(Figure 4.6a). Although this technique worked fairly well, it was difficult to interpret the cliff edge where the cliff slopes more gently, and where noise on top of the cliff from vegetation or development obscured the cliff edge. The gradient rendering of the gridded LiDAR data, while well suited to dune environments [21], made it difficult to interpret the edge of the cliff due to high slope gradients near the cliff edge from structures and vegetation (Figure 4.6b). As a result, the best and most consistent method of deriving a regional cliff edge from LiDAR data was to visually interpret the edge from hillshade renderings.

4.3.4.3 Calculation, Presentation, and Interpretation of Rates of Change

Rates of coastal cliff retreat were generated in a GIS with the DSAS [5]. This tool contains three major components that define a baseline, generate transects perpendicular to the baseline that intersect the cliff edges at a user-defined separation along the coast, and calculate rates of change. Baselines were constructed seaward of, and roughly parallel to, the general trend of the cliff edges. Using DSAS, transects were spaced at 20 m intervals. Transects were manually edited to assure that they were as orthogonal to the cliff edges as possible. This is an issue along crenulated coastlines and can result in erroneously high retreat rates (Figure 4.7). Rates of coastal cliff retreat were calculated at each transect using an end-point rate, which is the change from one time period to the next, applied to both cliff edges.

In this case study, cliff retreat rates are averaged over a ~70-year time period and the total amount of retreat is also presented. Cliff retreat rate data are frequently used for coastal zone management and can provide information on the spatial distribution of regional cliff retreat trends; they provide little information on specific hazard zones because of the highly episodic nature (both spatially and temporally) of the coastal cliff retreat process and response. For this reason, the total retreat

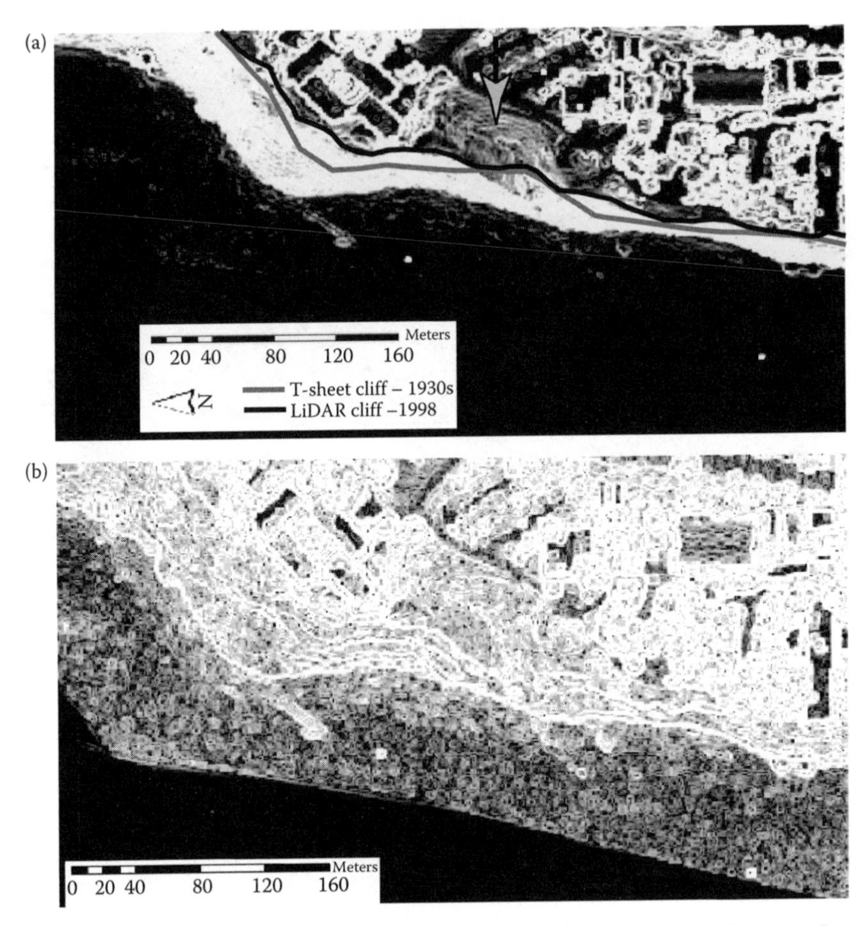

FIGURE 4.6 (a) Slope image derived from LiDAR data of the same general area as shown in Figure 4.1. The arrow indicates the location of the gully, which is somewhat ambiguous as a feature in the slope image. The cliff edges were digitized from the T-sheet and the slope image prior to the identification of the feature on oblique aerial photographs and the hillshade. In both cases, the gully was not identified correctly and the line was digitized across the mouth of the gully. (b) Gradient map of the gridded LiDAR data. The cliff edge is very difficult to identify due to buildings and vegetation near the cliff edge and the gully is not a readily identifiable feature.

amounts are also assessed. The dominant influences on the temporal variation of coastal cliff retreat are related to weather variations (storm intensity and frequency), climate variations (El Niño and Pacific Decadal Oscillation), and fluctuations in water level due to tides, storm waves, and eustatic sea-level rise. Spatial distributions in cliff retreat are related to the physical characteristics of the cliff-forming material (lithology and geologic structures), the orientation of the coastline with respect to the dominant wave direction, and anthropogenic impacts such as irrigation and the emplacement of protective structures.

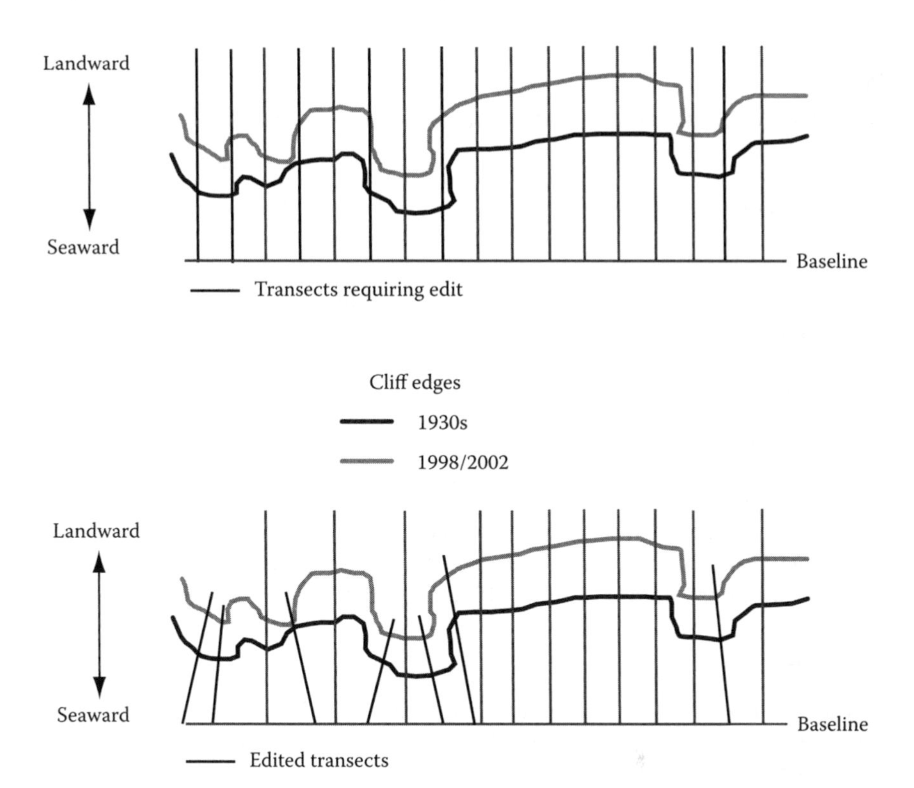

FIGURE 4.7 Schematic diagram showing required transect edits that are common on crenulated coastlines. If the transects are left orthogonal to the baseline (top diagram), they may result in under- or overestimation of the cliff retreat rate. The lower diagram shows the edited transects that more accurately reflect the retreat of the cliff edge.

4.3.4.4 Uncertainties and Errors

Documented trends and calculated rates of coastal cliff retreat are only as reliable as the measurement errors that determine the accuracy of each cliff edge position and statistical errors associated with compiling and comparing cliff edge positions. A variety of authors have provided general estimates of the typical measurement errors associated with mapping methods and materials for historical shorelines, registry of shoreline position relative to geographic coordinates, and shoreline digitizing [6–9]. Fewer reports have been published that document errors associated with coastal cliff retreat, although it has been addressed [22–24].

The largest errors in this analysis were positioning errors of ±10 m, which were attributed to scales and inaccuracies in the original T-sheet surveys. However, the influence of large position errors on long-term rates of change can be reduced if the rate is calculated over a sufficiently long period of time. Additional data source errors implicit in this analysis result from GPS positioning errors (±1 m) in the LiDAR data [10].

Estimates of the maximum measurement errors for this study are provided in Table 4.5 to show how each error contributes to inaccuracy in the cliff edge position.

TABLE 4.5
Estimated Positional Uncertainties for California Cliff Edges

	Time Period	
Measurement Uncertainties (m)	**1920s–1930s**	**1998–2002**
Georeferencing	4.0	—
Digitizing	1.0	1.0
T-sheet survey/T-sheet, DRG[a] position	10	—
LiDAR position uncertainty	—	1.0
Total position uncertainty	10.8	1.4
Annualized retreat rate uncertainty (m/yr)	0.2	

[a] DRG, Digital Raster Graphic.

The annualized error is calculated and subsequently incorporated into the cliff retreat rate calculations as outlined below. The uncertainty on the end-point rates, using a best estimate for California cliff edges, is ±0.2 m/yr (Table 4.5).

4.3.5 Results

This section presents the results of the California coastal cliff retreat analysis. Each California section (Northern, Central, and Southern) is subdivided into regions (Figure 4.4), which are based broadly on littoral cells and data coverage. Table 4.6 summarizes the average cliff retreat rates and amounts within each region.

The cliff retreat assessment for the broad divisions of the coast indicated that the average 70-year cliff retreat rates for California were highest in the Santa Monica region in Southern California (–0.3 m/yr, or 18 m of retreat), the San Francisco North region in Central California (–0.5 m/yr, or 36 m of retreat), and the Eureka region in Northern California (–0.7 m/yr, or 53 m of retreat). The maximum retreat in the state was 223 m of erosion in the Navarro region at the location of a large, deep-seated coastal landslide. The second highest amount of retreat, 210 m, was on the north-facing side of a large coastal headland in the San Francisco South region.

4.3.5.1 Northern California

Northern California extends from the Oregon border to Tomales Point (Figure 4.4), and for the presentation of the retreat rates, was divided into four regions: Klamath, Eureka, Navarro, and Russian River. Northern California is dominated by high-relief steep coastal slopes. The steep slope is interrupted where small streams and rivers drain to the coast. Marine terraces and wave-cut platforms occur sporadically along the coast.

The average amount of coastal cliff erosion measured over 70 years in Northern California was 28.8 m, and the average rate was –0.5 m/yr, as measured on 2325 transects. Many of the highest rates in Northern California were measured along headlands that lie in between embayments. The embayments occur either where there are small creeks draining the coastal slope, or in many cases they are deep-seated landslide complexes with wavelengths (distance from the center of one embayment to the next) on the order of 1 km.

TABLE 4.6
Number of Transects, Coastal Extents, and Average Cliff Retreat Rates for California

Region	Number of Transects	Length of Region (km)	Length of Measured Cliffs (km)	Average Retreat Rate (m/yr) ±0.2	Average Retreat Amount (m) ±10.9
Klamath	319	101	6	−0.5	−36.2
Eureka	135	154	3	−0.7	−53.4
Navarro	1441	142	29	−0.4	−28.9
Russian River	433	102	9	−0.2	−15.3
Northern CA	2325	499	47	−0.5	−28.8
San Francisco N	1092	119	22	−0.5	−36.2
San Francisco S	1551	99	31	−0.2	−16.4
Monterey Bay	1098	76	22	−0.4	−24.4
Big Sur	1929	145	39	−0.3	−17.2
Morro Bay	738	91	15	−0.2	−12.6
Santa Barbara N	3982	174	80	−0.2	−11.3
Central CA	10,390	704	208	−0.3	−17.3
Santa Barbara S	828	111	17	−0.2	−13.3
Santa Monica	1118	91	22	−0.3	−17.9
San Pedro	498	87	10	−0.2	−9.8
Oceanside	1993	86	40	−0.2	−12.0
San Diego	501	48	10	−0.2	−12.0
Southern CA	4938	400	99	−0.2	−13.3
State totals	17,653	1603	353	−0.3	−17.7

4.3.5.2 Central California

Central California is divided into six analysis regions including San Francisco North, San Francisco South, Monterey Bay, Big Sur, Morro Bay, and Santa Barbara North (Figure 4.4). The section begins just south of Tomales Point and extends to Point Conception. Central California has a more mixed geomorphology than Northern California in that there are areas of high-relief coast (i.e., the Big Sur coast), long stretches of well-developed, elevated marine terraces, and coastal lowlands that are typically associated with river mouths. The average retreat rate of cliff retreat was −0.3 m/yr, and the average amount of retreat was 17.3 m over the 70-year time period of the study. Numerous seawalls and revetments exist along this stretch of coast, especially in more heavily developed areas. These structures, built in response to cliff erosion threatening private homes and/or community infrastructure, act to reduce rates of cliff retreat.

4.3.5.3 Southern California

The Southern California section extends from Point Conception north of Santa Barbara to the Mexico border (Figure 4.4), comprising 400 km of coastline. The cliff retreat data for this section of the California coast is divided into five regions: Santa Barbara South, Santa Monica, San Pedro, Oceanside, and San Diego (Figure 4.7). Southern California is dominated geomorphically by long linear stretches of beach that in some areas are backed by low-to-moderate relief cliffs. There are a few areas of large deep-seated landslides, but where these do occur they are usually the locations of the highest rates for each region. This is also the most populous coast and, as a result, is also the most engineered coastline in the state. In addition to numerous harbors, breakwaters, jetties, and groins that disrupt the littoral flow of sand, large portions of the coast that are backed by cliffs have coastal protection structures. These efforts, which were employed to mitigate cliff retreat, have likely impacted the rates of cliff retreat, and likely contribute to the fact that the average retreat rate in Southern California is the lowest in the state (–0.2 m/yr).

4.3.6 Discussion

According to a recent study by the California Department of Boating and Waterways and the State Coastal Conservancy [25], the state of California has 1860 km of open ocean coastline. Of this, 1340 km has some type of coastal cliff. The remaining sections of the coast are generally coastal lowlands formed near the outlets of rivers and streams. In this report, long-term rates of coastal cliff retreat were provided for 353 km of the total length of cliffed coastline. For this analysis, gaps in either the LiDAR data or T-sheets, or the absence of a definable cliff edge in high-relief areas, resulted in a lack of two cliff edges over 74% of the coast characterized as cliffed or rocky. As a result, in this study long-term cliff retreat rates were calculated for 19% of the total California coast and 26% of California's cliffed coast.

The Eureka region in Northern California had the highest regionally averaged retreat rate (–0.7 m/yr) in the state, which translates to an average retreat amount of 53 m over the 70-year time period of this study. However, the rates were measured along a short length of coast (3 km) due to data gaps and may not be as representative as the other analysis regions. In general, the longer extents of continuous rates were measured in areas with well-developed and relatively continuous elevated marine terraces, such as the Santa Barbara North and Oceanside regions. Even in regions with lower average trends, there are clearly specific areas of coastline with high erosion rates, or "hotspots." In many locales, these hotspots present a high hazard to coastal development. Even in areas where the retreat rates are not exceptionally high, small amounts of cliff retreat may threaten homes and other community infrastructure.

Coastal cliff retreat rates are directly related to the geomorphology and geologic processing driving the retreat of the coast. As a result, the highest rates occurred along high-relief coastal slopes and were associated with large, deep-seated coastal landslide complexes. Thus, Northern California had the highest average retreat rates. In both the Santa Monica and San Pedro regions in Southern California, the highest rates were measured at specific sites of large landslides. For most other regions, the rates were highest where the cliff is composed of weaker geologic materials, such as Monterey Bay. Coastal protection structures likely influence the rates presented in this report. However, because most structures were emplaced in the time period covered by this analysis (1930s–1998/2002), it is not possible to quantitatively evaluate the extent of their effect.

The geomorphic influences on the rates of cliff retreat are also evident in the relationship between promontories and headlands and high rates of retreat. This relationship was more frequently true in Northern and Central California where the coastline is more crenulated, and thus has a higher density of headlands and embayments. The focusing of wave energy at headlands is likely driving these high rates, and underscores the importance of wave energy and water level on processes of coastal cliff retreat.

4.3.7 Summary

The average coastal cliff retreat rate for the State of California was –0.3 m/yr. This is based on retreat rates averaged along a total of 353 km of the coast, or about 20% of the state. Data gaps were numerous, and often a function of the lack of a discernible cliff edge on the historical maps or from gaps in the LiDAR data, which did not always extend inland far enough to capture the cliff edge.

The analysis found that the highest average rates were in Northern California, and for an individual region, the Eureka region had the highest average retreat rate in the state (–0.7 m/yr). Southern California had the lowest overall average cliff retreat rates, potentially because of the abundance of protective structures. In many cases, the regional trends in cliff retreat can be related to human intervention within the natural coastal system. Engineering structures, such as seawalls and revetments, have altered cliff retreat rates by lowering or even halting wave impact at the cliff base. Additionally, human activities that disrupt the natural sediment supply may ultimately result in increases in the cliff retreat rates by reducing the size of protective beaches.

The geomorphic influences on the rates of cliff retreat are evident in the relationship between promontories and headlands and high rates of retreat. In almost all of the analysis regions, the rates were consistently high in focused headland areas. This relationship was more frequently true in Northern and Central California where the coastline is more crenulated and thus has a higher density of headlands and embayments. The focusing of wave energy at headlands is likely driving these high rates, and underscores the importance of geology, wave energy, and water level on processes of coastal cliff retreat.

More information about this study, along with details of the methods and region-by-region results and interpretations, can be found in Hapke and Reid [24] and Hapke et al. [26]. An important note is that the change rates discussed in this case study represent change measured through the date that the LiDAR was collected (1998 or 2000) and thus may not reflect more recent trends in coastal cliff retreat. In addition, although retreat rates in some areas are reported as relatively low, even a small amount of local erosion may present serious hazards to the coastal resources and community infrastructure in a given area.

4.4 CONCLUDING REMARKS

The case studies presented in this chapter provide examples of how similar remote sensing and historical data sources can be used for a variety of coastal hazard mapping applications. The examples presented both utilize modern LiDAR data in an attempt to bridge a gap between traditional historical data (maps and aerial photographs) and allow the different datasets to be interfaced. In addition, both analyses used similar approaches and techniques, regardless of the temporal or spatial scale of the assessment.

In the first study presented, long-term shoreline change on a barrier island beach (Santa Rosa Island) was compared with the change caused by a series of severe storms to try to understand how the short-term (storm) record influences the historical rates of change. This analysis used shorelines derived from historical maps and a time series of modern LiDAR data. In many areas of Santa Rosa Island, the long-term rates of shoreline change were good predictors of where the shoreline eroded most rapidly during the storms. However, the analysis identified several locations that may have shifted in trend and may be areas of accelerated erosion in the future. Such analyses and results are important for coastal planners and managers of the National Park resources.

The analysis conducted on Santa Rosa Island covered a total of 25 km of coastline, a relatively small spatial scale. In contrast, the second case study along the California coast covered over 300 km of coastline and mapped a different geomorphic feature, but utilized similar data sources. The research developed a new method of deriving a cliff edge with regional coverage from LiDAR data and also utilized historical maps as a data source. The California cliff analysis helped to provide a regional perspective on coastal erosion hazards along the California coastline, and identified specific areas with very high long-term retreat rates. Detailed summaries of each case study are presented in the study descriptions above, and will not be repeated here.

The approaches and techniques utilized in both the case studies are highly applicable to other coastal regions, whether they are sandy beaches or rocky, cliffed coasts. Datasets that have commonly been used for coastal zone assessments can be integrated with newer data sources to modernize and update analyses. Such assessments will continue to be critical to coastal management and planning, especially with the currently predicted rates of sea-level rise through the twenty-first century.

REFERENCES

1. Stone, G.W., Liu, B., Pepper, D.A., and Wang, P., 2004. The importance of extratropical and tropical cyclones on the short-term evolution of barrier islands along the northern Gulf of Mexico, USA, *Mar. Geol.*, 210, 63–78.
2. Shalowitz, A.L., 1964. Shore and sea boundaries. Publication 10-1, U.S. Department of Commerce, Washington, DC.
3. Ruggiero, P., Kaminsky, G.M., and Gelfenbaum, G., 2003. Linking proxy-based and datum-based shorelines on a high-energy coastline: Implications for shoreline change analyses, *J. Coast. Res. Special Issue*, 38, p. 57–82.
4. Moore, L., Ruggiero, P., and List. J., 2006. Comparing mean high water and high water line shorelines: Should proxy-datum offsets be incorporated in shoreline change analysis? *J. Coast. Res.*, 22(4), 894–905.
5. Thieler, E.R., Himmelstoss, E.A., Zichichi, J.L., and Miller, T.L., 2005. Digital Shoreline Analysis System (DSAS) version 3.0; An ArcGIS© extension for calculating shoreline change, *U.S.G.S Open-File Report* 2005–1304.
6. Anders, F.J. and Byrnes, M.R., 1991. Accuracy of shoreline change rates as determined from maps and aerial photographs, *Shore and Beach*, 59(1), 17–26.

7. Crowell, M., Leatherman, S.P., and Buckley, M.K., 1991. Historical shoreline change; Error analysis and mapping accuracy, *J. Coast. Res.*, 7(3), 839–852.
8. Thieler, E.R. and Danforth, W.W., 1994. Historical shoreline mapping (1). Improving techniques and reducing positioning errors, *J. Coast. Res.*, 10(3), 549–563.
9. Moore, L.J., 2000. Shoreline mapping techniques, *J. Coast. Res.*, 16(1), 111–124.
10. Stockdon, H.F., Sallenger, A.H., List, J.H., and Holman, R.A., 2002, Estimation of shoreline position and change from airborne topographic lidar data, *J. Coast. Res.*, 18(3), 502–513.
11. Griggs, G.B. and Patsch, K., 2004, California's coastal cliffs and bluffs, in *Formation, Evolution, and Stability of Coastal Cliffs—Status and Trends*, Hampton, M.A. and Griggs, G.B., Eds., U.S.G.S. Prof Paper 1693, 53–64.
12. Dolan, R., Anders, F., and Kimball, S., 1985, Coastal erosion and accretion, in *National Atlas of the United States of America*, U.S. Geological Survey, Reston, VA, 336 pp.
13. Griggs, G.B. and Savoy, L.E., 1985, Living With the California Coast: Duke University Press, Durham, North Carolina, 393 pp.
14. Griggs G.B., Patsch, K., and Savoy, L., 2005, Understanding the shoreline, in *Living with the Changing California Coast*, G. Griggs and K. Patsch, Eds., University of California Press, Berkeley, 551 pp.
15. Moore, L.J., Benumof, B.T., and Griggs, G.B., 1999, Coastal erosion hazards in Santa Cruz and San Diego counties, California, *J. Coast. Res. Special Issue*, 28, 121–139.
16. Sallenger A.H., Krabill W., Brock J., Swift R., Manizade S., and Stockdon H., 2002, Sea-cliff erosion as a function of beach changes and extreme wave runup during the 1997–1998, *El Nino: Marine Geology*, 187(3), 279–297.
17. Collins, B.D. and Sitar, N., 2004, Application of high resolution 3D laser scanning to slope stability studies, *Proceedings of the 39th Symposium on Engineering Geology and Geotechnical Engineering*, Butte, Montana, May 18–21, 79–92.
18. Young, A.P. and Ashford, S.A., 2006, Application of airborne LIDAR for seacliff volumetric change and beach-sediment budget contributions. *J. Coast. Res.*, 22(2), 307–318.
19. Krabill, W.B., Wright, R.N., Swift, E.B., Frederick, E.B, Manizade, S., Yunkel, J.K., Martin, C.F. et al., 2000, Airborne laser mapping of Assateague National Seashore Beach, *Photogramm. Eng. & Rem. Sens.*, 66(1), 65–71.
20. Sallenger, A.H., Krabill, W.B., Swift, R.N., Brock, J., List, J., Hansen, M., Holamn, R.A., et al., 2003, Evaluation of airborne topographic lidar for quantifying beach changes, *J. Coast. Res.*, 19(1), 125–133.
21. Elko, N., Sallenger, A.H., Guy, K., Stockdon, H., and Morgan, K., 2002, Barrier island elevations relevant to potential storm impacts: 1. Techniques, *U.S.G.S. Open-File Report 2002-287*, 1 sheet.
22. Moore, L.J. and Griggs, G.B., Long-term cliff retreat and erosion hotspots along the central shores of the Monterey Bay National Marine Sanctuary, *Mar. Geol.*, 181, 1–3, 2002.
23. Hapke, C., 2004, The measurement and interpretation of coastal cliff and bluff retreat, in *Formation, Evolution, and Stability of Coastal Cliffs—Status and Trends*, Hampton, M.A. and Griggs, G.B., Eds., U.S.G.S. Prof Paper 1693, pp. 39–52.
24. Hapke, C.J., and Reid, D., 2007, The National assessment of shoreline change: Part 4, Historical coastal cliff retreat along the California coast: U.S. Geological Survey Open-file Report 2007–1133 (http://pubs.usgs.gov/of/2007/1133).
25. California Department of Boating and Waterways and State Coastal Conservancy, California Beach Restoration Study, 2002.
26. Hapke, C.J., Reid, D., and Richmond, B., 2009, Rates and trends of coastal change in California and the regional behavior of the beach and cliff system, *J. Coast. Res.*, 25(3), 603–615.

5 Coastal 3D Change Pattern Analysis Using LiDAR Series Data

Guoqing Zhou

CONTENTS

5.1 INTRODUCTION

Beach morphology fluctuations cover a wide range of time scales, varying from periods of hours (e.g., storms), to years and decades associated with long-term erosion trends. Different types of data collection methods have been used to compute the net deposition or erosion of sand on beaches. Coastal monitoring projects have employed different approaches for various monitoring purposes on spatial data acquisition, including bathymetric survey of the offshore sand banks, nearshore bathymetric surveys, elevation profiling of beaches, photogrammetric digital surface modeling, airborne laser-scanned cliff, and the beaches and surrounding coastline (Carter and Shrestha, 1997).

In the past several decades, different methods were employed to analyze the coastal change. For example, traditionally, investigators employed photogrammetry, classical surveying, or integration of aerial remote sensing with Global Positioning System (GPS) kinematic surveys to analyze 3D data changes of the coast along beach areas. These methods have inherent advantages, such as a full coverage of areas of interests and high accuracy (Moore et al., 2003; White and Wang, 2003). In the past several years, methods for coastal change analysis using airborne Light Detection And Ranging (LiDAR) have been increasingly interested in large-scale coastal mapping and monitoring including assessment of flood risk, storm damage, geomorphology, landfill monitoring, sediment transportation, and creation of nautical charts (Irish and Lillycrop, 1999; Schmitz-Peiffer and Grassl, 1990; Strom et al., 2001; Tralli et al., 2005), monitoring beach nourishment and evolution (Irish and White, 1998), coastline erosion and coastal structures change detection, nearshore and upland topography analysis, natural morphological changes and response to man-made alterations (Guenther et al., 2000), and emergency response to hurricanes and ship groundings (Parson et al., 1997), monitoring 3D sand dunes changes (Woolard and Colby, 2002), evaluating hurricane-induced beach erosion by analyzing volumetric change analysis of deposition or erosion (Meredith, 1999), and morphological change along the North Carolina coastline (White and Wang, 2003). Other researchers also have employed LiDAR data for coastal data analysis (Krabill et al., 2000b) and for shoreline, beach face, and headland change in response to El Nino and wave action (Sallenger, 2000; Sallenger et al., 1999). This chapter presents a case study that employed a series of LiDAR data for coastal 3D morphological change analysis.

5.2 STUDY AREA AND DATA

5.2.1 Study Area

The study area was the Assateague Island National Seashore (*Assateague Island*) in Virginia between 37.88°N and 38.02°N latitudes, and 75.38°W and 75.22°W longitudes (Figure 5.1). Historically, the study area was affected by northeasters and hurricanes and replenished by sand moving south along the Maryland coast until the establishment of the Ocean City Inlet at which time the erosion south of the inlet began to accelerate. Irish and Lillycrop (1999) demonstrated that the Assateague Island is exceptionally dynamic, experiencing average erosion rates as high as 3.1 m/yr in some areas. Dolan et al. (1997) and Dolan and Davis (1992) found that the Assateague Island, like other barrier islands, is constantly changing shape and geographical position. The important features of the Assateague Island are its fragile coastal elements characterized by sand dunes, maritime forests, inlets, lagoons, back-barrier marshes, and the vegetation. Increased urban development and natural processes have affected the morphology of these barrier islands. Observed loss of shoreline could be with rates of 5–12 m/yr in some particular section (Pilkey et al., 1998).

Section / Date	Study area data availability						Illustration
	Sect 1	Sect 2	Sect 3	Sect 4	Sect 5	Sect 6	
Oct. 9, 1996	√	√	√	√	√	√	
Oct. 11, 1996	√	–	–	–	–	–	
Sept. 15, 1997	√	√	√	√	√	√	
Sept. 16, 1997	√	√	–	–	–	–	
Sept. 18, 1997	√	–	–	–	–	–	
Feb. 9, 1998	√	√	√	√	√	√	
Feb. 10, 1998	√	√	√	√	√	√	
April 3, 1998	–	√	√	√	√	√	
Dec. 1, 1998	√	√	√	√	√	√	
Sept. 20, 2000	√	√	√	√	√	√	
Nov. 1, 2000	√	√	√	√	√	√	
Nov. 2, 2000	√	√	√	√	√	√	

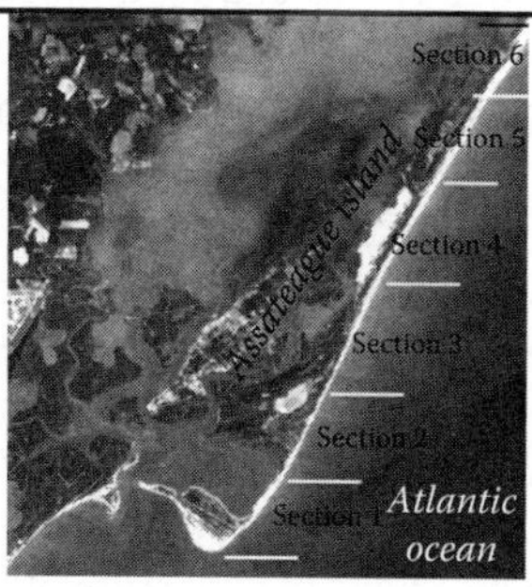

FIGURE 5.1 Downloaded LiDAR datasets and segmented intensive study areas.

5.2.2 LiDAR Dataset

A LiDAR system consists of three basic components, that is, a laser scanner, an inertial measuring unit (IMU), and an airborne GPS. They are mounted photogrammetrically at the bottom of an airplane, in a similar manner to an aerial camera system. The LiDAR laser scanner measures the distance between the laser emitter and the ground surface using the time between the pulses of laser light striking the surfaces of the earth and pulse return. The distance can be transformed into a foot printer's three coordinates using the onboard IMU and GPS. Robust data storage is required to process the return time for each pulse and to calculate the variable distances and 3D coordinates. This system can collect data at day or night, as long as clear flying conditions are present. Moreover, it is capable of distinguishing the canopy and bare ground, and collecting the 3D data for those areas, such as steep slopes, shadowed areas, and inaccessible areas of mud flats and ocean jetties.

The data for this study area were downloaded from the NOAA Coastal Services Center's LiDAR Data Retrieval Tool (LDART). The datasets were acquired on

October 11, 1996, on September 16–18, 1997, in February and December 1998, and in September and November 2000. While the 2000 data (NOAA, 2000) cover the entire study data, the data acquired on October 11, 1996 and September 16–18, 1997, cover only the southern end of the Assateague Island and the data acquired on April 3, 1998, cover only the eastern shoreline. The downloaded data were resampled into grid digital elevation model (DEM) using ArcGIS inverse distance weighting (IDW) methods with a planimetric resolution of 0.9×0.9 m^2. Such a cell resolution used in the grid is on the order of the spatial density from 1 point/m^2 to roughly 1.5 points/m^2. All the DEMs were georeferenced to the WGS84 spheroid and North American Vertical Datum (NAVD) of 1988, respectively.

The downloaded LiDAR data for the entire study area were partitioned into six sections due to the nature of the coastal condition and environment (see Figure 5.1). Limited capacity of computer processing memory was a factor for this section subseting. In each section, three representative sites, which were referred to as *Areas of Interest* (AOIs), were selected and created. Ancillary data, such as the spatial surface profiles of the DEMs, slope and relief data of the DEMs, Landsat-7 orthoimageries, and USGS color infrared (CIR) digital orthophoto quads (DOQs), were used to assist the identification and creation of the AOIs. The selected AOIs represent particular segments of the coastline, where the dune line and dry beach are obviously distinguished and are generally carried out in proximity to some highly valued coastlines. A number of environmentally sensitive sites were selected where the processes of erosion and deposition could be easily studied spatially. Each selected AOI covers almost exactly the same location and portion of coastline for the separated yearly analysis.

5.3 DATA ANALYSIS

5.3.1 Development of Analysis Tools

ArcObject software was employed to develop an analysis tool. The major modules include File, DataPre, CoastExt, MorphAnal, and DynView (Figure 5.2). The functions of each module are described as follows.

1. *File*: This module allows users to input and/or output different formats of data, such as raster, vector, text, images, and ASCII. These different formats of data, which are used for describing an identical object, can be converted from one to another. Output menu allows users to output traditional ArcView Layouts function, and the analyzed results into hardcopy datasheets or digital documents.
2. *DataPre*: This module allows users to prepare the data. This module is used to interpolate the original LiDAR data into DEM with different cell sizes, divide an entire study area into several sections, and coregister different formats of data such as LiDAR, Landsat, and USGS orthoimageries. In this module, an interface is designed for users to input the parameters.

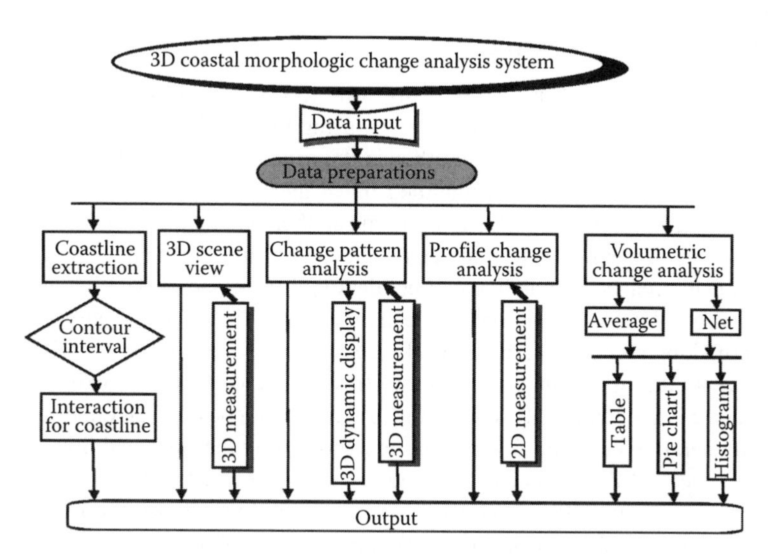

FIGURE 5.2 Coastal change analysis of LiDAR series data.

3. *CoastlineAnal*: This module is mainly for the coastline extraction and coastline change detection on the basis of LiDAR data. This module can provide information, for example, coastline position and changes, to support coastal 3D morphological analysis. The algorithm used for coastal changes is described in Section 3.2.
4. *MorphAnal*: This module includes the analysis of topographic change patterns, seasonal topographic change patterns, profile plot analysis patterns of each AOI, volumetric change patterns of net deposition and erosion and volume change per unit area (PUA) (m^3/m^2), and statistics of seasonal and annual average volume change (m^3). Each analysis module also allows users to manually measure the magnitudes and direction of changes in 2D or 3D form.
5. *DynView*: This module dynamically displays the changes of coastline, coastal topographic profile of AOI, deposition and erosion, and view of 3D scenes. Various change patterns can be dynamically animated using ArcGIS 3D functions. Seasonal and annual average erosion and/or deposition volume (m^3) and volume change PUA (m^3/m^2) are quantitatively and dynamically displayed through histograms, charts, tables, and pie charts. The dynamic display of the various types of data, such as shoreline, profile, erosion, TIN, and 3D scene provides an effective way to observe the coastal morphological changes from different points of views.

There are other software tools, such as Beach Morphology Analysis Package (BMAP) and Digital Shoreline Analysis System (DSAS), for shoreline analysis. The main characteristic of this tool developed in this paper is its capability of dynamically visualizing the changes of the coastline, coastal topographic profile, deposition and erosion, and 3D scene view.

5.3.2 Data Analysis Method

An algorithm developed in the above software tool for exacting the instantaneous coastline at imaging epoch was achieved by using LiDAR data in combination with Landsat 7 orthoimagery. The algorithm is described as follows. Contour lines at 0.05 m intervals are generated using the LiDAR data first. Multiple contour lines at elevations such as 0.0, 0.05, 0.10, and 0.15 m can then be found and an instantaneous coastline can be identified from the extracted contour lines using a tangent between Landsat orthoimagery. Coastline can be identified from one of the contours as the elevations of all points along a coastline are the same. This unique contour line with a smooth slope and curvature, along the coastline on the Landsat orthoimages, is identified as the instantaneous coastline. In fact, the contour line extracted from LiDAR data is not completely fitted to the instantaneous coastline on Landsat images. Manual operation has been designated for smoothening the extracted lines. In addition, because the LiDAR data do not include water wave information, three contours, whose elevations are 0.0, 1.2, and 1.8 m, are created to reference the low, average, and high wave coastlines, respectively (Zhou and Xie, 2008).

The calculation method used in the above software tool for the 3D morphological change pattern is for differentiating the *Z* coordinates of the second year from the first year values on each grid cell for each DEM pair. The volume change at each cell location can then be calculated. A positive, negative, or zero volumetric value (m^3) at a cell represents the amount of deposition, erosion, or no change. The morphological change of topography over the study area can be obtained through summing the positive and negative volumetric values (m^3) in each cell. The volumes of deposition for each section can be calculated by summarizing all positive volumetric values of cells. Similarly, the volumes of erosion for each section can be calculated by summarizing all negative values of the cells. The net change is calculated by differentiating the total deposition from the total erosion. Considering that each section does not cover exactly the same size of beach area in each yearly LiDAR data, the net volumetric change per square meter (m^3/m^2) is adopted for comparing the volumetric changes at various time intervals. Using the above methods, topographic change between selected years, and total volumetric change of beach and sand dunes can be obtained.

In addition, the algorithm used in the above software tool for beach transects and profiles was created to assist in analysis of beach changes in a specified time interval. By taking a close look at profiles that are made over a given time interval, it is possible to determine whether the beach is staying the same (Brock et al., 2002).

5.4 ANALYSIS RESULTS

5.4.1 Spatial Pattern of Topographical Change

Topographic changes of dune, berm, foreshore, and nearshore between 1996 and 2000 for each section were obtained and measured on the visualized Triangulated Irregular Network (TIN) data structure (Figure 5.3). As observed, the average width of dune in Section 5 is wider than that in Section 6. The berm in Section 4 narrows from north to south and disappeared in Section 3 as the dune transitions directly into

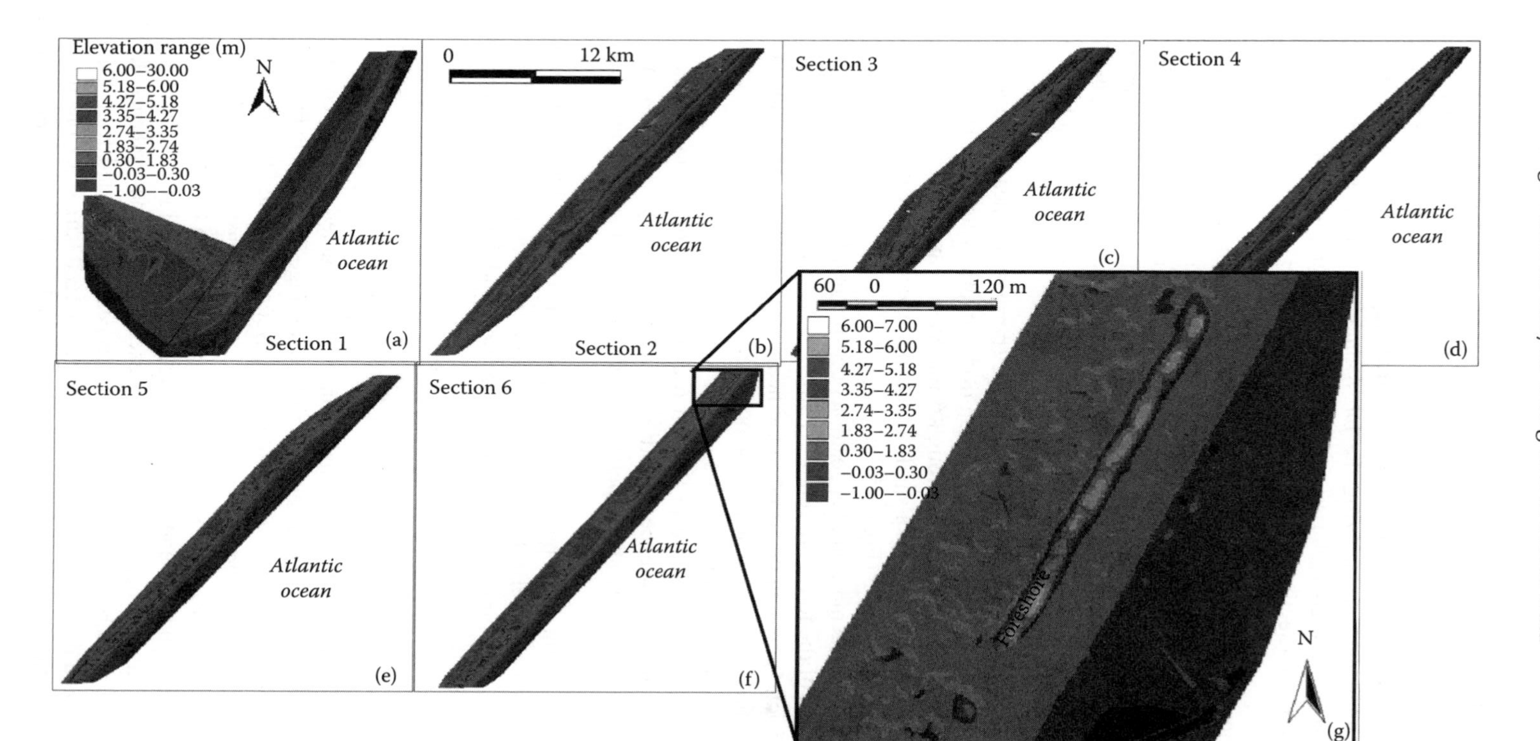

FIGURE 5.3 Dune line and dry beach of six sections in 1996. The magnified window is Section 6 segment of the coastline consisting of the primary portion of the dune line and dry beach.

the foreshore, forming a ridge. The ridge of Section 2 is narrower than that of the other sections, and its height is lower (Figure 5.3g). The topographic elevation in the southern Assateague Island (Section 1) has changed greatly between 1996 and 2000. Moreover, this change is irregular over the entire study area. We further analyzed the topographic changes via analyzing the DEM data pair between 1996 and 2000. The results reveal the characteristics of shoreline topographic change from south (Section 1) to north (Section 6) as follows.

1. A 1500-m-long shoreline in the southern end of the Assateague Island in Section 1 experienced significant deposition from 1996 to 2000 over an area that is 190 m wide and 1500 m long from nearshore to foreshore. Meanwhile, an erosion of 200 m width in the dune and a berm area also occurred (Figure 5.3a), and a more significant erosion of 100 m width and 980 m length in the berm and foreshore area of the northern end of Section 1 is evident (Figure 5.3a).
2. A 400-m-long shoreline on the west side (berm areas) near the middle of Section 2 has experienced deposition over a width of up to 100 m (Figure 5.3b), and from Sections 3 through 6 the erosion appears to have slowed from 1996 to 2000. Only a small area experienced severe erosion, such as Profile 1 in Section 3 (Figure 5.3c).
3. In Section 4, the most rapid deposition occurred in a 1200-m-long by 85-m-wide area of the foreshore (Figure 5.3d).
4. In Section 5, erosion occurred in the nearshore, foreshore, and dune areas facing the Atlantic, and deposition occurred in the berm and dune areas not facing the Atlantic (Figure 5.3e).
5. In contrast, the foreshore in Section 6 experienced deposition, while the berm and dune area experienced erosion. Additionally, a 40-m-wide by 415-m-long foreshore area experienced significant deposition (Figure 5.3f).

The analysis of the 3D topographic changes from 1996 to 2000 is dynamically displayed through color refreshing, as depicted in Figure 5.4, in which the blue color denotes erosion, the orange color denotes deposition, and the white color denotes no change (between −0.03 and 0.03 m).

5.4.2 Seasonal Topographic Change Pattern Analysis

Through observation of DEM seasonal changes, it was noted that the coastal area had generally grown in late summer and fall from average 15 m to 7 m from south to north. Most deposition occurred in winter from average 2200 to 998 m^3 from south to north (Table 5.1). This finding is consistent with what coastal morphologists have generally found. Discussions of seasonal changes have found that beaches grow during the less stormy summer and early fall when offshore sand from submerged bars are transported onto the beach face and berm except for summer and early fall hurricanes. During winters the beach materials are generally eroded from the beach face and berm and deposited on the nearshore submerged bars when the area is frequently affected by northeasters. Generally, during the winter and spring seasons

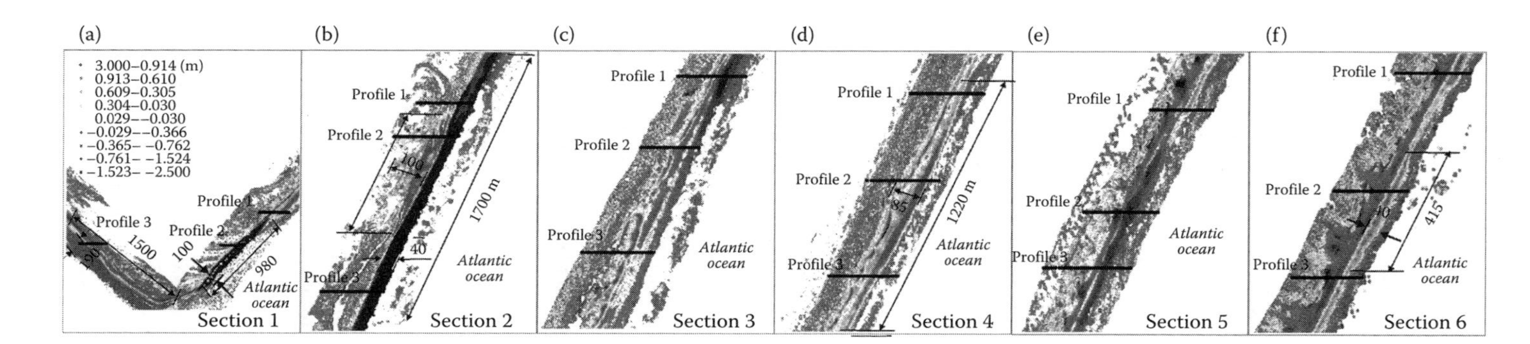

FIGURE 5.4 Dynamic refresh display of analyzed results of the topographic changes in six sections from 1996 to 2000 using different colors. The black lines indicate the location of a profile, which is used for the study of rapid changes.

TABLE 5.1
Summary of Annual Volumetric Change

		Net Change		Deposition		Erosion	
Period	**Sections**	**Sum (m^3)**	**PUA (m^3/m^2)**	**Sum (m^3)**	**PUA (m^3/m^2)**	**Sum (m^3)**	**PUA (m^3/m^2)**
1996–1997	Section 1	+879	+0.002	+8800 m^3	+0.347	−7920	−0.302
	Section 2	+3629	+0.015	+6794	+0.442	−3164	−0.422
	Section 3	+7798	+0.018	+10,929	+0.397	−3131	−0.273
	Section 4	+3723	+0.009	+8143	+0.339	−4419	−0.287
	Section 5	−188	−0.001	+4971	+0.301	−5159	−0.333
	Section 6	−1117	−0.003	+4329	+0.250	−5447	−0.262
	Total	+14,725	+0.006	+43,967	+0.349	−29,241	−0.302
1997–1998	Section 1	+2052	+0.004	+16,846	+0.574	−14,794	−0.665
	Section 2	−9169	−0.037	+3570	+0.477	−12,739	−0.828
	Section 3	−6601	−0.016	+2311	+0.189	−8912	−0.333
	Section 4	−1575	−0.004	+4499	+0.247	−6074	−0.286
	Section 5	−2474	−0.007	+4058	+0.303	−6531	−0.351
	Section 6	−3804	−0.009	+3654	+0.232	−7458	−0.335
	Total	−21,571	−0.009	+34,937	+0.362	−56,508	−0.447
1998–2000	Section 1	−5545	−0.010	+10,729	+0.644	−16,274	−0.466
	Section 2	−3962	−0.016	+4016	+0.465	−7978	−0.560
	Section 3	−3441	−0.008	+5650	+0.401	−9092	−0.365
	Section 4	−149	−0.000	+6684	+0.414	−6833	−0.293
	Section 5	−2155	−0.006	+6052	+0.483	−8207	−0.421
	Section 6	−4594	−0.011	+4748	+0.370	−9342	−0.370
	Total	−19,847	−0.008	+37,878	+0.468	−57,726	−0.406

Note: PUA—per unit area, which is calculated via the formula in text.

there are usually pronounced cusps in the beach where beach materials have been removed up to the high water mark during the storm. Such winter erosion not infrequently cuts into the face of the dune and, in some cases, has been known to breach the dunes. In the early spring and late fall, the coastal topographic change undulates; this may be caused by varying weather.

5.4.3 Profile Pattern Analysis

Analysis patterns of profile changes can provide a comprehensive understanding of the topographic changes in specified areas because topographic profiles provide the vertical and horizontal changes of the shoreline. Topographic profiles are representative of significant changes at nearshore, foreshore, berm, and dune. The profiles help determine if the sea level is rising or falling or if the beach is staying the same. In the developed software, a dialog box is designed for choosing a profile for dynamic displays using four years of data. Based on each profile, the erosion can be directly measured from the profile, and the erosion rate can be calculated. As an example, three profiles in each section are selected to demonstrate the topographic changes, as depicted in Figures 5.4–5.6.

As observed from Figures 5.4a and 5.5a, about 1500 m in the southern end of Section 1 have significant deposition of up to 1.37 m height and 190 m width from the nearshore to the foreshore. Significant erosion is observable for up to a maximum 0.8 m in the 200-m-wide dune and berm area. This results in the formation of a flat, elevated area of 0.6–0.9 m in the southern end as observed in 2000. Most of the 4267-m-long by 100-m-wide berm area along the eastern shoreline of Section 1 experienced severe erosion in the berm area like that shown in Profile 2 (Figure 5.5b) of Section 1 in the period 1996–2000. In only two years of time from 1998 to 2000, a 67-m-wide berm was eroded 2.4 m, resulting in the coastline shifting 67 m inland (Figure 5.5a and b). Additionally, a 700-m-long foreshore in Section 1 sustained deposition like that seen in Profile 1 of Section 1 (Figure 5.5a). An approximately 67-m-wide foreshore near the shore gained 1.0 m in height, resulting in shoreline movement outward by 67 m. An equally wide dune area that experienced erosion results in the formation of a berm area. As observed, a 1700-m shoreline in Section 2 has experienced severe erosion of 3.35 m depth and 40 m width from foreshore to berm (Figure 5.5b and f). However, the berm and dune areas of 400 m length and 100 m width at the middle of Section 2 had deposition of more than 0.3 m height (see Figure 5.4b). Most of the foreshore and dune areas have experienced more than 2 m erosion, resulting in the coastline shifting inland ~18–24 m (Figure 5.4d–f).

As observed from Figures 5.4a and 5.6a, the most significant deposition occurred in a 1220-m-long by 85-m-wide foreshore area (Figure 5.4d) in Section 4, resulting in a wrack line movement out by 27–30 m (Figure 5.6a–c). On the other hand, as observed from Profile 2 in Section 4, Profile 1 in Section 4, and Profile 3 in Section 5, the top of dunes increased in height from 0.7 to 1.5 m (Figure 5.6a–f). The mechanism for the growth in height of the dunes may result in natural dune building, which is generally brought about by aerolian processes over considerable periods. Another possible mechanism for this increase in dune height may be anthropogenic changes due to bulldozing of the dunes. In Sections 5 and 6, the areas near the shore, the foreshore, and the dune were all generally eroded, which caused the dune areas and the shoreline to retreat 6–15 m (Figures 5.4e, f, and 5.6d–i). The nearshore extended to foreshore by 30–45 m (Figure 5.6e and f). In Section 6, the tops of the dune decreased 0.8 m in the south of Profile 3, 1.5 m in the middle of Profile 2, and 1.8 m in the north of Profile 1 (Figure 5.6g–i).

Based on the above analysis, it can be concluded that

1. The elevation for most of the dunes has decreased, and the area for the dunes has retreated toward the west. This observation could explain why the entire island becomes narrower from 1996 to 2000.
2. The nearshore areas decreased about 0.3–1.0 m in height and migrated west into the foreshore from 1996 to 2000.
3. The berm that is connected with the foreshore experienced serious erosion and resulted in inland migration of the shoreline. Moreover, the slopes of foreshores in all six sections have grown steeper.

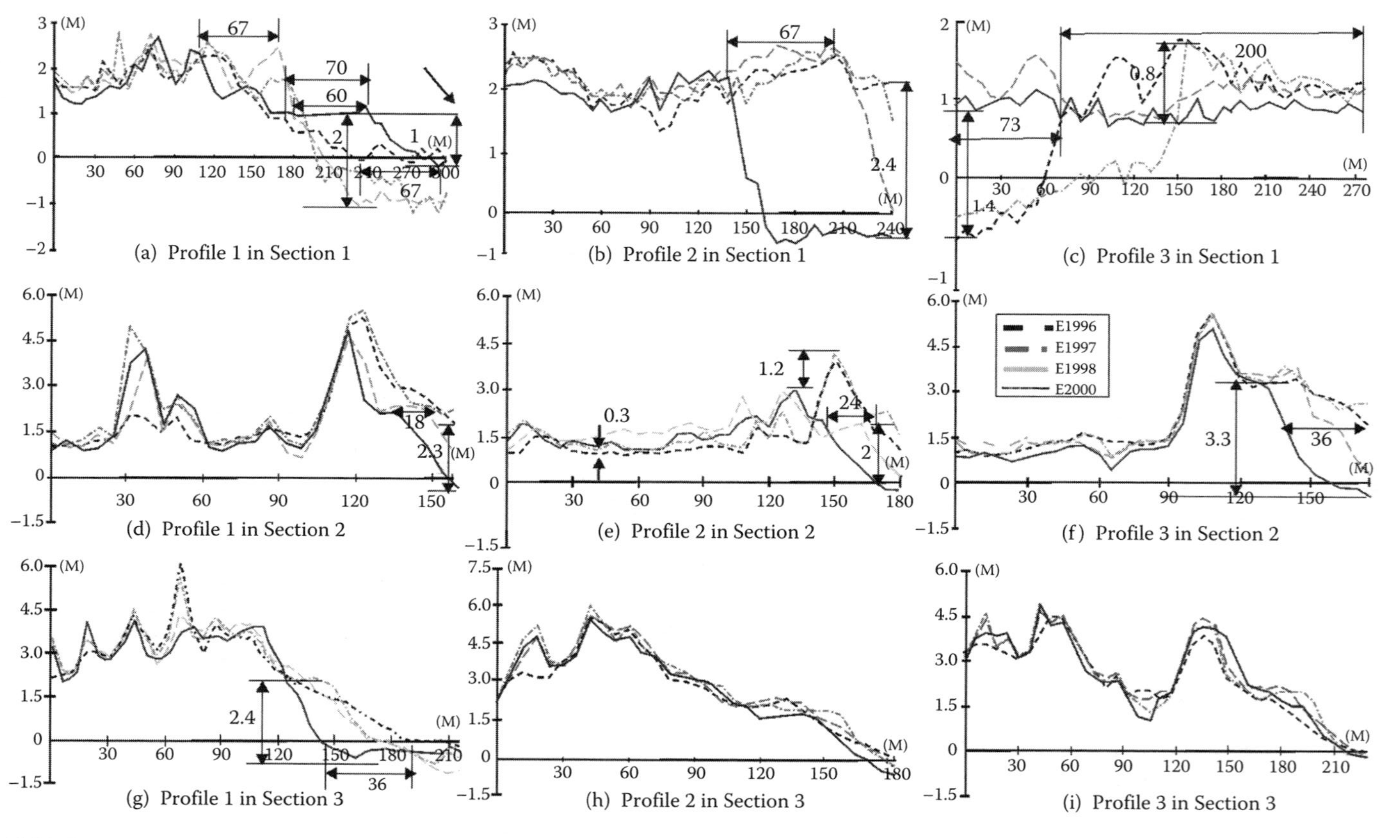

FIGURE 5.5 Profile analysis of topographic changes for Sections 1–3 from October 1996 to November 2000.

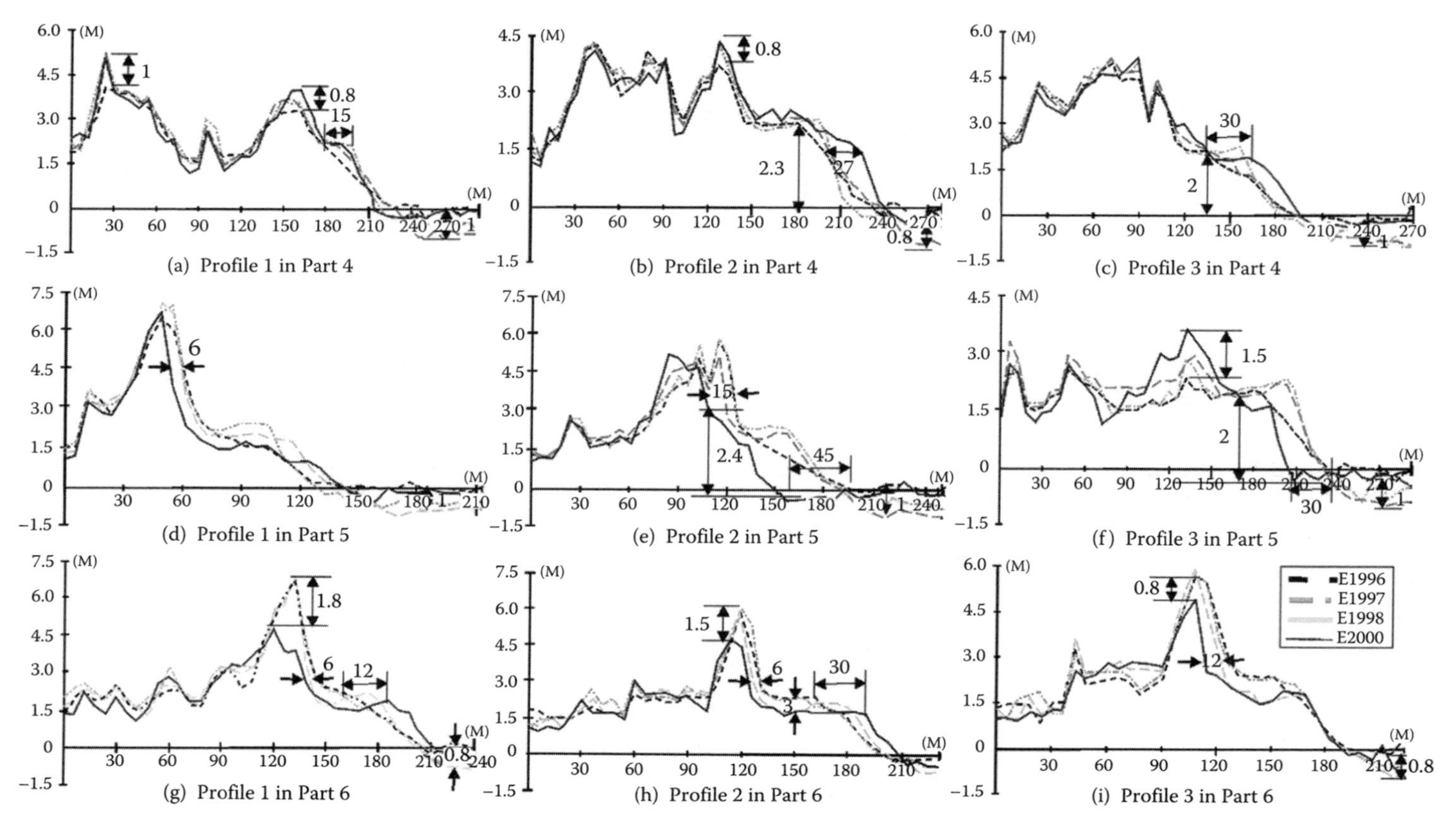

FIGURE 5.6 Profile analysis of topographic changes for Sections 4–6 from October 1996 to November 2000.

4. Changes of dunes differ in topographic profile and morphology over the course of the years. For example, the dunes from Sections 1 through 3 decreased in volume, accreted in Section 4, and retreated west in both Sections 5 and 6. Additionally, the rate of change was not the same. For example, the dunes from Sections 1–3 decreased rapidly between 1997 and 1998 (Figure 5.5a–i), while the dunes from Sections 5 to 6 retreated rapidly between 1998 and 2000 (Figure 5.6d–i). In contrast, the dunes in Section 4 built up rapidly between 1996 and 1997 (Figure 5.6a–c).
5. An approximately 2600-m-long shoreline has retreated inland from 2 to 40 m through different coastal conditions and environments. An example is the foreshore in Section 2, which retreated about 40 m from 1996 to 2000.

5.4.4 Patterns of Volumetric Change

The volumetric change can be calculated by differentiating the *Z* coordinates on each cell. Four-year topographic changes of the study area are summarized in Table 5.2. Section 1 had the largest amount of both deposition (14,647 m^3) and erosion (17,261 m^3). Section 4 experienced a positive net volumetric gain (1998 m^3). The gain in section might largely be attributed to beach nourishment during the time. The entire study area exhibited a net volumetric loss of 26,693 m^3 from 1996 to 2000 at an average loss rate of 0.011 m^3/m^2. Section 6 had the least amount of deposition (2835 m^3). The total erosion was 67,389 m^3. Among those, Section 1 contributed 17,261 m^3 to this loss at an average rate of 0.602 m^3/m^2. The net loss of Section 1 was 2613 m^3 at a rate of 0.005 m^3/m^2. Over the entire study area, the estimated average erosion rate was approximately 0.485 m^3/m^2, and the average rate of deposition was 0.005 m^3/m^2.

TABLE 5.2
Summary of 4-yr Volumetric Change for Each Section from 1996 to 2000

		Net Change		Deposition		Erosion	
Period	Sections	Sum (m^3)	PUA (m^3/m^2)	Sum (m^3)	PUA (m^3/m^2)	Sum (m^3)	PUA (m^3/m^2)
1996–2000	Section 1	−2613	−0.005	+14,647	+0.639	−17,261	−0.602
	Section 2	−9501	−0.038	+4961	+0.607	−14,463	−0.983
	Section 3	−2244	−0.005	+7140	+0.471	−9384	−0.394
	Section 4	+1998	+0.005	+7714	+0.420	−5715	−0.271
	Section 5	−4817	−0.014	+3396	+0.322	−8213	−0.383
	Section 6	−9515	−0.023	+2835	+0.321	−12,351	−0.422
	Total	−26,693	−0.011	+40,696	+0.485	−67,389	−0.485

Note: PUA—per unit area, which is calculated via the formula in text.

Annual volumetric changes are shown in Table 5.1. As observed, Section 1 had the largest amount of erosion (16,274 m^3) and, at the same time, it had the largest amount of deposition (10,729 m^3). The erosion and deposition in Section 1 were almost twice that of other sections. Section 2 exhibited the largest amount of net loss (3962 m^3). Section 4 demonstrated the largest amount of sand exchange between deposition and erosion with a loss of 6073 m^3 and deposition of 4498 m^3. Between 1998 and 2000, the entire study area experienced more erosion than deposition (Table 5.1). Section 3 had the largest amount of deposition (10,929 m^3), while Section 6 had the largest amount of erosion (5446 m^3) and the greatest net loss of 1117 m^3. In contrast, volumetric changes for the year between 1997 and 1998 showed that the entire study area experienced more erosion than deposition (Table 5.1). For example, Section 1 had the largest amount of erosion (14,794 m^3), and at the same time had the largest amount of deposition (16,846 m^3).

As observed from Table 5.1, volumetric losses increased each year from 29,241 to 57,726 m^3, while deposition varied. Net annual change from 1996 to 2000 varied dramatically from +14,725 to −21,571 m^3. Between 1998 and 2000 all sections had a net negative (erosive) result with losses ranging from 5545 to 149.3 m^3 in Sections 1–4. The observations likely show the effects that a series of stormy periods had upon the beaches of Assateague Island.

5.4.5 Volumetric Net Change

The volumetric net change per square meter (m^3/m^2) has been considered as an index to analyze quantitatively the change per square area. The calculation can be obtained by

$$\text{Volumetric net change per square area (m}^3\text{/m}^2\text{)} \\ = \text{total erosion/deposition volume (m}^3\text{)/erosion/deposition area.}$$

The calculated result for the volumetric net change per square area has been visualized using histograms illustrated in Figures 5.7 and 5.8. The largest and smallest net change rates PUA were 0.018 m^3/m^2 in Section 3 and −0.003 m^3/m^2 in Section 6, respectively (Figure 5.7a). The largest mean erosion and deposition rates are 0.422 and 0.442 m^3/m^2, both of which occurred in Section 2 (Figure 5.7a). Additionally, Section 1 experienced net deposition at a rate of 0.004 m^3/m^2, while other sections

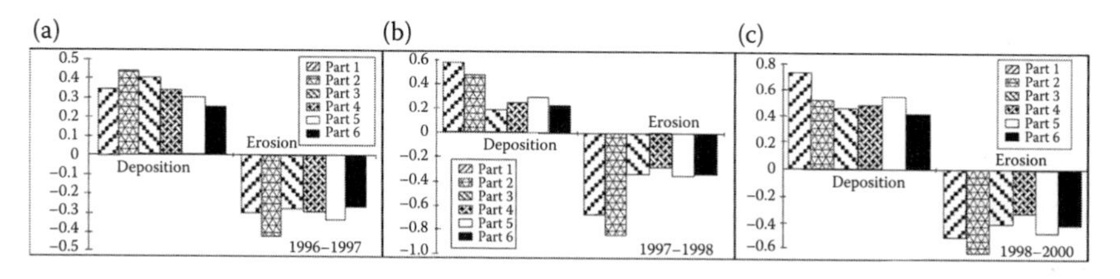

FIGURE 5.7 Average volumetric change PUA in the periods (a) 1996–1997, (b) 1997–1998, and (c) 1998–2000.

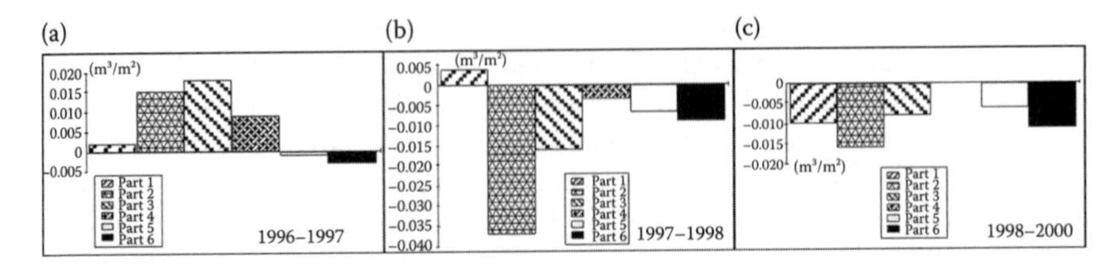

FIGURE 5.8 Net volumetric change PUA in the periods (a) 1996–1997, (b) 1997–1998, and (c) 1998–2000.

had net erosion at a rate of 0.004–0.037 m^3/m^2 throughout 1997–1998 (Figure 5.8b). The erosion rates gradually decreased from 0.016 to 0 m^3/m^2 from Sections 1–4, and then gradually increased from 0 to 0.011 m^3/m^2 from Sections 4–6 in the time period of 1998–2000 (Figure 5.8c). Comparing Figure 5.8a–c, the topographic changes at the two ends of the Island (Sections 1 and 6) experienced the most significant erosion and deposition. The changes gradually decreased toward the middle sections (Section 4) of the island. After 1997, the Assateague Island had a serious erosion, which can be demonstrated by the net volumetric change PUA at a rate of 0.037 (m^3/m^2) in Section 2 (Figure 5.8b).

5.5 CONCLUSION

This chapter presented coastal morphological change analysis at the study area, Assateague Island National Seashore. The results illustrated that the coastline change of Assateague Island is very complex and dynamic. Active coastal management, shoreline protection, and beach nourishment programs affected the topography of Assateague Island significantly. The results also demonstrated that LiDAR sensors can provide an extraordinary capability for capturing data upon which high-accuracy and high-density coastal DEMs can be created and used for quantitative analysis of coastal topographic morphology. Topographic morphology analysis can provide precise and reliable information for the effective planning and management of the coastal area. For the frequency of LiDAR data collection, it would be beneficial to have data collection before and after major impacting events such as hurricanes, which would have significant effects on deposition and erosion. A common knowledge indicates that beaches grow during less stormy summers and early fall. In winters, beach materials are generally eroded from beach face and berm and deposited on the nearshore submerged bars. Thus, a seasonal LiDAR data collection will not contribute to 3D morphological analysis. For a long-term strategy of coastline management, the annual LiDAR data collection will be helpful for active coastline management.

ACKNOWLEDGMENTS

This project was supported by a NASA research grant under the contract number NAG-13-01009 "Virginia Access." Carl Zimmerman, Chief of Division of Resource Management at Assateague Island National Seashore, provided essential support for topographic change data interpretation, such as deposition, erosion, and net change. Andrew Meredith of the Coastal Service Center at NOAA provided help in LiDAR data delivery and interpretation. Many thanks to Mark Weiss (the GIS System Engineer) and Dr Changqing Song for their help in ArcObject code development.

REFERENCES

Brock, J.C., Wright, C.W., Sallenger, A.H., Krabill, W.B., and Swift, R.N., 2002, Basis and methods of NASA airborne topographic mapper lidar surveys for coastal studies, *Journal of Coastal Research*, 18: 1–13.

Carter, W.E. and Shrestha, R.L., 1997, Airborne laser swath mapping: Instant snapshots of our changing beaches, *Proceedings of the International Conference on Remote Sensing for Marine and Coastal Environments*, March 17–19, Orlando, Florida (Environmental Research Institute of Michigan), pp. 298–307.

Dolan, R. and Davis, R.E., 1992, An intensity scale for Atlantic Coast Northeast storms, *Journal of Coastal Research*, 8(4): 840–853.

Dolan, R., Hayden, B., and Heywood, J., 1997, Atlas of environmental dynamics Assateague Island National Seashore. U.S. Dept. Interior, National Park Service, Natural Resources Report, 11.44 pp.

Guenther, G.C., Brooks, M.W., and LaRocque, P.E., 2000, New capabilities of the SHOALS airborne lidar bathymeter, *Remote Sensing of Environment*, 73: 247–255.

Irish, J. and White, T.E., 1998, Coast engineering applications of high-resolution LiDAR bathymetry, *Coastal Engineering*, 35: 47–71.

Irish, J.L. and Lillycrop, W.J., 1999, Scanning laser mapping of the coastal zone: The SHOALS system, *ISPRS Journal of Photogrammetry and Remote Sensing*, 54(2): 123–129.

Krabill, W., Wright, W., Swift, R., Frederick, E., Manizade, S., Yungel, J., Martin, C., Sonntag, J., Duffy, M., and Brock, J., 2000b, Airborne laser mapping of Assateague National Seashore Beach, *Photogrammetric Engineering and Remote Sensing*, 66(1): 65–71.

Meredith, A.W., 1999, An evaluation of hurricane induced erosion along the North Carolina coast using airborne LiDAR surveys. NOAA Coastal Services Center Technical Report, NOAA/CSC/99031-PUB/001.

Moore, R., Fish, P., Koh, A., Trivedi, D., and Lee, A., 2003, Coastal change analysis: A quantitative approach using digital maps, aerial photographs and LiDAR, *International Conference on Coastal Management, Coastal Management*, pp. 197–211.

NOAA, 2000, The potential consequences of climate variability and change on coastal areas and marine resources. A Report of the National Coastal Assessment Group for the U.S. Global Change Research Pro-gram, October 2000. NOAA Coastal Ocean Program Decision Analysis Series No. 21, NOAA Coastal Ocean Program, Silver Spring, MD, 163pp.

Parson, L.E., Lillycrop, W.J., Klein, C.J., Ives, R.P., and Orlando, S.P., 1997, Use of LiDAR technology for collecting shallow water bathymetry of Florida Bay, *Journal of Coastal Research*, 13(4): 1173–1180.

Pilkey, O.H., Neal, W.J., Riggs, S.R., Webb, C.A., Bush, D.M., Pilkey, D.F., Bullock, J., and Cowan, B.A., 1998. *The North Carolina Shore and its Barrier Islands, Restless Ribbons of Sand*, Duke University Press, Durham.

Sallenger, A.B., Krabill, W.B., Swift, J., Jansen, R., Manizade, S., Richmond, B., Hampton, M., and Eslinger, D., 1999, Airborne laser study quantifies El Niño-induced coastal change, *EOS, Transactions of the American Geophysical Union*, 80: 89 and 92.

Sallenger, A.H., 2000, Storm impact scale for barrier islands, *Journal of Coastal Research*, 16: 890–895.

Schmitz-Peiffer, A. and Grassl, H., 1990, Remote sensing of coastal waters by airborne lidar and satellite radiometer part 1. A model study, *International Journal of Remote Sensing*, 11(12): 2163–2184.

Strom, L., Tjernstrom, M., and Rogers, D., 2001, Observed dynamics of coastal flow at Cape Mendocino during coastal waves, *Journal of the Atmospheric Sciences*, 58(9): 953–977.

Tralli, D.M., Blom, R., Zlotnicki, V., Donnellan, A., and Evans, D., 2005, Satellite remote sensing of earthquake, volcano, flood, landslide and coastal inundation hazards, *ISPRS Journal of Photogrammetry and Remote Sensing*, 59(4): 185–198.

White, S.A. and Wang, Y., 2003, Utilizing DEMs derived from LiDAR data to analyze morphologic change in the North Carolina coastline, *Remote Sensing of Environment*, 85(2): 39–47.

Woolard, J.W. and Colby, J.D., 2002, Spatial characterization, resolution, and volumetric change of coastal dunes using airborne LiDAR: Cape Hatteras, North Carolina, *Geomorphology*, 48: 269–287.

Conference Proceedings, First International Symposium on Carbonate Sand Beaches, December 5–8, 2000, Westin Beach Resort, Key Largo, Florida, U.S.A. by Orville T. Magoon, Lisa L. Robbins, and Lesley Ewing, Eds. *Reston, VA: ASCE*, 0-7844-0640-5, 2002, 273 pp., ASCE (American Society of Civil Engineers) publisher, (doi 10.1061/40640(305)3).

Zhou, G. and Xie, M., 2008, Coastal 3-D Morphological Analysis Using LiDAR Series Data: A Case Study of Assateague Island National Seashore, *Journal of Coastal Research*, 25(2), 2009, 435–447.

Section II

Hyperspectral Remote Sensing

6 Mapping the Onset and Progression of Marsh Dieback

Elijah Ramsey III and Amina Rangoonwala

CONTENTS

6.1 INTRODUCTION

Along the Gulf of Mexico (GOM) coasts, vast wetlands inject valuable nutrients and suspended and dissolved materials into the coastal ocean. *Juncus roemerianus* (black needlerush) wetlands, dominating coastlines in the northeastern GOM, transition to the *Spartina alterniflora* (smooth cordgrass) coastline of Louisiana. Mixed marsh and mangrove barrier island systems occupy the southeastern and southwestern GOM [1,2].

Storms, fire, herbivory, and hydrologic changes, many times promoted by human activities and development, destroy the viability of wetlands to buffer against storm surges and shore erosion and diminish their capacity to provide coastal recreation, to aid in the maintenance of water quality, to act as sources of nutrients, and to serve as nurseries and habitats for fish, shrimp, various birds, and fur-bearing animals. Development continues as a large and growing proportion of the nation's population and facilities are being located along the GOM coasts. In consequence, these coastal ecosystems are beset with critical and complex problems that reflect the bleak prospect of diminished coastal wetlands worldwide [3]. Conservation of these coastal resources is urgently needed, and their continued viability depends on managing them in a sustainable fashion [4].

Sustainable management of coastal ecosystems requires an assessment of their biophysical aspects, both on the regional and community level. With this information, ecologists, resource managers, and coastal planners can implement a more rational decision-making system for allocating financial resources and directing responses to adverse changes in the coastal ecosystem. In addition, federal and state jurisdiction, administration, and protection rely on such information to determine wetland identification and delineation [5]. Remote sensing is capable of identifying landscape features over large areas on a repetitive basis by using a wide variety of spectral and spatial ranges and sensors [6,7]. By linking these features that are observed to the physical and biological processes, remote sensing can be used to improve the coverage, quality, and timeliness of biophysical data that are useful in ascertaining the status of coastal resources and understanding their trend dynamics.

All terrestrial remote sensing systems measure the combined effects of plants and background (e.g., mud, water, dead leaf material, and bacteria) components within the pixel. This mixture of components applies whether we are classifying species type, mapping biomass, determining change, or attempting to map differences in a single species canopy or subtle changes in a species undergoing adverse stress. In consequence, it is necessary to understand how the various components alter the different remote sensing measures in order to accurately map and monitor changes in coastal resources, especially if these changes are subtle or extend over long time periods. Even if the target contains only a single plant species, natural variability in the species' optical properties and its density brought about by environmental conditions must be estimated when attempting to detect abnormal change.

6.2 BACKGROUND

The optical signal from about 400 to 1300 nm collected by terrestrial remote sensing is a measure of the percent illuminating sunlight reflected from the composition of plants and background (referred to as the canopy reflectance) included in the pixel (ground spatial resolution element). Changes in the percent reflected sunlight can be used to detect and monitor changes in the plants and background. The optical leaf properties broadly include pigments and water content. Pigment concentrations can be determined via spectroscopic analyses primarily in the visible wavelength region (400–700 nm) (visible) (Figure 6.1), and in some cases in the near infrared (NIR) (700–1300 nm) (e.g., bacteriochlorophylls) [8]. Selective spectral absorptions (e.g., peak location and width) are used to determine the pigment type (e.g., chlorophyll and carotene). The reflected remote sensing signal contains information related to the pigment absorption via Kirchoff's radiation law as

$$\%\text{absorption}(\lambda) = 1 - [\%\text{reflectance}(\lambda) + \%\text{transmittance}(\lambda)]. \tag{6.1}$$

Percent reflectance is the corrected and normalized optical signal measured from an aircraft or satellite platform. Information directly related to leaf pigments is gained by first isolating from the reflectance the component that is solely associated with the plant-leaf material [9]. Although leaf reflectance is indirectly related to pigment absorption (Equation 6.1), features that are apparent in the reflectance spectra can be used as indicators of leaf absorption. Changes in the leaf pigments and their associated concentrations are the best indicators of biological changes in the plant that are obtainable via remote sensing, and these changes offer the best indication of abnormal change [9]. Even though they are used to describe leaf biochemical and biophysical properties, differences among and changes in optical leaf properties are often overlooked in studies detecting and tracking vegetation change [10–13].

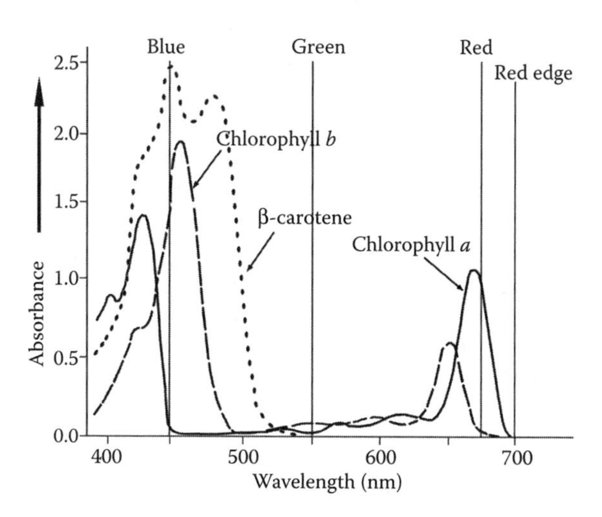

FIGURE 6.1 Absorption spectra of chlorophyll and carotene. Note that the blue and red bands are located near peak absorptions, and the green and red edge bands are located at the tails of these absorptions. (Adapted from Kirk, J., in *Light & Photosynthesis in Aquatic Ecosystems*, Chap. 8, pp. 229, 233, Cambridge University Press, Cambridge, 1994. With permission.)

Separate from the plant-leaf properties, the canopy structure combines the plant structure (leaf and stalk orientations) and the plant density. As a first-order approximation and from a remote sensing standpoint, the canopy structure can be effectively estimated as the percent plant cover from a planar viewpoint (looking straight down from above the canopy) by defining the plant structure orientation as spherical [14,15]. Although changes in canopy structure most often indicate later stages of impact than do pigment changes, natural variations in canopy structure and abnormal changes co-occurring with impact progression must be identified or minimized in order to isolate reflectance changes that are solely due to changes in plant-leaf pigmentation.

Within coastal marsh wetlands, plant-leaf optical properties and canopy structure can vary highly in response to changes in species type, environmental conditions, and health within a region and within a local area [9,16]. Changes in canopy structure associated with adverse impact would be observable when the impact progressed to a severity level at which the whole plant would die, leading in many cases to marsh conversion to mud and open water [17]. Even though the ability to detect these structural changes (presumably resulting from similar mechanisms) would provide one level of impact monitoring, the ability to consistently detect marsh loss through remote sensing would still be subject to the spatial and temporal resolutions of the remote sensing sensor and platform. In addition to these spatial and temporal constraints, the ability to discriminate marsh changes from background changes with remote sensing data is not absolute in these highly variable coastal environments (e.g., Figure 6.2).

A more informative method of mapping would supply estimates of the level of impact and recovery. Developing such a method requires creating a remote sensing model or set of protocols that allows estimation of the progression of healthy to dead marsh. The capacity to perform a more subtle discrimination would provide knowledge concerning the impact onset and progression before unanticipated and irrevocable loss of the coastal marsh. To provide this subtle discrimination, the separate optical influences of plant-leaf properties, canopy structure, and background variability on the remote sensing signal must be estimated or unwanted influences suppressed. If a standard metric of abnormal change based on consistent and relevant pre-event monitoring existed, the decoupling of canopy reflectance factors might be accomplished by occupying numerous sites featuring various levels of impact and documenting optical properties of plant leaf, canopy structure, and background. Without a metric to gauge marsh change, however, quantification of change related to marsh dieback is impossible. Without quantification of the change criteria, assertions of change are subjective, increasing the potential for confusion and misclassification while diminishing the capacity to provide consistent and relevant assessments and to determine impact progression.

6.2.1 Marsh Dieback in Coastal Louisiana

Areas of marsh dieback, commonly termed "brown marsh," were first observed in parts of Texas and Florida and throughout coastal Louisiana in the spring of 2000. The dieback was widespread and discontinuous, incorporating a broad range of areas

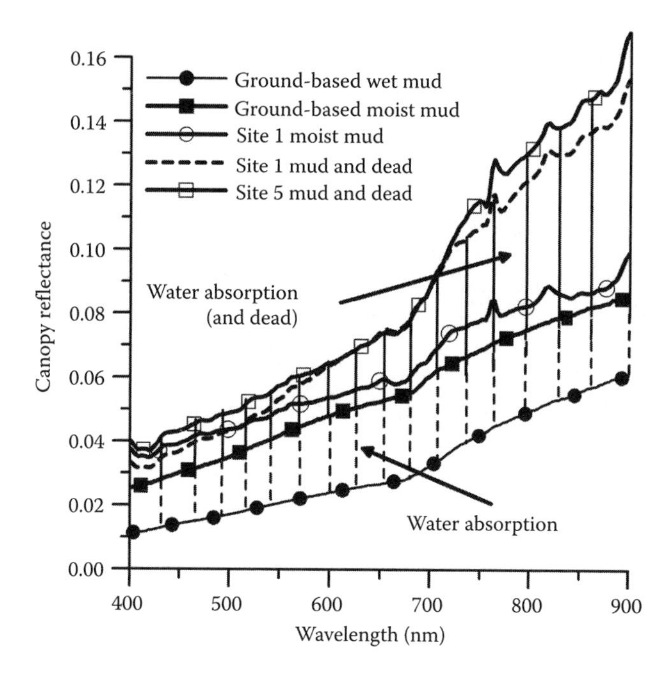

FIGURE 6.2 Canopy reflectance as a function of mud background water content. Note the high variability in spectra magnitude depending on the moisture content. The terms *wet* and *moist* refer to high and moderate mud water contents, respectively. "Mud and dead" refers to sites where mud water content was relatively low and dead marsh was present. Note the abrupt reflectance increase around 700 nm in the "mud and dead" spectra indicating that live leaf material was present even though visually the marsh was brown. Relative differences in water content were based on visual observations and were not quantitatively determined. Sites 1 and 5 spectra were obtained from helicopter-based recordings.

(Figure 6.3). At dieback sites, the normally dense and healthy intertidal salt marshes, mostly composed of smooth cordgrass, browned and, in many cases, ultimately died. The occurrence of dieback clearly suggested an elevated change in the marsh biophysical properties, and although some regeneration occurred, a portion of the dieback sites converted to mud flats or open water.

The dieback of Louisiana coastal marsh offered an example of a progressive detrimental change. Our investigation of the dieback phenomenon was directed at refining and developing remote sensing techniques to detect the spatial occurrence of marsh dieback and to determine the stage of dieback progression. To reach this goal, we needed not only to detect the dramatic change of a healthy salt marsh to a dead marsh but also to detect the more subtle stages of onset and progression, ahead of visible differences and before irrevocable damage had occurred. We initiated dieback studies at three resolutions, including optical properties of plant-leaf and top-of-canopy reflectance, as well as the spectral, spatial, and temporal resolutions of canopy reflectance associated with resource satellites such as Landsat Thematic Mapper (TM) [18].

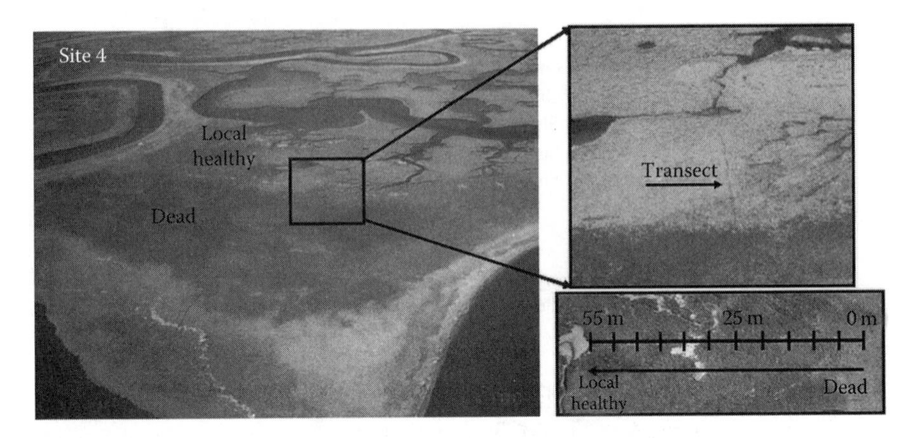

FIGURE 6.3 **(See color insert following page 206.)** Aerial photography of the marsh dieback at site 4. The severest dieback impact is in the center, and the local healthiest marsh is in the background. The right top segment shows a closeup of the transect, and the right bottom segment shows the transect with plant sample locations (m). (Adapted from Ramsey, E., III and Rangoonwala, A., *Photogrammetric Engineering and Remote Sensing*, 71, 299, 2005.).

6.2.2 Metric Development

The lack of information concerning the timing of onset and progression of marsh dieback and the disparate occurrence and noticeably high variability in dieback sizes and impact intensities necessitated the development of a new sampling methodology. In other words, there was no metric by which change in leaf optics or canopy structure could be compared in order to determine the relative time since onset or the dieback progression. Furthermore, the categories of marsh dieback (e.g., [19]) were clearly subjective (e.g., brown-green, green-brown), and simple occupation of sites that visually appeared impacted or nonimpacted did not provide an acceptable metric to quantify dieback progression (e.g., Figure 6.4). The metric model combined with the sampling methodology permitted estimation of the onset and progression of impact at each dieback site and provided the possibility of relating the metrics at one site to those at all other sites.

Field reconnaissance from the air and on the ground indicated that many dieback sites that were not abruptly curtailed by open water or some other barrier exhibited a central dead marsh zone surrounded by progressively denser and healthier marsh (local healthy marsh) (e.g., Figure 6.3). We hypothesized that, where this zonal pattern existed, the distance metric from the dead marsh to the adjacent, healthy marsh could approximate a progressive timeline simulating the elapsed time since dieback onset and, thereby, provide an indicator of dieback progression. This distance and time interrelation was our zonation-trend model (Figure 6.5, top). Although a portion of any uncovered trend along this conceptual transect could be due to natural variation, we believed that the superposition of the dieback trend signal would be in excess of any natural progression. Changes in the plant optical leaf properties along a transect from dead to local healthy marsh would provide an indication of the progressive impact through time (Figure 6.5, bottom).

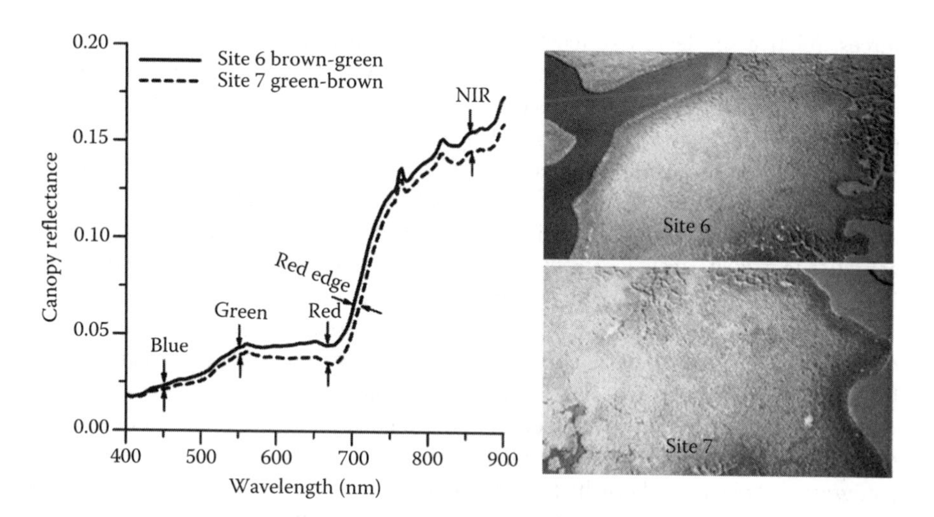

FIGURE 6.4 (**See color insert following page 206.**) Canopy reflectance associated with two sites that exhibited spectral variability near the onset of dieback progression. Note the visual dieback classifications (brown-green, green-brown) and the small change in canopy reflectance (600–700 nm) that caused the perceived visual change.

6.2.3 Study Area and Site Descriptions

Historically, human alterations have caused the most dramatic changes in coastal resources and wetlands [3]. This causal relationship is true of the GOM coast, particularly coastal Louisiana, and the lower Mississippi River Delta. Of the natural

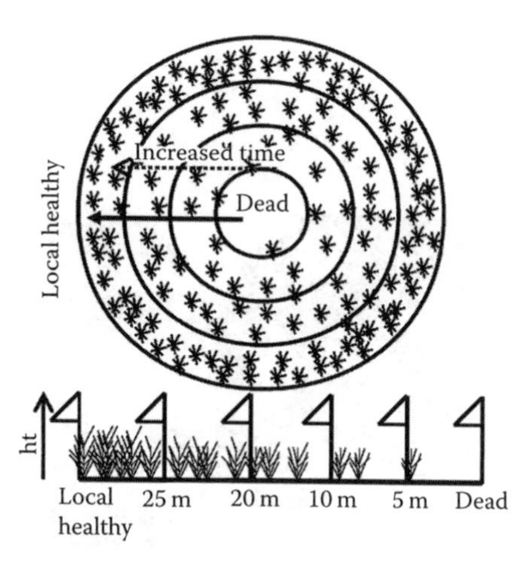

FIGURE 6.5 Conceptual model for dieback progression. The top portion shows that dieback progressively spread outward from an initial and central location. The bottom portion illustrates the plant sampling strategy and the progression of time related to the physical distance along the transect away from the dieback center.

forces, relative sea level rise (eustatic plus subsidence) holds the potential for the most widespread loss of these coastal wetlands. As relative sea level continues to rise, human protection mechanisms (e.g., coastal armoring and levees) aggravate this problem, causing unknown, but most likely detrimental, effects to coastal resources. While flood control acts to protect human lives and property and maintain inland ports, it disrupts natural flooding and channel switching, which act to preserve the coastal wetlands. The loss of sediment replenishment in addition to other human activities (e.g., waterway construction and fluid extractions) has resulted in dramatic losses of the coastal wetlands since 1956 [3,7,18,20].

Coastal wetland resources within the GOM include a diverse and extensive area of marshes and swamps that support the development of a number of floristic species [18,20]. Floristic marsh species in coastal Louisiana may be classified into four functional zones and dominant species that are partitioned on the bases of inundation frequency, salinity level, and edaphic factors [21]: (1) saline marshes (smooth cordgrass and black needlerush), (2) brackish marsh, which has moderate salinity and is generally found inland from saline marshes (*Spartina patens* and *Schoenoplectus americanus*), (3) intermediate marsh, which has salinity levels between those of brackish and freshwater marshes (*S. patens*), and (4) fresh marsh, existing outside of ocean salinity influences (*Panicum hemitomon*).

During the summer of 2000, four smooth cordgrass dieback sites and one nonimpacted, long-term field site farther inland were identified; radiometer collections from a helicopter platform were obtained at eight additional sites (Figure 6.6).

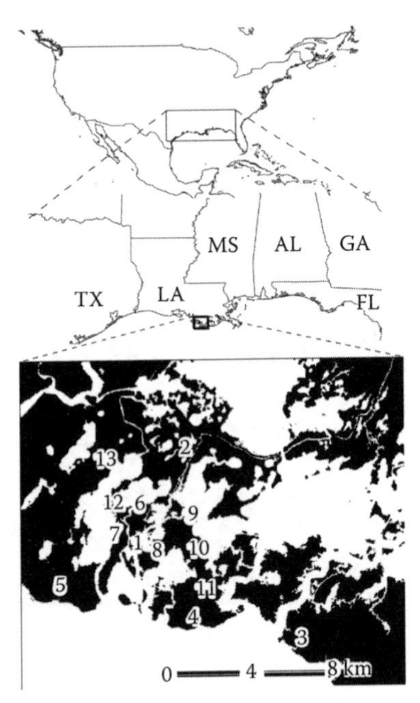

FIGURE 6.6 Site locations. Map showing the approximate locations of the 13 study sites where marsh dieback was analyzed along the GOM coastline. Plants were collected for leaf analyses along transects at sites 1, 3, 4, and 5, and at site 2, the reference site.

Samples were collected from September 19, 2000 to October 10, 2000. Location of all sites adjacent to tidal flushing conduits typified the situation of all marsh in this highly bifurcated coastal wetland environment. Although adjacent to tidal channels, site distances from the coast front differed. Reference site 2 was located farthest inland, and site 1 was located on an inland embayment. Sites 3 and 5 were located along tidal channels nearer the coast front, while site 4 was the farthest seaward, located nearly on the GOM coastal front. Impact patterns differed among the four impacted sites, primarily portrayed in the level and width of the impacted marsh. Site 1 was situated on the smallest marsh platform between tidal channels and the embayment. The narrow region of most severely impacted marsh contained mostly dead stubble and a few intermixed live plants with a scarcity of green leaves. The width of the most severely impacted marsh at site 3 was narrower than that at all other sites; however, other than being almost entirely composed of dead brown leaves, the impacted canopy structure at site 3 remained similar to that of the adjacent healthy marsh. This retention of marsh structure was documented by light attenuation measurements collected along the transect [16]. Site 5 was located more interior, and sites 4 and 5 were within larger contiguous expanses of marsh platform than were sites 1 and 3. As at site 1, the impact centers of sites 4 and 5 were almost devoid of plants but included dead marsh stubble. Progression from the impact center to the adjacent healthy marsh was indicated by increasing live plant density (Figure 6.5) and at site 4 by increasing plant height.

6.3 METHODS

Initially, on the basis of visual observations, we separated marsh into three broad classes: healthy, intermediate dieback, or dead (categories in Table 6.1). Although no quantifiable relation from site to site was implied in these classes, this categorization helped establish the collection strategy. The category of "healthy marsh" included marsh where there was a visual preponderance of green leaves in the plant canopy. In order to account for natural variability, we created a more specific category, "local healthy marsh," to describe healthy marsh relative to each transect site. "Intermediate dieback marsh" areas visually contained a higher amount of yellowing and dead leaves and, at times, whole plant removal opening the canopy. "Dead marsh" areas exhibited severe impact with loss of whole plants and little remaining green to yellow-green leaves. Even though microalgal growth on the exposed mud substrate can contribute substantially to canopy reflectance, microalgal growth was not observed, and therefore, this factor was not considered. No standing water was present at any field site during the radiometer collections.

6.3.1 LEAF REFLECTANCE

To discover whether or not the concept of the zonation-trend model would provide indicators of dieback progression, we first determined whether changes in optical leaf properties existed along transects established in the field. We hypothesized that distance from the dieback center would be accompanied by a progressive change in

TABLE 6.1
Percent Compositions of Marsh Based on Site-Specific 35 mm Slides Classification

Site Names	Mud	Green Marsh	Yellow-Green Marsh	Yellow-Brown Marsh	Brown Marsh
2 (reference)	5.4	59.1	23.6	8.7	3.0
1h1	14.9	52.1	21.6	11.2	0.0
1h2	26.3	37.4	0	0.0	36.2
1i	33.9	0.0	2.5	37.5	25.9
1d	33.2	0.0	0.0	13.3	53.6
3h	17.1	29.1	26.6	0.0	27.0
3i	56.2	1.1	0.0	17.4	25.2
3d	56.1	6.8	0.0	6.4	30.4
4h	17.0	29.3	18.0	7.2	28.3
4i2	34.6	16.7	5.8	32.3	10.3
4i1	29.9	1.5	0.0	26.5	41.5
4d	57.0	0.0	0.0	4.3	38.6
5h	31.7	31.8	0.0	0.0	36.3
5d	43.6	0.7	0.0	10.4	45.2
5i	29.3	0.3	0.0	23.6	46.6
6i	22.7	14.1	5.7	43.5	13.9
7i	19.1	21.0	5.5	41.5	12.7
8i	36.9	5.3	0.0	25.8	31.7
9i	23.7	9.8	7.4	35.3	23.5
10h	30.7	38.1	13.2	0.0	17.7
11i	31.7	3.0	10.3	32.8	22.0
12h	13.3	41.3	11.6	13.5	20.2
13h	4.6	57.7	24.2	7.5	5.8
GB1	Ground-based—moist mud flat with dead marsh				
GB2	Ground-based—wet mud surrounded by healthy marsh				

Visual observations: h = healthy, h1 = healthy1, h2 = healthy2, i = intermediate, i1 = intermediate1, i2 = intermediate2, d = dead. GB1 and GB2 were not classified. Moist and wet modifiers were not quantified.

plant stress expressed as a change in optical leaf properties above natural variability. Previous works have proved that optical leaf measurements are essential if detrimental marsh changes are to be detected before irreversible change has occurred [9]. In other words, if optical leaf properties and impact level (assessed by transect distance) were not related, there would be little hope that a clear remote sensing indicator of prelethal stress existed.

Leaf spectra were obtained from whole plants collected every 5 m along four transects (Figures 6.3 and 6.5). Techniques from Daughtry et al. [22] were adapted to accommodate leaves of various shapes and sizes, including the flat and narrow smooth cordgrass leaves [9,23]. A LICOR external integrating sphere [24] attached to the

Spectron field radiometer [25] was used to directly measure diffuse spectral properties of the leaf. Leaf reflectance and transmittance were calculated directly from the measurements, and leaf absorption was calculated via Equation 6.1. Spectra spanned the visible (VIS, about 400–700 nm) to most of the NIR (700–1000 nm) region (the VIS and NIR, referred to as VNIR) with operational bandwidths near 10 nm [26].

6.3.2 Broadband Analyses

Broadband analyses were conducted to elucidate associations between the indicators of dieback and bandwidths typifying most sensor systems that measure regional resources, such as the Landsat TM and Enhanced TM. Bandwidths incorporating the wavelengths 454–459 nm (blue), 545–550 nm (green), 670–675 nm (red), 695–705 nm (red edge), and 845–850 nm (NIR) were extracted from the set of leaf and canopy spectra (Figure 6.7). Reflectance band ratios amenable for remote sensing mapping via broadband satellites were generated to help reduce atmospheric contamination from the remotely sensed signal and to dampen influences of background variability in the canopy reflectance. Although many band ratios were tested, only NIR/red and NIR/green were found useful for these analyses. The NIR/red ratio has long been used to map and monitor live canopy biomass, and NIR/green has been documented by Gitelson et al. [27] to be more sensitive to leaf pigment changes than the NIR/red ratio [9,27]. Changes in leaf pigment concentrations (e.g., chlorophyll and carotene) were inferred from leaf properties following predicted relations [27]. Researchers studying flushing frequency and marsh stress (e.g., Mendelssohn et al. [28]) have shown that relative chlorophyll and carotene changes can indicate

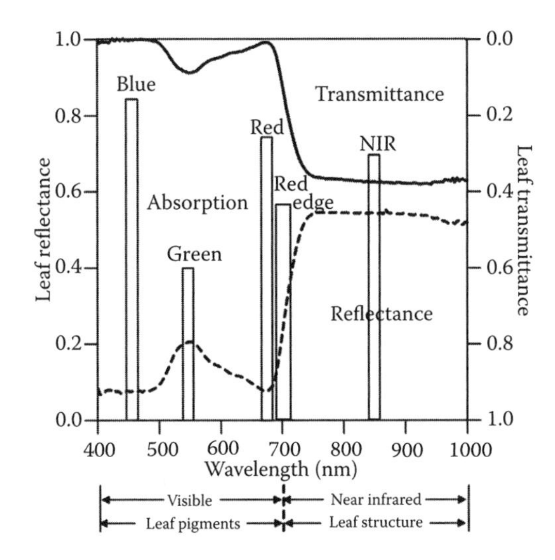

FIGURE 6.7 Measured leaf reflectance and transmittance and the calculated leaf absorption, as well as the locations of the blue, green, red, red edge, and NIR wavelength bands used in the study. (Adapted from Ramsey, E., III and Rangoonwala, A., *Photogrammetric Engineering and Remote Sensing*, 71, 299, 2005.)

stress levels of marsh plants. Spectral indicators of dieback onset and progression uncovered in the leaf analyses were transferred to canopy reflectance to simulate atmospherically corrected satellite image data.

6.3.3 Canopy Reflectance

The same Spectron Engineering radiometer [25] used in the leaf spectral measurements was used to collect upwelling radiance from a helicopter platform. An average aboveground level of about 190 m and a radiometer fixed entrance slit of 6° resulted in a ground instantaneous field of view (GIFOV) of about 20 m that approximated the spatial resolution of the EO-1 Hyperion hyperspectral and Landsat TM sensor systems. Up to seven upwelling radiance spectra collected with the Spectron radiometer were normalized by simultaneous ground-based collections with the LICOR radiometer of downwelling sunlight irradiance (300–1100 nm at 2 nm bandwidths) [24], thus creating mean reflectance spectra. Calibration of both the Spectron and LICOR field radiometers to the same bench-top light calibration instrument ensured compatibility of spectral intensity. Any artifacts in the resulting reflectance spectra were minimized in postnormalization processing [17].

Canopy reflectance spectra obtained from collections along the four marsh dieback transects and the reference site farther inland provided the spectral basis for linking the canopy reflectance changes to dieback progression. At the transects and the reference site, canopy reflectance spectra were computed at the dieback center, in the area of local healthy marsh surrounding the dieback, and at transect locations between the dead and healthy marsh containing marsh exhibiting intermediate dieback (Figure 6.8). Additional canopy reflectance spectra were obtained from collections at new sites containing various mixtures of healthy and impacted marsh (Figure 6.6).

6.3.4 Site-Specific 35 mm Slide Classification

Photographic 35 mm slides were collected along with the upwelling measurements collected with the Spectron radiometer from a helicopter. The slides were centered as much as possible at the location of upwelling recordings and covered an area of about 61 by 81 m at the ground level. Slides were scanned; separated into blue, green, and red images; and classified by using a supervised classification methodology described by Ramsey et al. [18] and Ramsey and Rangoonwala [17] (Table 6.1). Marsh composition classes included “mud” (shaded mud, mud, and bare ground), “green marsh” (including shadowed vegetation), “yellow-green marsh,” “yellow-brown marsh,” and “brown marsh” (dead marsh only). Classification applicability and precision were estimated by comparing biomass percentages of dead (totally brown) leaves and live (with any green) leaves obtained from ground-based transect-site measurements to composition indicators derived from slide classifications.

In order to better identify and assess associations among the marsh classes interpreted from the 35-mm slides and canopy reflectance, we applied a rotated orthogonal coordinate system to the marsh class variance (principal component analysis, PCA), SAS Institute Inc. [29]). The PCA removed intercorrelations among the marsh classes that interfere with establishing the strength of individual relations [17]. Scores

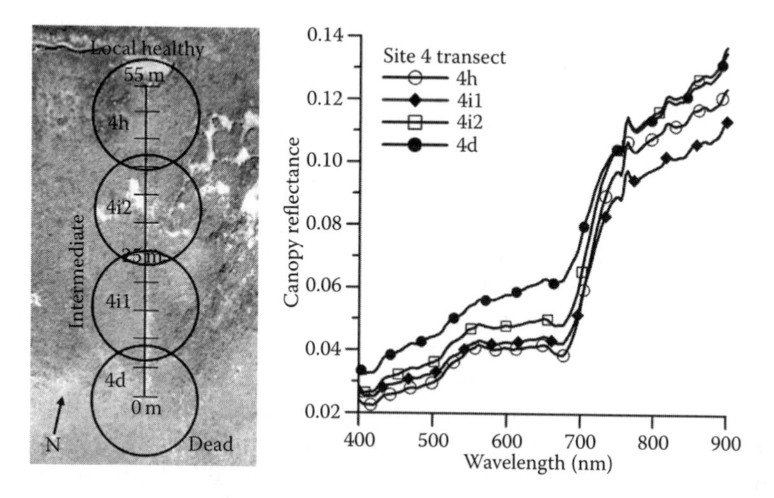

FIGURE 6.8 Canopy sampling strategy and reflectance spectra of transect 4. In the left portion, circles along the transect approximately represent the categories of marsh area recorded by the helicopter radiometer, from bottom to top: dead (4d); intermediate yellow-brown (4i1); intermediate yellow-green (4i2); and local healthy (4h) (see Table 6.1 for site codes). On the right, canopy reflectance at site 4 associated with the ground areas depicted by the circles and site codes shown on the transect. (Adapted from Ramsey, E., III and Rangoonwala, A., *Photogrammetric Engineering and Remote Sensing*, 72, 641, 2006.).

depicting the proportion of the original class variance (in the 35-mm classifications), as defined by each rotated orthogonal factor, were used as independent measures of marsh dieback progression. Dieback progression at all nontransect sites (sites 6–13) was estimated solely from PCA composition relations to whole-spectrum and broadband dieback indicators. "PCA-healthy" represented nonimpacted and dieback onset, "PCA-yellow-green" represented early dieback progression, "PCA-yellow-brown" represented later dieback progression, and "PCA-brown" represented the final stages of dieback.

6.3.5 Whole-Spectrum Analysis

In addition to the dieback analyses based on broadband reflectance, we implemented analytical methods pertinent to high spectral resolution sensors such as the EO-1 Hyperion, which collects data that simulate the whole reflectance spectrum obtained by our field radiometers (about 400–1000 nm) (e.g., Figure 6.8). Polytopic vector analysis (PVA), a multivariate analysis technique, was used to extract characteristic spectra [17,30–32]. In PVA, the form of the characteristic spectra types was not assumed a priori as in spectral libraries or extraction from homogeneous imaged areas. Rather, the characteristic spectra were extracted directly from the input set of whole canopy spectra, because a pertinent spectral library was unavailable and because characteristic spectra are not necessarily visibly identifiable by reconnaissance or distinguishable in the canopy reflectance spectra. Characteristic spectra extracted directly from the spectral data on canopy reflectance were used to de-emphasize the background variability and to isolate those influences related to leaf reflectance within the marsh canopy [17]. For these extractions, we restricted the characteristic spectra selection to reflectance spectra that were generated from the four transects and the one reference site. By concentrating on sites where dieback progression had been previously determined in the leaf analyses, we minimized the variability, or noise (data not pertinent to detecting the dieback progression), and maximized the extraction success without compromising the necessary generality of the extracted characteristic spectra.

To provide results that were more interpretable with respect to marsh dieback onset and progression, we used field observations to assemble the canopy reflectance spectra collected over the transect sites into two extreme groups of characteristic spectra. The "dead group" combined eight canopy spectra collected over the dead to intermediate end of each transect that represented composites of yellow to brown marsh and mud with little or no green marsh (sites included 1i, 1d, 3d, 4d, 5i, and 5d) (Table 6.1). To this group, we added one spectrum typifying a moist mud flat with dead marsh (GB1) and another typifying wet mud surrounded by healthy marsh (GB2) (Table 6.1); the latter was the only spectra that did not directly include any vegetation. The second component group contained five canopy reflectance spectra generated from the "local healthy marsh" at each transect (sites included 1h2, 3h, 4h, 5h, and reference site 2) (Table 6.1). Within the dead and healthy groups, we identified the minimum number of spectra that could be used to recreate all spectra in each input group; these spectra were referred to as "dead" and "healthy" characteristic spectra. As in the scores created within the PCA, the percent association of each canopy spectrum with the "dead" or "healthy" characteristic spectrum was defined by the composition weight.

Composition weights associated with the healthy and the dead characteristic spectra were used to identify the dieback progression at each transect location and marsh site. To combine characteristic composition weights for both healthy and dead marsh into a definition of dieback progression at each transect location and site, the composition weights were entered into a distance-clustering procedure [29].

6.3.6 Satellite Spectral Analyses

The results of the broadband canopy spectral analyses were implemented into the satellite detection and monitoring strategy. (A comparison of whole canopy spectra to TM canopy bandwidths is depicted in Figure 6.9.) We acquired six TM images collected on 08/14/99, 11/18/99, 01/05/00, 04/26/00, 09/18/00, and 12/22/00 that covered about 6 months to 1 year before and about 6 months after the first reported observation of dieback in coastal Louisiana in the spring of 2000. The final collection was near the end of the dieback episode. The comparison to the patterns identified in the canopy spectral analyses was limited to the blue, green, red, and NIR TM bands and band transforms (Figure 6.9). The 12/22/00 image was not used in this temporal assessment.

In preparation for implementing the broadband canopy reflectance results into the satellite analyses, the TM images were converted to absolute units, corrected for atmospheric influences, and transformed to apparent reflectance by using

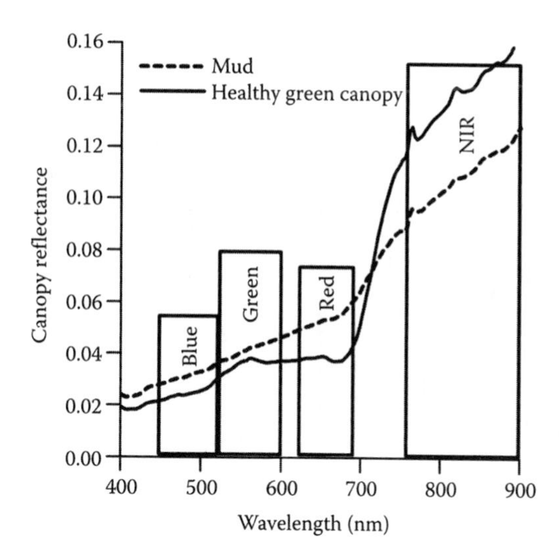

FIGURE 6.9 Landsat TM bandwidths overlaid on canopy reflectance spectra obtained from a healthy marsh and a mud flat.

methods similar to those described in Ramsey and Nelson [33]. By correcting and transforming the satellite data to reflectance, the image data were more closely comparable to the canopy reflectance obtained via helicopter and ground-based radiometer collections [14,15]. In addition, the reflectance estimates portrayed a closer and more accurate depiction of the marsh condition than could be obtained with more simplistic transforms based solely on the sun zenith as an atmospheric optical depth estimate [14,33].

The ability to accurately and consistently define changes in temporal and spatial spectral patterns is directly dependent on the correctness of surface reflectance determined via the atmospheric correction. The conversion of at-sensor brightness values to surface-reflectance estimates was a complex and intense effort. Numerous light and dark targets that exhibited fair stability over the image collection period were identified and used to calibrate the atmospheric correction. An optimization drove a radiative atmospheric program by incrementally changing the program inputs until there was convergence (minimization) of the summed differences between the predicted and expected values of the light and dark reflectances (e.g., [33]). The set of input parameters and variables at convergence was then used to correct the at-sensor intensity for atmospheric influences. Validation of the corrected image data was not completed, so only preliminary results are included from selected dieback sites.

6.4 RESULTS

6.4.1 Leaf Spectral Analyses

Comparison of NIR/red and NIR/green ratios from plants collected at the healthy ends of each transect (local healthy) confirmed the natural spectral variability of the

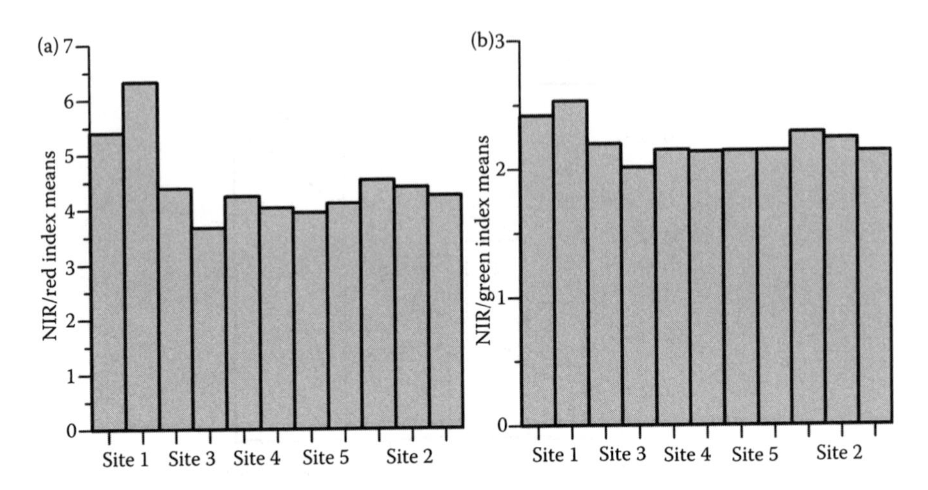

FIGURE 6.10 Spectral variability in areas of local healthy marsh, expressed as (a) NIR/red and (b) NIR/green indexes. Each site is characterized by the next-to-last two locations on each transect (e.g., site 1, as depicted on the chart, includes the first two bars). Site 2 (reference) includes the three last bars on the chart.

smooth cordgrass marsh (Figure 6.10). Even so, spectral properties of live leaves gathered from plant samples taken along transects spanning impacted coastal Louisiana marsh sites did depict changing plant status (with respect to dieback progression) that far exceeded the observed natural variability. We found that green and red edge reflectance trends generally represented the early stages and fairly well the later stages of dieback progression, while blue and red reflectance and absorption trends represented the later stages of marsh impact that were most closely related to visible signs of marsh impact (Figure 6.11a–d). NIR reflectance was not compatible with visual reflectance trends and did not covary with derived indicators of water stress. Predicted carotene concentrations tended to remain constant or increase relative to chlorophyll concentrations in stressed plants at the two least impacted sites (1 and 3), while the pigments covaried at the two most impacted sites (4 and 5) (Figure 6.11e–h). The NIR/red and NIR/green ratio magnitudes and ranges generally increased from the most to the least impacted site. The NIR/red trends indicated later stages of impact, while the NIR/green trends indicated impact onset and progression as well as generally indicated later stages of impact.

6.4.2 Canopy Spectral Analyses

6.4.2.1 Slide Classification

Covariation was high between the measured live and dead biomass and the slide classifications. There was significant ($p = 0.05$, significance level) correlation ($r = 0.98$, $n = 4$) between combined percentages of green and yellow-green classes with live biomass percentages and also ($r = 0.98$, $n = 4$) between brown and yellow-brown percentages with dead biomass percentages. Although this analysis was

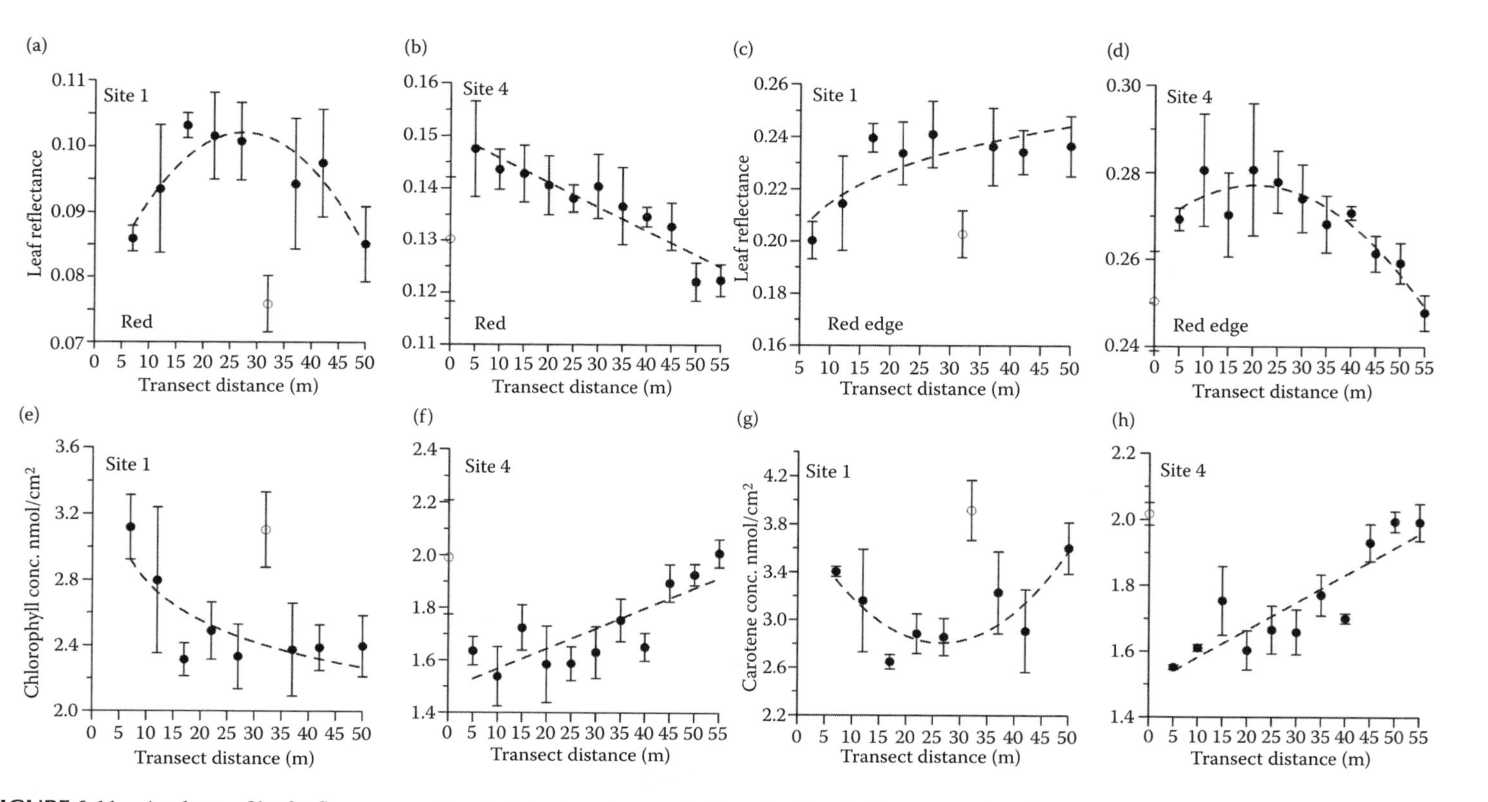

FIGURE 6.11 Analyses of leaf reflectance and trends in leaf carotene and chlorophyll levels. The top portion shows trends in mean values of leaf reflectance and standard errors associated with the (a and b) red and (c and d) red edge bandwidths at sites 1 and 4. Explained variances determined by using regression models (dashed line, open symbols not used, significant level $p < 0.05$) were 83% at site 1 and 88% at site 4 for red bands and 80% at site 1 and 87% at site 4 for red edge bands. The bottom portion illustrates predicted leaf (e and f) chlorophyll and (g and h) carotene trends. Explained variances determined by using regression models were 62% ($p < 0.10$) at site 1 and 76% ($p < 0.05$) at site 4 for chlorophyll and 81% ($p < 0.05$) at site 1 and 66% ($p < 0.05$) at site 4 for carotene. (Adapted from Ramsey, E., III and Rangoonwala, A., *Photogrammetric Engineering and Remote Sensing*, 71, 299, 2005.)

restricted to only four transect sites where the healthy marsh was sampled, these results suggested that our classifications of the 35-mm slides were adequate indicators of marsh composition. Transformed PCA-green plus PCA-yellow-green classes retained the significant ($p = 0.05$) relation to measured live biomass ($r = 0.93$, $n = 4$); however, the transformed PCA-brown plus the PCA-yellow-brown class was not significantly related to dead biomass. The realignment of the class variables, resulting in better separation of marsh classes, coupled with the inability to separate these classes within the field biomass measures (live was defined as a leaf with any green, while dead leaves were totally brown) may have led to the decreased correspondence between the PCA class and measured dead biomass. The high overall correspondence and the highly diminished significance of intercorrelation among slide classes, however, justified replacing the original slide-class percentages with the PCA output scores in all subsequent statistical analyses in order to alleviate problems inherent when using intercorrelated regressor variables [34].

6.4.2.2 Broadband Spectral Reflectances

Results of broadband covariations with marsh dieback indicators from the transformed slide classes showed that the blue, green, red, and red edge reflectance band variances were individually related to the PCA-transformed variance in the healthy, yellow-green, and brown marsh slide classification indicators. The blue, red, and red edge bands were significantly ($p = 0.05$) and positively related to PCA-brown ($r = 0.44$, 0.49, and 0.46, respectively) and negatively related to PCA-healthy ($r = -0.66$, -0.65, and -0.63, respectively), and the green band was negatively related to PCA-healthy ($r = -0.64$). The NIR band was not significantly related to any PCA-transformed class, while NIR/red and NIR/green ratios were negatively related to PCA-brown ($r = -0.43$ and -0.42, respectively) and positively related to PCA-healthy ($r = 0.67$ and 0.66, respectively).

6.4.2.3 Characteristic Spectra

On the basis of site-specific observations and the associations among the eight dead marsh spectra, PVA extracted two characteristic spectra for dead vegetation, one associated with site 1i (sparse dead and live marsh in a mud background) and the other with the ground-based site composed of a moist mud flat with dead marsh (GB1) (Figure 6.12a). The association and alignment of the six remaining characteristic spectra for dead composition weights corresponded to field observations and bare-ground percent compositions.

As with the spectral extractions for dead marsh, two spectra were ultimately selected as characteristic of healthy marsh (Figure 6.12b). The first characteristic spectrum for healthy vegetation was associated with site 5h (site 5 healthy), and the second was associated with the reference site (site 2). The composition weights associated with healthy marsh at each transect mirrored findings documented in the leaf optical analyses. Local healthy marsh occupying the more spatially extensive dieback sites was similar, while local healthy marsh at less extensive dieback sites was progressively aligned with the most inland and nonimpacted reference site 2.

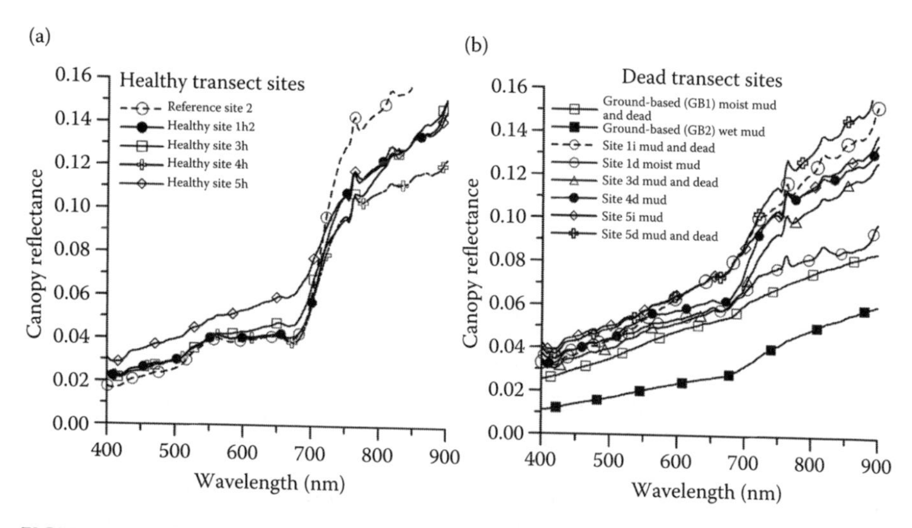

FIGURE 6.12 Canopy spectra used to identify the best spectra to characterize healthy and dead marsh. (a) The five background reflectance spectra used in PVA to identify spectra that best characterized healthy canopy. (b) The eight mean background spectra used in the PVA to identify spectra that best characterized dead canopy. Dashed lines denote healthy and dead characteristic spectra used to derived PVA composition weights. (Refer to Table 6.1 for site codes.) (Adapted from Ramsey, E., III and Rangoonwala, A., *Photogrammetric Engineering and Remote Sensing*, 72, 641, 2006.)

Subsequent regression indicated that the two best indicators of dieback onset and progression with respect to the orthogonal slide classifications were the characteristic spectra representing dead marsh (sparse live and dead marsh and mud) of site 1i and the healthy marsh of the reference site 2. Covariation between the characteristic spectra of healthy and dead marsh and the transformed slide classes illustrated that variability in the characteristic spectra was most sensitive to canopy changes associated with the PCA-healthy, PCA-yellow-green, and PCA-brown indicators but not necessarily with the PCA-yellow-brown indicator.

6.4.2.4 Whole-Spectrum Canopy Reflectance

Composition weights exceeding the expected 0–1 range at sites 12h and 13h suggested that these sites contained marsh with spectral characteristics outside the range of the original transects and reference site used for the creation of characteristic spectra [17]. Cluster analysis of the PVA compositions based on the dead and healthy characteristic spectra (site 1i and reference site 2) separated the whole-spectrum canopy reflectance into three groups: healthy, intermediate, and dead marshes (Figure 6.13). The clustering procedure further separated healthy marsh into three groups: (a) those most similar to marsh found at the reference site 2, (b) those with somewhat less green marsh, and (c) two sites (12h and 13h) visually containing the most green and probably healthiest marsh. The intermediate marsh class included both the local healthy marsh at sites 4h and 5h and sites of similarly impacted marsh. This category was separated into two site-based categories, with marsh dieback

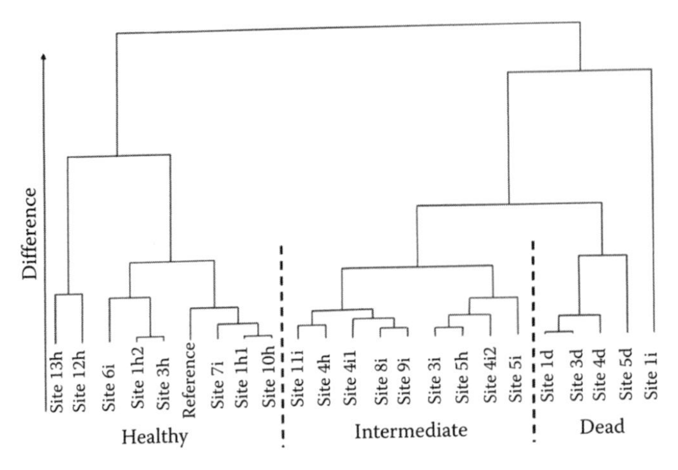

FIGURE 6.13 Similarities in dieback progression [17]. A distance-clustering procedure was used to accomplish the grouping (PROC CLUSTER—Method Average, SAS Institute Inc. [29]). Inputs were the PVA composition weights associated with all field top-of-canopy reflectance spectra as related to the dead and healthy characteristic spectra. The coordinates on the ordinate denote average distance between clusters. Note that site codes based on visual observations as defined in Table 6.1 are mixed in the whole-spectrum classification. (Adapted from Ramsey, E., III and Rangoonwala, A., *Photogrammetric Engineering and Remote Sensing*, 72, 641, 2006. With permission.)

progression aligned with marsh at site 4h and those more aligned with marsh at site 5h. The dead marsh class was divided into two major groups, one with higher portions of brown marsh than yellow-brown marsh and one solely related to moist mud and sparse live and dead marsh.

6.4.3 Satellite Spectral Analyses

Graphical comparisons proved that removal of atmospheric influences in the Landsat TM sensor signal was required in order to correctly represent spectral patterns in the marsh (Figure 6.14). Because a complete atmospheric correction was not accomplished, however, a normalized different vegetation index was used in place of the NIR/green vegetation ratio to help dampen any remaining atmospheric influences. Temporal patterns of the normalized index created from the nonatmospherically corrected (Figure 6.14a) and the corrected (Figure 6.14b) TM image data clearly show the benefit of atmospheric correction in monitoring of coastal resources. Once the atmospheric influence has been minimized, the temporal patterns of the biophysical variables provide timely detection and continued monitoring of adverse impacts, such as marsh dieback (Figure 6.14c and d).

6.5 DISCUSSION

Successful linkage of changes in optical leaf properties to dieback onset and progression provided a strategy for accomplishing a similar result with canopy reflectance

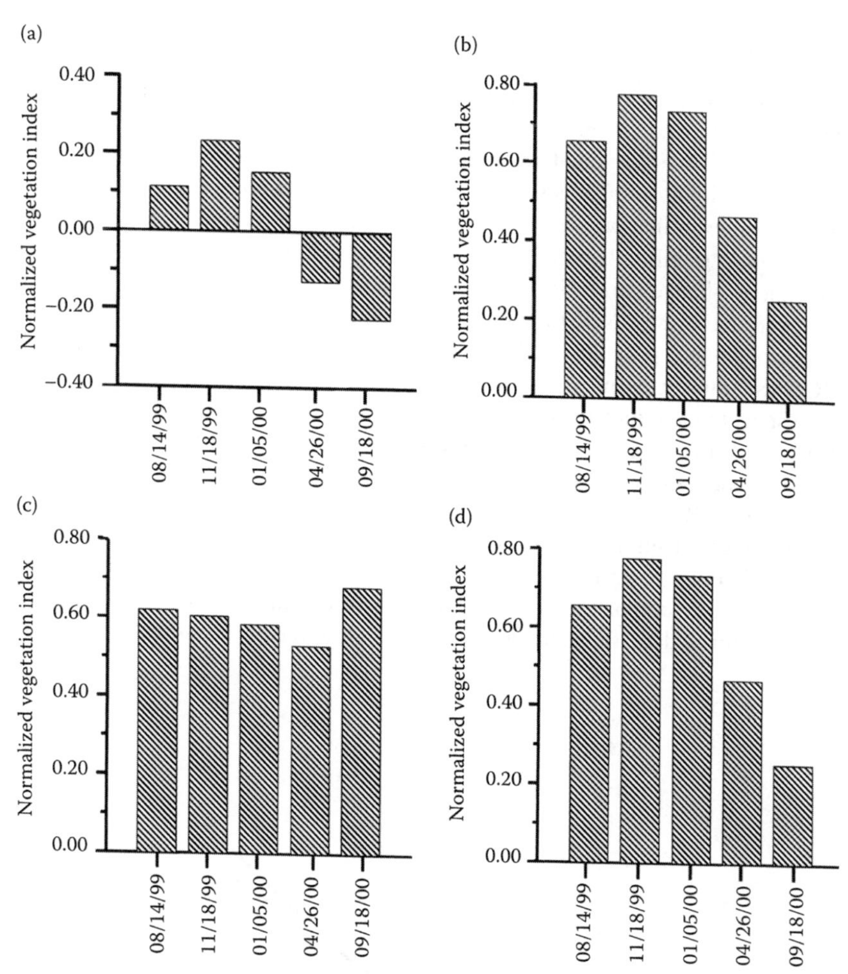

FIGURE 6.14 Dieback trends derived from Landsat TM with and without atmospheric correction. (a and b) The five dates of reflectance data from TM transformed to a normalized vegetation index at a dieback site. The transform is shown without atmospheric correction (a) and with atmospheric correction (b). (c and d) Vegetation indexes of healthy (c) and dieback (d) sites. Abscissa labels denote Landsat TM collection dates (the last collection date, 12/22/00, is not included).

spectra. The strategy generated canopy reflectance along the same transects used in the leaf optical analyses so that the results of analyzing these canopy spectra could be validated by comparison to known optical changes in the plant leaves along the transects. Once validated, model parameters were used to extend the analyses beyond the transect locations to the broader coastal region affected by the dieback. These extended results provided guidance and limits for temporal and regional satellite mapping and monitoring.

6.5.1 Broadband Reflectance

In contrast to associations of dieback progression that were found in the leaf spectral analyses, broadband canopy analyses based on comparisons (via regression analyses) to 35 mm slide classifications indicated few differences in blue, red, green, and red edge sensitivity to changes related to dieback onset and progression. As in the leaf spectral examination, variance in NIR reflectance explained less of the dieback progression variability than did the other reflectance bands and seemed more aligned with canopy changes in the later stages of dieback. In agreement with results from leaf spectral analyses, NIR/red and NIR/green transforms were good indicators of dieback onset and progression. Contrary to leaf spectral results, the NIR/red and NIR/green transforms provided similar sensitivity to canopy dieback progression with respect to the indicators derived from classification of the 35-mm slides. On the other hand, correspondence between the whole-spectrum healthy compositions and the broadband canopy NIR/green indexes was somewhat higher than those calculated versus NIR/red (Figure 6.15).

The higher sensitivity to dieback progression in the NIR/green versus NIR/red expressed in these comparisons to the whole-spectrum healthy compositions but not in the slide-based comparisons may lie in the nature of the latter classifications. Even when transformed to independent measures, slide-based classification indicators

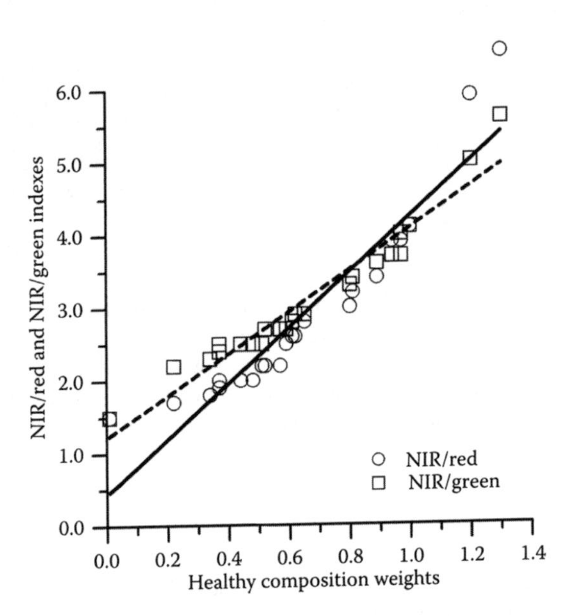

FIGURE 6.15 Correspondence of the vegetation ratios with the whole-spectrum composite weights for healthy marsh, as derived from the PVA. (solid line) The regression model for composite weights of the NIR/red versus the whole spectrum had an overall explained variance of 88% with a mean square error of 0.21. (dashed line) The regression model for the composite weights of the NIR/green versus the whole spectrum had an overall explained variance of 94% with a mean square error of 0.06.

provided only an estimate of dieback progression. The exact physical linkage between those classification indicators and the canopy reflectance could neither be fully quantified nor be considered completely consistent along transect locations or from site to site. In other words, the orthogonal transform ensured independence of the slide classes but did not ensure that the slide classes consistently represented exactly the same plant canopy characteristic. Although not offering subtle distinctions, the slide-based comparisons were more quantifiable than were visual comparisons. The slide-based indicators offered essential evidence for understanding what the derived broadband and whole-spectrum measures represented and for validating their derived linkage to marsh dieback onset and progression.

6.5.2 Whole-Spectrum Analyses

Relying on site-specific observations and known dieback progression at the transect sites, we found that application of the single pair of dead and healthy characteristic spectra to all canopy spectra and subsequent groupings broadly divided the canopy spectra into three categories: healthy marsh, intermediately marsh, and the most severely impacted marsh. The last group was further separated into marsh clearly exhibiting visible dieback and marsh dominated by dead plants and mud. Within this dead marsh group, some alignments (grouping of site-spectra similarities) contrasted with the leaf spectral results. For example, the spectral properties of leaves from sites 1d and 3d were most closely aligned with the leaf spectral properties of intermediate marshes at sites 4i2 and 5i. In contrast, in the whole-spectrum analyses, canopy spectra at sites 1d and 3d aligned with canopy spectra at sites 4d and 5d. The preponderance of background versus marsh canopy cover within the imaged area skewed the canopy spectral response, diminishing the ability to identify the relatively healthier but scarce marsh within the imaged area (e.g., sites 1d and 3d). In general, as the impact severity advanced in the dead marsh category, the canopy reflectance became progressively less a function of green-leaf content; ultimately, it became more a function of exposed mud and moisture content.

6.5.3 Whole-Spectrum and Broadband Canopy Spectral Analyses

There were fewer differences when comparing the NIR/green than when comparing the NIR/red bands to the whole-spectrum indicators (Figure 6.15). Although the evidence was limited, the associations and trends of NIR/green indicated more sensitivity to dieback onset and progression, as was expressed in the leaf optical. Even though this correspondence was moderately high, the characteristic spectra derived from the whole-spectrum analyses provided greater detail of relative dieback progression and dieback similarity among sites than did the NIR/green band analyses. For example, in the whole-spectrum analysis, local healthy marsh at sites 4h and 5h were separately defined into two major intermediate groups, thereby providing more dieback detail than did the simpler NIR/green indicator in the broadband spectral analyses. This enhanced detail in the whole-spectrum results provided evidence that this methodology performed better than did indicators solely based on the variability in selected wavelength regions.

6.5.4 Satellite Spectral Analyses

The ability to map the occurrence of diebacks by identifying patterns of change in the Landsat TM time series was partially determined. The determination relied first on atmospheric correction and then on subsequent conversion of the Landsat TM image data into reflectance estimates. As illustrated in the plant-leaf and canopy spectral analyses, the most convincing evidence of dieback impact or nonimpact was reflected in the temporal pattern of the normalized difference ratio, not evidence from a single TM image date. These positive correspondences were based on selected sites where impact history was known. Once the atmospheric corrections of the Landsat TM image data are validated, the impact patterns predicted from satellite data will be analyzed by comparing to marsh dieback establishment observed with photo-interpreted distributions, fixed-wing transect observations, and ground-based observations conducted during the image collection period. Upon completion of this accuracy assessment, pattern analyses will adapt techniques from Ramsey et al. [18,35] in order to categorize the historical development of the marsh at each specific location and thus the rate of dieback spread.

6.6 CONCLUSION

Techniques implemented in this study proved that remote sensing data can detect dieback onset and progression if spectral band indicators that accurately simulate leaf optical results can be extracted from the satellite data. To accomplish these developments, we first created a conceptual model that provided a physical portrayal of the onset and expansion of marsh dieback. That conceptual portrayal, combined with sample transects oriented according to the model, afforded an essential component to describe change—a metric. Trends of spectral indicators derived from the optical analyses of leaves collected from plant samples obtained along transects proved the validity of the model and the collection strategy. This leaf-based analysis also provided another essential result—that the impact progression was exhibited in the plant-leaf spectral properties. Without this confirmation, remote sensing methods would have only been useful in detecting perceptual changes in canopy structure that occur near the end of dieback progression.

Dieback indicators uncovered in the leaf analyses were transferred to the satellite analysis via canopy reflectance based on helicopter collections along the same transects where plant samples were collected. Coupling the leaf-level results to the canopy analyses and using the same sampling methodology at the same locations ensured compatibility between the two scales of monitoring and provided procedural guidance for the canopy analyses and cross-validation of the analytical results. A second validation was based on independent photographic classifications of the marsh sites targeted in the canopy collections. As found in the broadband and whole-spectrum site classifications and clearly illustrated in Figure 6.13, the canopy reflectance analyses demonstrated that remote sensing offered a more accurate depiction of dieback severity than is offered by perceptions relying on visible indicators of marsh change (e.g., visual site codes in Table 6.1).

High spectral resolution data formed the bases of all spectral analyses. The use of high spectral resolution data allowed us to determine (a) which reflectance data bandwidths and (b) which whole-spectrum indicators (from 400 nm to about 900 nm) were best in revealing the various stages of impact within the highly variable marsh landscape. Based on these analyses, spectral indicators amenable to operational broadband sensors, such as Landsat TM, and hyperspectral resolution sensors, such as EO-1 Hyperion, were constructed, and limitations were quantified. Results showed that changes in chlorophyll and carotene pigments were practical indicators of impact progression and that green and red edge reflectance bands, located on the absorption tails of the chlorophyll and carotene pigments, provided estimates of early impact onset. A new vegetation index based on NIR/green reflectance proved to be an adequate indicator of both the onset and the relative time since the impact onset; however, progression detail was lacking. In contrast, whole-spectrum impact indicators provided the broadest impact divisions as well as detailed progression and reasonable onset information. Neither the broadband nor the whole-spectrum remote sensing analyses provided the same level of detail as the leaf analysis did in places where the marsh had been severely impacted by dieback.

The final step transformed the integrated leaf and canopy results into operational satellite mapping tools for detecting and monitoring subtle spectral changes that indicate abnormal status in a spectrally and spatially complex marsh system. A temporal suite of Landsat TM images spanning the dieback episode were adjusted to surface reflectance values compatible with the top-of-canopy-based spectral indicators, and satellite broadband dieback indicators were derived. Although validation of the satellite-derived impact progression is not complete, results at known marsh sites suggested that temporal patterns portrayed by spectral indicators differentiate impacted and nonimpacted marshes.

Although directed at mapping and monitoring marsh dieback in coastal Louisiana, results of these studies proved that remote sensing technologies can supply accurate regional maps depicting subtle changes in marsh vegetation. Subtle optical changes provide vital information about detrimental long-term plant stress that can be used to prevent the ultimate outcome of acute plant stress, conversion of marsh to mud flats or open water. We hope that methods developed as part of this project will provide a template for mapping and monitoring subtle changes within native landscapes in different geographic areas.

ACKNOWLEDGMENTS

We thank Greg Linscomb of the Louisiana Department of Natural Resources, who first observed and recognized the importance of the marsh dieback. We thank Professor Anatoly Gitelson for suggesting the successful NIR and green ratio. We are grateful to Kristine Martella, a GIS Analyst at the Department of Information Technology, Solano County, California, for help in sample collection and analyses, Beth Vairin and Connie Herndon of the U.S. Geological Survey, and Victoria Jenkins of IAP World Services, Inc. for editing this manuscript. Partial funding for this work was provided through the Louisiana Department of Natural Resources Agreement Number 2512-01-11. Mention of trade names does not constitute endorsement by the U.S. Government.

REFERENCES

1. Seneca, E.D., Techniques for creating salt marshes along the east coast, in *Rehabilitation and Creation of Selected Coastal Habitats: Proceedings of a Workshop*, U.S. Fish and Wildlife Service, Biological Services Program, Washington, DC, FWS/OBS-80/27, 1980, p. 1.
2. Stout, J.P., The ecology of irregularly flooded salt marshes of the northeastern Gulf of Mexico: A community profile, U.S. Fish and Wildlife Service Biological Report 85(7.1), 1984, 98pp.
3. National Research Council (NRC), *Responding to Changes in Sea Level, Engineering Implications*. Committee on Engineering Implications of Changes in Relative Mean Sea Level, National Academy Press, Washington, DC, 1987.
4. Saenger, P., Hegeri, E., and Davie, J. (Eds), Global status of mangrove ecosystems, *The Environmentalist*, 3(Suppl. 3), 1983, p. 1–88.
5. Lyon, J., *Practical Handbook for Wetland Identification and Delineation*, Lewis Publishers, Boca Raton, FL, 1993.
6. Ramsey, E., III, Radar remote sensing of wetlands, in R. Lunetta and C. Elvidge (Eds), *Remote Sensing Change Detection: Environmental Monitoring Methods and Applications*, pp. 211–243, Ann Arbor Press, Michigan, 1998.
7. Ramsey, E., III, Remote sensing of coastal environments, in M.L. Schwartz (Ed.), *Encyclopedia of Coastal Science*, Encyclopedia of Earth Sciences Series, 797pp, Kluwer Academic Publishers, The Netherlands, 2005.
8. Gitelson, A., Stark, R., Dor, I., Michelson, O., and Yacobi, Y., Optical characteristics of the phototroph *Thiocapsa reseopersicina* and implications for real-time monitoring of the bacteriochlorophyll concentration, *Applied and Environmental Microbiology*, 65, 3392, 1999.
9. Ramsey, E., III and Rangoonwala, A., Leaf optical property changes associated with the occurrence of *Spartina alterniflora* dieback in coastal Louisiana related to remote sensing mapping, *Photogrammetric Engineering and Remote Sensing*, 71, 299, 2005.
10. Gates, D., Keefan, H., Schleter, J., and Weidner, V., Spectral properties of plants, *Applied Optics*, 4, 11, 1965.
11. Gausman, H., Allen, W., Cardenas, R., and Richardson, A., Relation of light reflectance to histological and physical evaluations of cotton leaf maturity, *Applied Optics*, 9, 545, 1970.
12. Carter, G., and Knapp, A., Leaf optical properties in higher plants: Linking spectral characteristics to stress and chlorophyll concentrations, *American Journal of Botany*, 88, 677, 2001.
13. Penuelas, J. and Filella, I., Visible and near-infrared reflectance techniques for diagnosing plant physiological status, *Trends in Plant Science*, 3, 151, 1998.
14. Ramsey, E., III and Jensen, J., Modeling mangrove canopy reflectance using a light interaction model and an optimization technique, in J. Lyon and J. McCarthy (Eds), *Wetland and Environmental Applications of GIS*, p. 61, CRC Press, Boca Raton, FL, 1995.
15. Ramsey, E., III and Jensen, J., Remote sensing of mangroves: Relating canopy spectra to site-specific data, *Photogrammetric Engineering and Remote Sensing*, 62, 939, 1996.
16. Ramsey, E., III, Nelson, G., Baarnes, F., and Spell, R., Light attenuation profiling as an indicator of structural changes in coastal marshes, in R. Lunetta and J. Lyon (Eds), *Remote Sensing and GIS Accuracy Assessment*, p. 59, CRC Press, New York, 2004.
17. Ramsey, E., III and Rangoonwala, A., Site-specific canopy reflectance related to marsh dieback onset and progression in coastal Louisiana, *Photogrammetric Engineering and Remote Sensing*, 72, 641, 2006.
18. Ramsey, E., III, Nelson, G., and Sapkota, S., Coastal change analysis program implemented in Louisiana, *Journal of Coastal Research*, 17, 55, 2001.

19. Michot, T., Ford, M., Rafferrty, P., Kemmerer, S., and Olney, T., Characterization of plants and soils in a *Spartina alterniflora* salt marsh experiencing "brown marsh" dieback in Terrebonne Parish, Louisiana, USA, in abstracts from coastal marsh dieback in the Northern Gulf of Mexico: Extent, causes, consequences, and remedies, in R. Stewart, C. Proffitt, and T. Charron (Eds), *Information and Technology Report*, USGS/BRD/ITR—2001-0003, 2001.
20. Ramsey, E., III and Laine, S., Comparison of Landsat Thematic Mapper and high resolution photography to identify change in complex coastal marshes, *Journal of Coastal Research*, 13, 281, 1997.
21. Chabreck, R., *Marsh Zone and Vegetative Types in the Louisiana Coastal Marshes*, PhD dissertation, Louisiana State University, Baton Rouge, 1970.
22. Daughtry, C., Ranson, K., and Biehl, L., A new technique to measure the spectral properties of conifer needles, *Remote Sensing of Environment*, 27, 81, 1989.
23. Ramsey, E., III and Rangoonwala, A., Determining the optical properties of the narrow, cylindrical leaves of *Juncus roemerianus*, *IEEE Geoscience and Remote Sensing*, 42, 1064, 2004.
24. LI-COR, *LI-1800UW Underwater Spectroradiometer Instruction Manual*, LI-COR Publication No. 8405-0037, LI-COR Inc., Lincoln, Nebraska, 1984.
25. Spectron Engineering, Inc., *Operating Manual: SE590 Fieldportable Data Logging Spectroradiometer*, Spectron Engineering, Denver, Colorado, n.d.
26. Markham, B., Williams, D., Schafer, J., Wood, F., and Kim, M., Radiometric characterization of diode-array field spectroradiometers, *Remote Sensing of Environment*, 51, 317, 1995.
27. Gitelson, A., Kaufman, Y., Stark, R., and Rundquist, D., Novel algorithms for remote estimation of vegetation fraction, *Remote Sensing of Environment*, 80, 76, 2002.
28. Mendelssohn, I. and McKee, K., *Spartina alterniflora* die-back in Louisiana: Time-course investigation of soil waterlogging effects, *Journal of Ecology*, 76, 509, 1988.
29. SAS Institute Inc., *SAS/STAT User's Guide: Version 6*, 4th edition, vol. 1, 943pp, SAS Institute Inc., Cary, NC, 1989.
30. Ehrlich, R. and Crabtree, S., *The PVA Multivariate Unmixing System, Self-training Classification*, Tramontane, Inc., and C & E Enterprises, Salt Lake City, UT, 2000.
31. Johnston, G., Ehrlich, R., and Full, W., Principal components analysis and receptor models in environmental forensics, in B.L. Murphy and R.D. Morrison (Eds), *An Introduction to Environmental Forensics*, p. 461, Academic Press, San Diego, 2002.
32. Ramsey, E., III, Rangoonwala, A., Nelson, G., and Ehrlich, R., Mapping the invasive species, Chinese Tallow with EO1 satellite Hyperion hyperspectral image data and relating tallow percent occurrences to a classified Landsat Thematic Mapper landcover map, *International Journal of Remote Sensing*, 26, 1637, 2005.
33. Ramsey, E., III and Nelson, G., A whole image approach for transforming EO1 Hyperion hyperspectral data into highly accurate reflectance data with site-specific measurements, *International Journal of Remote Sensing*, 26, 1589, 2005.
34. Hoerl, A. and Kennard, R., Ridge regression: Applications to nonorthogonal problems, *Technometrics*, 12, 69, 1970.
35. Ramsey, E., III, Sapkota, S., Barnes, F., and Nelson, G., Monitoring the recovery of *Juncus roemerianus* marsh burns with the normalized vegetation index and Landsat Thematic Mapper data, *Wetlands Ecology and Management*, 10, 85, 2002.
36. Kirk, J., The photosynthetic apparatus of aquatic plants, in *Light & Photosynthesis in Aquatic Ecosystems*, Chap. 8, pp. 229, 233, Cambridge University Press, Cambridge, 1994.

7 Estimating Chlorophyll Conditions in Southern New England Coastal Waters from Hyperspectral Aircraft Remote Sensing

Darryl J. Keith

CONTENTS

7.1 INTRODUCTION

Nationally, increased population and development have contributed significantly to environmental pressures along many areas of the U.S. coastal zone. These pressures have resulted in substantial physical changes to beaches, loss of coastal wetlands, declines in ambient water and sediment quality, and the addition of higher volumes of nutrients (primarily nitrogen and phosphorus) from urban, nonpoint source runoff. Some nutrient inputs to coastal waters are necessary for a healthy functioning coastal ecosystem. However, when nutrient concentrations from sources such as stream and river discharges, wastewater sewage facilities, and agricultural runoff are increased beyond the natural background levels of estuaries and other coastal receiving waters, algal growth is stimulated. These excess nutrients and the associated increased algal growth can also lead to a series of events that can decrease water clarity, cause benthic degradation, and result in low concentrations of dissolved oxygen.

The initial response of coastal systems to the addition of nutrients (especially nitrogen) is to increase phytoplankton biomass [1]. This biological response is routinely measured in terms of chlorophyll *a* (chl *a*), which is the dominant light-harvesting pigment and is universally present in eukaryotic algae and cyanobacteria [2]. Chl *a* is commonly measured in water-quality monitoring programs for coastal and freshwater systems [3–6], in surveillance programs for harmful algal blooms [7–10], and in ecological studies of phytoplankton biomass and productivity [11–13]. The concentration of chl *a*, when evaluated with other condition indicators such as water clarity and dissolved oxygen concentrations, provides information on the environmental quality of estuaries and coastal waters [14]. These indicators of environmental condition are also used to gauge the extent to which coastal habitats and resources have been altered. It must be noted that while these indices may not address all characteristics (societal and scientific) of estuaries and coastal waters, they do provide information on both ecological condition and human use of estuaries.

In the U.S. Environmental Protection Agency (EPA) National Coastal Assessment (NCA) program, the condition of estuarine and coastal water bodies is assessed using rating criteria based on existing legal statutes, federal and state guidelines, or interpretation of the scientific literature. In this framework, coastal condition ratings vary based on the conditions of individual sites within the region. These individual ratings are aggregated to create a regional assessment rating. The regional criteria are determined through surveys of local environmental managers, resource experts, and the general public. In the NCA program, the concentration of chl *a* is one of several environmental metrics that are used to develop a water-quality index of coastal condition. This and several other indices (e.g., sediment quality) are used to develop an assessment to inform the U.S. Congress and the public about general water-quality conditions in the United States as required by the National Water Quality Inventory (Section 305 (b) of the Clean Water Act). In the National Coastal Condition Report II [14], the NCA Program reported that, with respect to chl *a* condition, approximately 15% of the estuarine area along the Northeast Coast was rated poor during the summer of 2000. The Northeast Coast ranged from the Chesapeake Bay region to the border between the United States and Canada. When these assessments are subset for southern New England bays and estuaries, approximately 4% were rated as poor during this season. These comparisons suggest that these assessments are subject to spatial variability and bias due to seasonally limited data collection.

In this chapter, the ecological condition of numerous individual embayments and estuaries along the southern New England coast, as well as the adjoining coastal ocean, are determined over an annual cycle using remote sensing and the criteria for assessing chl *a* concentrations found in EPA NCA guidelines [14]. In Phase I, surface concentrations of chl *a* are calculated using an empirically derived, band ratio model developed from *in situ* hyperspectral data acquired during a multiyear monitoring program in Narragansett Bay, Rhode Island. In Phase II, the model was applied to remotely sensed, hyperspectral data passively acquired during aerial surveys of estuaries and bays along the southern New England coastline to estimate chl *a* concentrations and assess environmental condition. In Phase III, chl *a* condition assessments are made for individual coastal systems and are aggregated over the survey period (2006) to create site-specific annual assessments. The individual site assessments are further used to create regional scale assessments at monthly, seasonal, and annual scales.

7.2 METHODS

7.2.1 Field Measurements

In general, Narragansett Bay is a well mixed, temperate estuary whose upper reaches are occasionally stratified as measured by salinity gradients [15]. The mean depth of the bay is 7.8 m [16]. The Narragansett Bay ecosystem is significantly impacted by eutrophic conditions during the summer season and nonpoint runoff from intensely populated watersheds in Rhode Island and Massachusetts [15]. Narragansett Bay is a phytoplankton-dominated ecosystem that usually experiences a bay-wide, winter-early spring bloom; several localized short-term blooms throughout the summer; and a late summer bay-wide bloom [15]. In the study, *in situ* hyperspectral radiance and irradiance data, as well as seawater samples for the laboratory estimation of chl *a*, were collected along a salinity gradient from the upper to the lower bay at 10 stations in Narragansett Bay and Rhode Island Sound (Figure 7.1).

Between July 2005 and July 2006 and during July 2007, spectral data were acquired using the Satlantic, Inc. Ocean Profiler II system configured in the buoy mode and equipped with a pressure sensor (depth), a thermal probe (sea surface temperature), conductivity probe (salinity), and a WetLabs ECO-BB2F combination scattering meter and fluorometer.

At each station, the profiler was deployed to a depth of 1 m on the sunny side of the sampling vessel to avoid shading of the instrument. The profiler package was allowed to drift on a tether approximately 15 m aft of the stern in order to sample waters unaffected by the presence of the vessel. During the deployment, the profiler collected *in situ* upwelling radiance (L_u) using a Satlantic HyperOCR-I (in-water) hyperspectral digital radiometer. Downwelling irradiance (E_d) measurements were collected onboard the vessel using a Satlantic HyperOCR-I (in-air) hyperspectral digital radiometer. Spectral data were collected at a resolution of 2 nm from 350 to 800 nm. Sampling occurred between 10:00 and 14:00 hours with an average time on

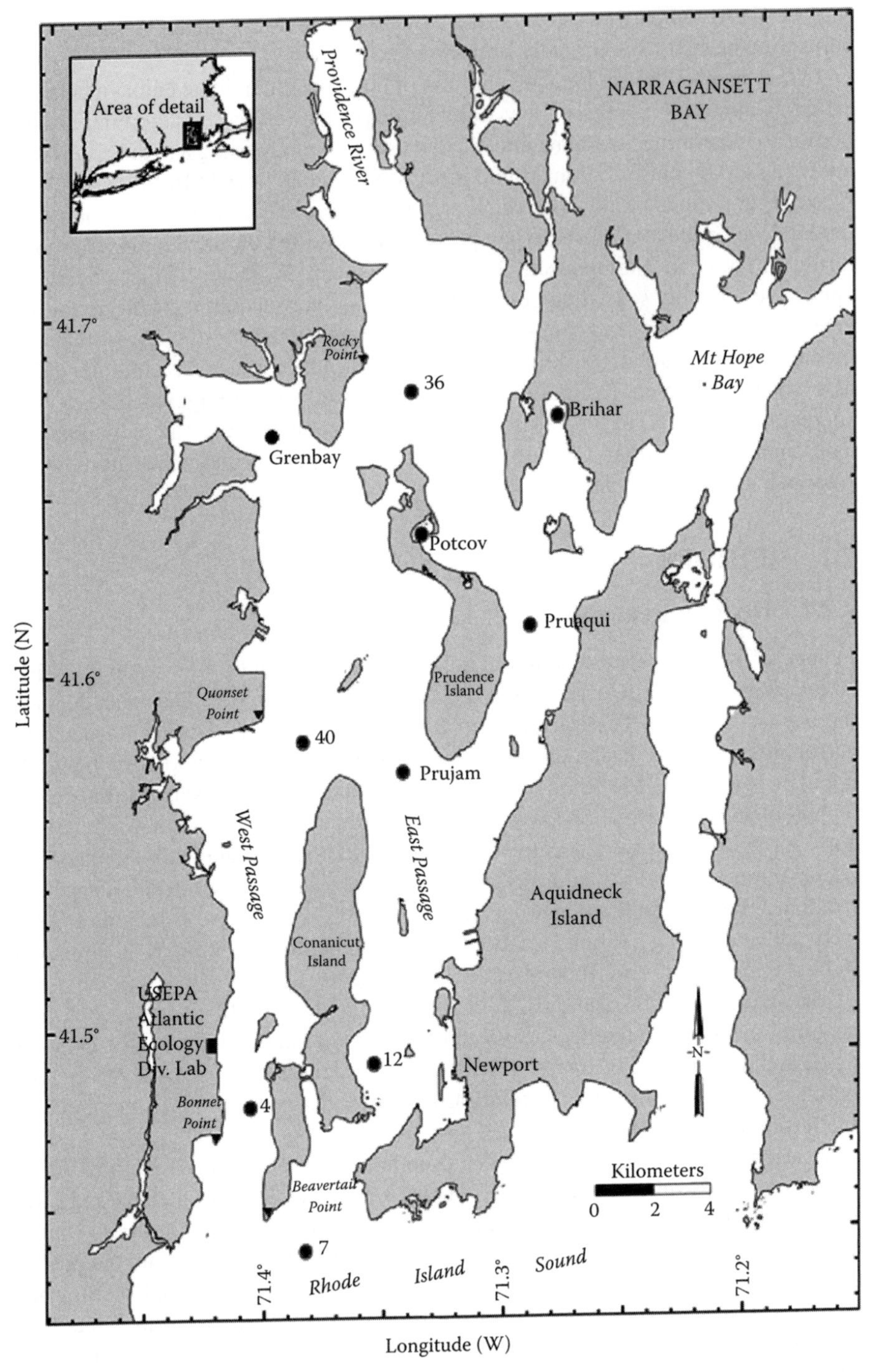

FIGURE 7.1 Location of sampling stations in Narragansett Bay used to collect *in situ* hyperspectral data and water samples for band ratio model development.

station of approximately 10 min. Data from each sensor were routed through a USB hub for input and storage on a laptop computer.

Concurrent with the hyperspectral data collection, seawater samples were taken at each station with a 5-l Go-Flo bottle lowered to the same depth as the Ocean Profiler system. Samples were stored on ice in the field and later analyzed for chl *a* concentrations in a laboratory setting using high-performance liquid chromatography (HPLC) methodology [17]. The HPLC chl *a* concentrations and the ECO-BB2F fluorometric estimates were used to groundtruth algorithm-derived chlorophyll estimates.

7.2.2 Development of a Regionally Tuned, Band Ratio Model for the Estimation of Chl *a* in Southern New England Waters

After each cruise, the hyperspectral data were processed using the Satlantic ProSoft data analysis package (version 7.7). At each wavelength (λ), remote sensing reflectances (R_{rs}) were derived from the ratio of the upwelling spectral radiance propagated through the sea surface upward radiance (L_w) to downward irradiance (E_d) using

$$R_{rs}(0^+; \lambda) = \frac{L_w(\lambda)}{E_d(\lambda)}. \tag{7.1}$$

$E_d(\lambda)$ is the downwelling spectral irradiance measured just above the sea surface. $L_w(\lambda)$ is expressed as

$$L_w(\lambda) = L_u(0^-; \lambda)[1 - \Delta(\lambda, \theta)/n_w^2(\lambda)], \tag{7.2}$$

where $L_u(0^-; \lambda)$ is the upwelling spectral radiance just below the sea surface, $\Delta(\lambda, \theta)$ is the Fresnel reflectance of seawater, and $n_w^2(\lambda)$ is the refractive index of seawater. Radiance was measured in units of $\mu W\ cm^{-2}\ nm^{-1}\ sr^{-1}$.

In Phase I, simple band ratio algorithms were developed by regressing the *in situ* remotely sensed reflectances from the red portion of the visible spectrum against the *in situ* chl *a* concentrations measured from Narragansett Bay and Rhode Island Sound. Conceptually, these algorithms use the spectral range that shows the maximum sensitivity to changes in chlorophyll concentration and the range with the minimum sensitivity to variation in chlorophyll concentration.

In order to test the ability of these models to accurately estimate chl *a* concentrations in waters of southern New England, an independent dataset with hyperspectral reflectance and *in situ* fluorometer data was collected from Narragansett and Greenwich Bays, Long Island Sound, Buzzards Bay, and the Providence River. These data were acquired during June 2005; March, September, and October 2006; May and July 2007; and May 2008. The proportion of variation explained by each model was estimated by calculating the coefficient of determination (R^2). The accuracy of the model was determined from the slope of the linear equation relating estimated and measured chl *a* concentrations, the Pearson correlation (r), and the root-mean-square error (RMSE) between predicted and measured chl *a* values. The slope was calculated using the Major Axis method of Model II linear regression analysis.

Model II linear regression was considered appropriate because the goal of the analysis was to estimate the functional relationship between the predicted and measured chl *a* concentrations and not to predict the measured concentrations as a function of the predicted chlorophyll values [18]. The linearity of the relationship between measured and predicted concentrations was calculated from the Pearson moment correlation coefficient. The RMSE was calculated from [19]

$$\text{RMSE} = \left[E \frac{(\text{chlorophyll } a \text{ predicted} - \text{chlorophyll } a \text{ measured})^2}{N - 1} \right]^{0.5}, \quad (7.3)$$

where N is the number of samples.

7.2.3 Application of the Chl *a* Model to Hyperspectral Aircraft Data Collected from Southern New England Coastal Waters

Multispectral and hyperspectral systems have been flown on low-flying aircraft to collect spectral data to estimate and map the distribution of chl *a* concentrations in several freshwater and estuarine systems and the coastal ocean at high-resolution spatial scales over long time periods [20–24]. In Phase II of this study, hyperspectral data were passively acquired during monthly surveys of 26 estuaries, bays, and coastal waters along the southern New England coast over one year (Figure 7.2). These surveys provided near-synoptic views of the distribution of surface chl *a* concentrations along coastal southern New England as well as provided data from bays and estuaries whose spatial dimensions are below the resolution capability of ocean color satellite sensors. The hyperspectral data were collected in a "pushbroom" scanning method with Satlantic HyperOCR digital sensors mounted on a low-flying (300 m altitude)

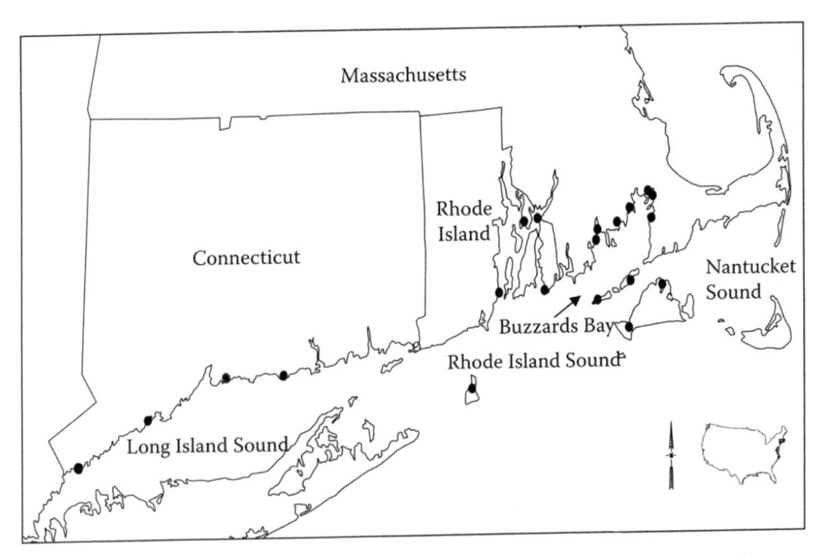

FIGURE 7.2 Location of estuaries and bays along the southern New England coast surveyed using hyperspectral aircraft remote sensing.

Cessna Skymaster. Pushbroom scanning is the technique of using the forward motion of the aircraft or sensing platform to sweep the detectors across the ground or sea surface. These sensors acquired total radiance from the sea surface [$L_t(\lambda, h)$] as a function of altitude (h), sky radiance (L_{SKY}), and downward irradiance just above the sea surface [$E_d(\lambda, 0+)$] from 350 to 800 nm with an 11.5° field of view. Flights were usually conducted between 9:00 and 15:00 hours and usually completed within 4–6 h of takeoff. Because the length of the solar day varied with season, flight time was adjusted on a monthly basis. Surveys were flown at an airspeed of approximately 100 knots, during clear sky conditions (visibility 5–10 miles with less than 30% cloud cover) with wind speeds less than 8 m/sec and wave heights less than 1 m. With a field of view of 11.5° at an altitude of 300 m, the instantaneous field of view of the sensors (or the pixel scale) had an area of approximately 60 m^2.

A simple radiative transfer equation that expressed the pathway that light traveled from the sea surface to the aircraft is [25]

$$L_t(\lambda, \theta_0, h) = L_P(\lambda, \theta_0, h) + t(\lambda, \theta_0, h)[L_G^{sun}(\lambda, \theta_0, h) + L_G^{sky}(\lambda, \theta_0, h) + L_w(\lambda, \theta_0)], \tag{7.4}$$

where L_t is the total spectral radiance reaching the sensor at a wavelength (λ) and at the altitude (h) of the aircraft's radiometer, θ_0 is the solar zenith angle, t is the atmospheric transmittance from the surface to h, L_w is water-leaving radiance, and L_G^{sun} and L_G^{sky} are surface sun and sky glint, respectively. L_P is the atmospheric path radiance and is expressed as

$$L_P(\lambda, \theta_0, h) = L_R(\lambda, \theta, h) + L_A(\lambda, \theta, h), \tag{7.5}$$

where

$$L_R(\lambda, \theta_0, h) = \frac{E_{sun}(\lambda, h)t(\lambda, \theta, h)\tau_R(\lambda, h)p_R(\theta)}{\cos\theta}, \tag{7.6}$$

$$L_A(\lambda, \theta_0, h) = \frac{E_{sun}(\lambda, h)t(\lambda, \theta, h)\tau_A(\lambda, h)p_A(\theta)}{\cos\theta}, \tag{7.7}$$

and E_{sun} is the direct solar irradiance incident at altitude h, τ_R and τ_A are the Rayleigh and aerosol optical thicknesses, θ is the viewing nadir angle, and $p_R(\theta)$ and $p_A(\theta)$ are the Rayleigh scattering and aerosol phase functions and are assumed to be unity.

Several important simplifications may be made for t and L_P:

1. For low altitudes (up to 300 m), the intervening atmospheric layer can be considered transparent [26] and, hence, $t(\lambda, \theta_0, h)$ can be taken as unity.
2. For these altitudes, aerosol content in the atmosphere is low [25,27,28] and, hence, $L_A(\lambda, \theta_0, h)$ may be assumed to be zero.

In addition, Rayleigh optical thickness (τ_R) is expressed as

$$\tau_R(\lambda, h) = 0.0088\lambda^{-4.15+0.2\lambda}[1 - \exp(-0.1188h - 0.0011h^2)]. \tag{7.8}$$

After these simplifications, Equation 7.4 may be rewritten as

$$L_t(\lambda, \theta_0, h) = L_R(\lambda, \theta_0, h) + L_G^{sun}(\lambda, \theta_0, h) + L_G^{sky}(\lambda, \theta_0, h) + L_w(\lambda, \theta_0). \quad (7.9)$$

L_G^{sun} values are identified in the remotely sensed data by an increase in radiance at all λs and are commonly filtered out during data processing. Rewriting Equation 7.8 yields

$$L_t(\lambda, \theta_0, h) = L_R(\lambda, \theta_0, h) + L_G^{sky}(\lambda, \theta_0, h) + L_w(\lambda, \theta_0). \quad (7.10)$$

L_G^{sky} is expressed as [25]

$$L_G^{sky} = \rho(0; W)L^{sky}(\lambda, \pi), \quad (7.11)$$

where $\rho(0; W)$ is surface reflectance at wind speed (W) [29] and $L^{sky}(\lambda, \pi)$ is the zenith sky radiance.

After calculation of $L_R(\lambda, \theta_0, h)$ and L_G^{sky} by Equations 7.6 and 7.10, water-leaving radiance spectra were calculated as

$$L_w(\lambda) = L_t(\lambda, \theta_0, h) - L_R(\lambda, \theta_0, h) - L_G^{sky}(\lambda). \quad (7.12)$$

$R_{rs}(\lambda)$ values were determined after each flight from the ratio of the water-leaving radiance $L_w(\lambda)$ to downwelling irradiance just above the sea surface $E_d(\lambda, 0+)$ using a data extraction and processing routine written in MATLAB, version 7.3.

7.2.4 Estimating Chlorophyll Conditions

In Phase III, chlorophyll condition assessments were conducted in a multistep process at site-specific and regional scales as well as over several seasons. Initially, after each flight, the spatial distribution of surface chl *a* concentrations along individual flight-lines was determined using the band ratio model and hyperspectral data. The surface chlorophyll concentrations along each flightline were translated into an assessment of condition for an area using the EPA/NCA criteria for chl *a* (Table 7.1). The criteria used in this study were designated for East/West coast and Gulf of Mexico areas [14]. These criteria classify the chlorophyll status of a water body as Good/Fair/Poor based on respective surface chl *a* thresholds (<5, 5–20, and >20 μg/L). These thresholds are modified for Hawaii, Puerto Rico, and Florida Bay in response to state criteria. By aggregating the area-specific chl *a* and condition assessments for the 2006 survey period, conditions can be assessed over an annual scale (i.e., annual site-specific assessments). The site-specific assessments could be aggregated to give a chlorophyll condition assessment (i.e., a regional assessment) for the southern New England region.

The temporal variation in the distribution of chlorophyll condition on a regional scale was also determined by aggregating the distribution of individual condition assessments on a monthly basis (i.e., monthly assessments of condition). The distribution of chlorophyll condition on a seasonal basis was created by combining the monthly assessments as a function of winter, summer, and fall seasons during 2006.

TABLE 7.1
Comparison of Band Ratio Combinations to Measured Chl *a* Values

Band Ratio Combination and Reference	Goodness of Fit (R^2) between Measured Chl *a* Concentrations and Band Ratio	Goodness of Fit (R^2) between Measured and Predicted Chl *a* Concentrations	RMSE (μg/L)
678/667 [50]	0.94	0.97	1.13
680/670 [48]	0.86	0.79	1.87
690/675 [51]	0.72	0.42	2.58
702/665 [31]	0.77	0.80	0.36
705/665 [31]	0.75	0.84	1.82
705/670 [52]	0.71	0.77	2.12

7.3 RESULTS

7.3.1 Hyperspectral Signature of Narragansett Bay Waters

A composite of reflectance spectra (spectral range from 350 to 800 nm) acquired from Narragansett Bay and Rhode Island Sound shows that these waters are characterized by strong absorption of light by colored dissolved organic matter (CDOM) and significant light scattering by algal cells and nonalgal particles [30] (Figure 7.3). These spectra were quite similar to reflectance spectra collected from turbid,

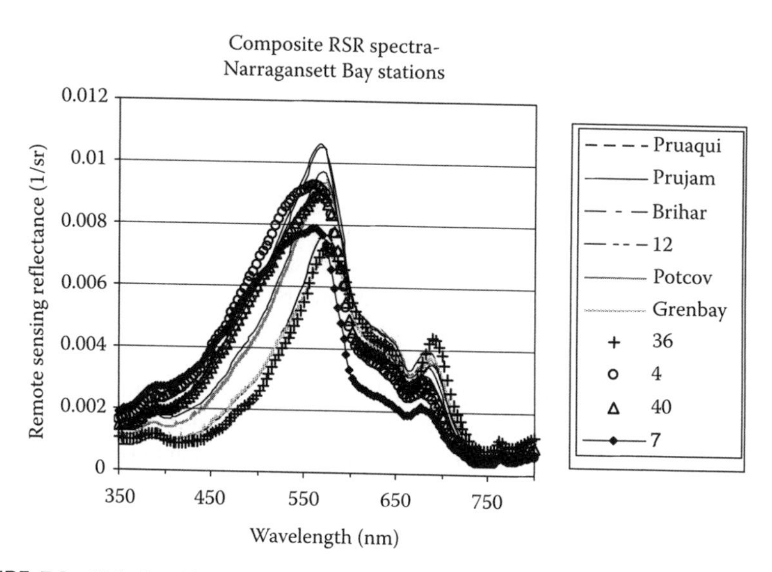

FIGURE 7.3 Relationship between chl *a* concentrations and R_{rs678}/R_{rs667} in Narragansett Bay, Rhode Island.

productive coastal waters as well as lacustrine systems [31–34]. These data also show a prominent peak around 700 nm whose position shifts to longer wavelengths with increased reflectance (Figure 7.3). The shift in reflectance peak position toward longer wavelengths has been found to coincide with increasing phytoplankton density and chl *a* concentrations [34,35]. Previous researchers have also suggested that the position of the peak represents the occurrence of chlorophyll fluorescence [36–38] and minimal absorption by constituents in the water combined with dominance by scattering due to inorganic and nonliving organic suspended matter [39–41].

7.3.2 Development of a Simple Band Ratio Algorithm for Determining Chl *a* Concentrations

The algorithms used to remotely monitor chl *a* concentrations in ocean waters are not valid in coastal waters due to the diversity of optically active constituents (e.g., CDOM and suspended sediments), which mask fundamental phytoplankton absorption and scattering relationships [42–46]. The complex interaction and highly variable temporal and spatial nature of these components have led to the development of regionally focused algorithms that establish a unique relationship between the remotely sensed signal and the variable to be estimated [47]. While chlorophylls and related compounds strongly absorb light in the blue portion of the spectrum, this characteristic is masked by the stronger absorption of light by CDOM, which makes it difficult to use this spectral region to estimate photosynthetic pigment concentrations. Studies of productive marine and freshwater ecosystems have demonstrated that light reflectances in the red and near-infrared (NIR) region can be strongly correlated with laboratory-measured chl *a* concentrations (correlation coefficients generally exceed 0.90) [21,35,48,49]. In freshwater and estuarine environments, the reflectance peak near 700 nm was found to be very sensitive to changes in chl *a* concentrations, while the reflectance at 670 nm was the least sensitive to changes in chl *a* concentrations [32].

For a northeastern U.S. estuarine system, Szekielda [50] found that a strong correlation also existed between the ratio of remotely sensed reflectances at 667 and 678 nm and chl *a* concentrations (over the range of approximately 1–106 mg/m^3) during noneutrophic to eutrophic conditions in the Peconic River, New York and in the Peconic Bay in Long Island Sound. When measured chl *a* data are plotted against R_{rs678}/R_{rs667} data ($n = 6$ stations) from the Peconic Bay estuary, the resulting scatter diagram was best fit using a power relationship that yielded an R^2 value of 0.98 (Figure 7.4).

Using this approach for Narragansett Bay, the R_{rs678}/R_{rs667} data, as well as data from other red/NIR band combinations, were plotted against *in situ* chl *a* concentrations ($n = 32$–35 sampling events) to determine the correlation between the spectral values and phytoplankton abundance in these waters. The results were very similar to the Peconic Bay study as the data for each band combination were best fitted using a power relationship that accounted for 71–94% of the variance in the data (Table 7.1). Comparisons were also made between predicted and measured chl *a* concentrations for each band combination as well as associated RMSEs. These results indicated that

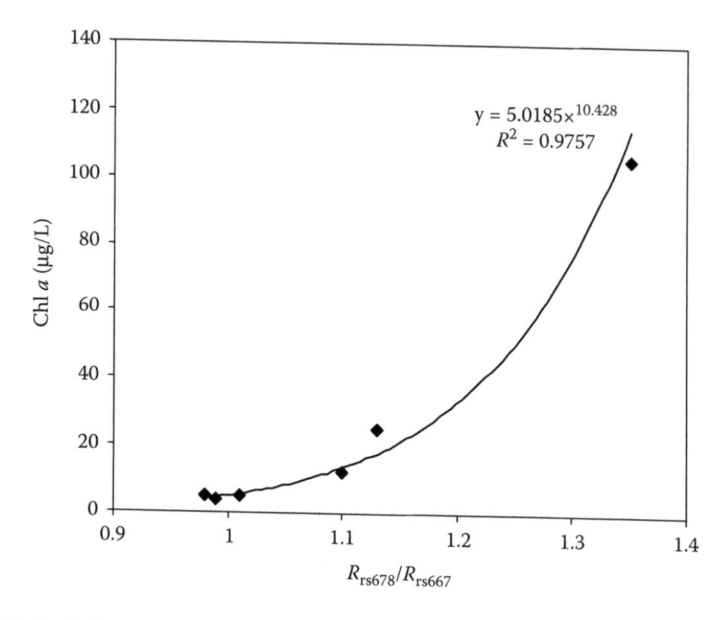

FIGURE 7.4 Relationship between chl *a* concentrations and R_{rs678}/R_{rs667} in the Peconic Bay estuary, Long Island. (Data from Szekielda, K.-H., *EARSeL eProceedings, European Association of Remote Sensing Laboratories*, 3, 261, 2004.)

the 678/667 band ratio is the preferred combination for southern New England waters because it had the highest R^2 value of any band combination for relating chl *a* concentrations to spectral ratios. The 678/667 band ratio also had the highest R^2 value when predicted chl *a* concentrations were plotted against measured chl *a* concentrations (Table 7.1). The 678/667 band also had one of the lowest RMSE values (Table 7.1).

The 678/667 band relationship is expressed (Figure 7.5) as

$$\text{Chl } a \text{ } (\mu\text{g/L}) = 1.919 \times \left(\frac{R_{rs678}}{R_{rs667}}\right)^{7.63}. \tag{7.13}$$

Model II linear regression of a scatterplot of 678/667 band-derived chlorophyll and measured chlorophyll indicated a linear slope for this relationship ($m = 1.1$) with a zero intercept ($b = -0.2$). The Pearson moment correlation coefficient (r) was 0.98. The difference between measured and predicted chl *a* concentrations resulted in an RMSE equal to 1.1 µg/L (Figure 7.6).

To test if the power relationship observed in Narragansett Bay varied seasonally, a scatter diagram was created of the R_{rs678}/R_{rs667} ratio and measured chl *a* data for available spring, summer, and fall cruises from 2005 to 2008. The results show that the 678/667 band combination accounted for approximately 85% of the variance in the spring, 98% of the variance in summer, and 90% of the variance in fall datasets (Figure 7.7).

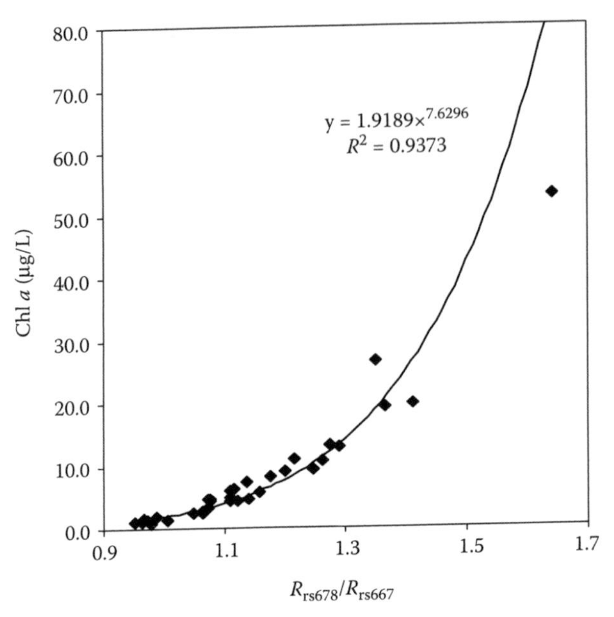

FIGURE 7.5 Relationship between chl *a* concentrations and R_{rs678}/R_{rs667} in the Narragansett Bay estuary, Rhode Island.

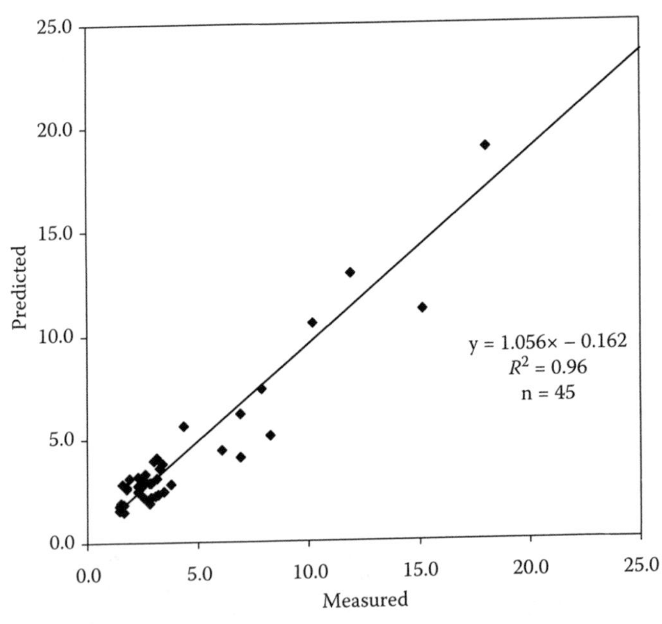

FIGURE 7.6 Model-derived chl *a* concentrations plotted against measured chl *a* from Narragansett Bay (Rhode Island), Buzzards Bay (Massachusetts), and Long Island Sound stations.

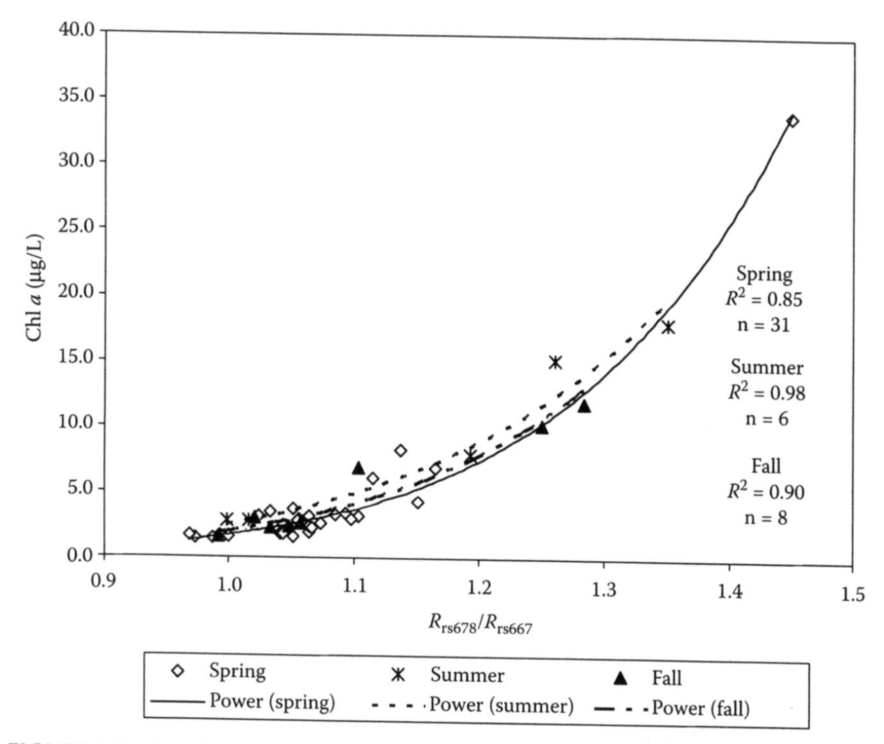

FIGURE 7.7 Relationship between chl *a* concentration and R_{rs678}/R_{rs667} as a function of season.

7.3.3 Assessment of Chlorophyll Condition in Southern New England Waters

During each flight, hyperspectral data were usually collected from 8–16 estuarine and coastal systems. These data were used to estimate, from the 678/667 band ratio model, the range of chl *a* concentrations along each flightline. The reliability of the remotely acquired data was assessed by comparing predicted chl *a* concentrations calculated from above-water reflectance measurements acquired during flights in summer 2006 with in-water spectral measurements and measured chlorophyll values collected during spring, summer, and fall 2006 field sampling (Figure 7.8). The resulting scatterplot showed close agreement and overlap between the datasets.

Generally, the estimated chl *a* concentrations showed the spatially variable nature of chl *a* concentrations characteristic of coastal systems (see Figure 7.9 for example). These data indicated a general trend in pigment concentration, with low concentrations usually at the seaward boundary of a system, which increased in an onshore direction. In addition, the data identified isolated phytoplankton blooms of varying spatial scales and concentrations along each flightline, which naturally occur or are generated in response to increased nutrient loadings. During this study, chl *a* concentrations ranged from less than 5.0 to greater than 20 μg/L. Highest concentrations

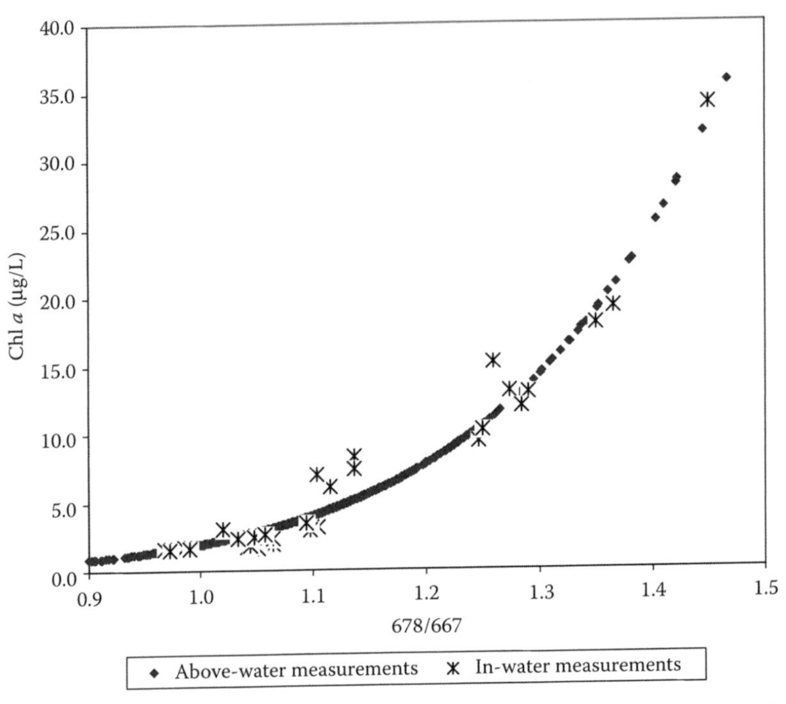

FIGURE 7.8 Comparison of 678/667 ratio above-water measurements and predicted chl *a* concentrations with in-water measurements and measured chl *a* concentrations.

(>20 μg/L) occurred in Mt. Hope Bay, Rhode Island (located in the northeast corner of the greater Narragansett Bay estuary) during September 2006.

Using NCA Site Criteria (Table 7.2), the chl *a* concentrations along each flight survey line were transformed into assessments of environmental condition. For a single system, these assessments would spatially range from Good to Poor reflecting the variability of chl *a* concentrations at that particular flight time. These condition assessments were aggregated for each system surveyed during 2006 to illustrate the variability in chl *a* condition over an annual period (Figure 7.10). The results showed that all 26 coastal systems along the southern New England coast were categorized as in good condition at least 50–100% of the survey time. Twenty-three of those systems were categorized as in fair condition from less than 5–50% of the time. Only one system (Mt. Hope Bay, RI) was classified as poor less than 1% of the time.

The annual site assessments were combined to illustrate chlorophyll condition on a regional scale (Figure 7.11). These data show that regionally the majority (80%) of the estuarine and coastal waters of southern New England were classified as good. Approximately 20% of the region's waters were classified as fair and a very small percentage (<0.1%) were categorized as poor. Based on the EPA/NCA criteria, the southern New England region is assessed in good condition because less than 10% of the coastal area is assessed in poor condition and more than 50% is assessed in good condition.

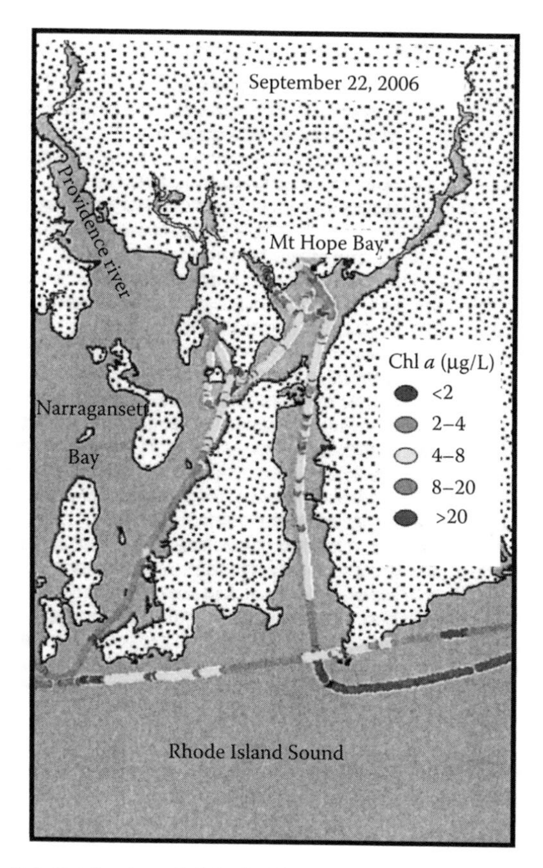

FIGURE 7.9 Spatial distribution of chl *a* in Mt. Hope Bay and Rhode Island Sound during September 22, 2006.

The chl *a* concentrations for all the survey sites were aggregated for each month between January and December 2006 to assess regional variability. These assessments indicated that chlorophyll condition at the regional scale generally ranged from fair to good. During January 2006, 50% of the survey sites were assessed as in either good or fair condition, which may reflect the presence of abundant winter

TABLE 7.2
U.S. EPA/NCA Criteria for Assessing Chl *a* Condition

Condition	Site Criteria	Regional Assessment
Good	<5 µg/L	<10% of coastal area in poor condition and >50% in good condition
Fair	5–20 µg/L	10–20% of the coastal area in poor condition or >50% in combined poor and fair condition
Poor	>20 µg/L	>20% of coastal area in poor condition

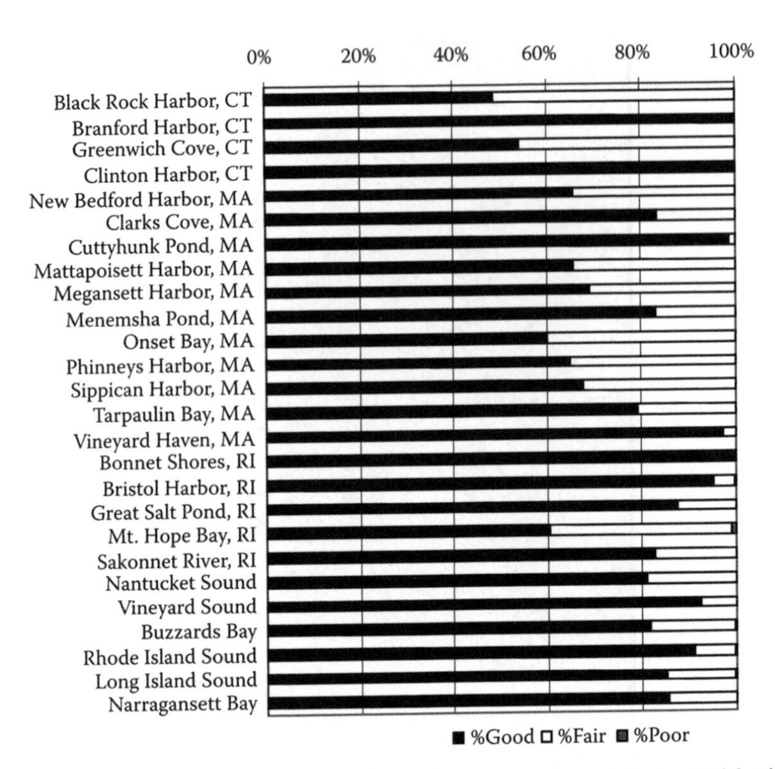

FIGURE 7.10 Site-specific assessments for each system surveyed during 2006 with the percentage of time the system was assessed at a particular condition.

phytoplankton blooms in local waters. From February to March, the percentage of sites assessed in good condition continued to increase to 80% of the survey sites as chl *a* concentrations declined. No flights were conducted during April and August because of weather-related conditions. The highest percentage of sites assessed in good condition occurred during June 2006 with 96% of the area surveys having

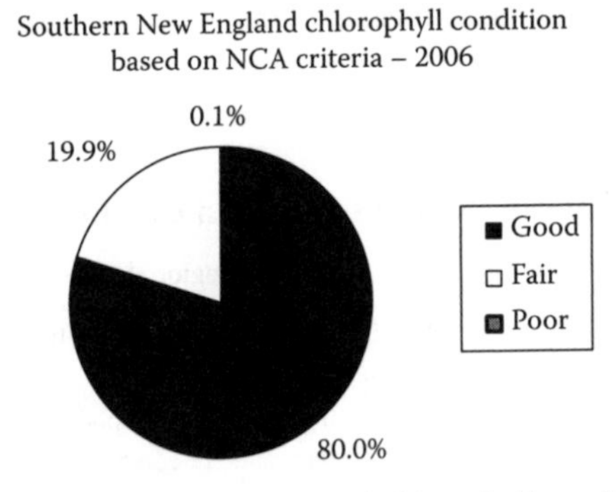

FIGURE 7.11 2006 regional scale assessment of chlorophyll condition for southern New England.

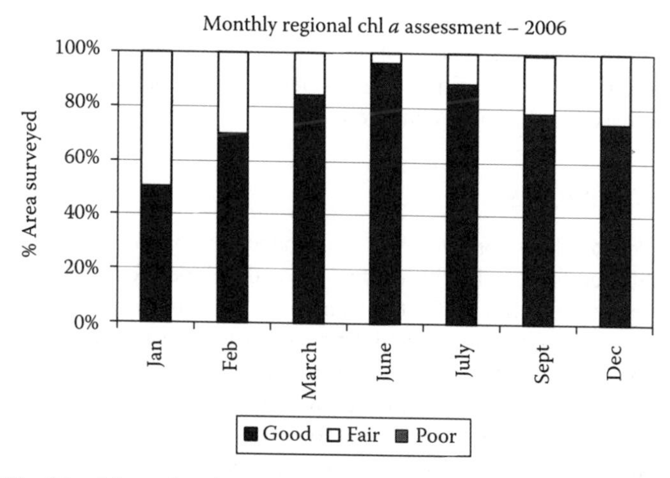

FIGURE 7.12 Monthly, regional scale chl *a* assessment of southern New England estuaries and coastal waters during 2006.

average chl *a* concentrations less than 5 μg/L. Four percent of the sites were assessed as fair and no sites were assessed as poor. After June, episodic and localized phytoplankton blooms during the summer season increased the percentage of fair assessments from 4% to 22% of the survey area and increased the amount of poor assessment to less than 1%. From September to December 2006, the percentage of sites assessed in fair condition continued to increase from 22% to 26% (Figure 7.12). During this period, less than 1% of the sites were assessed in poor condition.

The monthly assessments were combined to illustrate changes in condition as a function of season (Figure 7.13). At the seasonal level, assessments of condition showed a rhythmic nature that may reflect regional scale variability. The data show that the percentage of sites assessed in good condition was lowest during the winter season (68% of sites surveyed), highest during the summer season (88%), and

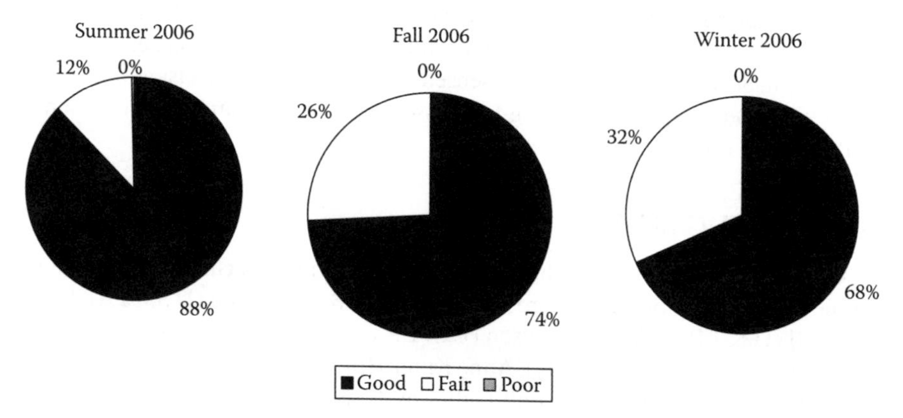

FIGURE 7.13 Assessment of chlorophyll condition for southern New England estuaries and coastal waters on a seasonal basis.

declined during the fall season to 74% of the sites surveyed. The percentage of sites assessed in fair condition was highest in the winter (32%) and lowest in the summer (12%). During winter and summer seasons, less than 1% of southern New England estuaries were assessed in poor condition.

7.4 CONCLUSIONS

The estimation of chl *a* concentrations in productive coastal and estuarine waters requires a different set of algorithms than those developed for ocean waters. Using the unique spectrum of light reflected from waters of estuaries and coastal systems along southern New England, a band ratio model was derived that retrieved chl *a* concentrations using spectral information from the red portion (600–700 nm) of the visible spectrum. These data were then used to monitor the temporal trends in chl *a* concentration and its spatial variability for 26 coastal locations in Connecticut, Rhode Island and Massachusetts. Using the EPA/NCA framework for chlorophyll, these estimates of phytoplankton concentrations were used to determine the environmental condition of these waters at regional and site-specific scales. Regionally, almost 80% of all sites along southern New England were categorized as in good condition for 2006. Site-specific assessments showed that chl *a* in any one coastal system might vary from 50% to 100% good condition. The data also suggested that seasonal variability is a factor since the highest percentage of sites categorized as in good condition occurred during the summer season with the lowest percentage occurring during the winter season, as phytoplankton concentrations increase. This chlorophyll increase, usually associated with diatom blooms, occurs during the winter/spring season as part of the seasonal productivity cycle that characterizes many New England estuaries and coastal areas.

Aircraft remote sensing provided near-synoptic, regional views of chlorophyll condition for coastal southern New England. These assessments would have been sample intensive and expensive if conducted over an annual period using traditional field-based monitoring. Routine aircraft-based monitoring resulted in surveys of numerous bays and estuaries in this region that have been excluded from satellite-based monitoring programs because their spatial dimensions may be smaller than the resolution capability of ocean color satellite sensors. From a management standpoint, the inclusion of these remotely sensed data into the EPA decision framework added value to the regional and long-term assessment of environmental condition of southern New England coastal waters.

ACKNOWLEDGMENTS

I acknowledge everyone who helped in the data collection and chl *a* analysis including the late John Ambroult (Ambroult Aviation), Barbara Bergen (U.S. EPA), Laura Winward (U.S. EPA), William Nelson (U.S. EPA), Saro Jayaraman (U.S. EPA), Ivy Ozmon (Old Dominion University), Anne Kuhn-Hines (U.S. EPA), and Donald Cobb (U.S. EPA). I especially thank Walt Galloway (U.S. EPA) and Marilyn ten Brink (U.S. EPA) for their support of this research. I also thank Dorsey Worthy (U.S. EPA), Walt Galloway (U.S. EPA), and Leonid Sokoletsky (U.S. EPA) for their review of the manuscript. I especially thank Larry Harding (UMCES-Horn Pt Laboratory), Leonid Sokoletsky (U.S. EPA), and John Schalles (Creighton University) for valuable discussions and information on the behavior of light in turbid lake and coastal systems and the retrieval of chl *a* concentrations from the spectral character of these waters. Although the research described in this chapter has been funded wholly (or in part) by the U.S. EPA under the NCA and Aquatic Stressors Programs, it has not been subjected to Agency-level review. Therefore, it does not necessarily reflect the views of the Agency. This is contribution AED-09-001 of the Atlantic Ecology Division, National Health and Environmental Effects Research Laboratory, Office of Research and Development, U.S. EPA. Mention of trade names or commercial products does not constitute endorsement or recommendation for use.

REFERENCES

1. Kelly, J.R., Nitrogen Effects on Coastal Marine Ecosystems, in R.F. Folett and J.L. Hatfield (Eds), *Nitrogen in the Environment: Sources, Problems, and Management*, Chap. 9, pp. 207–251, Elsevier Science, Amsterdam, 2001.
2. Rowan, K.S., *Photosynthetic Pigments of Algae*, Cambridge University Press, Cambridge, 1989.
3. Jordan, T.E., Cornell, D.L., Maklas, J., and Weller, D.E., Long-term trends in estuarine nutrients and chlorophyll, and short-term effects of variation in watershed discharge, *Marine Ecology Programming Series*, 75, 121, 1991.
4. Morrow, J.H., White, B.N., Chimiente, M., and Hubler, S., A bio-optical approach to reservoir monitoring in Los Angeles, *Limnology and Lake Management—Archive für Hydrobiologie Special Issues, Advances in Limnology*, 55, 171, 2000.
5. Casazza, G., Silvestri, C., and Spada, E., Classification of coastal waters according to the new Italian water legislation and comparison to the European Water Directive, *Journal of Coastal Classification*, 9, 65, 2003.
6. Gohin, F., Saulquin, B., Oger-Jeanneret, H., Lozac'h, L., Lampert, L., Lefebvre, A., Riou, P., and Bruchon, F., Towards a better assessment of the ecological status of coastal waters using satellite-derived chlorophyll-*a* concentrations, *Remote Sensing of Environment*, 112, 3329, 2008.
7. Paerl, H.W., Nuisance phytoplankton blooms in coastal, estuarine, and inland waters, *Limnology and Oceanography*, 33, 823, 1988.
8. Richardson, L.L., Remote sensing of algal bloom dynamics, *Bioscience*, 46, 492, 1996.
9. Kahru, M. and Mitchell, B.G., Spectral reflectance and absorption of a massive red tide off southern California, *Journal of Geophysical Research*, 103(21),601, 1998.
10. Pettersson, L.H., Durand, I., Johannessen, O.M., Svendsen, E., and Soiland, H., Satellite observations and predictions of toxic algae blooms in coastal waters, *Proceedings of the Sixth International Conference on Remote Sensing for Coastal and Marine Environments*, 1, 48, 2000.
11. Cole, B.E. and Cloern, J.E., An empirical model for estimating phytoplankton productivity in estuaries, *Marine Ecology Programming Series*, 36, 299, 1987.
12. Gallegos, C.L. and Jordan, T.E., Impact of the spring 2000 phytoplankton bloom in Chesapeake Bay on optical properties and light penetration in the Rhode River, Maryland, *Estuaries*, 25, 508, 2002.
13. Lefevre, N., Taylor, A.H., Gilbert, F.J., and Geider, R.J., Modeling carbon to nitrogen and carbon to chlorophyll *a* ratios in the ocean at low altitudes: Evaluation of the role of physiological plasticity, *Limnology and Oceanography*, 48, 1796, 2003.

14. USEPA, National Coastal Condition Report II. Office of Research and Development/ Office of Water, EPA-620/R-03/002, Washington, DC, 2004.
15. Kremer, J. and Nixon, S.W., *A Coastal Marine Ecosystem*, Springer, New York, Berlin, Heidelberg, 1978.
16. Chinman, R. and Nixon, S.W., Depth–Area–Volume relationships in Narragansett Bay, Graduate School of Oceanography, the University of Rhode Island, NOAA/Sea Grant Marine Technical Report, p. 87, 1985.
17. Wright, S.W., Jeffrey, S.W., Mantoura, R.F.C., Llewellyn, C.A., Bjornland, T., Repeta, D., and Welschmeyer, N., Improved HPLC method for the analysis of chlorophylls and carotenoids from marine phytoplankton, *Marine Ecology Programming Series*, 77, 183, 1991.
18. Laws, E., *Mathematical Methods for Oceanographers*, Wiley, New York, 1997.
19. O'Reilly, J., Maritorena, S., Siegel, D., O'Brien, M.C., Toole, D., Mitchell, B.G., Kahru, M., [21 co-authors]. Ocean color chlorophyll *a* algorithms for SeaWiFS, OC2 and OC4: Version 4, in S.B. Hooker and E.R. Firestone (Eds), *SeaWiFS Postlaunch Calibration and Validation Analyses, Part 3*, Vol. 22, SeaWiFS Postlaunch Technical Report Series, NASA/TM-2000-206892, 2000.
20. Harding, L.W., Jr., Itsweire, E.C., and Esaias, W.E., Determination of phytoplankton chlorophyll concentrations in the Chesapeake Bay with aircraft remote sensing, *Remote Sensing of Environment*, 40, 79, 1992.
21. Schalles, J.F., Gitelson, A.A., Yacobi, Y.Z., and Kroenke, A.E., Chlorophyll estimation using whole seasonal, remotely sensed high spectral-resolution data for a eutrophic lake, *Journal of Phycology*, 34, 383, 1998.
22. Keith, D., *Determination of Chlorophyll a Concentrations and Phytoplankton Primary Production in New England Waters Using Ocean Color Remote Sensing from Low-Flying Aircraft*, PhD dissertation, University of Rhode Island, 2004.
23. Montes-Hugo, M.A., Carder, K.L., Foy, R., Cannizzaro, J., Brown, E., and Pegau, S., Estimating phytoplankton biomass in coastal waters of Alaska using airborne remote sensing, *Remote Sensing of Environment*, 98, 481, 2005.
24. Churnside, J.H. and Wilson, J.J., Ocean color inferred from radiometers on low-flying aircraft, *Sensors 2008*, 8, 860, 2008.
25. Lazin, G., Harding, L.W., and McLean, S., Ocean color radiometry from aircraft: I. Low altitude measurements from light aircraft, in J.L. Mueller, G.S. Fargion, and C.R. McClain (Eds), *Ocean Optics Protocols for Satellite Ocean Color Sensor Validation, Revision 4, Volume VI: Special Topics in Ocean Optics Protocols and Appendices*, Chap. 4, p. 79, NASA/TM-2003-211621/Rev4-Vol.VI, 2003.
26. Lazin, G., *Correction Methods for Low Altitude Remote Sensing of Ocean Color*, Master of Science thesis, Dalhousie University, 1998.
27. Zibordi, G. and Maracci, G., Determination of atmospheric turbidity from remotely-sensed data: A case study, *International Journal of Remote Sensing*, 9, 1881, 1988.
28. Lazin, G., Davis, R.F., Ciotti, A.M., and Lewis, M.R., Ocean color measurements from low-flying aircraft: Atmospheric and surface glint correction, in S.G. Ackleson and R. Frouin (Eds), *Ocean Optics XIII*, Vol. 2963, p. 703, 1996.
29. Mobley, C.D., Estimation of remote sensing reflectance from above-surface measurements, *Applied Optics*, 38, 7442, 1999.
30. Keith, D.J., Yoder, J.A., and Freeman, S.A., Spatial and temporal distribution of coloured dissolved organic matter (CDOM) in Narragansett Bay, Rhode Island: Implications for phytoplankton in coastal waters, *Estuarine, Coastal and Shelf Science*, 55, 705, 2002.
31. Dall'Olmo, G. and Gitelson, A., Effect of bio-optical parameter variability on the remote estimation of chlorophyll-*a* concentration in turbid productive waters: Experimental results, *Applied Optics*, 44, 412, 2005.

32. Gitelson, A.A., Yacobi, Y.Z., Schalles, J., Rundquist, D.C., Han, L., and Stark, R., Remote estimation of phytoplankton density in productive waters, *Archiv für Hydrobiologie Special Issues Advance Limnology*, 55, 121, 2000.
33. Lee, Z., Carder, K.L., Hawes, S.K., Steward, R.G., Peacock, T.G., and Davis, C.O., Model for the interpretation of hyperspectral remote-sensing reflectance, *Applied Optics*, 33, 5721, 1994.
34. Schalles, J.F., Optical remote sensing techniques to estimate phytoplankton chlorophyll *a* concentrations in coastal waters with varying suspended matter and CDOM concentrations, in L.L.Richardson and E.F. LeDrew (Eds), *Remote Sensing of Aquatic Coastal Ecosystem Processes; Science and Management Applications*, Chap. 3, p. 27, Springer, Berlin, 2007.
35. Gitelson, A.A., Schalles, J.F., and Hladik, C.M., Remote chlorophyll-*a* retrieval in turbid, productive estuaries: Chesapeake Bay case study, *Remote Sensing of Environment*, 109, 464, 2007.
36. Gower, J.F. and Borstad, G., Use of the *in vivo* fluorescence line at 685 nm for remote sensing surveys of surface chlorophyll *a*, in: J.F. Gower (Ed.), *Oceanography from Space*, pp. 329–338, Plenum Press, New York, 1981.
37. Gower, J.F.R., Doeffer, R., and Borstad, G.A., Interpretation of the 685 nm peak in water-leaving radiance spectra in terms of fluorescence, absorption, and scattering, and its observation by MERIS, *International Journal of Remote Sensing*, 20(9), 1771, 1999.
38. Roesler, C.S. and Perry, M.J., *In situ* phytoplankton absorption, fluorescence emission, and particulate backscattering spectra determined from reflectance, *Journal of Geophysical Research*, 100, 13279, 1995.
39. Gitelson, A.A., The peak near 700 nm on reflectance spectra of algae and water: Relationships of its magnitude and position with chlorophyll concentration, *International Journal of Remote Sensing*, 13, 3367, 1992.
40. Vasilkov, A. and Kopelevich, O., Reasons for the appearance of the maximum near 700 nm in the radiance spectrum emitted by the ocean layer, *Oceanology*, 22, 697, 1982.
41. Vos, W.L., Donze, M., and Bueteveld, H., On the reflectance spectrum of algae in water: The nature of the peak at 700 nm and its shift with varying concentration, Technical Report, 1986.
42. Morel, A. and Prieur, L., Analysis of variation in ocean color, *Limnology and Oceanography*, 22, 709, 1977.
43. Carder, K.L., Steward, R.G., Harvey, G.R., and Ortner, P.B., Marine humic and fulvic acids: Their effects on remote sensing of ocean chlorophyll, *Limnology and Oceanography*, 34, 68, 1989.
44. Gallegos, C.L., Correll, D., and Pierce, J.W., Modeling spectral diffuse attenuation, absorption, and scattering coefficients in a turbid estuary, *Limnology and Oceanography*, 35, 1486, 1990.
45. Ritchie, J.C., Schiebe, F.R., Cooper, C., and Harrington, A.J., Jr., Chlorophyll measurements in the presence of suspended sediment using broad band spectral sensors aboard satellites, *Journal of Freshwater Ecology*, 9, 197, 1994.
46. Schalles, A.F., Sheil, A.T., Tycast, J.F., Alberts, J.J., and Yacobi, Y.Z., Detection of chlorophyll, seston, and dissolved organic matter in the estuarine mixing zone of Georgia coastal plain rivers, *Proceedings of the Fifth International Conference on Remote Sensing for Marine and Coastal Environments*, 2, 315, 1998a.
47. IOCCG, Remote Sensing of ocean color in coastal, and other optically-complex waters, in S. Sathyendranath (Ed.), *Reports of the International Ocean-Colour Coordinating Group*, 3, IOCCG, Dartmouth, Canada, 2000.

48. Szekielda, K.-H., Gobler, C., Gross, B., Moshary, F., and Ahmed, S., Spectral reflectance measurements of estuarine waters, *Ocean Dynamics*, 53, 98, 2003.
49. Schalles, J.F. and Hladik, C., Remote chlorophyll estimation in coastal waters with tripton and CDOM interferences, *Ocean Optics 17*, Fremantle, Australia, October, 2004.
50. Szekielda, K.-H., Spectral recognition of marine biogeochemical provinces with MODIS, *EARSeL eProceedings, European Association of Remote Sensing Laboratories*, 3, 261, 2004.
51. Dekker, A.G., *Detection of Optical Water Quality Parameters for Eutrophic Waters by High Resolution Remote Sensing*, PhD thesis, Vrije Universiteit, 1993.
52. Rundquist, D.C., Han, L., Schalles, J.F., and Peake, J.S., Remote measurement of algal chlorophyll in surface waters: The case for the first derivative of reflectance near 690 nm, *Photogrammetric Engineering and Remote Sensing*, 62, 2, 1996.

8 Mapping Salt Marsh Vegetation by Integrating Hyperspectral and LiDAR Remote Sensing

Jiansheng Yang and Francisco J. Artigas

CONTENTS

8.1 INTRODUCTION

Coastal wetlands or salt marshes are subjected to a high frequency of disturbances caused by hurricanes, storms, tidal cycles, ditching and draining which continuously affect the establishment, persistence, and distribution of plant species in marsh habitats. These areas retain sediments and provide the substrate that supports plants and creates buffer zones that help alleviate the impact of sea surges by reducing the impact of floods on coastal areas [1]. Disturbances have many ways of affecting how salt marshes look and function. In some cases, disturbances expose new sediment and create the opportunity for invasive species to establish and, as a consequence, the extent and type of the vegetation cover change [2]. Similarly, high-energy storms can significantly alter the topography by exposing buried sediments, rerouting creeks and creating new saline intrusions that modify the intrinsic chemical properties of the sediments, which in turn has an effect on the distribution of wetland plants. Moreover, toxic insoluble metals that lack mobility when buried under anaerobic conditions may be stirred-up during a storm and exposed to the atmosphere making metals soluble and available to plants and animals up the food chain [3]. For this and many other reasons, monitoring and sampling wetland vegetation is a common and important activity. Wetland managers continuously monitor for invasive and opportunistic species such as the common reed (*Phragmites australis*), which is known to take over coastal wetlands in the northeastern United States and completely replace well-adapted native species that support local wildlife [4]. Once established, these invasive species are difficult to remove and further alter the topography to promote their expansion. Wetland vegetation is also sampled to benchmark native species distribution at certain terrain elevations so that when restoring or establishing new wetlands, these benchmark species are used as references to recreate the natural environment at the correct elevations [5]. After restoration of new wetlands, monitoring continues for some years to make sure that there is adequate plant cover, there is no threat of invasive species, and ultimately to make sure that the community is self-sustaining. Newly established salt marshes are monitored by scientists that use field methods designed to capture the extent of plant cover and the diversity of species in a given area [6]. One way to measure diversity is to randomly toss a 1×1 m quadrant square over a representative area and record all new species that fall within the quadrant after each toss. After several tosses, the number of new species starts to level-off and the approximate number of species present in the area is determined. A good field biologist can describe wetland vegetation fairly accurately using these methods that are based on a series of samples measured in a given location within a site and extrapolated to characterize an entire area. Quadrants are usually laid out over points along transects where each one is divided into four subquadrants from where all species are counted and their density and percent cover estimated [7]. These field methods to determine plant community composition are fairly accurate but are costly and labor intensive.

In highly fragmented urban wetlands, remote sensors can be used to find small relict communities that hold invaluable genetic stock and at the same time serve as crystallization points for further expansion of these plant cover types. Differences in light reflectance are effective in locating nondesirable species (e.g., invasive species) where subsequent treatments for eradication can easily be planned and budgeted by calculating affected areas directly from images [8]. On the other hand, there are natural vegetation types that are good surrogates for assessing the ecological and health condition of wetland sediments. High marsh communities, for example, composed of low dense grass associations are usually good indicators of anaerobic sediments with low oxidation reduction potential that favor the immobilization of toxic metals. These vegetation types point to areas where toxic metals are less likely to move up the food chain and can be mapped accordingly.

Remote sensing has been increasingly applied to environmental monitoring and has become popular for the convenience of capturing detailed images of the vegetation without having to physically move around the difficult terrain [9,10]. Traditional remote sensors, such as those on Landsat and SPOT with a few spectral bands (i.e., 4–7 bands) and low horizontal resolution (i.e., 10–30 m), are able to give general information about large areas but no detailed information about the fine composition of plant communities. It was not until the emergence of hyperspectral sensors mounted on a fixed wing aircraft that has the ability to look at plant distribution at species level [11]. In recent years, airborne hyperspectral remote sensing has been used in coastal wetlands characterization due to its large number of narrow, contiguous spectral bands as well as high horizontal resolution. For example, it has been used to map the habitat heterogeneity [12], determine plant cover distribution in salt marshes [13,14], map the spread of invasive plant species [15], and map the spatial pattern of tree species abundances [16]. In a previous study, authors used hyperspectral image to separate vigor types by detecting slight differences in coloration due to stress factors, infestation, or displacement by invading species [17]. Certain stunted forms of marsh vegetation reflect low quantities of light compared to the more vigorous forms signaling poor sediment conditions and pointing to areas that may require man-made modifications for plant establishment and subsistence. Based solely on reflectance, a few individuals may show great vigor among similar stunted forms in other cases, indicating the existence of naturally occurring phenotypes that make ideal candidates for cloning and cultivating. While recent improvements in classification techniques have been made, the application of hyperspectral imaging to salt marshes is not without challenge. Salt marsh plant communities can be highly heterogeneous at small spatial scales and the relationship between heterogeneity and pixel size makes selection of the appropriate endmembers critical to the accuracy of the final vegetation classification [18].

A detailed characterization of salt marsh vegetation is particularly challenging because these sites have a complex net of small channels with low elevation relief, and much of the marsh is covered by dense vegetation that hides underlying terrain features [15]. The high vertical and horizontal precision and accuracy of airborne Light Detection And Ranging (LiDAR) makes it suitable for mapping surfaces with great detail [19,20]. Most LiDAR devices provide at least the first and last returns for every pulse emitted. The first return is assumed to originate from reflections from taller parts of objects on the ground such as vegetation surface, buildings, and roads, while the last return is assumed to come from lower parts [15]. LiDAR greatly contributes by generating accessory information in the form of terrain and surface models that help further separate plant communities and populations by the slight differences in elevation and has effectively been used to produce a three-dimensional characterization in forest ecosystems [21,22]. Therefore integrating light reflected from salt marsh vegetation from many narrow spectral bands and LiDAR-derived elevation offers enormous possibilities to effectively describe, monitor, and assess the value and services that these important ecosystems provide to all living organisms.

The objective of this study was to test the applicability of LiDAR-derived elevation in improving the classification accuracy of salt marsh vegetation using hyperspectral image in the New Jersey Meadowlands estuary.

8.2 METHODS

8.2.1 Site Description

The New Jersey Meadowlands District encompasses more than 34 km^2 of tidal salt marsh and mud flats along the lower reaches of the Hackensack River of New Jersey. Also known as the Hackensack Meadowlands, it has been logged, diked, drained, farmed, filled, and contaminated in the past. These low lands located a few miles west of New York City are surrounded by some of the most densely populated and developed areas in the United States. The remaining open spaces of this area include high marsh, low marsh, marsh reed, mudflats, and open water. The high marsh vegetation is dominated by *Spartina patens* (Aiton.) Muhl and *Distichlis spicata* (L.) Greene, whereas low marsh is dominated by *Spartina alterniflora* Loisel. The invasive marsh reed, *P. australis* ((Cav) Trin. ex Stued), occupies the higher elevation dredge spoil islands, tidal creek banks, and levees. Uniform marsh vegetation is interrupted by ditches, patches of bare and exposed mud, and/or water-filled depressions.

8.2.2 Hyperspectral Image Acquisition and Processing

Hyperspectral image of a 20-km^2 area within the New Jersey Meadowland District was collected on June 7, 2007, by an imaging company SpecTIR using Airborne Imaging Spectrometer for Applications (AISA). AISA is a solid-state, push-broom instrument that is capable of collecting data within a spectral range from visible to near-infrared in up to 286 spectral bands [23]. In this study the sensor was configured to 128 spectral bands from 430 to 1000 nm with average bandwidth of approximately 5 nm. The AISA sensor had a 36° field-of-view and a 3080-m altitude above sea level, which corresponds to a 1670-m swath width with overlay and a 2-m horizontal resolution. A total of three north–south flight strips were collected to cover the study area. Dates and times of flights were chosen to coincide with clear skies and the low tide.

Radiometric correction was conducted by the vendor to derive spectral reflectance images using the ATCOR/MODTRAN algorithm and then polished using a SpecTIR proprietary program based on a Savitsky–Golay algorithm with refined handling of atmospheric absorption features associated with CO_2 and water. Authors used a histogram matching technique to correct the brightness distortions between strips because of the tilt angle of the camera and sun angle. After histogram matching, three AISA strips were mosaicked into a single seamless image, which was then georectified to the New Jersey State Plane coordinates (NAD 83) using high-resolution orthophoto (Figure 8.1). Twenty-two ground control points (GCPs) were carefully selected on both hyperspectral image and orthophoto, which generated an average root-mean-square error (RMSE) of ±1.1 (~2.2 m) in georectification. A narrow-band normalized difference vegetation index (NDVI) image was derived from an analysis of AISA band 52 (~863 nm) and band 30 (~660 nm) [24]. A threshold value of $NDVI > 0.3$ was selected to build a mask to eliminate all nonvegetated areas including roads, open water, and mudflats of the study area. The masked vegetation-only image was used for further image classification. The Environment for Visualization

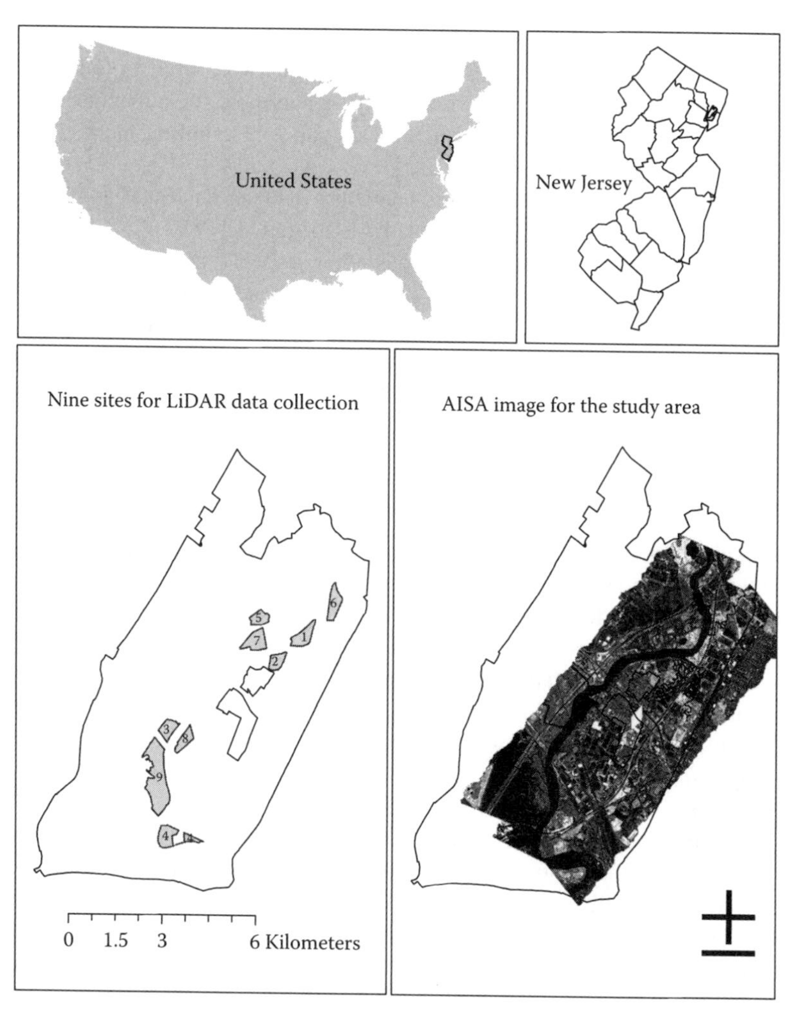

FIGURE 8.1 LiDAR data collection sites and hyperspectral AISA image for the study area in the New Jersey Meadowlands District.

Images (ENVI) software package (ENVI version 4.3, Research Systems, Inc.) was used to perform all image preprocessing and advanced image analysis.

8.2.3 LiDAR Data Acquisition and Processing

Airborne LiDAR data were collected for the nine target sites within the New Jersey Meadowlands District (Figure 8.1) by Photo Science, a company specializing in the design of laser-based ranging, mapping, and detection systems. They used a Leica ALS50 LiDAR sensor mounted on a fixed wing aircraft flown at an approximate 914-m altitude above sea level on May 5, 2007. The laser pulse rate was 74 Hz and the average point spacing was 0.56 m with a point sampling density of 3.12 points/m^2.

The field-of-view of the sensor was 28°, which resulted in a swath width of 455 m for each flight line. A total of 16 north–outh flight lines were used to cover the nine target sites. Mosaicked LiDAR image was georegistered to the hyperspectral image with the first-order polynomial and nearest neighbor resampling methods with an RMSE of 0.76 (~3.04 m) based on 20 GCPs.

Initial LiDAR data consisted of a point database in which each site had FirstReturn, LastReturn, BareEarth, and Intensity in LAS format. ENVI LiDAR Toolkit was used to convert original data into digital elevation models (DEMs) for each site. Two elevation models were derived in this study: a digital terrain model (DTM) based on the BareEarth return and a digital surface model (DSM) using the FirstReturn pulses. In addition, a digital vegetation height model was developed to calculate vegetation height for each laser point as the difference between the interpolated terrain elevation and corresponding surface elevation [25].

8.2.4 Endmember Selection

In the field, trained people can easily identify individual plant species according to their color, size, and shape. However, the hyperspectral image is a matrix of 2 × 2 m pixels of recorded reflectance values from 3000 m above sea level, which makes it difficult to identify plant species as easily as people do on the ground. In this study, the salt marsh vegetation was grouped into four dominant plant communities: (1) high marsh (HM), including *D. spicata*, *S. patens*, and Black grass; (2) low marsh (LM), including *S. alterniflora* and its stunted form; (3) tall *Phragmites* (TP), including *P. australis* higher than or equal to 1.8 m; and (4) short *Phragmites* (SP), including *P. australis* shorter than 1.8 m.

Because most adjacent spectral bands are highly correlated and provide similar information, a minimum noise fraction (MNF) rotation was used to reduce the data dimensionality as well as computational requirements for subsequent processing [26,27]. The first 15 eigenimages (MNF bands) were derived from the original hyperspectral image, which accounted for more than 98% of the variance (Figure 8.2). These MNF bands were then used to find the purest spectral curves (i.e., endmembers) by calculating the pixel purity index (PPI) through repeatedly projecting *n*-dimensional scatter plots onto a random unit vector. The extreme pixels in each projection are recorded and the total number of times each pixel is marked as extreme is noted [28]. In this case, 16 endmember candidates were first interactively selected from the PPI pool and seven endmembers were finally obtained by merging some similar curves. In addition to the automated endmember extraction, polygons of representative pixels of each class (i.e., endmember training sites) were also manually selected from the AISA image based on high-resolution orthophoto and field verification for each of the four dominant marsh vegetation communities.

8.2.5 Image Classification

Several algorithms, including the spectral angle mapper (SAM), linear spectral unmixing, spectroscopic library matching, and specialized hyperspectral indices, have been developed and can be used in hyperspectral image classification [24].

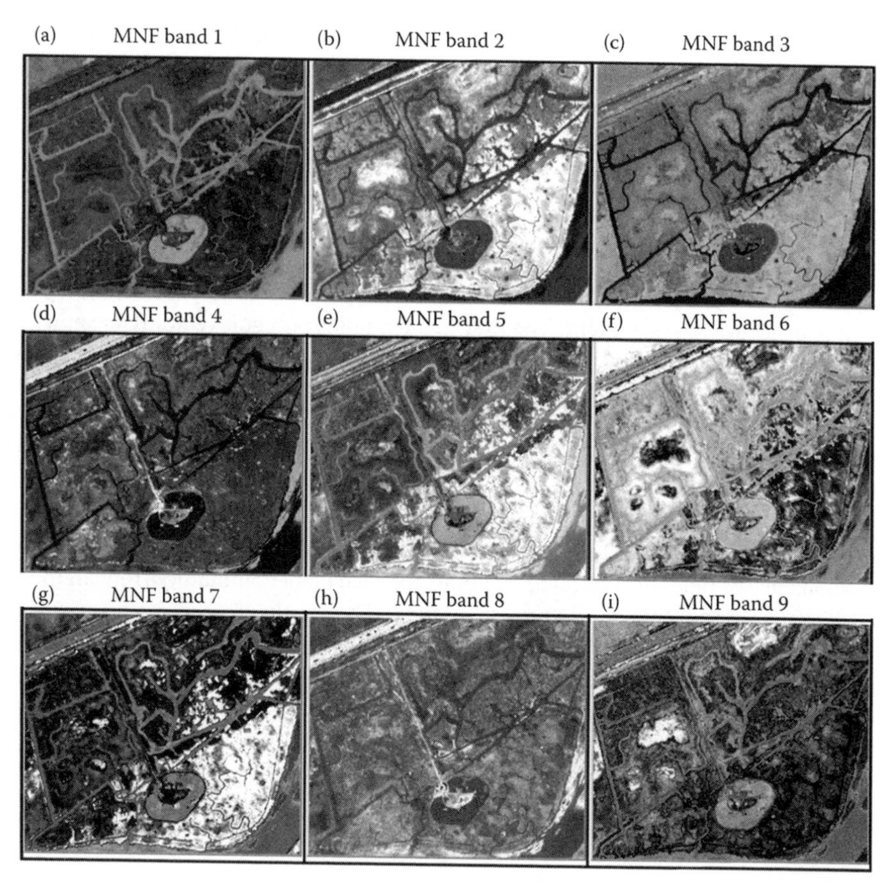

FIGURE 8.2 Nine of the first 15 eigenimages (MNF bands) extracted from the Marsh Resources Meadowlands Mitigation Bank (site 1) in the New Jersey Meadowlands District.

Some traditional classification algorithms have also been applied to hyperspectral image with varying degrees of success [27].

The first classification algorithm used in this study was SAM. SAM determines the similarity between two spectra by calculating the "spectral angle" between them, treating them as vectors in a space with dimensionality equal to the number of bands [29]. The angle between the reference spectrum and the hyperspectral image pixel was compared and the reference spectrum class that yields the smallest angle was assigned to the image pixel. Because SAM uses only the "direction" of the spectra and not their "length," this method is insensitive to the unknown gain factor, and all possible illuminations are treated equally. The second classification method we used is minimum distance to means (MDM). MDM is a supervised classifier that first analyzes the training sites and calculates a mean vector for each prototype class, described by the class center coordinates in feature space [24]. The algorithm then determines the Euclidean distance from each unclassified pixel to the mean vector of each prototype class. The unknown pixel will be assigned to the vector class that has the shortest distance to its mean. The MDM method works best in applications where spectral classes are dispersed in feature space and have similar variance. Based on ground reference data in the Riverbend (site 4), three training areas for each of the four plant communities (HM, LM, TP, and SP) were selected for this classification.

8.2.6 Accuracy Assessment

Accuracy assessment for wetlands study is always a challenge due to its accessibility. In this study, three sources of field-collected data were used to select reference points in accuracy assessment. The first was the point/polygon vector data showing the location of salt marsh species collected with the aid of a Global Positioning System (GPS) since 2005; the second was the reflectance spectra of individual salt marsh species collected using a hand-held spectroradiometer since 2004; and the third was images taken by a digital camera mounted on a low-flying tethered balloon in 2006.

Since there is a consensus among researchers that a minimum of 50 samples for each category is reasonable, both statistically and in practical terms for accuracy assessment, we randomly selected 50 points for each plant community from all those available ground reference data in this study. The result of accuracy assessment was reported in an error (or confusion) matrix, which includes overall accuracy, producer's accuracy, and user's accuracy for each category and a more rigorous κ statistic obtained by a statistical formula that utilizes information in the error matrix [24].

8.3 RESULTS

8.3.1 Hyperspectral Image Classification

Reflectance spectra in the visible and near-infrared portions of the spectrum for the dominant salt mash plant communities are shown in Figure 8.3. We found that the high marsh species have much higher reflectance in the near-infrared region and can be separated from the other associations. Tall *Phragmites* has the lowest reflectance and is separable from stunted *Phragmites*, which grows in a mixture interspersed with *S. patens*. The low marsh species is difficult to separate from the stunted *Phragmites* and *S. patens* mixtures due to their similar spectral reflectance.

The error matrix of the hyperspectral image classification on the site Riverbend using SAM and MDM are presented in Tables 8.1 and 8.2, respectively. The overall accuracy from the SAM algorithm was 56.5% with a κ coefficient of 0.42. Tall *Phragmites* had the highest accuracy of 68%, while low marsh had the lowest accuracy of 42%. For results from MDM algorithm, the producer's accuracy for high marsh increased from 62% to 96%, although there were no significant changes in overall accuracy and producer's accuracies for the other three classes. This shows that MDM is better in distinguishing high marsh from other salt marsh plants than SAM. The species that is difficult to separate from others was low marsh, which had similar reflectance spectra with short *Phragmites* occurring in mixtures with *S.patens*. Proportionally, tall *Phragmites* and high marsh are the two dominant

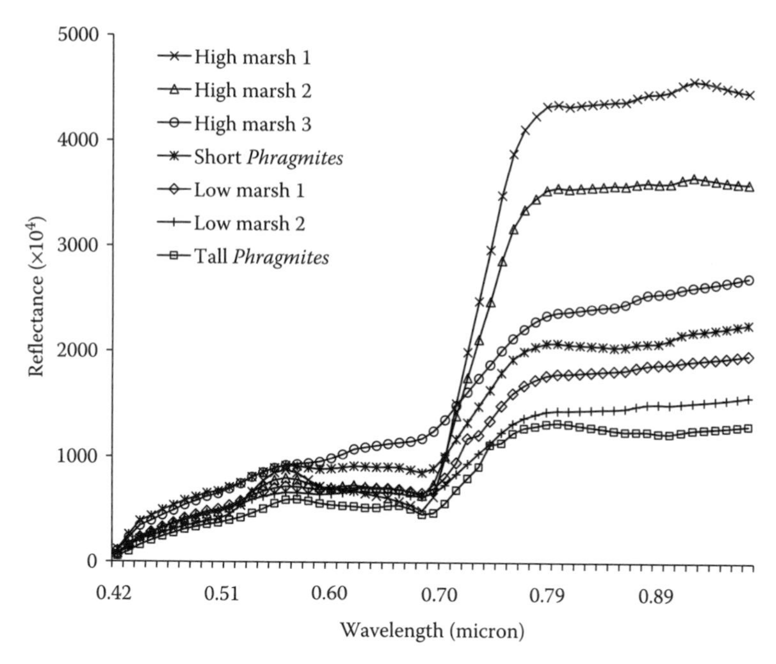

FIGURE 8.3 Endmember spectra of dominant salt marsh vegetation communities in the New Jersey Meadowlands District extracted through the MNF/PPI process.

categories in the Riverbend, which occupy approximately 13% and 10% among all surfaces and 35% and 28% among the vegetation-only marshes, respectively. Low marsh makes up approximately 8% of all surfaces and 23% of the vegetation-only marshes, and short *Phragmites* is the least in abundance: 5% for all surfaces and 14% for vegetation-only mashes.

TABLE 8.1
Error Matrix of the Classification Map Derived from Hyperspectral Image of the Riverbend in the New Jersey Meadowlands District Using SAM Algorithm

	Reference Data					
Classification	**HM**	**TP**	**SP**	**LM**	**Row Total**	**User's Accuracy (%)**
HM	31	5	2	5	43	72
TP	19	34	7	8	68	50
SP	0	0	27	16	43	62
LM	0	11	11	21	43	48
Column total	50	50	47	50	197	
Producer's accuracy (%)	62	68	57	42		
Overall accuracy = 56.5%				κ coefficient = 0.42		

TABLE 8.2
Error Matrix of the Classification Map Derived from Hyperspectral Image of the Riverbend in the New Jersey Meadowlands District Using MDM Algorithm

	Reference Data					
Classification	**HM**	**TP**	**SP**	**LM**	**Row Total**	**User's Accuracy (%)**
HM	48	1	0	2	51	94
TP	2	27	6	8	43	62
SP	0	1	21	18	40	52
LM	0	21	20	22	63	34
Column total	50	50	47	50	197	
Producer's accuracy (%)	96	54	44	44		
Overall accuracy = 59.9%				κ coefficient = 0.46		

8.3.2 LiDAR-Derived Marsh Topography

Plant covers in most vegetated plots in the Meadowlands are dense, thus LiDAR pulses have difficulty penetrating through the entire plant canopy to reach the ground, especially in high marsh, which occurs as a dense math of low-lying grasses. In a similar study conducted in southern California, they found that of the total number of vegetation plots, LiDAR penetrated the vegetation canopy in only 3% of cases, and the average depth into the canopy that LiDAR capable to penetrate was 3–10 cm [18].

LiDAR-derived DSM for the Riverbend showed a relatively flat topography in May. Most of the area southwest and northwest of the Riverbend is 0.6 m above sea level where *Phragmites* patches tend to occur. Lower canopies below 0.3 m are sparsely distributed in the middle and northeastern portions where high marsh patches tend to occur. According to the LiDAR profile, 70% of the marsh canopy in the Riverbend in May was found to lie between 0.6 and 0.9 m elevation (Figure 8.4).

8.3.3 Integration of Hyperspectral and LiDAR Data

LiDAR-derived DEM, including maximum, mean, minimum surface elevation as well as intensity, were layer-stacked with hyperspectral data as distinct bands; SAM and MDM were then applied to the stacked image, respectively. The classification maps were assessed using the same 50 ground reference points. The results showed that the accuracies were not improved significantly except tall *Phragmites*, whose producer's accuracy was increased from 54% to 70%.

There was a considerable amount of variation in average canopy height between salt marsh categories. Consistent with their individual growth habits in the Meadowlands, high marsh usually occurs on areas where ground elevation is above 0.5 m while low marsh occurs where ground is below 0.5 m. In addition, it seems possible to separate tall *Phragmites* (≥1.8 m) and short *Phragmites* (<1.8 m) based

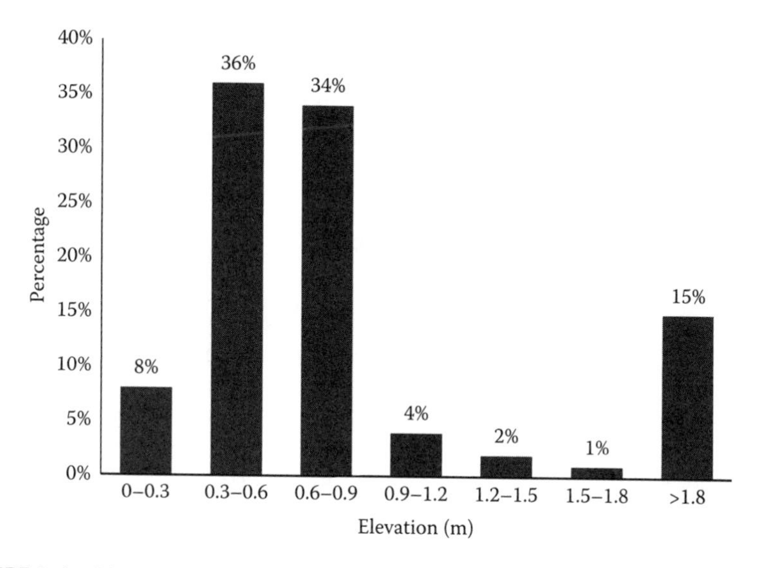

FIGURE 8.4 Pixel distribution with surface elevation in the Riverbend in the New Jersey Meadowlands District.

on the canopy surface height. Therefore a reclassification tree was first established to use the LiDAR-derived DTM to separate high marsh from low marsh and the LiDAR-derived DSM to distinguish tall *Phragmites* from short *Phragmites* (Table 8.3a). The result of this classification was evaluated using the same ground reference points and no increase in accuracy was obtained. This indicated that the most difficult species to separate in our study might be between low marsh and *Phragmites*. This is consistent with the results in the error matrix Tables 8.1 and 8.2, where the low producer's accuracies are mainly due to the misclassification among low marsh, short *Phragmites*, and high *Phragmites*. Then we reclassified the map based on the revised criteria in

TABLE 8.3
LiDAR-Derived Elevation Criteria in Reclassifying the Maps Obtained from Hyperspectral Image

	Hyperspectral Classes	Reclassified Classes	
a	TP	TP (if DSM > 1.8 m)	SP (if DSM < 1.8 m)
	SP	TP (if DSM > 1.8 m)	SP (if DSM < 1.8 m)
	HM	HM (if DTM > 0.5 m)	LM (if DTM < 0.5 m)
	LM	HM (if DTM > 0.5 m)	LM (if DTM < 0.5 m)
b	LM	SP (if DTM > 0.5 m)	LM (if DTM < 0.5 m)
	TP	TP (if DTM > 0.5 m)	LM (if DSM < 0.5 m)
	SP	SP (if DSM > 0.5 m)	LM (if DSM < 0.5 m)

TABLE 8.4
Error Matrix of the MDM Classification Map of the Riverbend in the New Jersey Meadowlands District by Integrating Hyperspectral and LiDAR Data

	Reference Data					
Classification	**HM**	**TP**	**SP**	**LM**	**Row Total**	**User's Accuracy (%)**
HM	48	1	1	2	52	92
TP	2	35	9	7	53	66
SP	0	6	23	11	40	58
LM	0	8	16	30	54	56
Column total	50	50	49	50	199	
Producer's accuracy (%)	96	70	47	60		
Overall accuracy = 68.3%				κ coefficient = 0.60		

Table 8.3b, which states that *Phragmites* will be classified as low marsh as long as they occur in the areas where ground elevation is below 0.5 m, and low marsh on areas where ground elevation is above 0.5 m will be classified as short *Phragmites*. The reclassified map was evaluated using the same ground reference points and the error matrix is presented in Table 8.4. The producer's accuracies of tall *Phragmites* and low marsh increased from 54% to 70% and 44% to 60%, respectively. The overall accuracy also increased to 68.3% with a κ coefficient of 0.60. The final classification map was finally overlaid on the hillshade image extracted from LiDAR-derived DSM with 50% transparency (Figure 8.5), which gives a unique perspective on salt marsh vegetation mapping in coastal wetlands.

8.4 DISCUSSION AND CONCLUSIONS

Results of this study showed that hyperspectral AISA image can be used to distinguish high marsh from other salt marsh communities due to its high reflectance on the near-infrared region of the spectrum. On the contrary, the separation among *Phragmites* and low marsh is still challenging because of their similar reflectance in June. This study further demonstrated that the integration of hyperspectral image and LiDAR-derived elevation did improve the accuracy in mapping salt marsh vegetation compared to using either dataset alone. Although SAM is the most popular algorithm that has been used in hyperspectral image classification in the past, we found that SAM is not well suited for our study. The classifier MDM resulted in higher map accuracy than SAM in this study, especially for high marsh with a good amount of ground reference information available. These ground reference data ensure high-quality training areas so that the accuracy of areas from where training sites were obtained are high, however this does not guarantee high accuracies in other high marsh communities where no representative training sites were obtained. On the other hand, although the endmembers extracted through the automate process

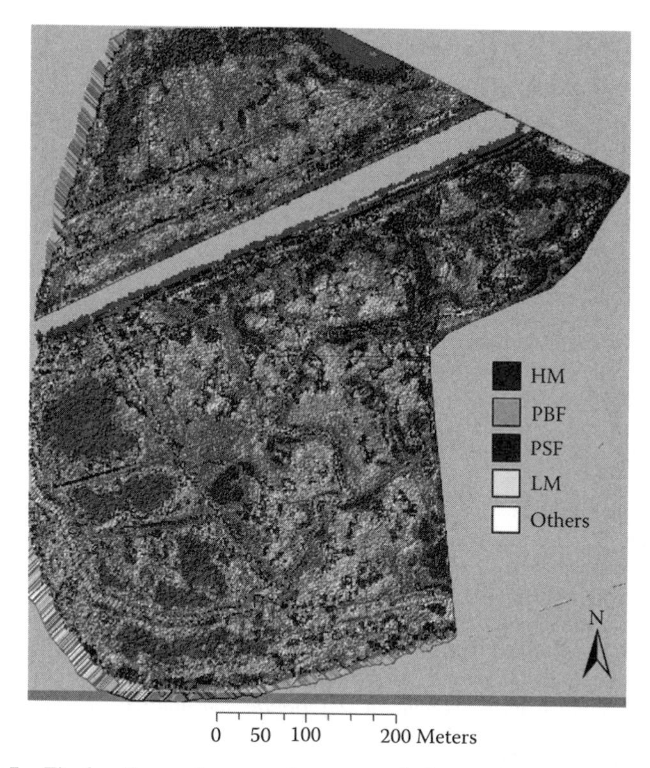

FIGURE 8.5 Final salt marsh vegetation map of the Riverbend in the New Jersey Meadowlands overlaid on the Hillshade image with 50% transparency.

in SAM resulted in low accuracy for high marsh, they are not related to the training sites and can be applied to the entire study area once they are confirmed.

Endmembers selected from AISA image manually based on field observation resulted in higher map accuracy than those derived directly either from the field or the image using an automated process. While a LiDAR-derived surface model is useful in separating tall *Phragmites* from salt marshes with low elevation, a LiDAR-derived vegetation height model is not as useful in studying dense and short salt marsh vegetation as it is for forest structure [16]. LiDAR-derived elevation has limitation in mapping ground topography under low and dense canopy vegetation; additional topographic surveying and vegetation sampling to develop plant species-specific corrections are necessary in the future study. Overall accuracy of the vegetation map in this study is acceptable and within the range found for studies done in environments with a high degree of spatial heterogeneity in plant distribution [14,30]. Several factors may have contributed to the overall accuracy: the number of MNF bands retained, the training areas selected, georeferencing inaccuracy, and ground reference points selected for accuracy assessment. This study confirms that the integration of hyperspectral and LiDAR remote sensing is an effective method for characterizing salt marsh vegetation, which holds significant potential for improving the inventory and monitoring of salt marshes in fragmented coastal wetlands.

ACKNOWLEDGMENTS

The authors would like to thank Ildiko Pechmann for collecting reflectance spectra and balloon images of dominant salt marsh species in the New Jersey Meadowlands. This research was supported through a grant from the U.S. EPA (grant no. CD97270201-1).

REFERENCES

1. Mitsch, W.J. and Gosselink, J.G., *Wetlands*, Wiley, New York, 2000.
2. Lathrop, R.G., Windham, L., and Montesano, P., Does *Phragmites* expansion alter the structure and function of marsh landscapes? Patterns and processes revisited, *Estuaries*, 26, 2003.
3. Weis, J.S. and Weis, P., Metal uptake, transport and release by wetland plants: Implications for phytoremediation and restoration, *Environment International*, 30, 685, 2004.
4. Windham, L., Microscale spatial distribution of *Phragmites australis* (common reed) invasion into *Spartina patens* (salt hay)—dominated communities in Brackish Marsh, *Biological Invasions*, 1, 131, 1999.
5. Daubenmire, R.F., *Plant Communities: A Textbook of Plant Synecology*, Harper & Row, New York, 1968.
6. Tiner, R.W., *Wetland Indicators: A Guide to Wetland Identification, Delineation, Classification, and Mapping*, Lewis Publishers, Boca Raton, FL, 1999.
7. Kent, M. and Coker, P., *Vegetation Description and Analysis. A Practical Approach*. CRC Press, Boca Raton, FL, 1992.
8. Coops, N., Stanford, M., Old, K., Dudzinski, M., Culvenor, D., and Stone, C., Assessment of Dothistroma needle blight of *Pinus radiata* using airborne hyperspectral imagery, *Phytopathology*, 93, 1524, 2003.
9. Cochrane, M.A., Using vegetation reflectance variability for species level classification of hyperspectral data, *International Journal of Remote Sensing*, 21, 2075, 2000.
10. Schmidt, K.S. and Skidmore, A.K., Spectral discrimination of vegetation types in a coastal wetland, *Remote Sensing of Environment*, 85, 92, 2003.
11. Silvestri, S., Marani, M., Settle, J., and Benvenuto, F., Salt marsh vegetation radiometry: Data analysis and scaling, *Remote Sensing of Environment*, 2, 473, 2002.
12. Artigas, F.J. and Yang, J., Hyperspectral remote sensing of habitat heterogeneity between tide-restricted and tide-open areas in New Jersey Meadowlands, *Urban Habitat*, 2, 1, 2004.
13. Li, L., Ustin, S.L., and Lay, M., Application of multiple endmember spectral mixture analysis (MESMA) to AVIRIS imagery for coastal salt marsh mapping, a case study in China Camp, CA, USA, *International Journal of Remote Sensing*, 26, 5193, 2005.
14. Belluco, E., Camuffo, M., Fererari, S., Modenese, L., Silvestri, S., Marani, A., and Marani, M., Mapping salt marsh vegetation by multispectral and hyperspectral remote sensing, *Remote Sensing of Environment*, 105, 54, 2006.
15. Rosso, P.H., Ustin, S.L., and Hastings, A., Use of lidar to study changes associate with *Spartina* invasion in San Francisco Bay marshes, *Remote Sensing of Environment*, 100, 295, 2006.
16. Anderson, J.E., Plourde, L., Martin, M.E., Braswell, B.H., Smith, M.L., Dubayah, R., and Hofton, M., Integrating waveform lidar with hyperspectral imagery for inventory of a northern temperate forest, *Remote Sensing of Environment*, 112, 1856, 2007.
17. Artigas, F.J. and Yang, J., Hyperspectral remote sensing of marsh species and plant vigor gradient in the New Jersey Meadowland, *International Journal of Remote Sensing*, 26, 5209, 2005.

18. Sadro, S., Gastil-Buhl, M., and Melack, J., Characterizing patterns of plants distribution in a southern California salt marsh using remotely sensed topographic and hyperspectral data and local tidal fluctuations, *Remote Sensing of Environment*, 110, 226, 2007.
19. Rango, A., Chopping, M., Ritchie, J., Havstad, K., Kustas, W., and Schmugge, T., Morphological characteristic of shrub coppice dunes in desert grasslands of southern New Mexico derived from scanning lidar, *Remote Sensing of Environment*, 74, 26, 2000.
20. MacMillan, R.A., Martin, T.C., Earle, T.J., and McNabb, D.H., Automated analysis and classification of landforms using high-resolution digital elevation data: Application and issues, *Canadian Journal of Remote Sensing*, 29, 592, 2003.
21. Lefsky, M.A., Cohen, W., Parker, G.G., and Harding, D.J., Lidar remote sensing for ecosystem studies, *Bioscience*, 52, 19, 2002.
22. Riaño, D., Meier, E., Allgöwer, B., Chuvieco, E., and Ustin, S.L., Modeling airborne laser scanning data for the spatial generation of critical forest parameters in fire behavior modeling, *Remote Sensing of Environment*, 86, 177, 2003.
23. AISA, http://www.specmap.com/rem-hyper-aisa.html, accessed May 2008.
24. Jensen, J.R., *Introductory Digital Image Processing*, 3rd edition, Prentice-Hall, Upper Saddle River, NJ, 2005.
25. Bork, E.W. and Su, J.G., Integrating LIDAR data and multispectral imagery for enhanced classification of rangeland vegetation: A meta analysis, *Remote Sensing of Environment*, 111, 11, 2007.
26. Boardman, J.W. and Kruse, F.A., Automated spectral analysis: A geological example using AVIRIS data, North Grapevine Mountains, Nevada, in *Proceedings of the Tenth Thematic Conference on Geological Remote Sensing*, Environmental Research Institute of Michigan, San Antonio, TX, pp. 407–18, 1994.
27. Underwood, E., Ustin, S., and DiPietro, D., Mapping nonnative plants using hyperspectral imagery, *Remote Sensing of Environment*, 86, 150, 2003.
28. Research Systems, Inc., *ENVI User's Guide*, Research Systems, Inc., Boulder, CO, 2008.
29. Kruse, F.A., Lefkoff, A.B., and Dietz, J.B., Expert system-based mineral mapping in northern Death Valley, California/Nevada using the Airborne Visible/Infrared Imaging Spectrometer (AVIRIS), Special issue on AVIRIS, *Remote Sensing of Environment*, 44, 309, 1993.
30. Thomson, A.G., Huiskes, A., Cox, R., Wadsworth, R.A., and Boormanet, L.A., Short-term vegetation succession and erosion identified by airborne remote sensing of Westerschelde salt marshes, The Netherlands, *International Journal of Remote Sensing*, 25, 4151, 2004.

Section III

High Spatial-Resolution Remote Sensing

9 Mapping Salt Marshes in Jamaica Bay and Terrestrial Vegetation in Fire Island National Seashore Using QuickBird Satellite Data

Yeqiao Wang, Mark Christiano, and Michael Traber

CONTENTS

9.1 INTRODUCTION

The Fire Island National Seashore (FINS) and the Gateway National Recreation Area (NRA) are two of the protected National Parks managed by the Northeast Coastal and Barrier Network (NCBN) of the National Park Service (NPS)'s Inventory and Monitoring (I&M) Program (see Chapter 16). These two parks consist of critical coastal habitat for many rare and endangered species, and serve as migratory corridors for wildlife. They also protect vital coastal wetlands essential to water quality, fisheries, and the biodiversity of coastal, nearshore, and terrestrial environments. In this chapter we report two case studies that applied QuickBird-2 high spatial resolution satellite remote data to map the salt marsh in the Jamaica Bay of the Gateway NRA and the terrestrial vegetation of the FINS as a step toward long-term coastal resource inventory and monitoring.

Jamaica Bay is one of the three units of the Gateway NRA and one of the largest open space areas in New York City. The Bay played an important role in the development of the city and its surrounding environment. The diversity and distribution of vegetation associations at Gateway NRA resulted from the geologic history of the region, such as glaciations and outwash sand depositions; the temperate climate on the northeastern U.S. coastal plain; and its proximity to the ocean and maritime ecological processes. The Jamaica Bay possesses typical coastal combination of salt marsh, grassland, woodlands, maritime shrublands, brackish, and freshwater wetlands, and open water. The maritime plant associations at Gateway NRA are heavily influenced by coastal processes, such as diurnal tides, dune deposition, shifting, storm overwash, salt spray, and high winds; and its setting within an urban landscape with an intensive human land-use history. In particular, the surrounding urban setting brings in problems of pollution, channel dredging, land filling, as well as invasive species, fragmentation, isolation, and loss of critical habitats (Edinger et al., 2008).

Productivity of salt marshes is a primary indicator of coastal ecosystem health. Salt marshes, one of the Jamaica Bay's most critical habitats, have been rapidly eroding and disappearing. Interpretation of historical aerial photographs shows that 51% of salt marshes in the Bay, largely *Spartina alterniflora*, had been lost between 1924 and 1999. Nationwide, emergent salt marshes declined by an estimated 5850 ha in 10 years from 1986 to 1997 (Dahl, 2000; Warren et al., 2001). Although the causes of the change are not fully understood, salt marsh change in the Jamaica Bay is similar to the trends found in other salt marshes elsewhere in the northeastern United States (Hefner et al., 1994; Moulton et al., 1997). Extensive studies have been conducted for monitoring and quantifying salt marsh dynamics (Niering and Warren, 1980; Peteet, 2001; Roman et al., 2002). Methods for examining salt marsh changes vary with project goals, compliance requirements, organization priorities, and financial limitations (Boumans et al., 2002; Neckles et al., 2002).

The FINS was officially designated in 1964, resulting from a commission from the U.S. Congress in the mid-1950s, which requested the NPS to study seashore and lakeshore areas for possible inclusion in the National Park system. The vegetation communities and spatial patterns on the FINS are dynamic with the impacts from forces such as sand deposition, storm-driven overwash, salt spray, and surface water. Mapping the vegetation communities and tracking their changes are among the important tasks that always challenge the resource managers.

Remote sensing and geographic information systems (GIS) have been applied in wetland and salt marsh habitat monitoring (e.g., Zhang et al., 1997; Ritter and Lanzer, 1997; Ozesmi and Bauer, 2002; Schmidt and Skidmore, 2003; Schmidt et al., 2004). The U.S. Fish and Wildlife Service Natural Wetland Inventory used aerial photography imagery and direct on-the-ground observations to record and monitor wetland changes over time. Remote sensing has been identified as one of the primary data sources for conducting change analysis and monitoring (Coppin and Baure, 1996; Hansen and Rotella, 2002; Griffith et al., 2003; Rogan et al., 2002; Turner et al., 2003; Wilson and Sader, 2002; Wang et al., 2005). Updating the vegetation mapping product on a regular basis is one of the challenges to ensure consistent data in the monitoring of the dynamics and changes of the vegetation communities in barrier islands and bay areas. Given that salt marsh and barrier island vegetation monitoring require repeatable and reliable updates of land-cover maps, exploration of new data and approaches that could efficiently update the salt marsh maps is necessary. Therefore, timely and repeated data acquisition, a quick and simple way of information extraction, low cost in data processing, and available baseline referencing data are key considerations in any protocol development.

The recent development of high spatial resolution satellite remote sensing data can improve significantly the capacity for salt marsh and coastal vegetation mapping. For example, Space Imaging's IKONOS satellite data, consisting of 1 m spatial resolution for the panchromatic band and 4 m spatial resolution for the multispectral bands (visible to near-infrared), have been applied in resource mapping. Digital Globe's QuickBird-2 satellite data possess 0.61 m spatial resolution for the panchromatic band and 2.5 m spatial resolution for the multispectral bands (visible to near-infrared). The capability for repeated data acquisition by high spatial resolution satellite imageries and the relatively low cost can facilitate the routine practice in salt marsh and coastal terrestrial vegetation change detection.

To help reach the goal of involving high spatial resolution remote sensing in coastal resources inventory and monitoring, the objectives of the case studies included to (1) map terrestrial vegetation within the FINS boundary; (2) map salt marsh within the Jamaica Bay, both using high spatial resolution QuickBird-2 satellite remote sensing data; and (3) compare and validate agreement between satellite-derived vegetation maps and the delineation result from conventional aerial photograph interpretation so that repeated monitoring of vegetation change using satellite remote sensing can be achieved.

9.2 METHODS

9.2.1 STUDY AREAS

The Jamaica Bay comprises several disjunct parcels in the Boroughs of Brooklyn and Queens in New York City (Figure 9.1). The central piece of the bay has been designated as the Jamaica Bay Wildlife Refuge. As an oasis of nature surrounded by metropolitan urban structures, the Bay provides a critical stopover area along the Eastern Flyway migration route. Although they have been protected since 1972 as part of the Gateway NRA, salt marshes in the Jamaica Bay have been declining significantly from 0.4% per year between 1924 and 1972, to 1.4% per year between 1974 and 1994, and to 3.0% per year between 1994 and 1999 (Hartig et al., 2002). Increasingly, those losses have occurred within the interior of marsh islands (Figure 9.1). As tidal pools increase in size, marsh areas become more and more fragmented. Marsh

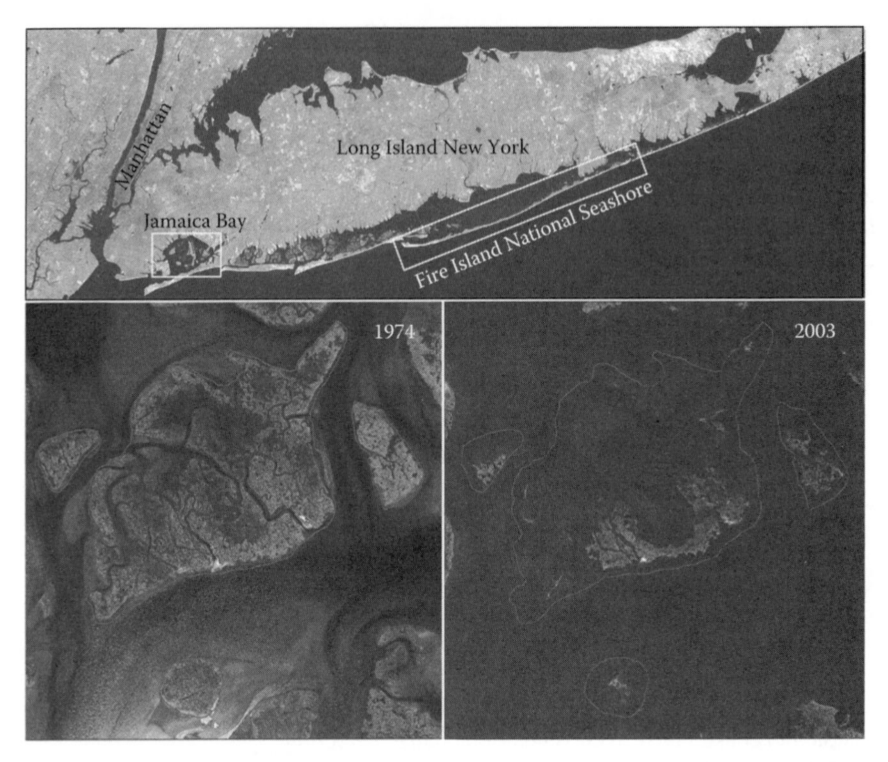

FIGURE 9.1 Jamaica Bay and FINS are two units managed by the Northeast Coastal Barrier Network of the NPS I&M Program. The comparison of a 1974 aerial photograph and the 2003 QuickBird imagery illustrates the change of salt marsh in the Duck Point site of the Jamaica Bay.

vegetation becomes inundated, withers, and dies. On some islands, up to 78% of the vegetation has disappeared in the past three decades (Gateway NRA, 2003).

The FINS is a member of Long Island barrier island system located in Suffolk County in the State of New York. Fire Island extends about 52 km from Fire Island Inlet to Moriches Inlet (Figure 9.1). The administrative boundary of FINS extends 305 m into the Atlantic Ocean to the south, and 1219 m north into the Great South Bay. In 1976, the 2.6 km^2 of William Floyd estate, the home of New York's Continental Congress delegate and signatory of the Declaration of Independence, became part of the FINS. In 1980, approximately 16 km^2 stretching about 11 km from Watch Hill to Smith Point County Park were designated as the Otis Pike High Dune Wilderness, which is the only federally designated wilderness area in the state of New York.

The vegetation communities and their spatial patterns on Fire Island are a direct result of dynamism. The dune morphology of Fire Island is typical of a barrier island. Maritime associations of the vegetation contain stunted "salt-pruned" trees and shrubs with contorted branches and wilted leaves and usually have a dense vine layer. These communities often occur as narrow bands parallel to the shoreline and contain plants that are adapted to these conditions (Edinger et al., 2008). Several zones can be readily identified along the gradient. The primary vegetation gradient extends from the Atlantic Ocean toward the Great South Bay roughly parallel to both along the entire island. Immediately adjacent to the open ocean is nonvegetated sand extending to the base of the primary dune. Sparse herbaceous plants can be found at the base of the primary dune and the dune face exposed to the ocean. Grass vegetation typically increases in cover from the crest of the primary dune and into the interdune swale area. These swales are often a mosaic of shrub and grass types. Different types of grass, dwarf-shrub, woody shrub, vine, and tree communities begin to appear in the swales. Occasionally depressions are present with near-surface water available to the vegetation. Shrubs tend to increase in density toward the secondary dune and the Bay salt marshes although many areas of Fire Island do not have a well-defined secondary dune. When a well-formed secondary dune is present, larger trees often replace shrubs. Trees can be over 10 m in height. Most of the Bay side of the island is salt marsh that gradually tapers into the shallows of the Great South Bay (Klopfer et al., 2002). The zonal patterns of vegetation distribution are observable on QuickBird-2 satellite image as confirmed by field photographs (Figure 9.2). The zonal pattern of

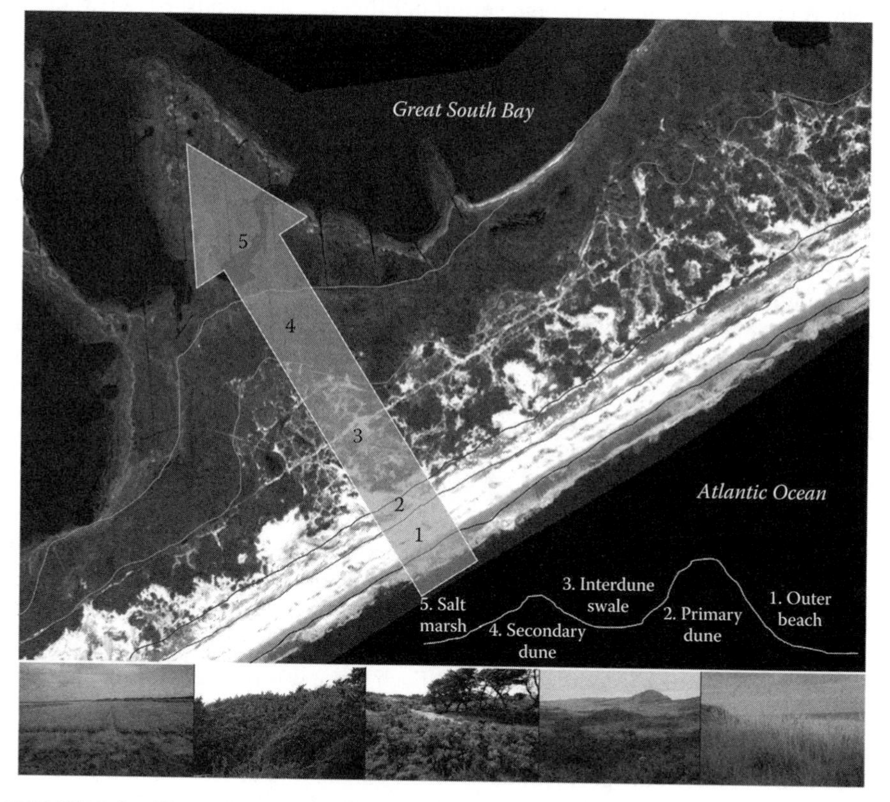

FIGURE 9.2 **(See color insert following page 206.)** Vegetation communities on the FINS barrier island follow the zonal pattern of outer beach, primary dune, interdune swale, secondary dune, and salt marsh from the Atlantic Ocean to the Great South Bay. The zonal pattern is identifiable on the QuickBird-2 satellite image.

vegetation distribution provided guidance in assisting with image interpretation as well as digital image classification (Wang et al., 2007).

9.2.2 Imagery and Reference Data

For the Jamaica Bay site, the QuickBird-2 satellite data were acquired on September 10 and October 6, 2003, respectively. For the FINS site, the high spatial resolution QuickBird-2 satellite imagery data were acquired on April 8, April 16, May 4, and May 6, 2004, respectively. The QuickBird-2 data possess 2.5 m spatial resolution for the multispectral bands that cover the visible lights and near-infrared portion of the spectrum, and a panchromatic band with 0.6 m spatial resolution.

We conducted a thorough review and compiled existing information, such as historical aerial photographs, interpretation maps from previous studies, and other GIS data. For the FINS site we obtained a set of true color digital orthophotograph acquired by the New York Department of State and NOAA Coastal Services Center in 2002. We mosaicked the orthophotographs to provide a precise and accurate reference dataset for rectification of QuickBird-2 images and other rasterized GIS data. A previous vegetation mapping effort in FINS was accomplished as part of the NPS Vegetation Mapping Project (Klopfer et al., 2002). The vegetation types were visually interpreted and manually delineated from a set of 1996 color infrared and true color 1:1200 scale aerial photographs. The hardcopy aerial photographs employed were digitally scanned at 600 dpi and georeferenced for head-up digitizing. The classification was based on a classification system developed in conjunction with the mapping project (Sneddon and Lundgren, 2001). The classification system has five broadly defined vegetation groups including salt marsh, dune grassland, dune shrublands, interdunal swales, and forests/shrublands, which is further separated into 39 subcategories. The minimum mapping unit was 0.25 ha. The project provided the baseline information for the FINS. We refer this accomplished vegetation map in vector GIS format as the Phase I map. It was the most relevant and up-to-date data source for this study. The baseline data allowed us to make a further improved mapping using QuickBird-2 satellite image.

9.2.3 Initial Image Processing

We registered each scene of QuickBird images independently by image-to-image geometric rectification using 2002 true color aerial orthophotograph as the reference base. The reported accuracy of the orthophotograph was documented as 5.5 m. We selected from 10 to 20 ground control points between each QuickBird-2 image scene and the reference true-color digital image. Each scene was registered to an accuracy of 0.5 pixels, determined from the root-mean-square error. This would yield a combined accuracy of 6.75 m between our registration and the reported accuracy of the base image. Finally, the georectified scenes were mosaicked into a single seamless imagery dataset for the entire FINS and the Jamaica Bay study areas. To reduce the spectral effects of water surface on terrestrial vegetation and vice versa, we clipped the registered image using areas of interest (AOI). This process masked out the pixels of Bay water areas so that the classification could be focused on salt marsh and terrestrial vegetation types.

Spatial resolution is one of the main considerations when deciding which type of imagery to be used for a mapping project. Since the Phase I map of the FINS adopted 0.25 ha as the minimum mapping unit, it takes about 400 pixels at 2.5 m spatial resolution to cover the 0.25 ha area. Therefore the spatial resolution of multispectral bands of QuickBird-2 satellite imagery data was sufficient for vegetation mapping and in comparison with the Phase I map. For the Jamaica Bay salt marsh classification, we merged the 2.5-m spatial resolution multispectral image data and the 0.6-m panchromatic image data from the original QuickBird-2 image to create a new dataset that possesses 0.6 m spatial resolution and four spectral bands. The new dataset took advantage of high spatial resolution of the panchromatic band and broader spectral coverage from the multispectral data.

9.2.4 Field Observation and Classification Schemes

We conducted field observations for both the Jamaica Bay and FINS to support mapping efforts. We used the Global Positioning System (GPS) technique including a Trimble® ProXR GPS unit to record locations of field transects and points of interest. We recorded the general characteristics of the landscape and associated information using a data dictionary uploaded to the ProXR. We also examined vegetation types on the Phase I vegetation map as the reference. Another effort was to record georeferenced site photographs along transects and at points of interest. The georeferenced field photographs identified both geographic locations by latitude/longitude coordinates and the general compass bearings. When associated with QuickBird-2 satellite image, the field photographs helped in the identification of the locations and characteristics of vegetation types as well as the landscape of the sites. The fieldwork and observations provided essential independent reference data for identifying vegetation types to be mapped.

The vegetation classification scheme used for this study was modified from the Phase I mapping project. The Phase I classification system included 39 classes (Sneddon and Lundgren, 2001). For keeping the classification consistent we adapted the same classification scheme. For the aim of this study, we grouped and renamed categories associated with urban land-cover types into a single category of impervious surface cover. We kept the terrestrial vegetation types in the Phase I classification scheme unchanged. Some of the characteristics of the classes referenced in the Phase I study are described as follows (Klopfer et al., 2002).

Maritime Holly Forest type has a definite forest signature. It occurs just behind the backdune and can be distinguished from younger shrub lands by visible tree canopies, tall vegetation, and dominance of evergreen holly.

Old Field Red-Cedar Forest type was found on William Floyd Estate. Individual trees are smaller-crowned and scattered in with hardwoods. Distinct shadows are visible through the canopy.

Maritime Post Oak Forest type was limited to the edge of waterways on the Floyd Estate. Only one stand was identified through ground survey. This stand appears very similar to the other oak-dominated types found on the property.

Coastal Oak-Heath Forest type covers a large portion of the William Floyd Estate. Individual green *Pinus rigida* is often present but not in high enough proportion to classify as coniferous forest.

Acidic Red Maple Basin Swamp Forest type is found on both the Floyd Estate and Fire Island. It appears as a light grayish tree canopy over very dark understory and usually near water or other wetland types.

Pitch Pine Oak Forest type is very similar to *Quercus coccinea/Quercus velutina/Sassafras albidum/Vaccinium palladium* Forest in color and texture except that the coniferous *Pinus rigida* is more dominant.

Japanese Black Pine Forest (*Pinus thunbergii*) is found in many isolated patches on Fire Island. It is often used to stabilize the fore dune, especially on the eastern end of the island and around residential communities. This type appears very similar to the Pitch Pine Dune Woodland but the density is usually higher and the canopy more continuous.

Pitch Pine Dune Woodland type is found throughout the Fire Island behind the primary dune. Polygons appear splotchy due to varying canopy heights within one polygon signifying open canopy of *Pinus rigida* with shrubs/herbaceous layer below. Bright patches of sand are often seen within the polygon.

Maritime Vine Dune type is difficult to identify from photographs as the *Toxidendron radicans* is often found growing over shrubs. It is perhaps best described as what it is not, for it does not appear like *Myrica pennsylvanica* usually found around it. This type does appear as a brown color highly interspersed with open sand when occurring in larger polygons.

Northern Salt Shrub type is often found between the more obvious, taller shrub types on and around the backdune. It is always near wetlands. *Baccharis halimifolia* has a uniform gray color lighter than the surrounding wetter areas and very fine speckling can be seen. Individual plants are impossible to discern.

Maritime Deciduous Scrub Forest vegetation type is found on the bay side, often behind a large primary dune on wider parts of the island. It appears as a mix of brown, gray, and green with variable-sized discernable crowns in most cases.

Northern Dune Shrubland type is very common and dominates the interdune areas on Fire Island. It varies in height between 0 and 1 m. It is often interspersed with lighter-colored signatures of *Ammophila brevigulata* and *Hudsonia tomentosa*. This type has a dark brownish-red color and is typically uniform in texture throughout the polygon. Little sand shows through.

Highbush Blueberry Shrub Forest is a wet shrub type found on both Fire Island and the William Floyd Estate. It is a brown color finely dissected with dark lines, which is the water showing through the canopy. It is seen in noticeable depressions or swales throughout the interdune area.

Northern Interdunal Cranberry Swale association is found in the interdune zone as small, pond-like bodies of shallow water. They appear dark, like open water, but do have a partially submerged layer of *Vaccinium macrocarpon*. These small polygons are found mostly in the Otis Pike Wilderness Area of Fire Island.

Beach Heather Dune type is widespread on Fire Island and is found from Fire Island Inlet to the Moriches Inlet predominantly, although not exclusively, in the interdune zone. The color is a very dark green with equal amounts of bright sand showing through. It is often intermingled with *A. brevigulata*, which has a light tan color. Because of this mixing, this type is also found in complex with the Northern Beachgrass Dune association.

Overwash Dune Grassland type is difficult to identify from photography alone, although it occurs on recent overwash areas (breach) near the foredune. The vegetation is typically more sparse and therefore "washed out" on the photography.

Reedgrass Marsh type is found in and around most wetland areas on both the Floyd Estate and Fire Island. It has variable color and texture on the photographs ranging from a very "soft" light brown/tan to a dark, "dirty"-looking brown at the high ends of the salt marshes. There is often a readily identifiable band of horizontal debris associated with the bay side of these polygons, which can aid in identifying the taller *Phragmites australis* stands.

Low-salt Marsh is one of the two common salt marsh types found on both the William Floyd Estate and Fire Island. This type is associated with more regularly flooded parts of the marsh. This type is usually adjacent to water bodies, mosquito ditches, and salt pannes. The *Spartina patens* types are found in proximity but tend to be lighter in color and more tan.

High-salt Marsh association dominates the salt marshes of Fire Island and the Floyd Estate. It is found in proximity to *S. alterniflora* on the less frequently flooded portions of the salt marsh. This type has a uniform tan color and is often bisected by mosquito ditches.

Northern Sandplain Grassland type was rare and apparently limited to the wider parts of the Otis Pike Wilderness Area. The codominance of *Myrica pensylvanica* in this type makes it difficult, if not impossible, to separate from the Northern Dune Shrubland type.

Cultivated Pasture types are found on the William Floyd Estate where past land-use practice has left openings in the forest. Individual fields differ in species composition due primarily to management.

Brackish Meadow type is found uncommonly near the highest portions of the salt marsh on the bay side of Fire Island. It typically occurs in small polygons or in thin bands on the edge of the Northern Salt Shrub type.

Northern Beachgrass Dune type is perhaps the most prevalent on Fire Island. It is found on the ocean side of the interdune area from the crest through the high-salt marsh. This association appears as a uniform tan when cover is high, but can have streaks of bright sand showing through. It is often found in small polygons interspersed with the Beach Heather Dune type and with Northern Dune Shrubland polygons.

Brackish Interdunal Swale type is found behind primary and secondary dunes where saline surface water is found. When these wetlands are not dominated by *Phragmites australis* they appear dark and closely match open water. The vegetation is emergent and is difficult to identify specifically from the photography.

For the Jamaica Bay site, we focused on several categories to characterize the overall salt marsh compositions, including (1) *Spartina* area coverage in excess of 50%, (2) *Spartina* area coverage between 10% and 50%, (3) mudflat areas, and (4) high marsh for specific islands. The area with less than 10% salt marsh coverage was on the edge between open water and the classes between 10% and 50% *Spartina* cover (Figure 9.3b). The class of 10–50% *Spartina* cover is made up of mixed mudflat marsh areas. Interior areas have started to degrade but coverage is still mostly complete (Figure 9.3c). Areas classified as greater than 50% cover represent dense expanses of *S. alterniflora*. There are no breaks in the vegetation (Figure 9.3d). High marsh does

FIGURE 9.3 (See color insert following page 206.) (a) This QuickBird satellite image illustrates the locations of salt marsh islands in the Jamaica Bay. Field photographs show the example of salt marsh coverage in <10% (b), between 10% and 50% (c), >50% (d), high marsh (e), and mudflat (f).

not exist on all salt marshes in the Jamaica Bay but is typically found on marshes that have upland components such as Joco, Black Bank, and Ruffle Bar (Figure 9.3a and e). The high marsh areas can have a variety of different plants, such as *S. patens*, *Phragmites*, and *Distichlis spicata*. Areas classified as mudflat have little to no vegetation cover. Mudflats are typically inundated with water for longer periods of time than the areas covered by vegetation. There are often subtle elevation differences between the mudflat and vegetated areas (Figure 9.3f). We chose those classes for the reason that they represent the process of marsh loss within the Jamaica Bay. Marsh vegetations degrade proceeding along the path from greater than 50% *Spartina* coverage to between 10% and 50% coverage, and break into smaller isolated marsh islands. As the density of marsh patch decreased, marsh areas break up, turning from marsh vegetation to mud, and finally to open water. This path is not reversible for growth. Growth of marsh vegetation tends to follow a different pattern with solid piece of marsh expanding outward. Those generalized classes allowed us to compare the new data with existing historical salt marsh data of the park for a possible change analysis.

9.2.5 Digital Image Classification

For the FINS site, we used a stratified digital image classification technique to extract terrestrial vegetation types from the QuickBird-2 imagery data. This protocol requires the support from an existing vegetation map in GIS format, which defines approximate boundaries of vegetation types from previous interpretation delineations on aerial photographs. We referenced the vegetation communities identified by the Phase I map for this purpose. The stratified classification helped reduce errors caused by the similar spectral features among common terrestrial vegetation types on the FINS. The procedures are as follows.

First we rasterized the vector format vegetation data from the Phase I map. The resampled pixel size matched with the 2.5-m pixel size of QuickBird-2 satellite multispectral imagery data. We rectified the rasterized GIS vegetation map by image-to-image geometric rectification with the georegistered QuickBird-2 image as the base. Upon finishing, the cells of rasterized Phase I vegetation map matched with QuickBird pixels in the corresponding locations. We then used one vegetation type defined in the rasterized vegetation map as the mask to extract pixels from QuickBird image that fell into the areas. Those pixels would most likely belong to the type of masking vegetation. We repeated this procedure to extract pixels that correspond to each vegetation type and land-cover category defined by the Phase I map. This masking process separated the QuickBird-2 imagery data into subset images for further classifications.

We employed unsupervised classifications on those masked subset images. For labeling the clusters resulting from unsupervised classification, the dominant category should be the vegetation type defined as the masking type from the Phase I GIS vegetation map data. Pixels that had distinct spectral differences than the masked type were labeled as other corresponding types. We repeated this procedure until all the masked subset images were classified and correctly labeled. Upon finishing, we examined the agreements between GIS vegetation map and classification results. We then mosaic all classified subset images to create the final updated vegetation map.

For the Jamaica Bay site, as the salt marsh islands are relatively simple in species composition and the spectral properties, we employed unsupervised classification to extract information about salt marsh coverage from QuickBird-2 images.

9.3 RESULTS

9.3.1 Vegetation Mapping of the FINS

The stratified classification mapped the vegetation based on the classification scheme adopted from previous mapping efforts and supported by the existing GIS data. Figure 9.4 illustrates the comparison between a Phase I map and the result from stratified vegetation classification. We used randomly selected pixel locations in comparison with the Phase I vegetation map and our own ground verification data for accuracy assessment. A total of 1500 pixels, 50 from each category, were used. The overall accuracy was about 82% with the overall κ statistics as 0.81 (Table 9.1).

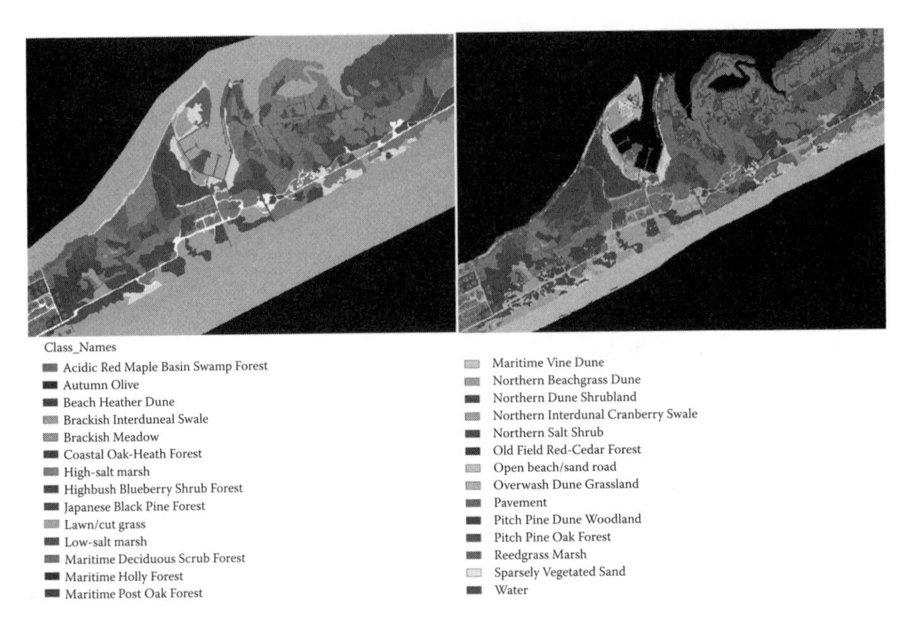

FIGURE 9.4 **(See color insert following page 206.)** Comparison of the Watch Hill section of the vegetation maps derived from manual delineation by the NPS vegetation mapping project and from stratified classification on QuickBird-2 satellite image.

The most significant error came between the categories of Coastal Oak-Heath Forest and Pitch Pine Oak. Spectral similarities between the two categories contributed to the misclassifications. Another major error occurred between categories of Northern Beachgrass Dune and Sparsely Vegetated Sand. These two categories were simply too dynamic and shared common features such as exposed sand that caused the confusion. Since the areas of Old Field Red-Cedar Forest and Maritime Post Oak Forest were too small, we eliminated the two types from accuracy assessment and the statistics. This does not necessarily mean that these vegetation types are not existing on the FINS. It needs to be identified by land managers who know the locations of these types. An update can then be done with postclassification editing if specific locations of these types are known and can be identified. Given that we used stratified classification with Phase I map data as the control, the resulting map could be considered as a new map updated from Phase I map with enhanced pixel level of details.

9.3.2 Salt Marsh Mapping of the Jamaica Bay

Classification results provided up-to-date coverage of the salt marsh categories in the Jamaica Bay. The total areas show that at the time of QuickBird-2 satellite image acquisition, the areas for >50% and between 10% and 50% *S. alterniflora* cover are about 170 and 120 ha, respectively, in the Bay. The high marsh areas are about 63 ha. The mudflat areas are about 89 ha. Figure 9.5 illustrates the salt marsh mapping for

TABLE 9.1
Accuracy Assessment of Vegetation Mapping at the FINS

Class	Reference Totals	Class Totals	Number Correct	Accuracy (%) Producers	Accuracy (%) Users	κ
Water	60	50	49	81.67	98.00	0.9791
Pavement/road	54	50	45	83.33	90.00	0.896
Open beach/sand	41	50	34	82.93	68.00	0.6703
Lawn/cut grass	42	50	38	90.48	76.00	0.7526
Sparse vegetation	48	50	34	70.83	68.00	0.6686
Autumn olive	46	50	45	97.83	90.00	0.8966
Maritime Holly	46	50	45	97.83	90.00	0.8966
Old Field Red-Cedar Maple	0	0	0	—	—	0
Maritime Post Oak Forest	0	0	0	—	—	0
Coastal Oak-Heath Forest	83	50	48	57.83	96.00	0.9575
Japanese Black Pine	53	50	45	84.91	90.00	0.8961
Pitch Pine Oak	17	50	15	88.24	30.00	0.2914
Pitch Pine Dune	41	50	38	92.68	76.00	0.7528
Northern Dune Shrub	57	50	44	77.19	88.00	0.8749
Maritime Deciduous Forest	72	50	39	54.17	78.00	0.7681
Acidic Red Maple Basin Swamp	35	50	34	97.14	68.00	0.6718
Maritime Vine Dune	46	50	44	95.65	88.00	0.8759
Highbush Blueberry Shrub Forest	53	50	40	75.47	80.00	0.7921
Northern Salt Shrub	42	50	34	80.95	68.00	0.6701
Beach Heather Dune	59	50	43	72.88	86.00	0.8538
Northern Interdunal Cranberry Swale	44	50	44	100.00	88.00	0.8761
Northern Beachgrass Dune	75	50	42	56.00	84.00	0.8309
Overwash Dune Grassland	50	50	49	98.00	98.00	0.9793
Brackish Interdunal Swale	44	50	41	93.18	82.00	0.8142
Brackish Meadow	38	50	38	100.00	76.00	0.7533
Reedgrass Marsh	63	50	47	74.60	94.00	0.9372
Low-salt Marsh	47	50	42	89.36	84.00	0.8344
High-salt Marsh	54	50	44	81.48	88.00	0.8752
Northern Sandplain Grassland	47	50	47	100.00	94.00	0.9379
Interdune Beachgrass Heath	43	50	41	95.35	82.00	0.8143

Overall classification accuracy = 82.07%.
Overall κ statistics = 0.8141.

the Jamaica Bay. Accuracy assessment based on 500 reference pixels selected by stratified random strategy indicates that the overall classification accuracy achieved was 84.6%. Most of the classification error came between the 10% and 50% *Spartina* cover and the mudflat categories (Table 9.2). The errors occurred between the 10% and 50% and >50% *Spartina* cover categories as well.

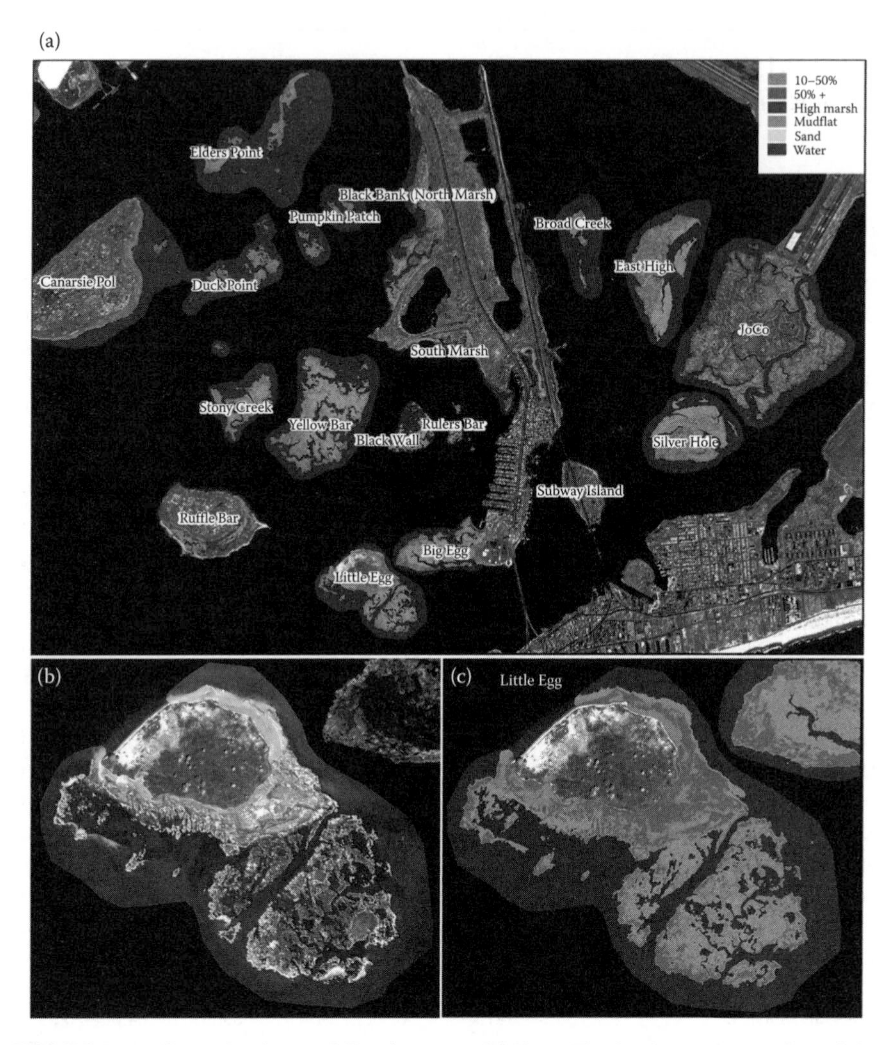

FIGURE 9.5 **(See color insert following page 206.)** (a) Final salt marsh mapping of the Jamaica Bay based on the classification of the QuickBird-2 satellite image. (b, c) The zoomed views illustrate the QuickBird-2 image and the salt marsh mapping of the Little Egg marsh island site.

9.4 CONCLUSIONS AND DISCUSSIONS

Monitoring of vegetation cover is a critical component for protecting coastal habitats of the National Parks. The barrier island ecosystem in the FINS is dynamic. Changes in dune topography, water availability, and salinity drive the community structure of the island. Periodic disturbances further change the vegetation communities. This dynamism, so integral to the function of the barrier island ecosystem, is often at odds with social and economic goals of the surrounding human communities. All of these factors contribute to the diversity and complexity of the vegetation found on Fire Island (Klopfer et al., 2002). As the barrier island is constantly changing, updating

TABLE 9.2
Confusion Matrix for Classification of Salt Marsh in the Jamaica Bay

Class Name	Reference Totals	Class Totals	Number Correct	Accuracy (%)	
				Producers	Users
Mudflat	120	101	95	79.2	94.1
10–50% *Spartina*	117	136	99	84.6	72.8
>50% *Spartina*	197	192	170	86.3	88.5
High marsh	66	71	59	89.4	83.1

Overall accuracy = 84.6%.

of vegetation data products on a regular basis presents a challenging and a necessary task. In order to accomplish a mapping task effectively, it is necessary to develop a protocol that can produce results quickly, accurately, and consistent with previous mapping efforts. The routine mapping should be conducted as a way to discover potential problem areas where critical vegetation communities are being lost due to a variety of human-induced or natural forces. Making these discoveries in a timely fashion means the difference between mitigation and recovery or just documenting the loss. This study that engaged high spatial resolution satellite remote sensing data and intensive field campaigns represents a successful effort to update the terrestrial vegetation map. With the existing Phase I terrestrial vegetation map data, this project produced the new pixel-based map data that should be valuable as a reference for future mapping efforts in this National Seashore.

In this study, changes of vegetation types that occurred after the Phase I project at the FINS were identified by classification of QuickBird-2 satellite images. The stratified and unsupervised classifications allowed the pixel locations with changed vegetation cover types to be identified as new categories, while unchanged pixels remained in the same category as the Phase I vegetation map. The advantages of this protocol include that it valued the previous vegetation mapping efforts, kept the classification scheme consistent, met the goal of updating vegetation map, and was valid for change analysis and monitoring. Because digital image classifications were pixel-based, more detailed spectral features within each delineated polygon from Phase I data were extracted. The GIS vegetation map played an important role as the control data layer. Combination of the GIS and QuickBird data helped achieve more comprehensive information on vegetation communities. Comparison of the two datasets allowed us to identify the locations and communities that were changed since the last mapping time period. Although simple and practical, this protocol requests the existence of vegetation maps in GIS format to guide the classification process. It means that this may not be an appropriate protocol to map areas where there is no existing vegetation map available in required level of details for both thematic information and spatial distributions.

Another consideration of satellite data acquisition is the orbit orientation. Most imaging satellites are in sun-synchronous orbit, which means that they track from north to south. With each pass the satellite has a limited imaging capability in the east–west orientation. To cover a study area that is predominantly oriented in the east–west direction, such as the FINS, it requires multiple days to acquire image data for the entire study area. The changing sky, tidal, and water column properties may result in having to classify the various images separately.

For the Jamaica Bay case study we concluded that unsupervised classification using high spatial resolution QuickBird-2 satellite imagery is a rapid, cost–effective, and accurate approach to map salt marshes. The protocol provides detailed steps to achieve the goal of salt marsh information extraction. With comparable spatial resolution as aerial photographs, the salt marsh information extracted from classification of QuickBird-2 image can be used for change detection against existing digital GIS format data derived from interpretation of aerial photographs. However, validation and justification must be conducted when a comparison is to be done between data from different sources and using different methodologies. Details of digital satellite imagery data involved in change detection, such as minimum mapping unit, classification system, projection and registration, purpose of original map data, and so forth, need to be considered.

These case studies demonstrate that high spatial resolution satellite remote sensing data provide a successful alternative data source for mapping both terrestrial vegetation in a barrier island setting and salt marsh in a bay setting. The map data and protocols developed provide the references for a long-term monitoring of a dynamic coastal ecosystems and aid in management decisions.

ACKNOWLEDGMENTS

The studies were funded by the NPS under the NCBN of the NPS's I&M Program through the NPS grant no. H4525037056 ("Remote Sensing of Terrestrial and Submerged Aquatic Vegetation in Fire Island National Seashore: Towards Long-term Resource Management and Monitoring") and by North Atlantic Coast Cooperative Ecosystem Studies Unit through grant no. J1770040405 ("Development of Salt Marsh Change Detection Protocol Using Remote Sensing and GIS"). We appreciate the contributions of Mike Bilecki and Diane Abell of the NPS FINS, Kathryn Mellander and George Frame of the NPS Gateway NRA, and Nigel Shaw, Charley Roman, Bryan Milstead, Sara Stevens, and Beth Johnson of the NPS for their expertise, insights, guidance, and logistical support. Fred Mushacke of the New York State Department of Environment Conservation helped greatly at the field. Our special thanks to Raymond Keyes for providing vessel facilities during our underwater video exercises as part of the fieldwork.

REFERENCES

Boumans, R.M.J., Burdick, D.M., and Dionne, M., 2002, Modeling habitat change in salt marshes after tidal restoration, *Restoration Ecology*, 10(3): 543–555.

Coppin, P.R. and Bauer, M.E., 1996, Digital change detection in forest ecosystem with remotely sensed imagery, *Remote Sensing Review*, 13: 207–234.

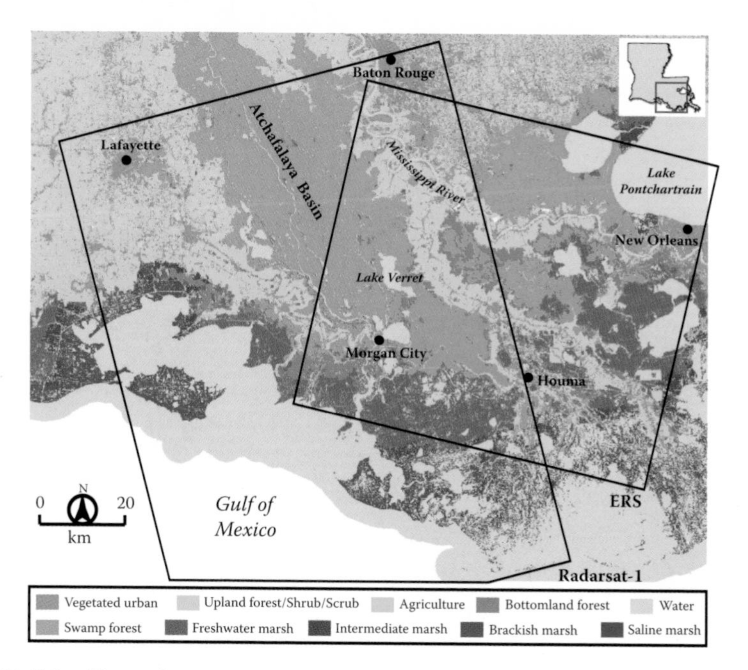

FIGURE 2.1 Thematic map. Modified from GAP and 1990 USGS National Wetland Research Center classification results, showing major land-cover classes of the study area. Polygons represent extents of SAR images shown in Figure 2.3 for the ERS-1/ERS-2 and RADARSAT-1 tracks.

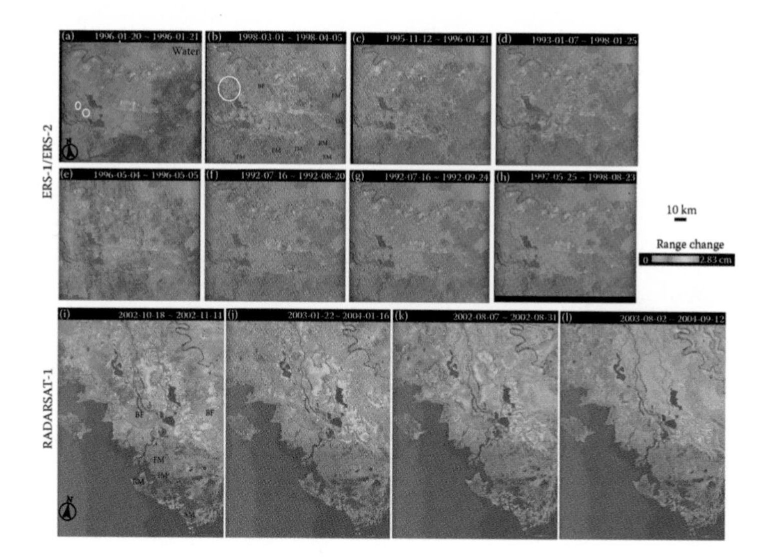

FIGURE 2.10 (a–d) ERS-1/ERS-2 InSAR images with different time separations during leaf-off seasons. (e–h) ERS-1/ERS-2 InSAR images with different time separations during leaf-on seasons. (i, j) RADARSAT-1 InSAR images during leaf-off seasons. (k, l) ERS-1/ERS-2 InSAR images during leaf-on seasons. Each fringe (full color cycle) represents 2.83 cm of range change between the ground and the satellite. The transition of colors from purple, red, yellow and green to blue indicates that the water level moved away from the satellite by an increasing amount in that direction. Random colors represent loss of InSAR coherence, where no meaningful range change information can be obtained from the InSAR phase values. AG: agricultural field, SF: swamp forest, BF: bottomland forest, FM: freshwater marsh, IM: intermediate marsh, BM: brackish marsh, SM: saline marsh, OW: open water.

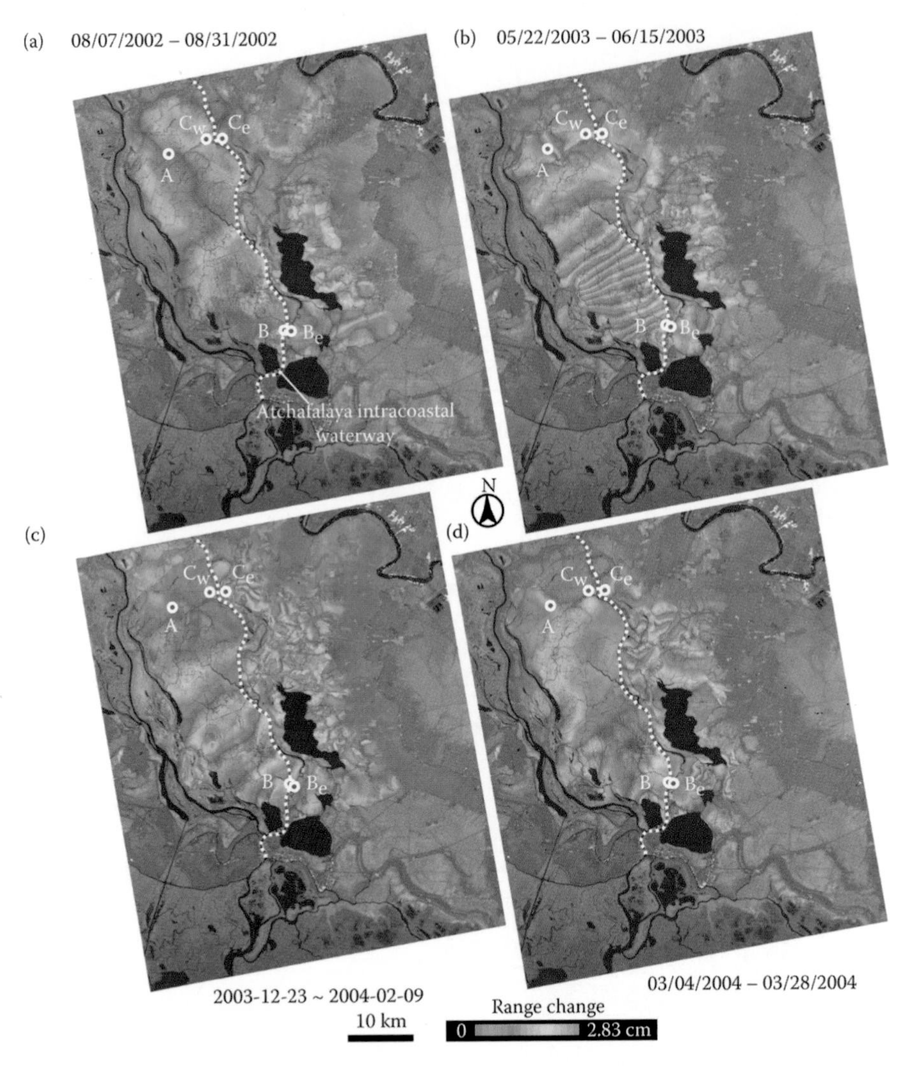

FIGURE 2.13 Unwrapped RADARSAT-1 images of the Atchatalaya Basin are used to quantify water-level changes over Atchafalaya Basin Floodway. InSAR-derived water-level changes at the selected locations are compared with gage readings (see Table 2.3 for details).

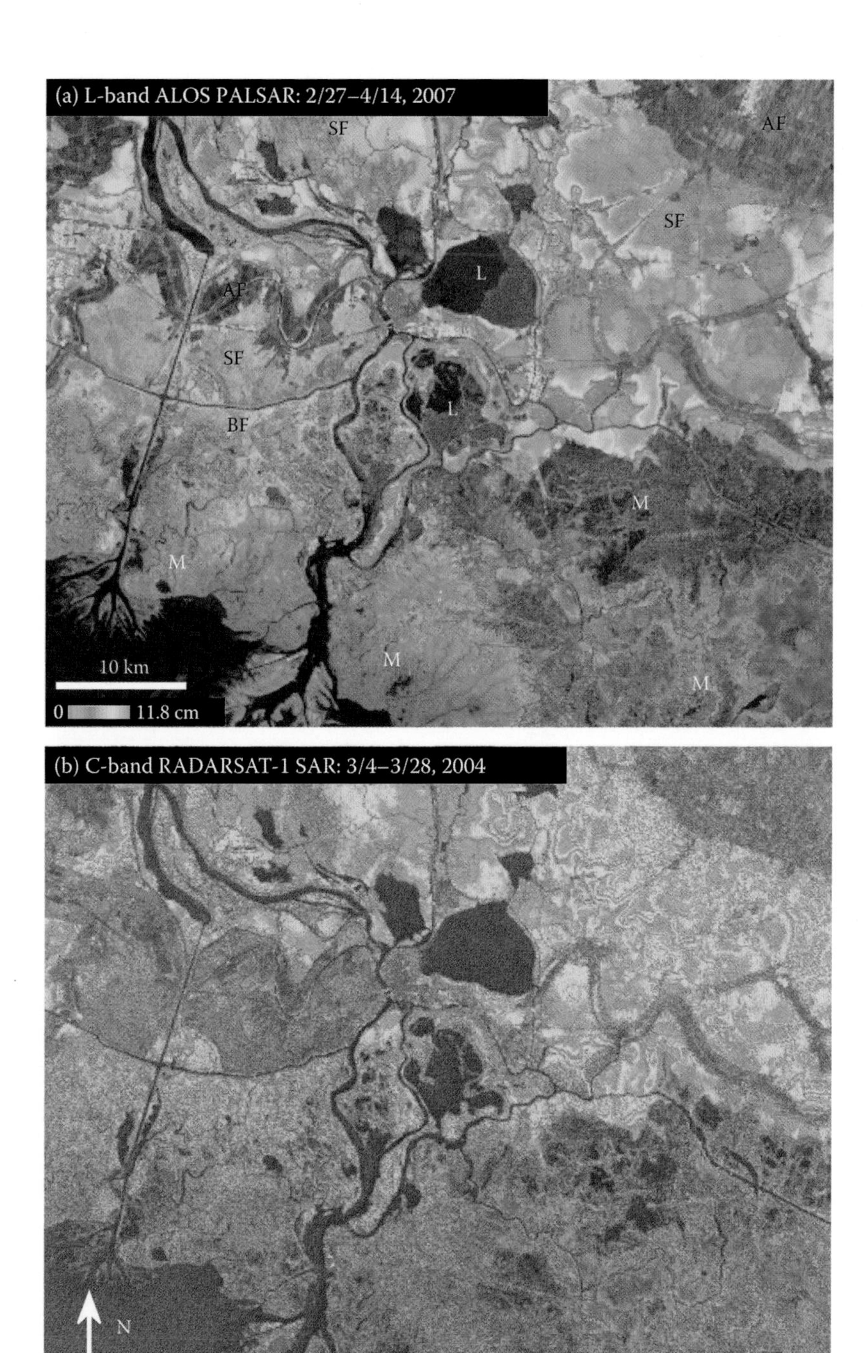

FIGURE 2.15 (a) L-band ALOS and (b) C-band RADARSAT-1 InSAR images showing water-level changes in coastal wetlands over southeastern Louisiana. Each fringe (full color cycle) represents a line-of-sight range change of 11.8 and 2.83 cm for ALOS and RADARSAT-1 interferograms, respectively. Interferogram phase values are unfiltered for coherence comparison and are draped over the SAR intensity image of the early date. Areas of loss of coherence are indicated by random colors. M: marshes (freshwater, intermediate, brackish, and saline marshes); L: lake, SF: swamp forest; BF: bottomland forest; AF: agricultural field.

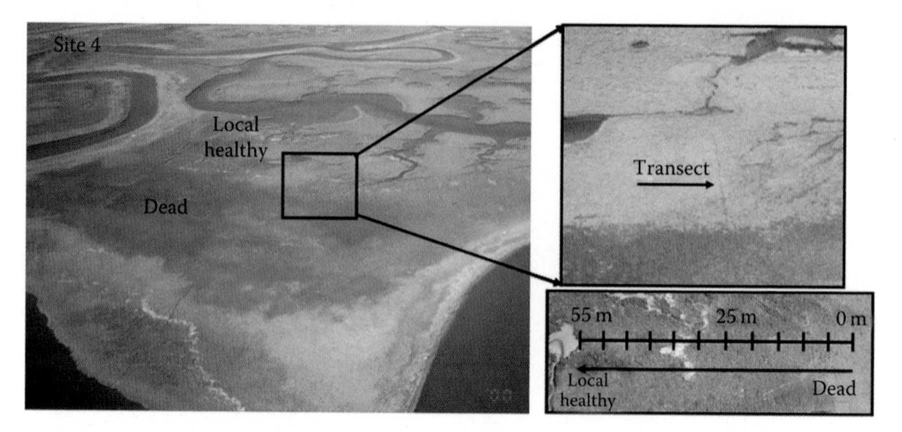

FIGURE 6.3 Aerial photography of the marsh dieback at site 4. The severest dieback impact is in the center, and the local healthiest marsh is in the background. The right top segment shows a closeup of the transect, and the right bottom segment shows the transect with plant sample locations (m). (Adapted from Ramsey, E., III and Rangoonwala, A., *Photogrammetric Engineering and Remote Sensing*, 71, 299, 2005.).

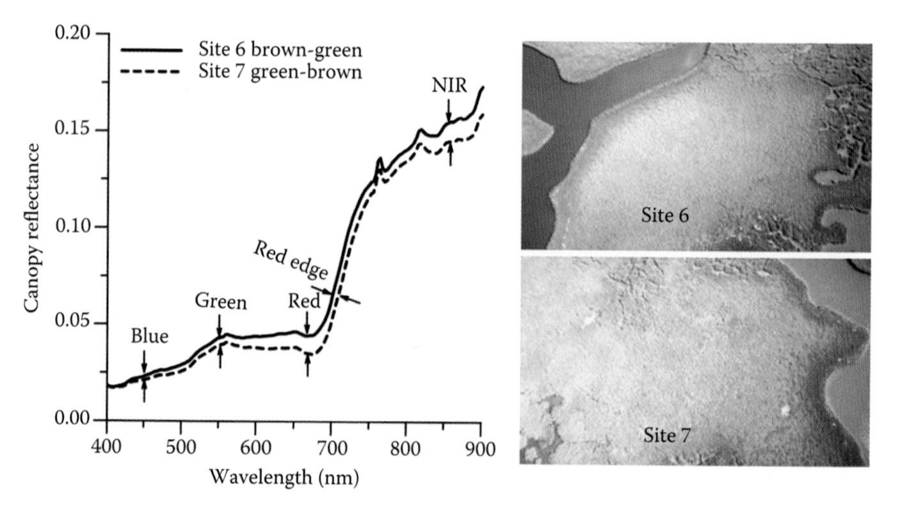

FIGURE 6.4 Canopy reflectance associated with two sites that exhibited spectral variability near the onset of dieback progression. Note the visual dieback classifications (brown-green, green-brown) and the small change in canopy reflectance (600–700 nm) that caused the perceived visual change.

FIGURE 9.2 Vegetation communities on the FINS barrier island follow the zonal pattern of outer beach, primary dune, interdune swale, secondary dune, and salt marsh from the Atlantic Ocean to the Great South Bay. The zonal pattern is identifiable on the QuickBird-2 satellite image.

FIGURE 9.3 (a) This QuickBird satellite image illustrates the locations of salt marsh islands in the Jamaica Bay. Field photographs show the example of salt marsh coverage in <10% (b), between 10% and 50% (c), >50% (d), high marsh (e), and mudflat (f).

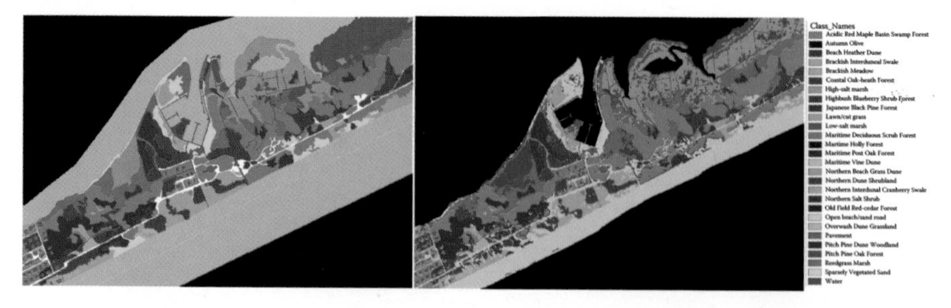

FIGURE 9.4 Comparison of the Watch Hill section of the vegetation maps derived from manual delineation by the NPS vegetation mapping project and from stratified classification on QuickBird-2 satellite image.

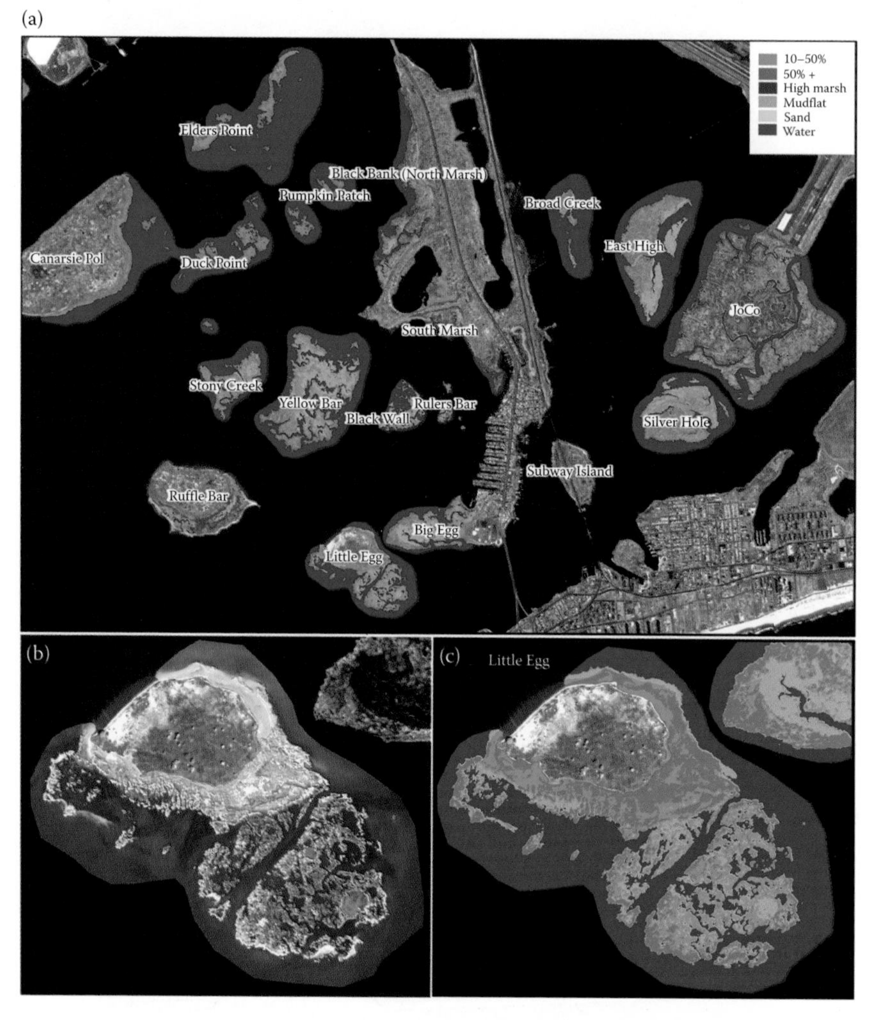

FIGURE 9.5 (a) Final salt marsh mapping of the Jamaica Bay based on the classification of the QuickBird-2 satellite image. (b, c) The zoomed views illustrate the QuickBird-2 image and the salt marsh mapping of the Little Egg marsh island site.

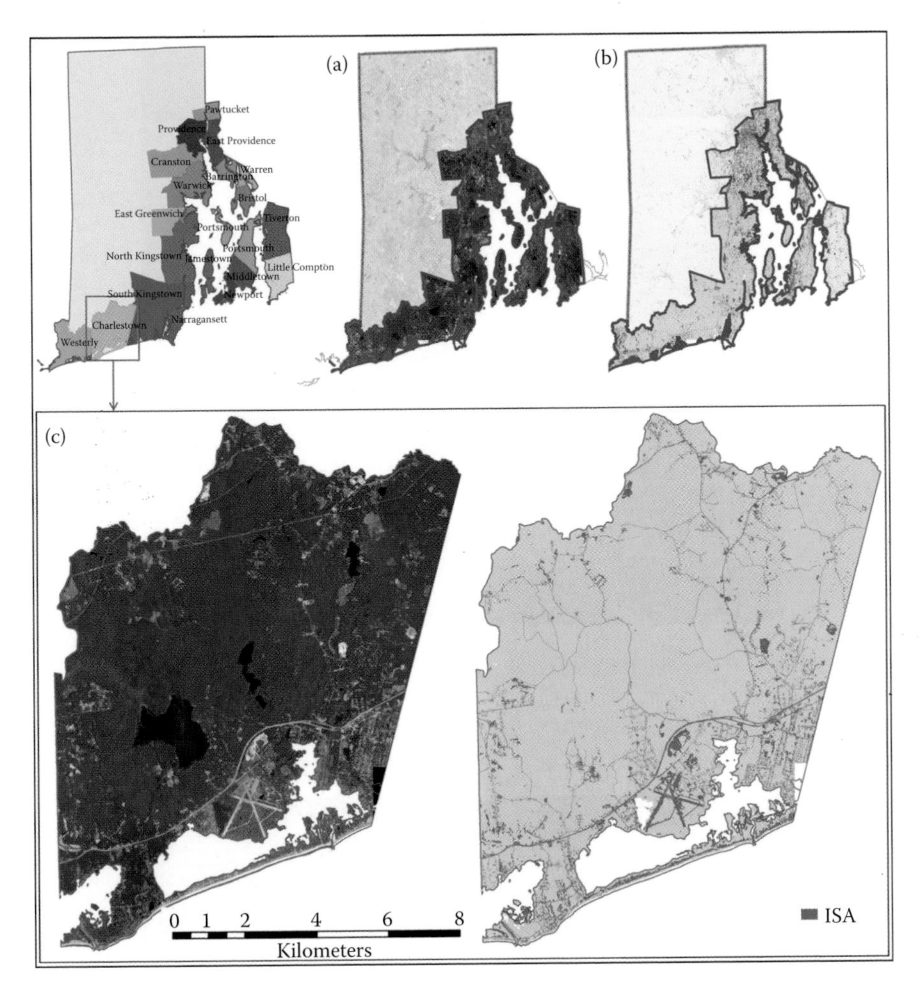

FIGURE 11.2 True-color orthophotography image data (a) and the extracted ISA (b) for coastal communities of Rhode Island. The zoomed example shows spatial details of the orthophotography data and ISA cover in Charlestown (c).

FIGURE 15.3 Underwater videography included (a) a video camera mounted on a slide, (b) GPS tracking of underwater camera, (c) VCR/TV monitoring equipment, (d) underwater footage, (e) and (f) direct field observation, and (g) example locations of underwater video transects.

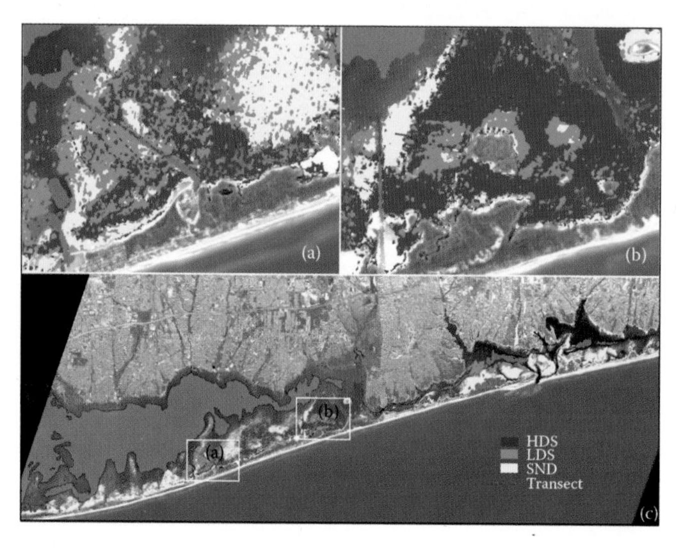

FIGURE 15.5 Zoomed display for selected sections of (a) and (b) and the SAV classification map displayed on top of the ALI image (c).

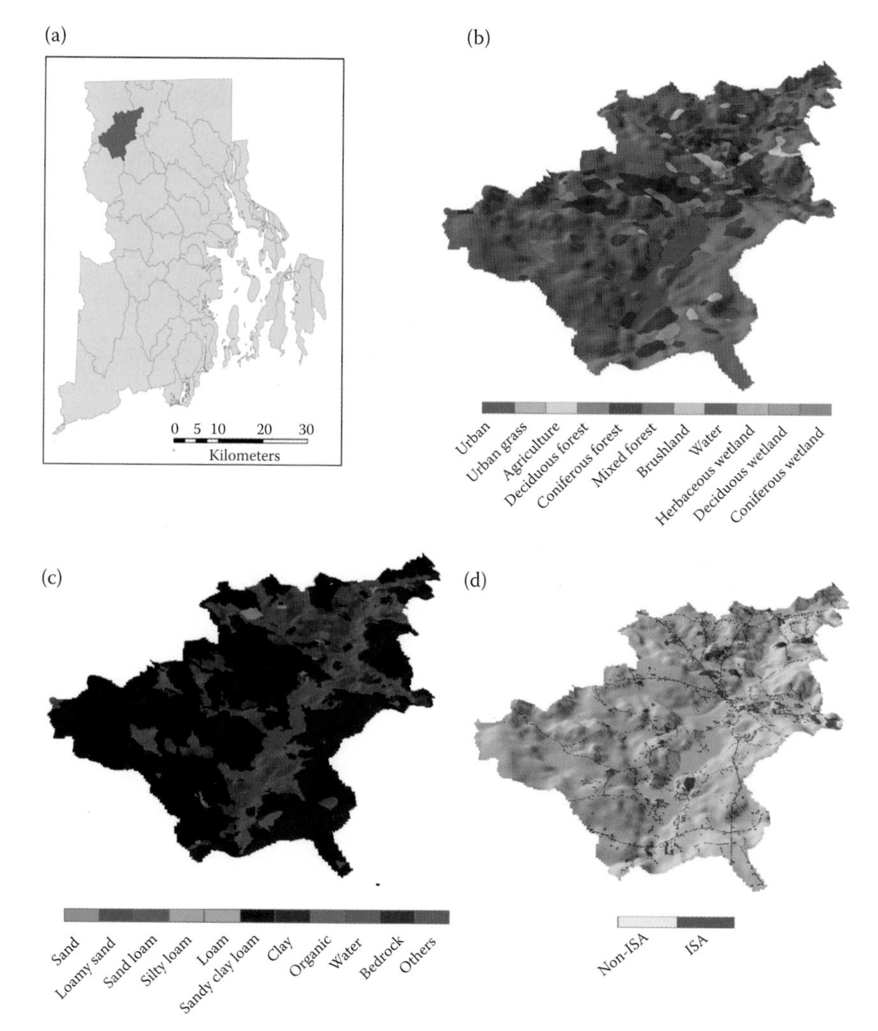

FIGURE 18.2 Location of an example watershed (a) and some of the selected data layers involved in modeling, including land-cover types (b), soil types (c), and high spatial resolution (1 m) ISA (d).

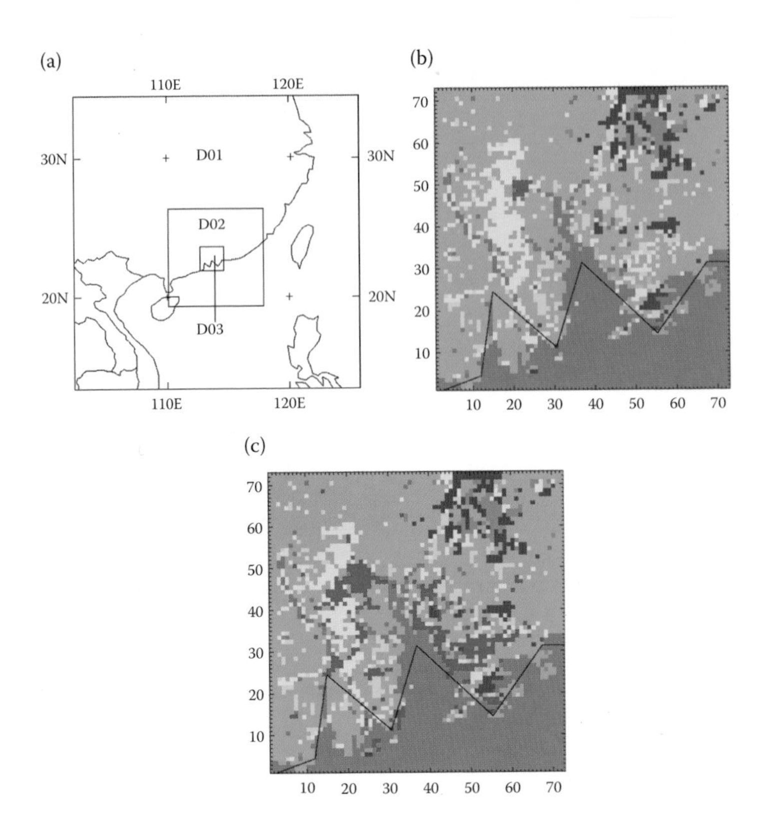

FIGURE 19.3 Computation domains and land-cover datasets used for MM5 simulations: (a) Computed domains. D01: 27 km grid spacing; D02: 9 km grid spacing; D03: 3 km grid spacing. (b) Original LC1980 data in the domain D03. (c) LC2004 data in the domain D03. The only change between (b) and (c) is the urban expansion.

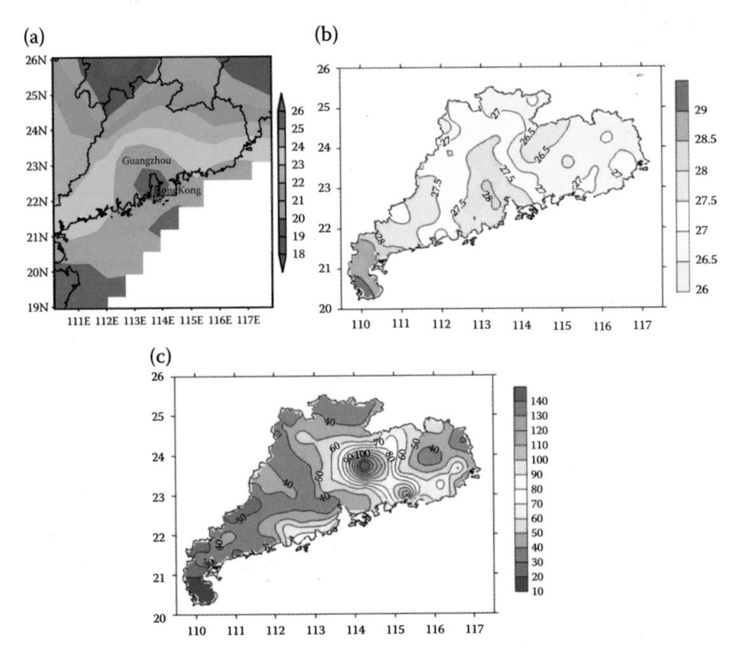

FIGURE 19.4 (a) Observed monthly mean 2 m temperature (unit: °C) of October 2004. (b) Observed monthly average 2 m temperature (unit: °C) of June 2005. (c) Observed monthly total precipitation (unit: cm) of June 2005. All data are provided by the Beijing Climate Center.

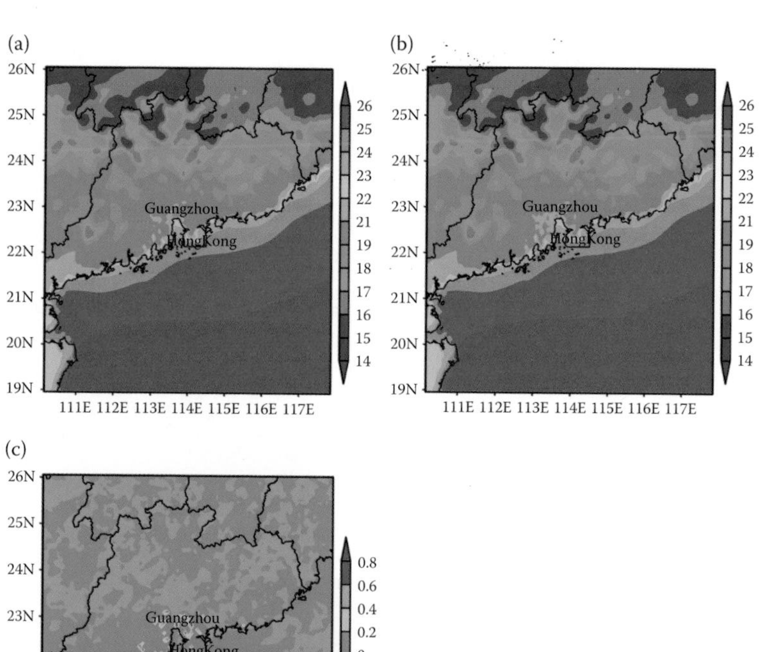

FIGURE 19.5 Simulated monthly mean 2 m temperature (unit: °C) of October 2004 in the domain D02 (a) with LC1980 data, (b) with LC2000 urban expansion data, (c) and the difference between (b) and (a).

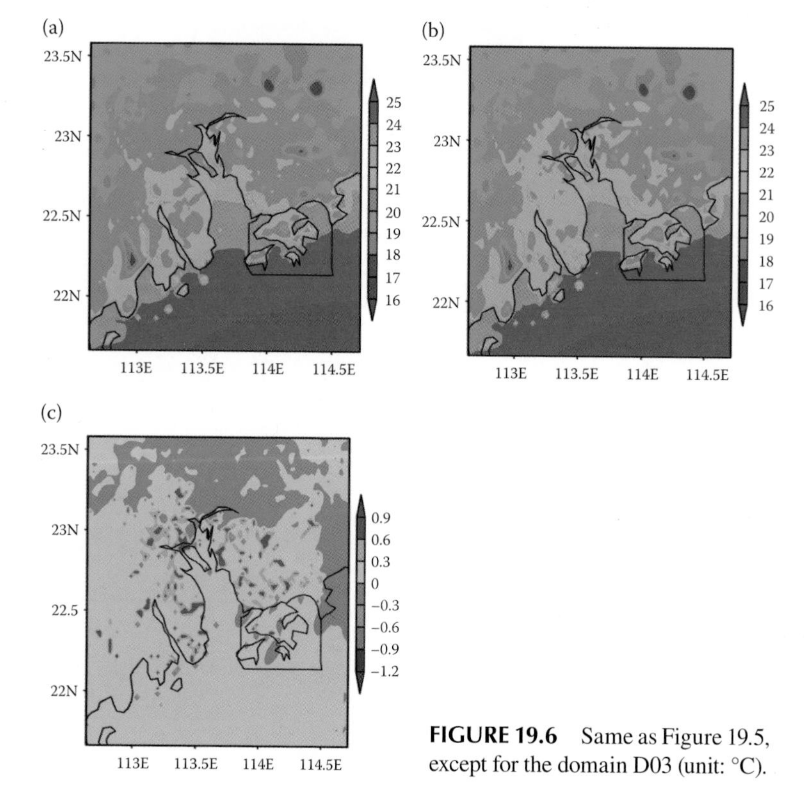

FIGURE 19.6 Same as Figure 19.5, except for the domain D03 (unit: °C).

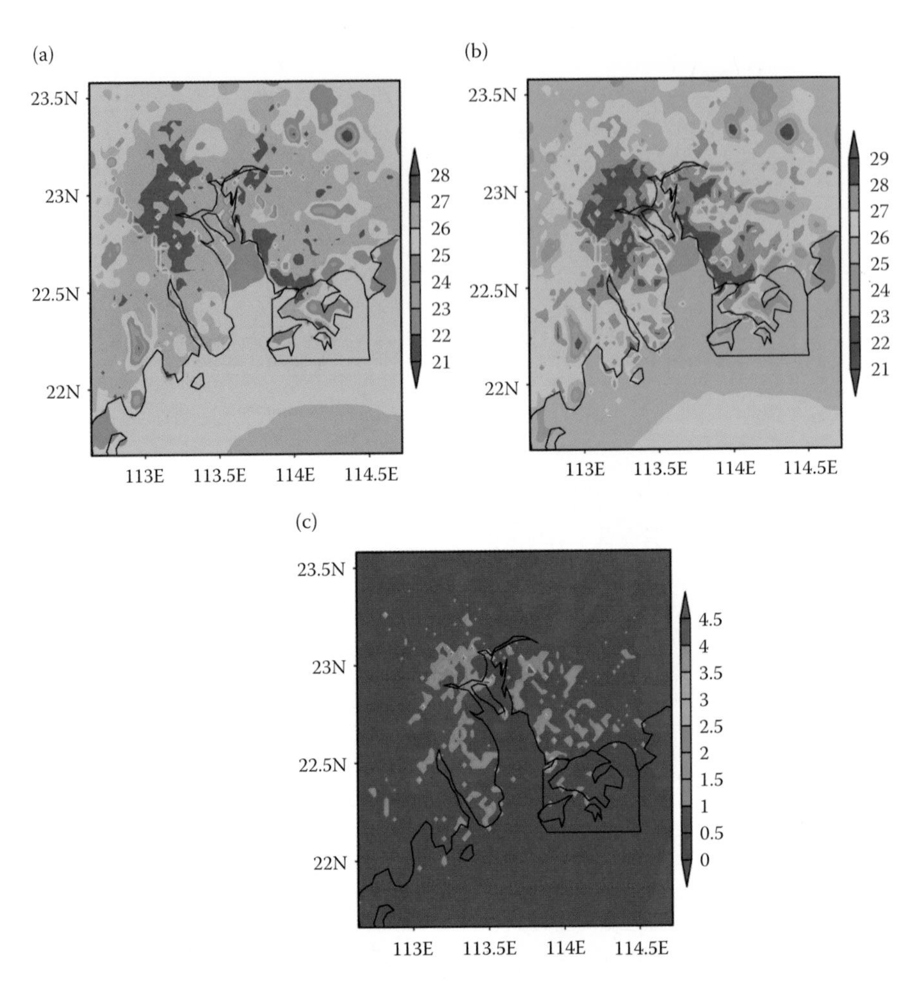

FIGURE 19.7 Same as Figure 19.5, except for daily maximum 2 m temperature (unit: °C).

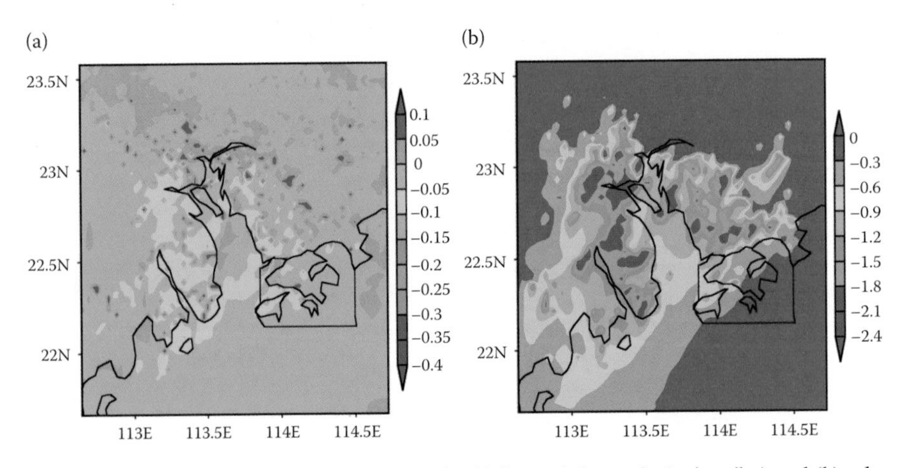

FIGURE 19.8 Same as Figure 19.5, except for (a) 2 m mixing ratio (unit: g/kg) and (b) relative humidity at level $\sigma = 0.995$ (unit: %).

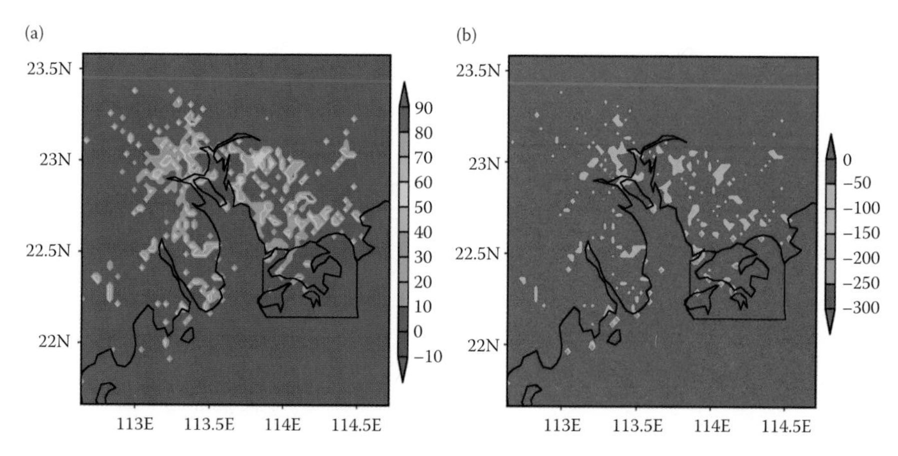

FIGURE 19.9 Same as Figure 19.5, except for (a) sensible heat flux (unit: W/m^2) and (b) latent heat flux (unit: W/m^2).

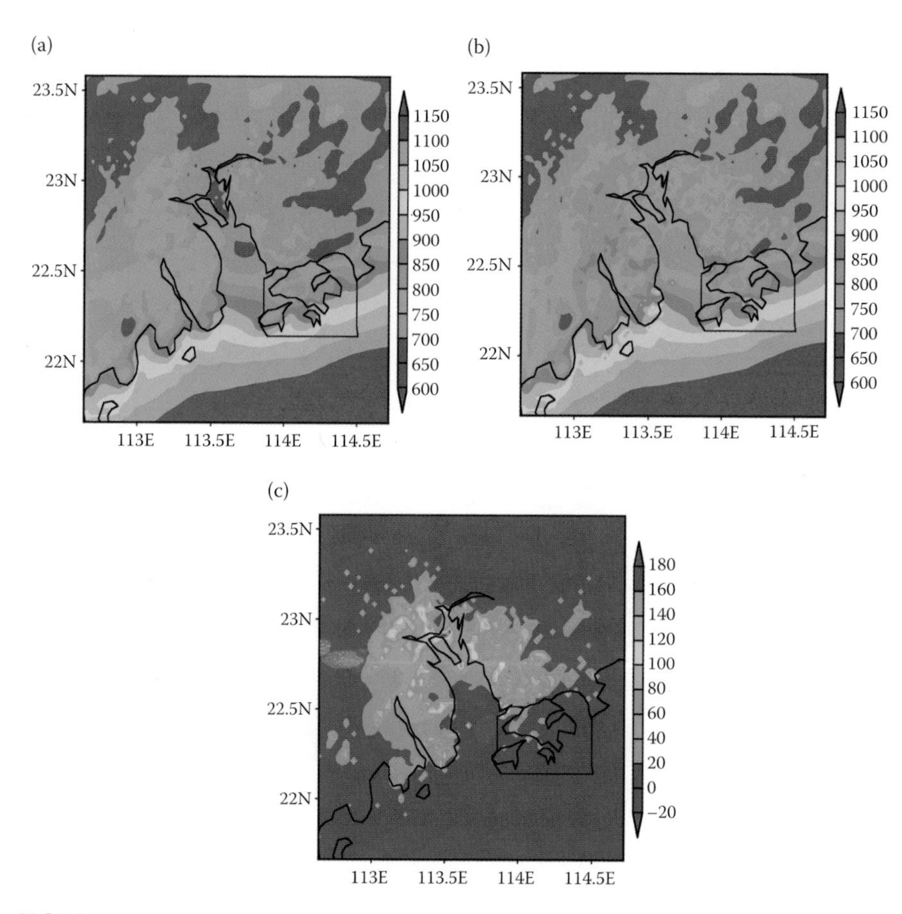

FIGURE 19.10 Same as Figure 19.5, except for PBL height (Unit: m).

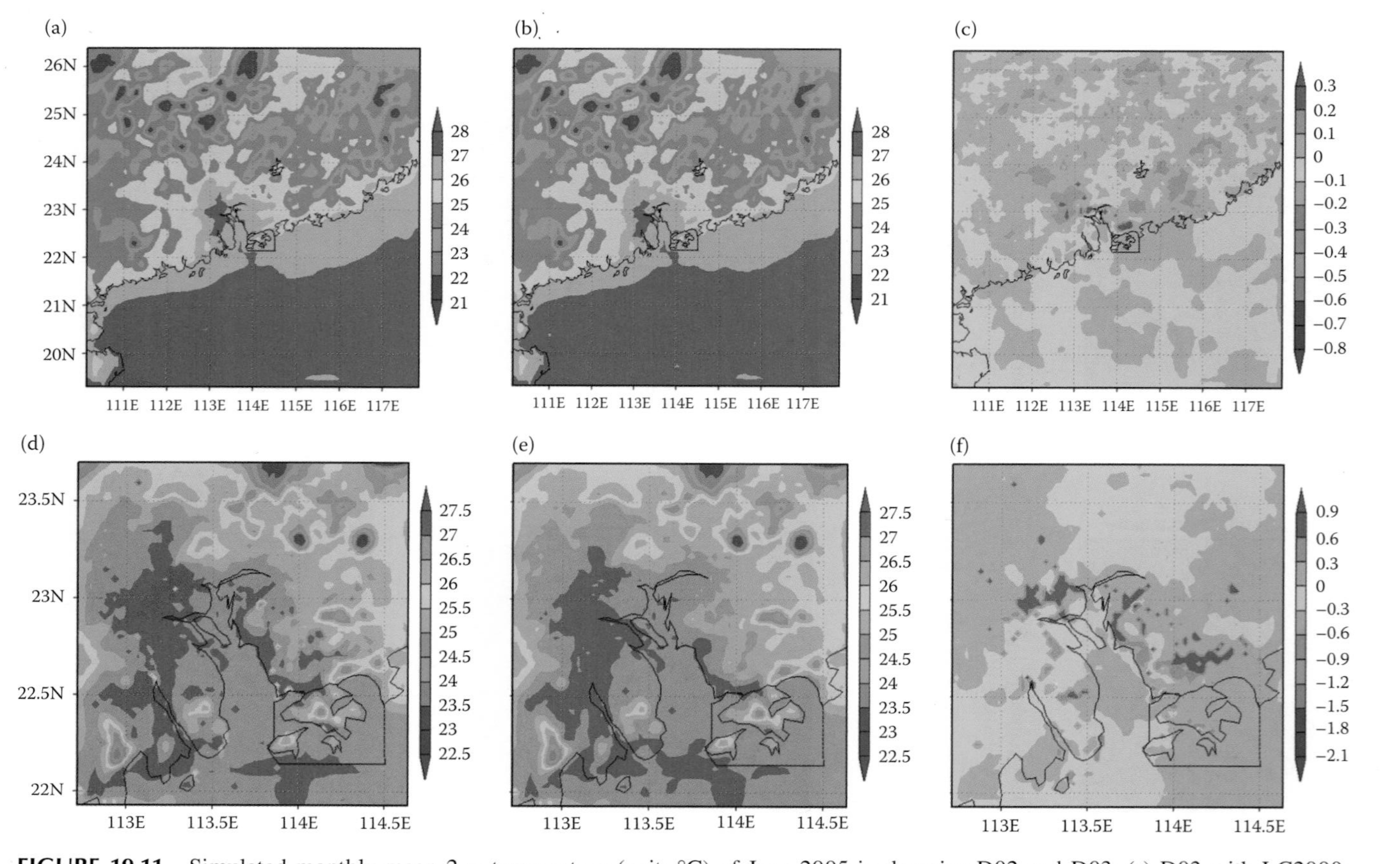

FIGURE 19.11 Simulated monthly mean 2 m temperature (unit: °C) of June 2005 in domains D02 and D03: (a) D02 with LC2000; (b) D02 with LC1980; (c) the difference between (a) and (b); (d) D03 with LC2000; (e) D03 with LC1980; and (f) the difference between (d) and (e).

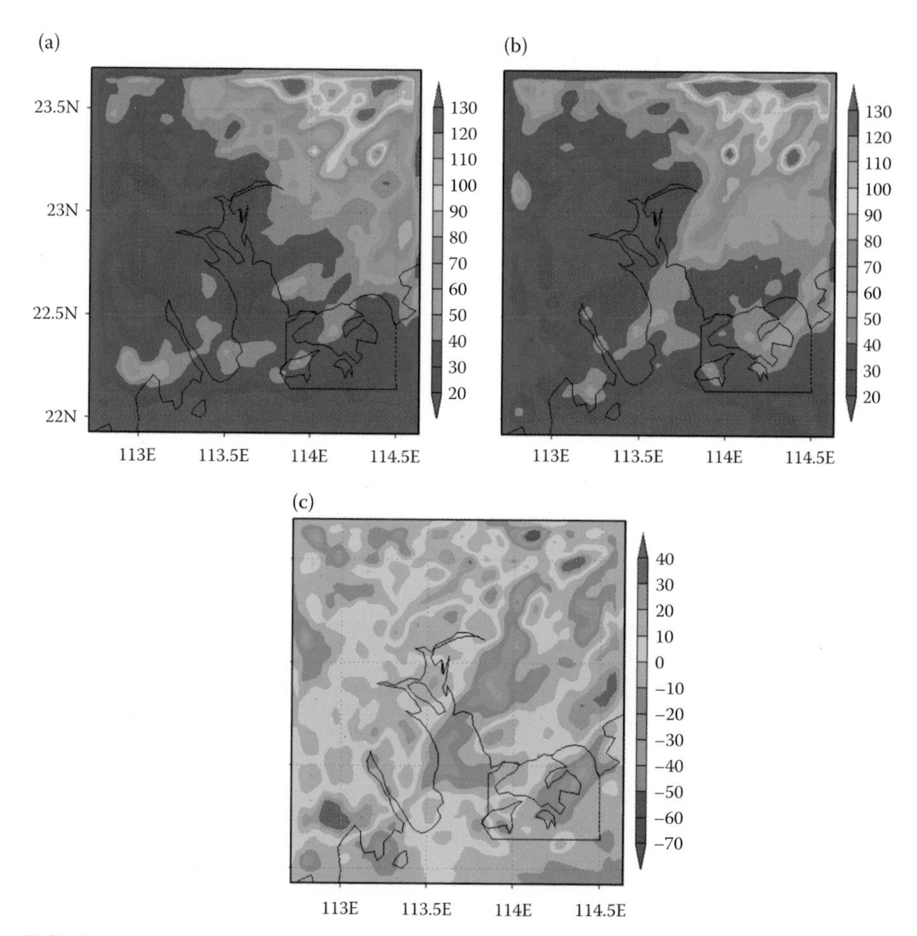

FIGURE 19.12 Simulated monthly total precipitation (unit: cm) of June 2005 in domain D03: (a) with LC2000; (b) with LC1980; and (c) the difference between (a) and (b).

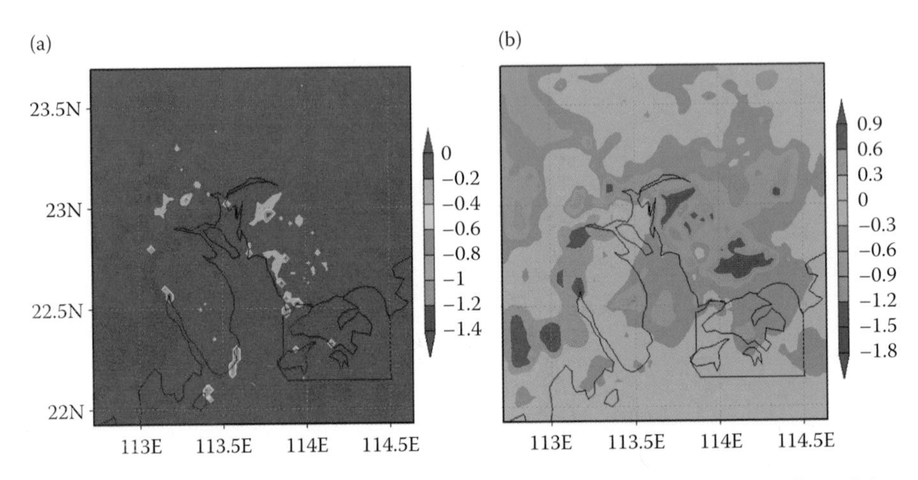

FIGURE 19.13 Differences between LC2000 and LC1980 in (a) monthly mean 2 m mixing ratio (unit: g/kg); (b) monthly mean relative humidity (unit: %).

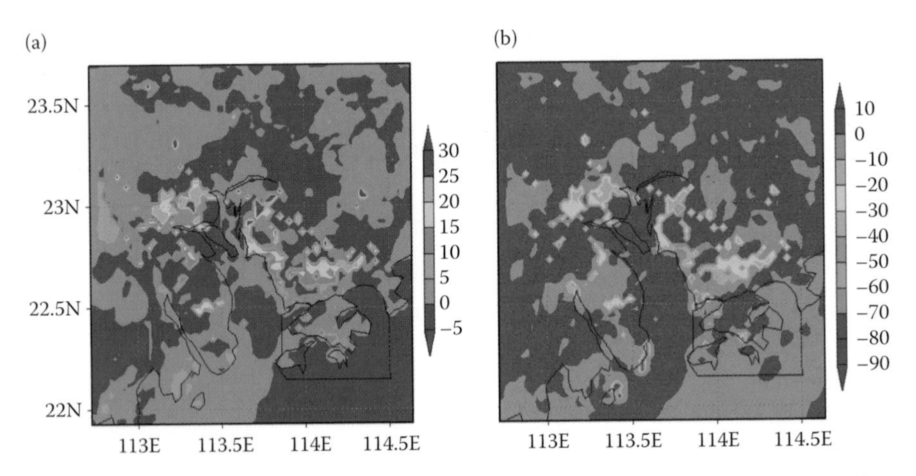

FIGURE 19.14 Differences between LC2000 and LC1980 in (a) monthly mean sensible heat flux (unit: W/m^2); (b) monthly mean latent heat flux (unit: W/m^2).

Dahl, T.E., 2000, *Status and Trends of Wetlands in the Conterminous United States 1986 to 1997*, U.S. Department of the Interior, Fish and Wildlife Service, Washington, DC, 82pp.

Edinger, G.J., Feldmann, A.L., Howard, T.G., Schmid, J.J., Eastman, E., Largay, E., and Sneddon, L.A., 2008, Vegetation classification and mapping at Gateway National Recreation Area, Technical Report, NPS/NER/NRTR—2008/107, National Park Service, Northeast Region, Philadelphia, PA.

Gateway National Recreation Area, 2003, A Report on Jamaica Bay, February 2003, National Park Service, U.S. Department of the Interior.

Griffith, J.A., Stehman, S.V., Sohl, T.L., and Loveland, T.R., 2003, Detecting trends in landscape pattern metrics over a 20-year period using a sampling-based monitoring programme, *International Journal of Remote Sensing*, 24: 175–181.

Hansen, A.J. and Rotella, J.J., 2002, Biophysical factors, land use, and species viability in and around nature reserves, *Conservation Biology*, 16: 1112–1122.

Hartig, E.K., Gornitz, V., Kolker, A., Mushacke, F., and Fallon, D., 2002, Anthropogenic and climate-change impacts on salt marshes of Jamaica Bay, New York, *Wetlands*, 22: 71–89.

Hefner, J.M., Wilen, B.O., Dahl, T.E., and Frayer, W.E., 1994, *Southeast Wetland; status and trends, mid-1970's to mid-1980's*, Department of the Interior, U.S. Fish and Wildlife Service, Atlanta, GA, 32pp.

Klopfer, S.D., Olivero, A., Sneddon, L., and Lundgren, J., 2002, Final report of the NPS vegetation mapping project at the Fire Island National Seashore, Conservation Management Institute, Virginia Tech, Blacksburg, VA, 205pp.

Moulton, D.W., Dahl, T.E., and Dall, D.M., 1997, Texas coastal wetlands-status and trends, mid-1950's to early 1990's, United States Department of the Interior, U.S. Fish and Wildlife Service, Albuquerque, NM, 32pp.

Neckles, H., Dionne, M., Burdick, D.M., Roman, C., Buchsaum, R., and Hutchins, R., 2002, A monitoring protocol to assess tidal restoration of salt marshes on local and regional scales, *Restoration Ecology*, 10(3): 556–563.

Niering, W.A. and Warren, R.S., 1980, Vegetation patterns and processes in New England salt marshes, *Bioscience*, 30: 301–307.

Ozesmi, S.L. and Bauer, M.E., 2002, Satellite remote sensing of wetlands, *Wetlands Ecology and Management*, 10: 381–402.

Peteet, D.M., 2001, Millennial climate and land use history from Jamaica Bay marshes, New York, The Geological Society of America (GSA) Annual Meeting Paper No. 187-0, Boston, MA, November 5–8.

Ritter, R. and Lanzer, E.L., 1997, Remote sensing of nearshore vegetation in Washington State's Puget Sound, *Proceedings of 1997 ASPRS Annual Conference*, Seattle, WA, Vol. 3, pp. 527–536.

Rogan, J., Franklin, J., and Roberts, D.A., 2002, A comparison of methods for monitoring multitemporal vegetation change using Thematic Mapper imagery, *Remote Sensing of Environment*, 80: 143–156.

Roman, C.T., Raposa, K.B., Adamowicz, S.C., James-Pirri, M.J., and Catena, J.G., 2002, Quantifying vegetation and nekton response to tidal restoration of a New England salt marsh, *Restoration Ecology*, 10: 450–460.

Schmidt, K.S. and Skidmore, A.K., 2003, Spectral discrimination of vegetation types in a coastal wetland, *Remote Sensing of Environment*, 85: 92–108.

Schmidt, K.S., Skidmore, A.K., Kloosterman, E.H., van Oosten, H., Kumar, L., and Janssen, J.A.M., 2004, Mapping coastal vegetation using an expert system and hyperspectral imagery, *Photogrammetric Engineering and Remote Sensing*, 70(6): 703–715.

Sneddon, L. and Lundgren, J., 2001, Vegetation classification of Fire Island National Seashore and William Floyd Estate, TNC/ABI Vegetation Mapping Program, Final Report.

Turner, W., Spector, S., Gardiner, N., Fladeland, M., Sterling, E., and Steininger, M., 2003, Remote sensing for biodiversity science and conservation, *Trends in Ecology and Evolution*, 18(3): 306–314.

Wang, Y., Tobey, J., Bonynge, G., Nugranad, J., Makota, V., Ngusaru, A., and Traber, M., 2005, Involving geospatial information in the analysis of land cover change along Tanzania Coast, *Coastal Management*, 33(1): 89–101.

Wang, Y., Traber, M., Milstead, B., and Stevens, S., 2007, Terrestrial and submerged aquatic vegetation mapping in Fire Island National Seashore using high spatial resolution remote sensing data, *Marine Geodesy*, 30(1): 77–95.

Warren, R.S., Fell, P.E., Grimsby, J.L., Buck, E.L., Rilling, G.C., and Fertik, R.A., 2001, Rates, patterns, and impacts of *Phragmites australis* expansion and effects of experimental *Phragmites* control on vegetation, macroinvertebrates, and fish within tidelands of the lower Connecticut River, *Estuaries*, 24: 90–107.

Wilson, E.H. and Sader, S.A., 2002, Detection of forest type using multiple dates of Landsat TM imagery, *Remote Sensing of Environment*, 80: 385–396.

Zhang, M., Ustin, S.L., Rejmankova, E., and Sanderson, E.W., 1997, Monitoring Pacific Coast salt marshes using remote sensing, *Ecological Applications*, 7: 1039–1053.

10 Object-Based Data Integration and Classification for High-Resolution Coastal Mapping

Jie Shan and Ejaz Hussain

CONTENTS

10.1 INTRODUCTION

Coastal areas refer to regions close to water, such as lake, sea, or ocean. They are considered to be very suitable for residential, recreational, and economic development. As a result, most coastal regions are subject to regular and intensive resource management, environmental protection, land development, and reclamation. The success of all such activities relies on good planning, for which it is essential to gather all relevant geospatial data. Acquisition of land-use and land-cover information through conventional field visits and surveying is very tedious, time-consuming, and costly. As an increasingly promising alternative, the use of high-resolution satellite and airborne remote sensing imagery provides a rapid and cost-effective means for the planning and management of coastal zones.

To acquire up-to-date land-use and land-cover information, various image classification techniques are utilized. They classify image pixels into a number of distinct categories or thematic classes. Although all such methods do perform well in many scenarios, a common problem encountered in image analysis is the separation of spectrally similar but actually distinct objects (Fisher, 1997). This difficulty can be due to the low spectral resolution (small number of bands) and high spatial resolution (small pixel size). Sensors of low spectral resolution can only record limited radiometric characteristics of the objects, based on which similar targets may not be differentiable. The improvement in the spatial resolution increases the within-class (internal) spectral variability and decreases the spectral variability between different classes (Alpine and Atkinson, 2001; Carleer et al., 2005; Bruzzone and Carlin, 2006). As the spatial resolution increases, the finer details may resolve features not recorded at coarser detail, thereby increasing rather than decreasing the proportion of mixed pixels, which are a source of errors and confusion (Campbell, 2002). This problem is minimized for images with relatively large homogeneous land-cover features like agriculture fields, large water bodies, wooded areas, but intensified in high-resolution images of urban areas due to spectrally similar small objects. Such a problem is tackled by the object-based image classification technique, which is reportedly effective and superior compared to the traditional pixel-based one (Benz et al., 2004; Wang et al., 2004), although there exist certain disadvantages, such as the difficulty in selecting segmentation parameters for objects of required size, necessary knowledge of the characteristics of different ground objects to form classification rule set, and more computationally intensive classification process. With the availability of high-resolution imagery and the development of object-based multiscale image analysis techniques, the research trend turns to explore the potential of the object-based image classification. Many studies in recent years were focused on the use and assessment of object-based classification techniques with reference to the traditional pixel-based classification (Niemeyer and Canty, 2001; Blaschke et al., 2005; Wei et al., 2005).

This study applies classification techniques to images as well as other geospatial data for coastal mapping. It compares the results of the pixel- and object-based mapping over a coastal area by Lake Michigan in the Lake County of Indiana, USA. The 1-m resolution color infrared (CIR) (three bands) orthoimages serve as the primary data source. Other data include the digital terrain model (DTM), digital surface model (DSM), and roads and streets data as a geographic information systems (GIS) layer. The topographic features to be mapped include coastal lines, sandy beach, sand dunes, water bodies including Lake Michigan and other small lakes and ponds, urban areas (roads and houses), open areas, and vegetation. The overall objective of this study is to examine the classification performance of high-resolution images with low spectral variability and the contribution of DTM, DSM, and road map in land-use and land-cover mapping of coastal areas.

The rest of the chapter consists of the following contents. Section 10.2 presents an overview of typical classification methods. Section 10.3 describes the geospatial data used in this study. Section 10.4 shows the classification results and discusses the joint use of auxiliary geospatial data besides images. The results from pixel- and object-based classifications are compared and evaluated. The study is summarized in Section 10.5.

10.2 PRINCIPLES OF IMAGE CLASSIFICATION

The objective of image classification is to automatically assign land-use and land-cover class categories to every pixel of the image (Lillesand et al., 2004). This section will address the principles of image classification, including classification unit, image segmentation, and several representative classification methods.

10.2.1 Classification Unit

Classification can be applied to different units in the image, either to an individual pixel or to a collection of connected pixels, called a segment. The original images are directly used as input for the pixel-based classification. The maximum likelihood classifier is one of the most commonly used pixel-based image classification methods. It is based on Bayes' decision theory, and makes a full use of the mean, variance, and covariance statistics from the selected training data. This classifier often assumes that the training data are Gaussian distributed. It classifies a pixel to the class that yields the maximum probability (Lillesand et al., 2004).

The object-based classification is based on image segmentation. The image is first segmented into spatial regions or objects at different resolutions, that is, multiresolution segmentation. The segmentation operates as a heuristic optimization procedure, which minimizes the average heterogeneity of image objects for a given resolution over the whole scene (Baatz and Schape, 2000). The objective is to construct a hierarchical net of image objects, in which finer objects are subobjects of coarser ones and each object "knows" its context, neighborhood, and subobjects. Thus, it is possible to define relations between the objects. The user-defined local object-based context information can then be used together with other features (spectral, spatial, and contextual) for classification. One of the popular schemes for object-based classification is the supervised fuzzy membership function approach. Object-based image analysis is often carried out in two sequential steps, that is, image segmentation and object classification, which are discussed below.

10.2.2 Image Segmentation

Image segmentation is the division of an image into spatially contiguous, disjoint, and homogeneous regions. It is the prerequisite for object-based image classification, in which an object or segment other than a pixel is used for classification. The aim of segmentation is to create meaningful objects by using appropriate methods and parameters to maintain the geometry of the objects as closely as possible (Blaschke and Strobl, 2001). There are many segmentation methods in practice for different applications. For land-use and land-cover mapping, the desired segmentation method should work equally well for all variable image contents and produce the segments that closely resemble the actual objects. Traditional image segmentation methods are pixel based (gray level), edge based (edge detection), and region based (splitting and merging).

Pixel-based segmentation is a sort of unsupervised classification. It uses the gray values of the pixels to segment an image with no regard to the pixel's neighborhood. It results in spectrally similar segments but not necessarily spatially contiguous regions. The edge-based method exploits the intensity variations at the edges and boundaries of image objects and searches for the edges discontinuities that exist between homogeneous areas. For segmentation, it looks for local maximum gradients and then the algorithm follows these gradient maxima until it reaches the starting pixel. The detected edges are linked to form the segments. The region-based method starts from a single seed pixel and then iteratively grows by grouping to neighboring pixels with similar intensity. Connected groups are then merged based on the homogeneity criteria such as color, intensity, shape, and texture. It continues until certain limits of the parameter set are met or all pixels are assigned to groups. Splitting is opposite to merging where it starts with large segments and iteratively splits them to smaller ones based on certain criteria.

Segmentation can be implemented at multiple resolutions. It can take a number of data layers (either raster or vector) in this process and the contribution of each data layer can be weighted according to its suitability for segmentation. The higher a layer's weight, the more its contribution to the segmentation. Thematic layers (such as roads and river vector data) can also be included in this process to control object splitting and restrict objects from crossing over the thematic boundaries. Segmentation of an image into meaningful objects is based on some heterogeneity criteria that include user-defined scale, color, shape (smoothness and compactness), and parameters (Benz et al., 2004). Analysts are always interested in getting the objects that closely resemble the real world, not just the object primitives. No specific, fixed, or optimal values for these parameters have yet been found that can be applicable to all situations. It mostly relies on trial and error to get the desired objects. Since not all objects can be well represented at one resolution, several hierarchical segmentation levels are often used to represent different objects of interest (Moeller et al., 2004).

Several parameters can be used in the object-based segmentation. Scale is an abstract value to determine the maximum possible change of heterogeneity caused by fusing several objects. The value of the scale parameter affects the size of the resulting objects. A higher scale will allow for merging more objects and resulting larger objects, while a lower scale will result in small size objects. The heterogeneity criterion is based on shape and color. Assigning a higher weight to shape results in a lower spectral or color contribution and more spatial homogeneity. Similarly, a higher weight to spectral component will lead to a lower contribution of shape in the segmentation task (Benz et al., 2004). The shape criterion is composed of smoothness and compactness. The smoothness describes the shape heterogeneity as a ratio of similarity between the image object border and a perfect square. It optimizes the image objects with respect to the smoothness of their borders. The compactness describes how an object is close to a circle (Baatz and Schape, 2000). As a result, the image can be segmented into homogeneous objects of different sizes.

10.2.3 Classification Methods

Image classification can be carried out in either unsupervised or supervised manner. The unsupervised classification is the identification of natural grouping of spectral values in the multispectral image. It first converts the image into inherent groups (also called clusters), followed by assignment of class categories to these groups. It requires no prior knowledge of the area at the clustering stage. However, without this knowledge it is difficult to identify and assign informative categories to these groups at the second stage. These spectrally homogeneous clusters can be very useful in selecting training samples for supervised classification (Schowengerdt, 1997). The *k*-means and iterative self-organizing data analysis (ISODATA) methods are two widely used methods for unsupervised classification. The *k*-means clustering starts with the specified number of clusters to be created. It assigns an arbitrary vector mean to the specified clusters and then each pixel is assigned to the cluster with the closest mean vector. It iteratively calculates the new cluster means and the pixel assignment is updated to the relevant cluster. This iterative assignment process continues until there is no substantial change between the successive iterations (Lillesand et al., 2004). The ISODATA algorithm starts with an initial threshold on the number of pixels to form a cluster. It assigns each pixel to a cluster with the closest mean and iteratively continues the process until a predefined threshold is approached. It is more suitable for recovering compact clusters (Theodoridis and Koutroumbas, 1999). It additionally merges the clusters that have less number of pixels than a threshold or if the center of two clusters is too close. It can also split a big cluster if the cluster standard deviation exceeds a predefined value. The ISODATA method can result in any number of clusters as against to the *k*-means method, which usually assumes the known number of clusters in the data.

Supervised classification exploits a priori knowledge about the data to identify which of the predefined informational classes closely resembles a classification unit. This prior knowledge typically is in the form of training samples, that is, a collection of representative data points, for which the class labels are known to the analyst. These training samples are then used by the classifier to identify the class labels of the unknown data points. As the classification process is based on the training data, they must be homogeneous and represent the classes as closely as possible (Landgrebe, 2003). Also, the training samples for a class should represent within-class spectral variations in the image. An image with high within-class and low between-class spectral variability makes it hard to select truly representative class samples to achieve satisfactory classification results.

The most commonly used supervised classification algorithms are parallelepiped, minimum distance to means, and maximum likelihood classifiers (Lillesand et al., 2004). Both parallelepiped and minimum distance to means algorithms are nonparametric, whereas the maximum likelihood classification algorithm is parametric. Nonparametric classifiers assume no statistical distribution, while the parametric ones assume statistical distributions of the data points (Schowengerdt, 1997). Training samples for parallelepiped classification algorithm require the lowest and highest pixel values of all bands, which form the lower and upper bounds of parallelepiped (Richards and Jia, 2006). Firstly, the mean vector and standard deviations are calculated for each class from the training data. Then, each pixel is classified under the class it falls. If the pixel does not fall into any of the parallelepipeds, it is labeled as unclassified. In the minimum distance to means classification, the means of each training class for all used bands are calculated. Then, the distance from each unknown pixel to every class means is calculated and the pixel is classified under the one that yields the closest distance. The Euclidean distance is the most commonly used distance measure out of a variety of methods for this classifier (Mather, 2005).

Technically, the object-based techniques can use any of the above methods for classification. However, a type of new method has been introduced, in which a classification unit can belong to more than one class at different likelihoods. This is the so-called fuzzy or soft classification. Typical soft classification methods include the fuzzy logic nearest neighbor classification and the fuzzy membership function approach. The nearest neighbor classifier uses the minimum distance to means for classification. The object-based nearest neighbor approach is similar to the supervised classification as it requires the selection of representative training samples for each class to train the classifier. The training sample objects may be just a few as every sample object can contain a number of pixels. After a representative set of the sample objects has been selected for each class, the algorithm searches for the closest sample object in the feature space for every unknown object. It assigns the unknown object to the class whose sample object is closest. The fuzzy nearest neighbor classifier assigns a membership value between 0 and 1 based on the object's distance to its nearest neighbor. If the image object is similar to a sample, it gets a membership value of 1, otherwise between 0 and 1, depending on the distance to the nearest sample object. The fuzzy membership function classification is based on the fuzzy logic principle that uses specific rules, vague, and even linguistic description of classes for the classification (Benz et al., 2004). It works on the possibilities that an object may belong to one or more classes at the same time. It best suits the situation where classes have high spectral, spatial, and geometrical variability. For this purpose, one needs to find and select appropriate features, and define membership functions for every class of interest. The classification results depend on these input features and the degree of membership value of a class. The closer the membership values to 1 with no or less alternative assignments, the better the classification results.

10.3 TEST DATA

The selection of data for mapping the coastal areas depends on the nature of the applications. Such applications may vary among coastline change, bathymetry, cartography, coastal wetlands, coastal vegetation, harbor limits, and general land use and land cover. In many of these applications, a variety of field surveys, old maps, satellite and aerial images, and GIS data are available. For this study, orthorectified aerial images, DSM, DTM, and road data are used. These data are available through Indiana spatial data portal (http://www.indiana.edu/~gisdata/). The CIR images were acquired from multiple flight lines via aerial photography during the leaf-off season in March and April 2005. They consist of green, red, and infrared (IR) 3 bands at a spatial resolution of 1 m. Their radiometric resolution is 8 bits per pixel. They were orthorectified with DSM. The DTM and DSM are at a spatial resolution

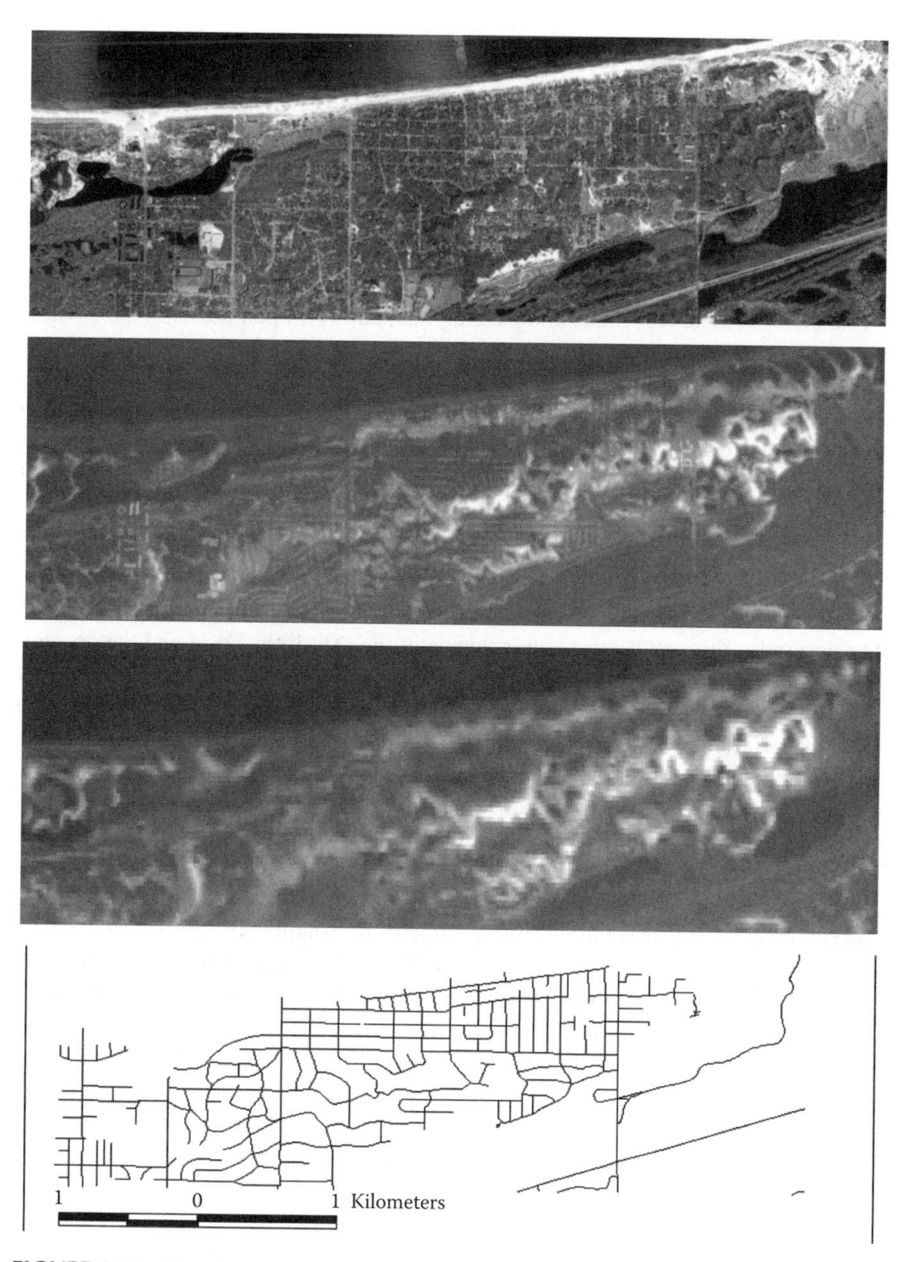

FIGURE 10.1 Data for the study area. From top: CIR orthoimage, DSM, DTM, city streets, and roads. The water at the top of the figures is Lake Michigan.

of 1 m. Besides these raster data, we also have road and street data over the mapping area. All data are under the UTM Zone 16 NAD 1983. The data used are shown in Figure 10.1. The software tools used in this study are ERDAS Imagine, ENVI, and Definiens Developer.

10.4 CLASSIFICATION RESULTS AND EVALUATION

The study area contains a variety of geographical features such as water bodies, sandy coastal area, deciduous trees and shrubs, open bare areas, and mixed urban areas. Visual analysis of the images leads to five very broad informative classes: water, sandy areas, open land, vegetative areas, and built-up areas. However, to best exploit the potential of the high-resolution aerial images, we further divide them into 11 more detailed classes: (1) Lake Michigan, (2) coast sand, (3) sand dunes, (4) lakes and ponds, (5) dry lakes and ponds, (6) trees, (7) grassy areas, (8) deciduous tree areas, (9) buildings, (10) roads and parking lots, and (11) open areas. Some of these classes are spectrally very similar and confusing, for example, roads and buildings. Buildings with multiple roof colors have high spectral variability, thus a high tendency of mixing up with other classes in the dataset. It should be noted that the images were acquired from multiple flight lines, which results in high spectral variations, noise, and artifacts. These factors badly influence the classification results of the ground objects that appear spectrally different but actually the same.

10.4.1 Pixel-Based Classification

The design for this test is based on inclusion of the DTM, the DSM and the normalized digital surface model (nDSM, i.e., DSM–DTM) along with the image data within the classification process. The classification is performed using a maximum likelihood classifier on image only, image and DTM, image and DSM, image and nDSM, and the combination of all. Maximum likelihood classifier estimates the means and variance–covariance matrix of the classes. The strength of these statistical properties depends on the size and representativeness of the training samples. Therefore, well-spread, homogeneous, highly probable, and adequate numbers of pixels for these classes should be selected for training (Mather, 2005; Landgrebe, 2003). The classifier uses these training samples to calculate the class statistics and assign each pixel in the image to a highly likely class. Along with these training samples, a number of test samples are also collected, which do not contribute to the classification process but are used for the evaluation of the classification results. The classification is performed interactively. The first run is to evaluate the performance and contribution of training samples to the classification accuracies. By excluding training samples with low performance, two to three such attempts are made to improve the classification and to achieve better results. The resultant supervised classification images are shown in Figure 10.2.

To assess the accuracy and contribution of the additional data layers in the pixel-based classification, the error matrices of the mapped classes relative to the reference data are generated. The error matrices for the test samples for the five data combinations are given in Tables 10.1a through e. These matrices provide statistical information about the classification results, including the user accuracy, producer accuracy, overall accuracy, and κ value (Campbell, 2002) for testing samples. It gives a fair idea about the class separability, between-class confusion, and misclassification in the form of omission and commission errors. The omission error indicates the number of pixels of a class wrongly omitted and assigned falsely to another class.

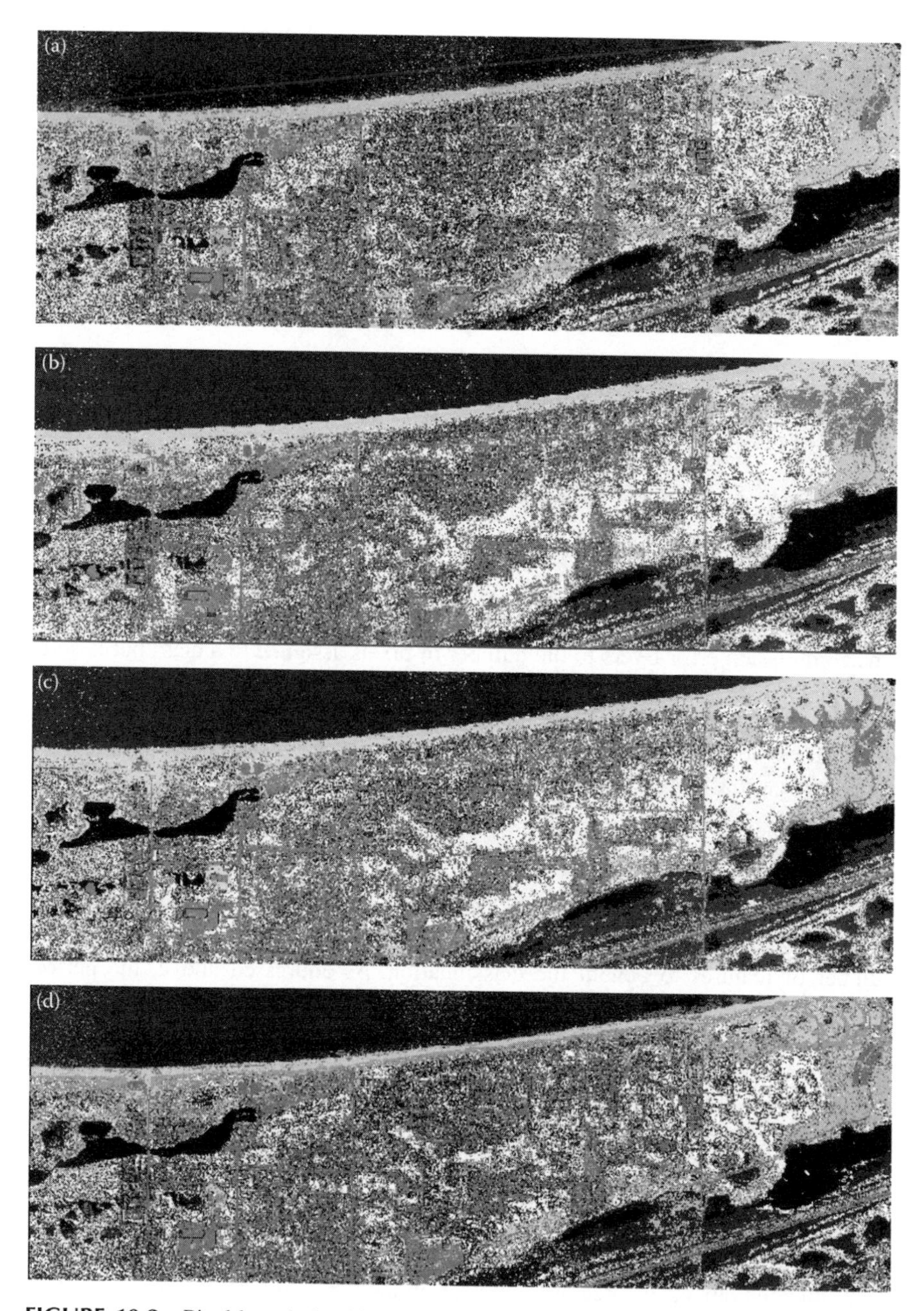

FIGURE 10.2 Pixel-based classification results with the maximum likelihood approach. From top: image only (a); image and DTM (b); image and DSM (c); image and nDSM (d); image, DTM, DSM, and nDSM (e).

FIGURE 10.2 (Continued).

The commission error refers to the number of pixels assigned to a class but is actually not a true representative of this class. All diagonal elements of the error matrix indicate the number of pixels correctly classified under a proper class. The off-diagonal elements in a column represent the omission error, while the off-diagonal elements in a row represent the commission error.

10.4.2 Object-Based Classification

In the object-based classification, DTM, DSM, nDSM, and roads data layers are used in combination with the images. The use of the combined data provides a large number of features to benefit the classification. As addressed above, the performance of the object-based classification depends on the quality of segmentation. Homogeneous and compact objects can be achieved by the selection and use of suitable parameters. The size of the objects is directly related to the type of features defined in the classification. To analyze and capture the small variations within the dataset, multilevel segmentation with varying parameters, that is, scale of 100, 35, and 25, shape of 0.3, and compactness and smoothness of 0.5 are used. This results in different numbers of objects at each level. All raster data layers are assigned equal weights, while the vector road layer is included as a boundary constraint in segmentation. Out of the selected 11 informational classes, buildings comprise smaller units; thus the objects created at segmentation level 2 with a scale parameter of 25 are used for subsequent classification. A subset of the multilevel segmentation results is shown in Figure 10.3.

After segmentation, a number of features need to be defined for objects of different classes. Since such features will be used for classification in the later stage, they should be selected so that they can uniquely characterize the corresponding classes with as little redundancy as possible. When selecting features, one needs to consider

TABLE 10.1a
Error Matrix of Pixel-Based Classification: Image Only

No.	Test Samples Class Name	Sample No.	Number of Pixels in Class 1	2	3	4	5	6	7	8	9	10	11	Producer's Accuracy
1	Lake Michigan	32,814	23,051	2	7	5212	615	1907	13	150	1592	242	23	70.2
2	Coast sand	3315	1	3113	194	0	0	0	0	1	2	4	0	93.9
3	Sand dune	892	2	319	410	0	11	2	9	17	2	34	86	46
4	Lakes/pond	8798	195	0	0	8379	112	0	0	0	112	0	0	95.2
5	Dry lakes and ponds	3147	31	0	23	44	1989	18	1	144	137	267	493	63.2
6	Tree	616	30	0	7	18	25	422	73	28	7	3	3	68.5
7	Grass	1450	0	0	34	0	3	45	1212	12	3	16	125	83.6
8	Deciduous tree	3218	18	0	78	22	842	196	56	1556	83	16	351	48.4
9	Building	2026	150	6	7	535	272	60	0	20	496	452	28	24.5
10	Roads/parking lots	3334	34	23	54	1	139	4	4	8	249	2779	39	83.4
11	Open areas	899	0	0	39	0	24	0	2	31	3	80	720	80.1
	Total	60,509	23,512	3463	853	14,211	4032	2654	1370	1967	2686	3893	1868	
	User accuracy		98	89.9	48.1	59	49.3	15.9	88.5	79.1	18.5	71.4	38.5	

Average user's accuracy = 59.63% Overall accuracy = 72.9% $\kappa = 0.634$ Average producer's accuracy = 68.81%

TABLE 10.1b
Error Matrix of Pixel-Based Classification: Image and DTM

No.	Test Samples Class Name	Sample No.	Number of Pixels in Class 1	2	3	4	5	6	7	8	9	10	11	Producer's Accuracy
1	Lake Michigan	32,814	31,978	3	52	7	0	687	0	47	21	19	0	97.5
2	Coast sand	3315	1	3176	136	0	0	0	0	2	0	0	0	95.8
3	Sand dune	892	9	286	559	0	0	3	0	11	1	10	13	62.7
4	Lakes/pond	8798	11	0	0	8537	61	125	0	0	64	0	0	97
5	Dry lakes and ponds	3147	0	0	3	34	2008	214	4	32	183	248	421	63.8
6	Tree	616	0	0	7	14	15	460	62	30	23	3	2	74.7
7	Grass	1450	0	0	0	0	7	97	1189	4	2	16	135	82
8	Deciduous tree	3218	0	0	28	14	480	182	68	1970	100	12	364	61.2
9	Building	2026	0	0	4	511	339	53	0	8	594	489	28	29.3
10	Roads/parking lots	3334	0	0	12	0	137	88	7	11	238	2801	40	84
11	Open areas	899	0	0	6	0	18	0	3	57	7	73	735	81.8
	Total	60,509	31,999	3465	807	9117	3065	1909	1333	2172	1233	3671	1738	
	User accuracy		99.9	91.7	69.3	93.6	65.5	24.1	89.2	90.7	48.2	76.3	42.3	

Average user's accuracy = 72.00% Overall accuracy = 83.9% $\kappa = 0.842$ Average producer's accuracy = 75.45%

TABLE 10.1c
Error Matrix of Pixel-Based Classification: Image and DSM

	Test Samples		Number of Pixels in Class											
No.	Class Name	Sample No.	1	2	3	4	5	6	7	8	9	10	11	Producer's Accuracy
1	Lake Michigan	32,814	31,996	2	4	78	0	526	7	92	1	66	42	97.5
2	Coast sand	3315	0	3269	32	0	0	9	0	0	0	4	1	98.6
3	Sand dune	892	0	8	751	0	4	16	7	14	1	40	51	84.2
4	Lakes/pond	8798	33	0	0	8651	52	24	0	36	0	2	0	98.3
5	Dry lakes and ponds	3147	0	0	20	90	2400	37	1	50	3	277	269	76.3
6	Tree	616	1	0	75	40	12	193	103	118	26	31	17	31.3
7	Grass	1450	0	0	5	4	15	109	1075	9	33	69	131	74.1
8	Deciduous tree	3218	0	0	80	41	318	221	109	1917	174	48	310	59.6
9	Building	2026	0	0	16	330	32	248	222	208	810	134	26	40
10	Roads/parking lots	3334	43	0	67	3	165	81	189	132	296	2178	180	65.3
11	Open areas	899	0	0	11	1	16	3	13	24	3	70	758	84.3
	Total	60,509	32,073	3279	1061	9238	3014	1467	1726	2600	1347	2919	1785	
	User accuracy		99.8	99.7	70.8	93.6	79.6	13.2	62.3	73.7	60.1	74.6	42.5	

Average user's accuracy = 70.00% Overall accuracy = 89.2% $\kappa = 0.841$ Average producer's accuracy = 73.60%

TABLE 10.1d
Error Matrix of Pixel-Based Classification: Image and nDSM

No.	Test Samples Class Name	Sample No.	Number of Pixels in Class 1	2	3	4	5	6	7	8	9	10	11	Producer's Accuracy
1	Lake Michigan	32,814	25,290	4	2	4183	604	1647	17	72	527	451	17	77.1
2	Coast sand	3315	1	3245	60	0	0	0	0	0	0	4	5	97.9
3	Sand dune	892	0	176	680	0	0	1	0	12	15	1	7	76.2
4	Lakes/pond	8798	117	0	0	8241	405	35	0	0	0	0	0	93.7
5	Dry lakes and ponds	3147	26	1	0	27	2123	22	3	55	31	327	532	67.5
6	Tree	616	39	1	3	50	54	320	49	44	24	22	10	51.9
7	Grass	1450	7	6	8	3	7	116	1082	5	3	60	153	74.6
8	Deciduous tree	3218	39	2	15	34	720	346	109	1416	74	36	427	44
9	Building	2026	35	2	50	215	50	41	2	380	1146	102	3	56.6
10	Roads/parking lots	3334	115	40	16	12	164	23	30	32	207	2637	58	79.1
11	Open areas	899	0	2	1	0	57	0	6	23	0	111	699	77.8
	Total	60,509	25,669	3479	835	12,765	4184	2551	1298	2039	2027	3751	1911	
	User accuracy		98.5	93.3	81.4	64.6	50.7	12.5	83.4	69.4	56.5	70.3	36.6	

Average user's accuracy = 65.18% Overall accuracy = 77.5% $\kappa = 0.689$ Average producer's accuracy = 72.36%

TABLE 10.1e
Error Matrix of Pixel-Based Classification: Image, DTM, DSM, and nDSM

No.	Test Samples Class Name	Sample No.	Number of Pixels in Class											Producer's Accuracy
			1	2	3	4	5	6	7	8	9	10	11	
1	Lake Michigan	32,814	32,532	3	0	67	0	179	2	6	1	20	4	99.1
2	Coast sand	3315	0	3269	31	0	0	10	0	0	0	5	0	98.6
3	Sand dune	892	0	0	840	0	0	22	0	4	1	22	3	94.2
4	Lakes/pond	8798	0	0	0	8558	140	100	0	0	0	0	0	97.3
5	Dry lakes and ponds	3147	0	0	0	59	2335	99	21	78	7	177	371	74.2
6	Tree	616	0	0	36	30	11	226	98	85	28	88	14	36.7
7	Grass	1450	0	0	1	3	33	40	991	44	100	137	101	68.3
8	Deciduous tree	3218	0	0	0	47	348	500	223	1398	194	133	375	43.4
9	Building	2026	0	0	0	79	127	55	296	124	1197	133	15	59.1
10	Roads/parking lots	3334	0	0	10	0	205	110	347	95	132	2251	184	67.5
11	Open areas	899	0	0	0	0	19	22	130	12	0	40	676	75.2
	Total	60,509	32,532	3272	918	8843	3218	1363	2108	1846	1660	3006	1743	
	User accuracy		100	99.9	91.5	96.8	72.6	16.6	47	75.7	72.1	74.9	38.8	
	Average user's accuracy = 71.45%		Overall accuracy = 89.7%				$\kappa = 0.847$				Average producer's accuracy = 74%			

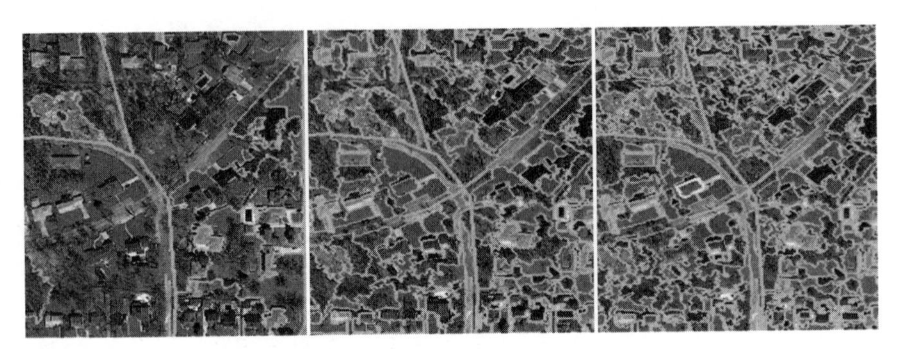

FIGURE 10.3 Segmentation results at scales of 100, 35, and 25 (left to right).

the spectral, contextual, textural, shape, and elevation properties of the classes and their relative relationships. In this study, we select features as band means, DTM means, DSM means, and nDSM (height) means, all in terms of objects. Band ratios, such as Normalized Difference Vegetation Index (NDVI) and the like, are used as features for some classes. In addition, shape parameters, such as area and length, thematic attributes, such as road names, and adjacent relationships, such as the distance to neighbors, are also selected as features in the object-based classification. Table 10.2 summarizes the features defined for objects of different classes in this study.

A proper selection of features should allow for a better classification. The spectral response over Lake Michigan water surface is not homogeneous rather it is varying over certain areas and near the shoreline. These areas are spectrally similar to buildings, trees, and their associated shadow, roads, and parking lots. As shown in Figure 10.4, none of the spectral feature values of different bands can provide satisfactory separation and classification. However, the feature space plots in the IR band and DSM values show a clear separation of lake water from other ground classes. Lake being low in elevation as compared to land, corresponding DSM and DTM

TABLE 10.2
Features Defined for Different Object Classes

Class	Object Features
Lake Michigan	Mean IR, mean DTM, mean DSM
Lakes and ponds	Brightness, mean IR, mean DTM, relation to neighboring objects
Dry lakes and ponds	Mean IR, ratio of IR and green, mean DTM, mean DSM
Sand (coast and dunes)	Brightness, means DTM, mean DSM, distance to class
Vegetation (trees, grass, and deciduous trees)	NDVI, ratio of red and green, mean IR, max difference
Buildings	nDSM, area, rectangular fit
Roads and parking lots	Length, ratio of IR and green, mean IR, thematic attributes, existence of neighbor classes, asymmetry
Open area	Brightness, mean difference to neighbors

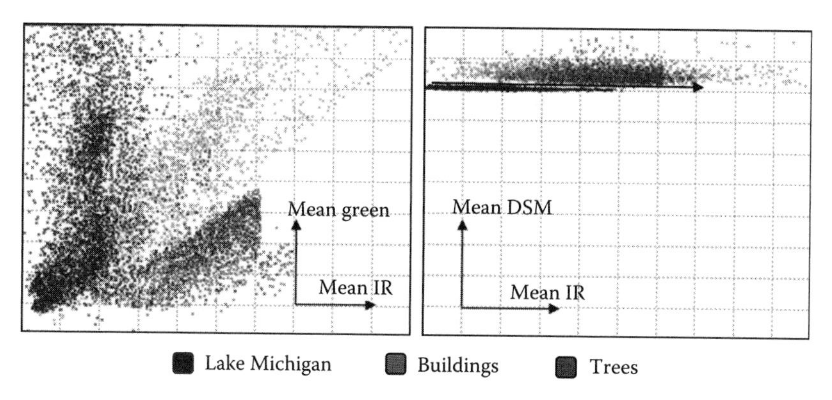

FIGURE 10.4 Feature space plots of Lake Michigan, buildings, and trees in IR-green (left) and IR-DSM (right).

values are also low and uniform. Including such features into classification can therefore result in a correct and homogeneous Lake Michigan classification.

The classification is carried out hierarchically. Initially, Lake Michigan and Land are selected as parent classes and the remaining class hierarchy is developed progressively. NDVI is used for vegetation classification, which later acts as the parent class for grass, trees, deciduous tree areas, and dry lakes and ponds. Due to the presence of vegetation over and near the dry lakes, this is included as a child class to vegetation class. NDVI supplemented with mean IR and the ratio of IR and green bands are used for the classification of these classes. Sandy areas, lakes and ponds, dry lakes, buildings, and roads and parking lots are classified under the parent class "Land." Because of the complexity of spectral similarity between and within classes, these are split into more than one subclass, for example, buildings and buildings-1, roads 1, 2, 3, and so on. This subdivision helps achieve better class separation and classification results. During this hierarchical classification process, contiguous class objects are merged to form larger objects, which helps filter wrongly assigned small objects. Figure 10.5 shows the class hierarchy in the object-based classification. The final classification result is shown in Figure 10.6.

The calculation of the classification accuracy is based on the statistics of the image objects assigned to classes. The assessment is carried out by generating 150 random points on the thematic image. The resultant matrix in Table 10.3 shows the mean, standard deviation, and minimum and maximum membership values. The higher the mean, the lower the standard deviation, and the closer the membership values to 1, the more reliable the classification results of a class.

10.4.3 Evaluation

The pixel-based classification using only imagery results in the lowest overall accuracy of 72.9% as shown in Table 10.1a. The least accurate classes are Lake Michigan, sand dunes, buildings, and roads and parking lots. Water class is divided into two subclasses: (1) Lake Michigan and (2) lakes and ponds. The confusion matrix shows

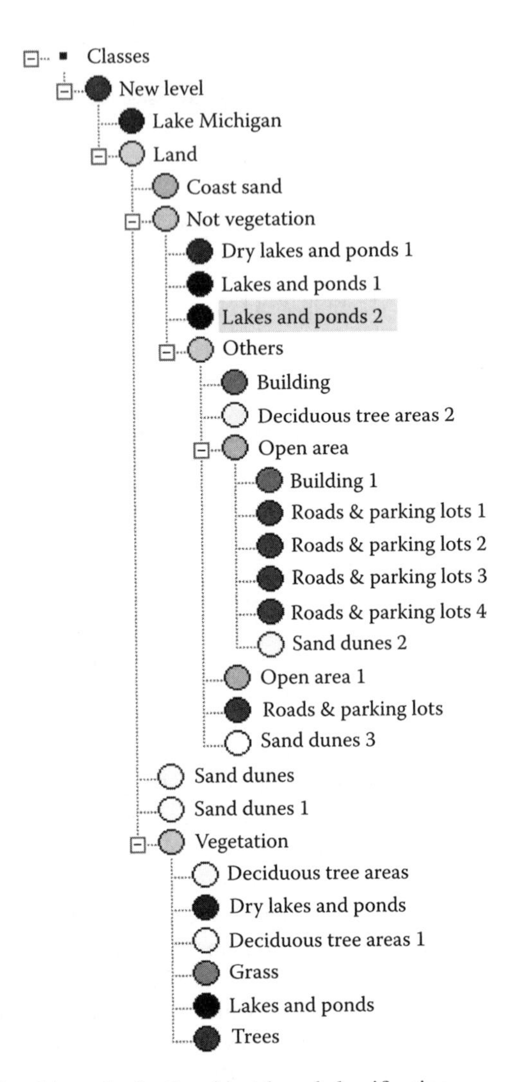

FIGURE 10.5 Class hierarchy for the object-based classification.

FIGURE 10.6 Results of the object-based classification.

TABLE 10.3
Quality Measures of the Object-Based Classification

Class	Mean	Standard Deviation	Minimum	Maximum	Accuracy (%)[a]	
					User	Producer
Lake Michigan	0.89	0.16	0.11	1	100	100
Coast sand	0.94	0.12	0.13	0.99	100	100
Sand dunes	0.88	0.12	0.56	0.99	85	90
Lakes and ponds	0.69	0.26	0.16	0.98	100	100
Dry lakes and ponds	0.73	0.24	0.11	0.98	87.5	80
Tree	0.54	0.3	0.10	0.99	100	85.7
Grass	0.71	0.3	0.10	0.99	100	62.5
Deciduous tree areas	0.82	0.09	0.7	0.99	81	98
Building	0.5	0	0.5	0.5	100	75
Roads/parking lots	0.88	0.12	0.6	1	87.5	96
Open areas	1	0	1	1	83.3	90
Average accuracy					85.5	81.4
$\kappa = 0.87$					Overall = 88.6%	

[a] Accuracy from randomly generated points.

some prominent misclassification statistics for water, sand, and built-up areas. 16% of the Lake Michigan water pixels are misclassified as lakes and ponds, 6% as trees, and 5% as buildings. This is because of the heavy wind on some parts of the Lake Michigan surface and the surfaces of a few other lakes and ponds along the same flight lines. These wavy water areas are very bright due to reflection and spectrally very different from the adjacent Lake Michigan and other lakes and ponds water, but similar to building roofs with multiple colors. Therefore, some of the water pixels, mostly water waves, along the coastalline are misclassified as buildings or roads, while calm lake water surface areas with no waves are correctly classified. In general, lakes and ponds are classified well except for a few pixels that are mistakenly labeled as Lake Michigan due to spectral similarity. These specific areas are shown in Figure 10.7a.

Sandy areas consist of the coast sand and the dunes mostly along the coastline. There are also a few such areas in other parts of the image. The coast sand is wet near the water line, and becomes dry away from the water. This dry coast sand appears spectrally very bright, smooth, and distinct as compared to all other classes thus classified accurately. Sand dunes are spectrally very similar to coast sand except at places where they have dry vegetation cover. As these two sand classes have very low separability due to similar spectral response, an interclass misclassification occurs. About 36% of pixels of sand dunes are wrongly classified as coast sand. The sand dunes with dry vegetation cover have spectral resemblance to open areas and are thus misclassified as open areas.

Another major source of low accuracy and misclassification are buildings, roads and parking lots classes. Spectral response of roads and parking lots from concrete

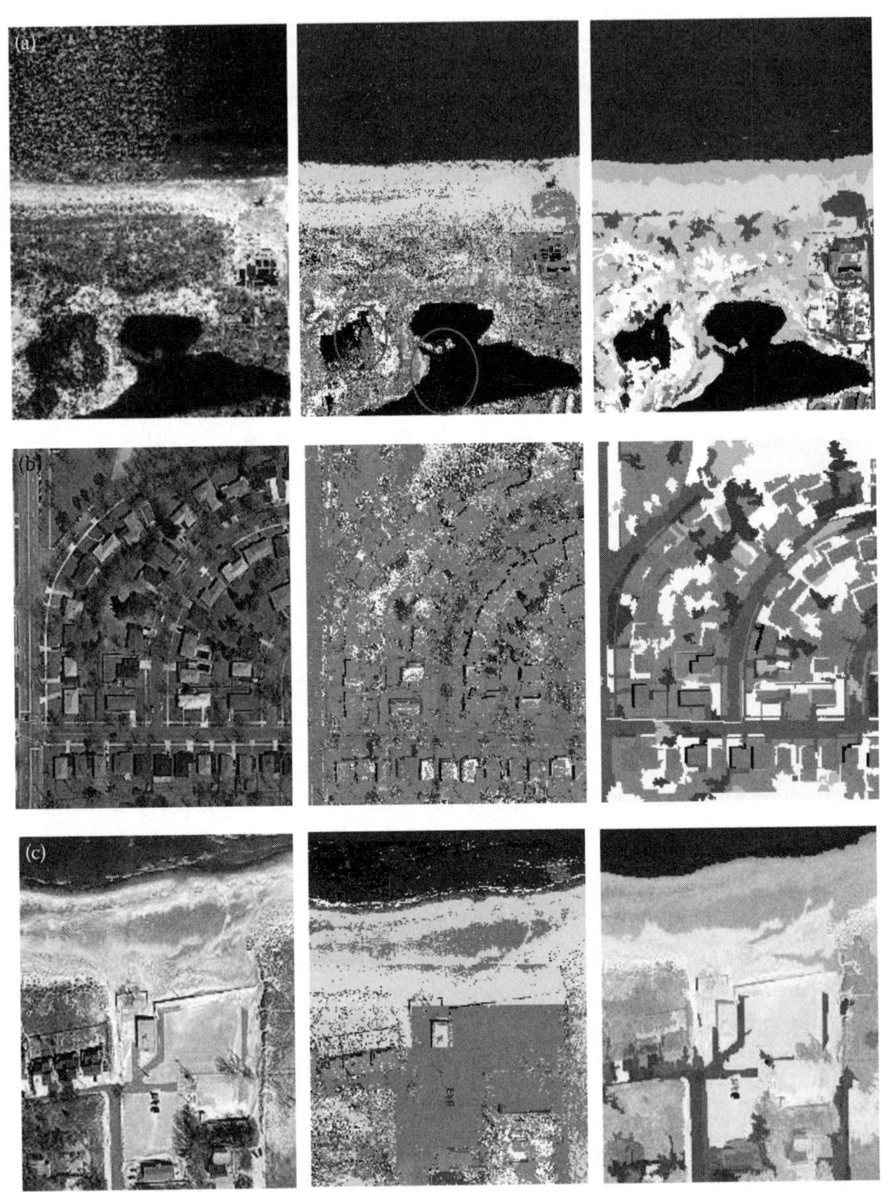

Left column: Original CIR image, Middle column: Combined data pixel-based classification image, Right column: Object-based classification image.

Lake Michigan | Buildings | Lakes and ponds

Sand dunes | Roads and parking lots | Coast sand

FIGURE 10.7 (a) Misclassification of water areas into buildings, trees, and roads. (b) Misclassification of buildings and roads, nonhomogeneous building footprints for pixel-based and homogeneous building for the object-based classification. (c) Mixture of wet coast sand, sand dunes, and roads and parking lots.

and bituminous surfaces results in misclassification of their pixels as dry lakes and ponds and buildings. Similarly, building roofs with multiple colors are misclassified as Lake Michigan, lakes and ponds, roads and parking lots, and dry lakes and ponds. This misclassification results in low accuracies (24.5%) and higher ranges of omission and commission errors (75–81%).

Vegetation grassy areas are well separated from other classes except the misclassification of a few pixels to open areas that might have some light vegetation, and result in an accuracy of 84%. On the other hand, tree class results in an accuracy of 69% with misclassification of its pixels to grass. It gets a high commission error of 84%, as a large number of Lake Michigan pixels are included in this class. Deciduous trees cover a large image area, which appears darker due to the associated shadow and visible bare ground underneath in the leaf-off season. A number of deciduous tree pixels are misclassified under most of the other classes, prominently as dry lakes, as trees probably due to the emergence of new leaves, and as open areas because of the visible ground underneath. This misclassification results in a low accuracy of 48% for this class. Dry lakes and ponds with both darker and brighter spectral response and presence of deciduous trees in the vicinity are misclassified as deciduous trees, roads and parking lots, buildings, and open areas. Open areas are classified well except for some confusion with roads and parking lots.

The use of DTM with the image considerably improves the classification results as compared to the image only case. It improves the accuracy of every class and especially those that have the lower accuracies in the image only case. This considerable improvement in the accuracy of Lake Michigan, sand dunes, trees, and deciduous tree classes is mainly due to the significant elevation difference among these classes. A clear elevation difference between Lake Michigan, lakes and ponds, dry lakes, trees, and buildings offsets the spectral similarity problems and improves Lake Michigan class accuracy from 70% to 95.5%, and also improves the accuracy of other classes as well. Another prominent improvement due to DTM is the increase in the accuracy of sand dune class, that is, from 46% to 63%. The contribution of DTM in the classification results in an increase in both the user and producer accuracies of all classes and achieves an overall accuracy of 84% as against 73% from the image only case. It produces slightly better homogeneous thematic map especially for Lake Michigan.

The addition of DSM to the image only classification improves the accuracy of most of the classes that have spectral confusion but considerable surface elevation differences. It increases the accuracies of some classes to that from DTM but decreases the accuracy of trees, grass, and roads and parking lot classes. Its major contribution to the increased accuracies is to Lake Michigan, sand dunes, dry lakes and ponds, deciduous tree, and building classes. The accuracy of building class improves from 24.5% to 60% mainly due to the building elevation being included in DSM. These classes have spectral confusion with others but the elevation differences help remove this confusion and improve the results. The decrease in the accuracy of trees is due to the misclassification of its pixels as grass and deciduous trees because of the presence of very less and scattered green leave trees, emerging leaves on deciduous trees, and confusion of "tree" with "grass." Thus, the spectral similarity and absence of any considerable surface elevation differences result in lower accuracy. The lower accuracy of grass and the misclassification of its pixels as open areas are due to their spectral and surface elevation similarities. The lower accuracy of roads and parking lots is because of its spectral similarity to many other classes and absence of any considerable surface elevation differences. The use of DSM improves those classes with some difference in surface elevation and decreases those that are similar in both spectral response and surface elevation. Overall, it yields a very high accuracy of 89.2% with an increase of about 16% and 6% as compared to the results of image only and image with DTM, respectively.

The use of nDSM for the pixel-based classification presents an accuracy increase for Lake Michigan, coast sand, sand dunes, and dry lakes and ponds as compared to the image only case, but a decrease in comparison to the use of DTM or DSM. It also results in the lower accuracy for tree and grass, roads and parking lots, deciduous trees, and open areas. The prominent contribution of nDSM is the improved accuracy of the buildings class as compared to DTM or DSM. Another improvement is the tree class. The main misclassifications in this case are under Lake Michigan, dry lakes and ponds, and deciduous trees. It achieves an overall accuracy of 77.5%, 4.5% more than the image only case, but 7–11% less than the DTM or DSM case.

The last combination for the pixel-based classification is that of image, DTM, DSM, and nDSM. The use of this combined data for classification results in an increase accuracy of six, and decrease in five classes as compared to image alone. The main increase is to Lake Michigan, sand dunes, and building classes, and the decrease is to the vegetation-related classes. The reason for the increased accuracy is due to the differences in terrain, surface elevation, or height. Wherever there is such a difference in elevation or height, the accuracy increases. Overall, the combined classification result carries the affects and contribution of individual data in the combination. The classification accuracy with the combined data is almost the same as from DSM except for slight variations for a few classes. The overall classification accuracy in the case of the combined data is 89.7%, almost 15% higher than the image only case.

None of the above dataset combinations is able to restrict the misclassification of Lake Michigan pixels as deciduous trees and trees classes. The combined data can reduce but not remove this misclassification. As a result, none of the data combinations result in homogeneous Lake Michigan class thematic map. The wavy and reflection-prone areas of lakes and ponds along one of the flight lines are misclassified as Lake Michigan, buildings, and roads and parking lots in all the above tests. No additional data alone or combined could resolve this spectral confusion as well as elevation similarities. Trees and grass are well classified in image only and the DTM case, but the DSM and nDSM could degrade their results. Also, buildings, and roads and parking lots could not achieve higher accuracies because of their spectral and elevation confusions with many other classes. Although the building footprints can be correctly captured, the within-class variations could not be removed to get noise-free and homogeneous thematic map.

Accompanying the results in Figure 10.7, Table 10.3 shows the object-based classification quality measures. Compared to the results of pixel-based classification in Tables 10.1a–e, the object-based classification provides higher mean membership values in all classes. The use of the object-based classification method improves the separability between most of the spectrally similar classes and produces more compact and homogeneous class results. In the object-based approach, the segmentation of image into homogeneous objects reduces the local spectral variations and allows adding shape and context information, which further improves the classification results and reduces the confusion. Some of the pixels of Lake Michigan far away and near coastline areas have spectral, terrain, and surface similarity and thus are misclassified in the pixel-based method as lakes and ponds, trees, and buildings even with the added data layers. However, the object-based approach correctly classifies these areas due to homogeneous segmented objects added with the additional information from the DTM, DSM, and nDSM and produced 100% classification accuracy and a very homogeneous Lake Michigan surface. The object's neighborhood feature very precisely delineates the coastline and sandy beach with 100% accuracy and disregards the wet or dry sand conditions. Some areas within lakes and ponds with wavy and refection surface due to spectral similarity to buildings and road sand parking lots are misclassified in the pixel-based method, but correctly classified in the object-based method with an accuracy of 100%, as illustrated in Figure 10.7a. The image segmentation includes all such area pixels and the adjacent pixels collectively to form the objects that are more homogeneous within themselves. The roads and parking lots, and buildings are spectrally very similar to each other, and have an accuracy of 67.5% and 59%, respectively, in pixel-based analysis. They are improved to a significant level of 96% and 75%, respectively, in the object-based classification. The improvement in buildings class accuracy is because of its homogeneous objects, and the use of additional object features such as length, rectangular fit, and height. The classification results show very compact and homogeneous building footprints as compared to the ones from the pixel-based method, as shown in Figure 10.7b. This improvement is attributed to the use of additional object features, such as length, rectangular fit, and height. It is noticed that some of the parking lots near the coastline and along sand dunes are mislabeled as sand dunes. This confusion is due to the presence of sand cover over the parking lots surface, light dark color of parking surface, and the same elevation as that of dunes as shown in Figure 10.7c. A few other parking places in urban areas are also mislabeled as sand dunes or open areas. The trees and grassy areas are well classified in the pixel-based image only case, except for some mix-up with deciduous tree areas. The object method yields an accuracy of 85.7% and 62.5% for these classes, although the object features used for the classification are also spectrum-based such as NDVI and ratio of bands. The spectral similarity problem of dry lakes and ponds with roads and many other classes is well tackled by the object-based method. The creation of homogeneous objects and the use of spectral and surface features help improve its separability and produce an accuracy of 80% as against 74% from the pixel-based method. The segmentation of image into objects and the option of use of shape and context in addition to the spectral features help to correctly classify all those classes that are misclassified in the pixel-based method. The use of objects coupled with the information from added data layers yields better accuracy and produces very sharp and smooth thematic map for most of the classes, as illustrated in Figure 10.6.

The above analysis shows that for classification of high-resolution images, object-based classification methods are comparable (1% lower overall statistical accuracy) with the combined data pixel-based classification methods, yet the classified thematic map is more homogeneous, visually appealing, and qualitatively clearer. Pixel-based classification by using only spectral data has lower accuracy and extensive salt-and-pepper affects. Such affects are more pronounced in the high-resolution imagery because of increased variance in the same land-cover types. The resultant thematic maps from high-resolution imagery with extensive noise do not provide satisfactory information to the users. The use of additional data layers in pixel-based classification removes some misclassification between spectrally similar classes (Lee and Shan, 2003), and improves the statistical accuracy by about 10%. Nevertheless, it cannot contribute much to the separation of classes with both spectral and elevation similarities. In the pixel-based method, an area of Lake Michigan adjacent to the coastline is classified as buildings even with the additional data layers. Such a spot gives a wrong impression about the presence of built-up area just close to the coastline. This problem is eliminated in the object-based classification approach with the use of segmented homogeneous objects and the additional object features. The segmentation of image into homogeneous objects coupled with the additional data layers provides a framework to use spatial, spectral, contextual, and geometric features to aid in the classification and avoid inherent misclassification in the pixel-based solutions. Better classification results (average user, producers, and overall accuracy of 85.5%, 81.4%, and 88.6%) are achieved through the object-based method that uses multilevel segmentation, carefully formed class hierarchy, precisely selected objects features, relevant fuzzy membership functions, and sequential classification strategy. Conversely, the trial-and-error process for segmentation and selection of suitable features takes much more time to perform the entire object-based classification. Even such elaborate effort in certain cases cannot totally remove the chance of misclassifying certain complex similar objects.

10.5 CONCLUSIONS

With the pixel-based method, it is difficult to capture any natural variations within a group of pixels of the same class due to the independency of pixels. It will classify every pixel based on its spectral response irrespective of its neighborhood, thus resulting in speckled appearance. Traditional pixel-based classification methods provide adequate results in many applications where the dataset contains spectrally homogeneous features. However, for inhomogeneous areas, for example, a residential area with several multicolor roofs, it is hard to collect homogeneous spectral signatures. A classifier totally relying on spectral characteristics of classes is thus unable to correctly label the classes and results in very low class accuracy and high omission and commission errors, for example, building class in this study (25%, 75%, and 79%). Such situations warrant the need for additional ancillary data and for finding some relationships between the additional data features and spectral bands to achieve better classification results. The use of additional data in pixel-based classification improves the building class accuracy (from 25% to 59%) and reduces the omission and commission errors by 41% and 28%, respectively. It well captures the building footprints but the spectral variation within the building footprints still exists and is mostly misclassified. The reduced accuracy of vegetation-related classes with additional data layers highlights the use of spectral-based vegetation indices for their classification in the pixel-based method. The use of additional data layers in the pixel-based method improves the classification results and reduces the degree of confusion between and within classes, if there exists a significant difference in the terrain, surface, and height among spectrally similar classes. The object-based classification method has the advantage of using objects for classification rather than pixels and utilizing the spatial, spectral, and shape characteristics of features. The class assignment of an object is more robust and homogeneous as all the pixels within an object are classified into one class even in the presence of spectral variations. The higher average user and producer accuracies of 85.5% and 81.4% as compared to 71.45% and 74% from the pixel-based method and the qualitatively better thematic map support the use of the object-based method for the classification of high spectral variations data. The better object-based classification outcome is attributed to the collective effect of both, the objects and the added data layers.

Mapping of coastal areas using high-resolution images can provide an accurate and elaborate land-use and land-cover details. This detailed land-use and land-cover information is helpful for preservation, restoration, and development of coastal resources. The complexity of extracting specific fine features can be effectively addressed by the integration of multiple geospatial data to the image classification process. The object-based classification is able to enhance the scope of the analysis because of its ability to integrate and process data of different characteristics with many data-specific features, including the relations among different data layers. Its classification results are satisfactory with relatively high accurate, smooth, and homogeneous thematic representation of the classes. These results can be directly used to update the existing high-resolution land-use and land-cover database.

REFERENCES

Alpine, P. and Atkinson, P.M., 2001, Sub-pixel land cover mapping for per-field classification, *International Journal of Remote Sensing*, 22(14): 2853–2858.

Baatz, M. and Schäpe, A., 2000, Multiresolution segmentation—an optimization approach for high quality multi-scale image segmentation, In: Strobl, J., Blaschke, T., Griesebener, G. (Eds), *Angewandte Geographische Informationsverarbeitung XII. Beiträge zum AGIT-Symposium Salzburg* 2000, Herbert Wichmann Verlag, Karlsruhe, pp. 12–23.

Benz, U.C., Hofmann, P., Willhauck, G., Lingenfelder, I., and Heynen, M., 2004, Multi-resolution, object-oriented fuzzy analysis of remote sensing data for GIS ready information, *ISPRS Journal of Photogrammetry & Remote Sensing*, 58: 239–258.

Blaschke, T., Lang, S., and Moller, M.S., 2005, Object-based analysis of remote sensing data for landscape monitoring, Recent developments, *Anais XII Simpósio Brasileiro de Sensoriamento Remoto*, Goiânia, Brasil, 16-21, INPE, pp. 2879–2885.

Blaschke, T. and Strobl, J., 2001, What's wrong with pixels? Some recent developments interfacing remote sensing and GIS, *GIS Zeitschrift für Geoinformationssysteme*, 6: 12–17.

Bruzzone, L. and Carlin, L., 2006, A multilevel context-based system for classification of very high spatial resolution images, *IEEE Transactions on Geoscience and Remote Sensing*, 44: 2587–2600.

Campbell, J.B., 2002, *Introduction to Remote Sensing*, 3rd edition, Guilford Press, New York.

Carleer, A., Debeir, O., and Wolff, E., 2005, Comparison of very high spatial resolution satellite image segmentation, In: *Proceedings of the SPIE Conference on Image and Signal Processing Remote Sensing IX*, Vol. 5238, pp. 532–542.

Fisher, P., 1997, The pixel: A snare and a delusion, *International Journal of Remote Sensing*, 18: 679–685.

Landgrebe, D., 2003, *Signal Theory Methods in Multispectral Remote Sensing*, Wiley, New York.

Lee, D.S. and Shan, J., 2003, Combining lidar elevation data and IKONOS multispectral imagery for coastal classification mapping, *Marine Geodesy*, 26(1): 117–127.

Lillesand, T.M., Kiefer, R.W., and Chipman, J.W., 2004, *Remote Sensing and Image Interpretation*, 5th edition, Wiley, New York.

Mather, P.M., 2005, *Computer Processing of Remotely-Sensed Images*, 3rd edition, Wiley New York.

Moeller, M.S., Stefanov, W.L., and Netzband, M., 2004, Characterizing land cover changes in a rapidly growing metropolitan area using long term satellite imagery, In: *Proceedings of the ASPRS Annual Conference*, Denver, USA, May 23–28.

Niemeyer, I. and Canty, M.J., 2001, *Object-Oriented Post-Classification of Change Images* In: *Proceedings of SPIE's International Symposium on Remote Sensing*, SPIE Vol. 4545, Toulouse, France, 17–21 September 2001.

Richards, J.A. and Jia, X., 2006, *Remote Sensing Digital Image Analysis*, 4th edition, Springer, New York.

Schowengerdt, R.A., 1997, *Remote Sensing, Models and Methods for Image Processing*, 2nd edition, Academic Press, New York.

Theodoridis, S. and Koutroumbas, K., 1999, *Pattern Recognition*, Academic Press, New York.

Wang, Z., Wei, W., Zhao, S., and Chen, X., 2004, Object-oriented classification and application in land use classification using SPOT-5 PAN imagery, *IEEE International Geoscience and Remote Sensing Symposium, IGARSS Proceedings*, pp. 3158–3160.

Wei, W., Chen, X., and Ma, A., 2005, Object-oriented information extraction and application in high-resolution remote sensing image, In: *Proceedings of the IGARSS 2005 Symposium*, Seoul, Korea, July 25–29.

11 True-Color Digital Orthophotography Data for Mapping Coastal Impervious Surface Areas

Yuyu Zhou and Yeqiao Wang

CONTENTS

11.1 INTRODUCTION

The incursion of residential and commercial development into terrestrial habitats is resulting in measurable changes to the composition and pattern of landscape and affects watershed hydrology. Monitoring of land-cover change and study of the effects of increasing impervious surface areas (ISAs) and declining forest cover as a result of suburban development are critical to the quantification of the state of, changes in, and anthropogenic threats to surface runoff—a major contributor to coastal waters. ISA is defined as any impenetrable material that prevents infiltration of water into the soil. Urban pavements, such as rooftops, roads, sidewalks, parking lots, driveways, and other man-made concrete surfaces, are among impervious surface types that feature the urban and suburban landscape. ISA is a critical factor in cycling of terrestrial runoff and associated materials to and within ocean margin waters. Increasing ISA impacts watershed hydrology in terms of influencing the runoff and baseflow. Urban runoff, mostly over the impervious surface, is the leading source of pollution in the Nation's estuaries, lakes, and rivers (Arnold and Gibbons, 1996; Booth and Jackson, 1997). Runoff frequently contains high concentrations of nutrients and pollutants, which can be harmful to freshwater and marine organisms, and therefore, it has significant impacts on sensitive tidal creeks and estuary systems (Schiff et al., 2002). The ISA has emerged as a key indicator to explain and predict ecosystem health in relation to watershed development. A published watershed-planning model predicts that most stream quality indicators decline when the watershed ISA exceeds 10% (Schueler, 2003). Assessment of the quantity of ISA in landscapes has become increasingly important with growing concern of its impact on the environment (Weng, 2001; Civco et al., 2002; Dougherty et al., 2004; Wang and Zhang, 2004). This is true even for coastal watersheds where continental shelves, inland or partially enclosed seas, estuaries, bays, lagoons, beaches, and terrestrial and aquatic ecosystems within them that drain into coastal waters. Human activities in urbanized coastal areas deliver sewage, solid wastes, refuse, sediments, dust, pesticides, and hydrocarbons to coastal rivers, estuaries, and the oceans. It is estimated that about 80% of all marine pollution originates from land-based sources and activities (Costa-Pierce, 2006). Assessing land-based sources of pollution to coastal oceans and how to mitigate and rehabilitate the impacts on coastal ecosystems are receiving growing attention. Precise data on ISA in spatial coverage and distribution patterns in association with landscape characterizations are critical for providing the key baseline information for effective coastal management and improved decision-making.

Conventionally, labor-intensive manual delineation through aerial photography has been used to extract ISA coverage (Draper and Rao, 1986). Moderate spatial resolution remote sensing data, such as those from Landsat sensors, have long been used in land-cover mapping and classifications. Most of the urban land category derived from digital image classification on Landsat data could be directly related to urban ISA. Due to limitations of spatial resolutions and corresponding spectral mixing within a given pixel size, the accuracy of ISA extraction has always been challenging through the conventional pixel-based classification process. Different methods have been developed to extract ISA from moderate spatial resolution remote sensing data. For example, subpixel methods have been developed for urban land classification (Lu and Weng, 2004). Wang and Zhang (2004) developed an artificial neural network-based SPLIT model to extract ISA information through subpixel extraction by integration of Landsat TM and high spatial resolution digital multispectral videography data. The subpixel method can extract the ISA percentage within mixed pixels. However, it is difficult to retrieve a precise spatial pattern of ISA. Extraction of ISA from high spatial resolution remote sensing data at meter or submeter level is in demand particularly by the planning agencies.

With improved capacity of remote sensing technologies, high spatial resolution data are becoming available for more precise measurements and quantification of landscape configurations. Digital true-color orthophotography data are being widely used and can be employed for extraction of ISA. The shortcomings are evident in using conventional pixel-based classification methods for extraction of ISA from high spatial resolution digital true-color orthophotography. The spatial information such as neighborhood, proximity, and homogeneity cannot be used sufficiently in the conventional classification processes (Burnett and Blaschke, 2003). The "salt-and-pepper" effect in classified images is significant due to increased spatial resolution and reduced spectral homogeneity among neighboring pixels. Object-based classifications have been developed in classification of high spatial resolution images (Baatz and Schäpe, 2000; Blaschke and Strobl, 2001; Shackelford and Davis, 2003; Carleer and Wolff, 2006; Yu et al., 2006; Jacquin et al., 2008; Walker and Blaschke, 2008). Object-based methods simulate the process of human image understanding in feature extraction. In addition, other spatial information can be integrated into the modeling process. The commercial software, such as eCognition (at Definiens Imaging, Germany), has been used for classification of high spatial resolution images.

Successful image segmentation, which defines regions in an image corresponding to objects in a ground scene, is the most important prerequisite (Baatz and Schäpe, 2000). Burnett and Blaschke (2003) discussed the application of multiscale segmentation in landscape analysis. Available segmentation algorithms include texture segmentation, watershed transformation, and mean shift among others (Woodcock and Harward, 1992; Li et al., 1999; Comanicu and Meer, 2002; Blaschke et al., 2004; Hu et al., 2005). As pointed out by Baatz and Schäpe (2000), a few of those methods lead to qualitatively convincing results that are robust and operationally applicable. In segmentation, objects are grouped into large homogeneous ones according to clustering cost functions adopted. In many region growing algorithms, the entire process of image segmentation is completed in a single pass. These methods may produce large differences between pixels at the opposite ends of a region. Woodcock and Harward (1992) developed a multiple-pass algorithm to extract forest information from Landsat TM data. This algorithm improved the segmentation of images and achieved reasonable accuracies with the consideration of local best fitting and merging coefficient. This approach can control the rate of region growing to avoid decision errors that occur as region centroids change due to premature segmentation merges. The mutual nearest neighbor principle employed in the multiple-pass technique can produce more accurate region boundaries. However, this algorithm aims at classifying remote sensing data at 30 m spatial resolution. Modification is necessary when dealing with high spatial resolution digital aerial photography data at meter or submeter levels.

In such an attempt we developed a synthetic *multiple agent segmentation and classification* (MASC) algorithm to obtain information on ISA from high spatial resolution digital true-color orthophotography imagery data. We then quantified the ISA coverage along the Narragansett Bay and the southern coast of the Rhode Island and analyzed the spatial distribution and pattern of ISA along the coastal communities. The data revealed spatial patterns and provided precise aerial information on ISA for coastal management and scientific research.

11.2 METHODS

11.2.1 Study Area and Data

Rhode Island is a heavily settled state, especially in the coastal areas, in the southern New England region. Narragansett Bay extends into the capital city of Providence and connects the developed coastal shoreline open to the Atlantic Ocean. Quantification

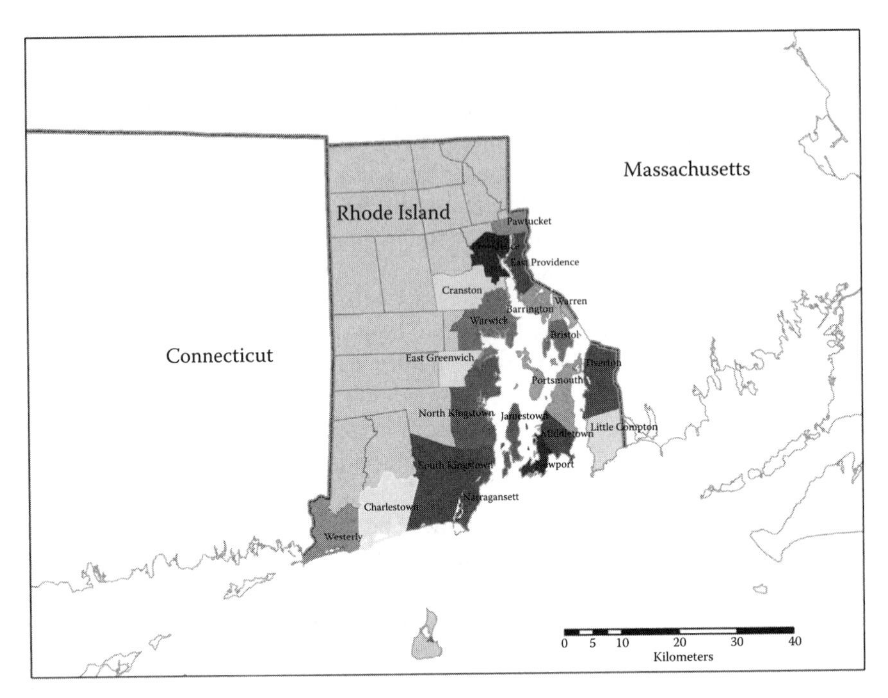

FIGURE 11.1 Coastal communities of Rhode Island.

of impervious surface coverage and revealing its spatial distribution are an important step toward understanding the coastal environment. Moderate spatial resolution ISA data of Rhode Island can be extracted from existing land-cover maps derived from classification of Landsat data (Novak and Wang, 2004). However, there is a lack of information at finer spatial resolution for a precise measurement of ISA distribution in coastal communities. In this study, therefore, we focused on 20 towns that cover the coastal areas of the state (Figure 11.1).

The digital true-color orthophotographs used in this study were acquired by the Rhode Island Statewide Planning Program between 2003 and 2004 through the National Agricultural Imagery Program (NAIP). This orthorectified imagery dataset with 1 m ground sample distance has a horizontal accuracy of within ±3 m of reference digital ortho quarter quads (DOQQS) from the National Digital Ortho Program. The orthoimages are projected into the Rhode Island State Plane Coordinate System with Zone 3800 in U.S. Survey feet. The resulting spatial resolution of the dataset is 1 m. The dataset has spectral bands of red, green, and blue light and was distributed in GeoTIFF format (Figure 11.2a). As texture information can be helpful for spatial information definition, we used a 3 × 3 pixel window to extract the variance as one of the features in the image segmentation process.

11.2.2 MASC Algorithm for ISA Extraction

In this study, we employed an object-oriented MASC algorithm for extraction of ISA from 1 m spatial resolution digital true-color orthophotography data. The MASC

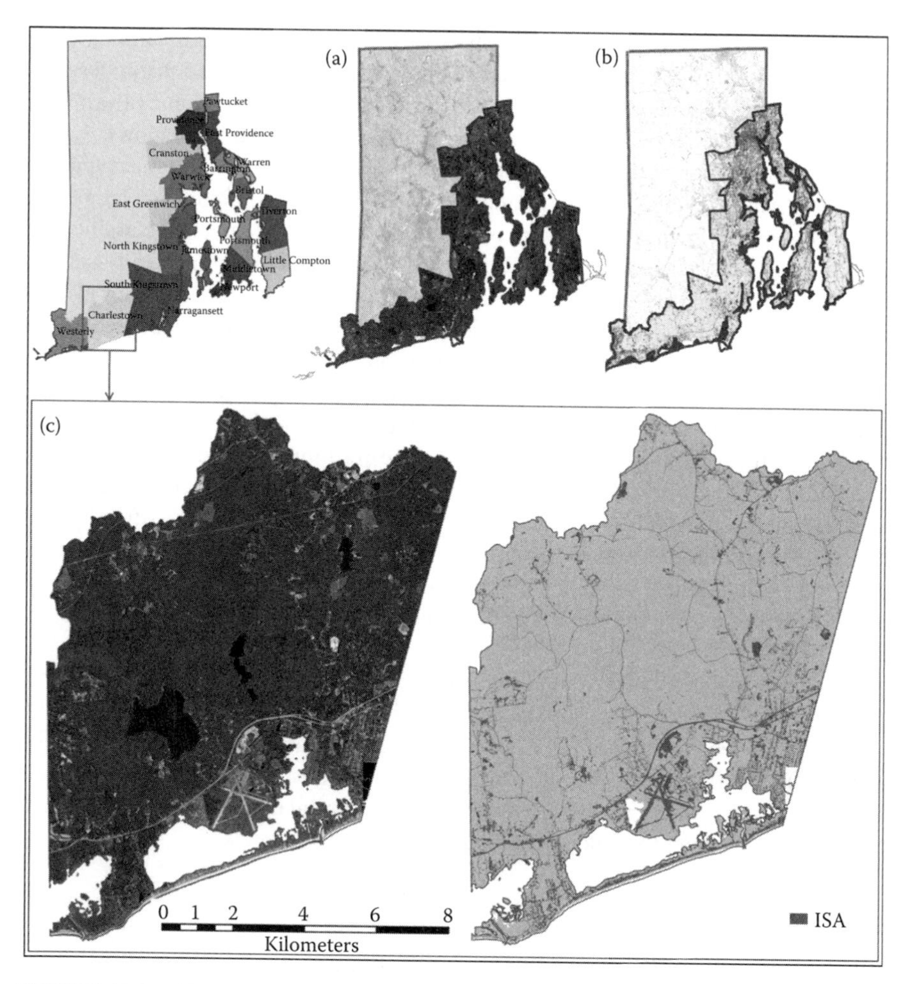

FIGURE 11.2 (See color insert following page 206.) True-color orthophotography image data (a) and the extracted ISA (b) for coastal communities of Rhode Island. The zoomed example shows spatial details of the orthophotography data and ISA cover in Charlestown (c).

algorithm includes submodels of segmentation, shadow-effect, MANOVA-based classification, and postclassification (Zhou and Wang, 2008). Segmentation defines regions in an image corresponding to objects in a ground scene. In the segmentation submodel, the added shape information enhanced the performance of the multiple-pass segmentation algorithm that has been used in object-oriented classifications. Shadow effects are typical in high spatial resolution remote sensing images. Therefore, the shadow-effect submodel used a split-and-merge process to separate shadows and the objects that cause the shadows. In order to conduct an information-enhanced image classification, the MASC algorithm employed a MANOVA-based classification submodel to incorporate the relationship between spectral bands and the variability in the training objects and the objects to be classified. For suburban areas, dense tree canopies unavoidably cover some ISAs such as the road segments in orthophotographs, which makes direct extraction of ISA extremely difficult. With a geographic information systems (GIS)-supported postclassification process, we employed the 1995 transportation data from the Rhode Island Geographic Information System (RIGIS) to recover the missing segments of ISA caused by shadows.

The segmentation and classification processes were extended to all the orthophotography scenes involved through a batch process. First, the algorithm segmented those photograph scenes using a set of the parameters. Second, a preclassification stratification was performed according to the 1995 land-cover and land-use data from the RIGIS and the training objects of land-cover types were selected in each subset from the preclassification stratification. The training samples included five main categories of ISA, forest, grassland, bare soil, and water. Each of the five categories contained several subcategories. For example, black asphalt and concrete pavement are both important urban ISA but with different spectral features. Those types were treated as subcategories of ISA. The training data were compared and checked using the covariance matrix. The samples with smaller covariance were selected in the MANOVA-based classification. Upon finishing classification the results were recoded to two categories of ISA and non-ISA for the final map. We then used the postclassification submodel and a rasterized GIS transportation data as a reference to identify the road networks and to obtain the high spatial resolution ISA cover.

Upon finishing the extraction of high spatial resolution ISA, we used the random point sampling method to evaluate the classification accuracies. We selected 200 samples in the coastal towns and examined the classification accuracies for the ISA and non-ISA only. We examined overall accuracy and the Producer's and User's accuracy. We then extracted ISA for each of the 20 coastal towns to reveal the spatial patterns of ISA along the Rhode Island coastal lines.

11.3 RESULTS

The results for ISA extracted in coastal areas are illustrated in Figure 11.2b. Enlarged example of the orthophotograph and extracted ISA for the town of Charlestown illustrates the spatial details of ISA distributions (Figure 11.2c). The MASC algorithm achieved a 91.5% overall accuracy for the extracted ISA with a 0.83 κ coefficient

TABLE 11.1
Error Matrix of Extracted ISA in the Coastal Area of Rhode Island

	Reference Totals	Classified Totals	Number Correct	Producer's Accuracy (%)	User's Accuracy (%)
ISA	83	96	81	97.6	84.4
Non-ISA	117	104	102	87.2	98.1
Totals	200	200	183		

Overall accuracy = 91.5%.
κ value = 0.83.

TABLE 11.2
Percentage of ISA for Coastal Towns in Rhode Island

Town Name	Area (ha)	ISA (%)	Town Name	Area (ha)	ISA (%)
Little Compton	5854	6	Westerly	7962	16
Charlestown	9900	6	Warren	1619	16
South Kingstown	15,880	7	Middletown	3420	18
Tiverton	7863	8	Cranston	7492	19
Jamestown	2505	9	Bristol	2559	20
East Greenwich	4226	11	East Providence	3625	20
Narragansett	3691	12	Warwick	9300	24
Portsmouth	6115	13	Pawtucket	2296	26
North Kingstown	11,444	14	Newport	2096	30
Barrington	2227	14	Providence	4873	37

Average ISA cover = 14%.

(Table 11.1). The Producer's and User's accuracies are 97.6% and 84.4% for the ISA and 87.2% and 98.1% for the non-ISA categories, respectively.

The results reveal that three out of the 20 coastal towns in Rhode Island have highest ISA cover over 25%, including major population centers of Pawtucket (26%), Newport (30%), and Providence (37%). Another three towns have ISA cover between 20% and 25%, including Bristol (20%), East Providence (20%), and Warwick (24%). There are nine coastal towns that have ISA cover between 10% and 19%. Only five of the coastal towns in the state have ISA cover less than 10%. Overall, 14% of the coastal areas in Rhode Island are covered by ISA (Table 11.2).

11.4 CONCLUSION AND DISCUSSION

This study demonstrates that the MASC algorithm is effective and capable of extracting ISA from high spatial resolution digital true-color aerial orthophotography data. The result of this study revealed that 14% of the land in coastal towns of Rhode Island is covered by ISA as of 2004. Compared with 7% of overall ISA coverage for the state (Zhou and Wang, 2007), the coastal communities are more heavily urbanized with ISA covers. Increasing ISA in the landscape of this coastal state is a major concern in environmental and coastal management. Another study indicates that ISA increased 43% between 1972 and 1999 in the state, six times faster than population growth (Rhode Island Economic Policy Council, 2006). The increase of ISA consumed large areas of forests and caused fragmentation of forest habitats as well. The altered landscape would have significant effects on watershed hydrology (Chapter 18 of this book).

Ecosystem modeling and management decisions require precise information on ISA distribution. The results from this study provide the needed information on ISA for land management, planning, and ecological and hydrological modeling. As true-color orthophotography data are available among state agencies and are widely used by the general public, the MASC algorithm can be easily adapted and extended to other coastal areas to provide more comprehensive information on ISA as an indication of the impacts of human-induced land-cover change in the coastal regions.

ACKNOWLEDGMENT

This research was funded by the Rhode Island Agricultural Experimental Station (project no. RI00H330).

REFERENCES

Arnold, C.A., Jr. and Gibbons, C.J., 1996, Impervious surface coverage: The emergence of a key urban environmental indicator, *Journal of the American Planning Association*, 62: 243–258.

Baatz, M. and Schäpe, A., 2000, Multiresolution segmentation: An optimization approach for high quality multi-scale image segmentation, In: J. Strobl and T. Blaschke (Eds), *Angewandte Geographische Informationsverarbeitung XII*, pp. 12–23, 542pp, Wichmann-Verlag, Heidelberg, Germany.

Blaschke, T., Burnett, C., and Pekkarinen, A., 2004, New contextual approaches using image segmentation for object-based classification, In: S.M. de Jong and F.D. van der Meer (Eds), *Remote Sensing Image Analysis: Including the spatial domain*, pp. 211–236, Kluver Academic Publishers, Dordrecht.

Blaschke, T. and Strobl, J., 2001, What's wrong with pixels?: Some recent developments interfacing remote sensing and GIS, *GeoBIT/GIS*, 6: 12–17.

Booth, D.B. and Jackson, C.R., 1997, Urbanization of aquatic systems: Degradation thresholds, stormwater detection, and the limits of mitigation, *Journal of American Water Resources Association*, 35: 1077–1090.

Burnett, C. and Blaschke, T., 2003, A multi-scale segmentation/object relationship modelling methodology for landscape analysis, *Ecological Modelling*, 168(3): 233–249.

Carleer, A.P. and Wolff, E., 2006, Urban land cover multi-level region-based classification of VHR data by selecting relevant features, *International Journal of Remote Sensing*, 27(5–6): 1035–1051.

Civco, D.L., Hurd, J.D., Wilson, E.H., Arnold, C.L., and Prisloe, S., 2002, Quantifying and describing urbanizing landscapes in the Northeast United States, *Photogrammetric Engineering and Remote Sensing*, 68: 1083–1090.

Comanicu, D. and Meer, P., 2002, Mean shift: A robust approach toward feature space analysis, *IEEE Transactions on Pattern Analysis and Machine Intelligence*, 24(5): 603–619.

Costa-Pierce, B., 2006, Can we have sustainable coastal cities, *41° North: A publication of Rhode Island Sea Grant & the University of Rhode Island Coastal Institute*, 3(2): 3–4.

Dougherty, M., Randel, L.D., Scott, J.G., Claire, A.J., and Normand, G., 2004, Evaluation of impervious surface estimates in a rapidly urbanizing watershed, *Photogrammetric Engineering and Remote Sensing*, 70: 1275–1284.

Draper, S.E. and Rao, S.G., 1986, Runoff prediction using remote sensing imagery, *Water Resources Bulletin*, 22(6): 941–949.

Hu, X.Y., Tao, C.V., and Björn, P., 2005, Automatic segmentation of high resolution satellite imagery by integrating texture, intensity and color features, *Photogrammetric Engineering and Remote Sensing*, 71(12): 1399–1406.

Jacquin, A., Misakova, L., and Gay, M., 2008, A hybrid object-based classification approach for mapping urban sprawl in periurban environment, *Landscape and Urban Planning*, 84(2): 152–165.

Li, W., Benie, G.B., He, D.C., Wang, S.R., Ziou, D., Gwyn, Q., and Hugh, J., 1999, Watershed-based hierarchical SAR image segmentation, *International Journal of Remote Sensing*, 20(17): 3378–3390.

Lu, D. and Weng, Q., 2004, Spectral mixture analysis of the urban landscape in Indianapolis with Landsat ETM+ Imagery, *Photogrammetric Engineering and Remote Sensing*, 70(9): 1053–1062.

Novak, A. and Wang, Y., 2004, Effects of suburban sprawl on Rhode Island's forest: A Landsat view from 1972 to 1999, *Northeastern Naturalist*, 11: 67–74.

Rhode Island Economic Policy Council, 2006, Community Development Scorecard, available at http://www.ripolicy.org/2010, accessed on April 2007.

Schiff, K., Bay, S., and Stransky, C., 2002, Characterization of stormwater toxicants from an urban watershed to freshwater and marine organisms, *Urban Water*, 4: 215–227.

Schueler, T., 2003, *Impacts of Impervious Cover on Aquatic Systems*, 142pp, Center for Watershed Protection (CWP), Ellicott City, MD.

Shackelford, A.K. and Davis, C.H., 2003, A combined fuzzy pixel-based and object-based approach for classification of high-resolution multispectral data over urban areas, *IEEE Transactions on Geoscience and Remote Sensing*, 41(10): 2354–2363.

Walker, J.S. and Blaschke, T., 2008, Object-based landcover classification for the Phoenix metropolitan area: Optimization versus transportability, *International Journal of Remote Sensing*, 29(7): 2021–2040.

Wang, Y. and Zhang, X., 2004, A SPLIT model for extraction of subpixel impervious surface information, *Photogrammetric Engineering and Remote Sensing*, 70: 821–828.

Weng, Q.H., 2001, Modeling urban growth effects on surface runoff with the integration of remote sensing and GIS, *Environmental Management*, 28: 737–748.

Woodcock, C.E. and Harward, V.J., 1992, Nested-hierarchical scene models and image segmentation, *International Journal of Remote Sensing*, 13: 3167–3187.

Yu, Q., Gong, P., Clinton, N., Biging, G., Kelly, M., and Schirokauer, D., 2006, Object-based detailed vegetation classification with airborne high spatial resolution remote sensing imagery, *Photogrammetric Engineering and Remote Sensing*, 72(7): 799–811.

Zhou, Y. and Wang, Y., 2007, An assessment of impervious surface areas in Rhode Island, *Northeastern Naturalist*, 14(4): 643–650.

Zhou, Y. and Wang, Y., 2008, Extraction of impervious surface areas from high spatial resolution imageries by multiple agent segmentation and classification, *Photogrammetric Engineering and Remote Sensing*, 74: 857–868.

12 FORMOSAT-2 Images in Mapping of South Asia Tsunami Disaster

Ming-Der Yang, Tung-Ching Su, and An-Ming Wu

CONTENTS

12.1 INTRODUCTION

Natural disasters are considered to be destined, but nowadays limited prevention and reduction approaches can be adopted due to a great deal of coordination. An efficient action of disaster mitigation and rescue relies on timely and effective information about hazard areas. Unfortunately, natural disasters are often widespread and the devastated areas are sometimes unreachable so that an overall on-site investigation becomes difficult or even impossible. With the rapid technology development of space platform and digital imagery, remote sensing data have been broadly applied to monitoring the Earth's surface in the last decades, such as agriculture monitoring [1,2], environmental change [3], and water pollution assessment [4–6]. Several studies applying remote sensing to natural hazard investigation and management were reviewed thereafter. Landsat5 TM images were applied to identify the neotectonic features of the September 7, 1999 Athens earthquake [7]. Gupta and Joshi [8] and Lin et al. [9,10] used remote sensing and geographic information system (GIS) techniques in assessing landslides and debris flows. The U.S. National Oceanic and Atmospheric Administration monitors droughts on a large area by developing an Advanced Very High Resolution Radiometer (AVHRR)–based Vegetation Condition Index (VCI) derived from the Normalized Difference Vegetation Index (NDVI) and the Temperature Condition Index (TCI) derived from the AVHRR-measured radiances [11]. SPOT satellite images were used to analyze the NDVI for comparison and evaluation of the vegetation recovery rates before and after the Chi-Chi Earthquake [12]. Aerial photographs and SPOT satellite images were coupled to assess the damage level and potential risk of collapsed spots on the Tsao-Ling landslide caused by the Chi-Chi Earthquake [13]. A synthetic probability map of the Tsao-Ling landslide reoccurrence was produced that provided efficient information for planning emergency response and making a rehabilitation strategy. Also, automated classifications were developed and applied to SPOT satellite images for the recognition of landslide hazards [14,15]. The integration of GIS, remote sensing, and global positioning system (GPS) can provide a low-cost and rapid methodology of disaster management as well as critical information for decision support by emergency managers and the disaster response community [16,17]. A decision support system (DSS) for open mining areas during the formulation of a restoration plan was developed by using thematic maps derived from observation data from Earth [18].

Satellite images have several advantages for hazard monitoring, such as digital data beneficial for processing; large area coverage; and routine, repetitive, continuous, and economical image acquisition. Compared with aerial photography, however, satellite images have a lower spatial resolution that limits their applications in urban areas. With the launching of IKONOS-1 in 1999 and QUICKBIRD in 2001, a new era of commercial satellite imaging with high spatial resolution was begun by offering resolutions of 1 and 0.6 m panchromatic images as well as 4 and 2.44 m multispectral images, respectively. The land cover map derived from IKONOS satellite images was used as input for the flood simulation model LISFLOOD-FP to produce a Manning roughness factor map of inundated areas [19]. Intensity-Hue-Saturation (IHS) was used to integrate hyperspectral and radar data into a single image of an urban area by image fusion that provided rich information necessary for assessment and mitigation of hazards in urban disaster management [20]. High-resolution digital elevation models (e.g., from InSAR, Lidar, and digital photogrammetry) and image spectroscopy (e.g., using ASTER, MODIS, and Hyperion) are believed to significantly contribute to hazard management of natural disasters, such as earthquakes, volcanos, floods, landslides, and coastal inundations [17].

However, high spatial resolution images do not fulfill the demand of a complete disaster management. A better temporal resolution (revisit capability) is desirable during hazard response and recovery due to their emphasis on "real-time" information about a disaster area. In this chapter, a civilian satellite with both high spatial and temporal resolutions, FORMOSAT-2, was employed for remote monitoring of the severe coastal disasters from the tsunami-devastated areas of the Sumatra catastrophe. FORMOSAT-2 offers images with a high spatial resolution of 2 m panchromatic and 8 m multispectral as well as a high temporal resolution of 1 day. FORMOSAT-2's superior temporal frequency of remote sensing data acquisitions makes it possible to apply satellite images to response and recovery of disaster management.

12.2 DISASTER MONITORING USING SATELLITE IMAGES

Hazard management involves four phases in a cycle, which should consist of mitigation, preparedness, response, and recovery [16]. Remote sensing has been playing a key role in disaster mitigation activities, including assessing risk and reducing potential effects of disasters. Spatial and temporal anomalies of the Earth's emitted radiation in the thermal infrared (TIR) spectral range measured from a satellite were analyzed for monitoring a seismically active region [21]. Geospatial information products derived from remote sensing data are addressing the operational requirements of the hazard DSS, which is helpful in risk assessments, mitigation planning, disaster assessment, and response prioritization for policy makers, emergency managers, and responders [17]. Various useful information, such as land use, land cover, infrastructures, human settlements, topography, vegetation recovery index, morphology, and geology for risk analysis, and hazard maps can be extracted from remote sensing data [12,13,17,19,22,23]. Preparedness consists of planning actions in a disaster and trying to increase available rescue resources. The information about risk analysis extracted from high spatial resolution satellite imagery can be introduced into GIS to establish a disaster-relief database for preparedness [22]. Response relates to all possible emergency activities immediately following a disaster occurrence. When a disaster occurrence passes away, short-term and long-term recovery continues until all life-support systems are back to normal [16]. The recovery rate of an ecosystem can be successfully estimated by comparing satellite images before and after a disaster [12].

Among these four phases of hazard management, satellite imagery provides limited support during the short period of disaster response due to its deficient temporal resolution for instant reactions immediately following a disaster. However, the response phase is the most critical portion of the disaster cycle because it provides emergency assistance to victims of an event and reduces the likelihood of secondary damage. Thus, routine, repetitive, and continuous observations of the Earth's surface from high temporal resolution satellite imagery are very useful for acquiring rapid assessments to take rapid and effective action in the disaster response phase [3,16]. The application of satellite images to disaster response includes several steps, including orbit determination, image acquisition, data preparation, image processing, image interpretation, and decision supports for disaster response. Through the analysis process, high temporal satellite imagery can efficiently assist authorities in making disaster response decisions and further recovery plans. In the phase of recovery, routine, repetitive, and continuous observations of the Earth's surface from operational satellites with high temporal resolution imagery are useful in assessing hazard impact and making recovery plans [3,16]. Also, satellite images are suitable for a long-term assessment of environment and ecosystem recovery.

12.3 FORMOSAT-2 SATELLITE

As other developed countries in the world, Taiwan faces dynamic changes of land use due to a rapid economic growth and a frequent attack of natural disasters such as earthquakes and typhoons. In the aftermath, the need for timely remote sensing data to monitor and assess damage becomes urgent. Thus, in 1991 the National Space Organization (NSPO) was founded by the National Science Council (NSC) as a governmental agency to execute the national space program [24]. Currently, two satellites are operated by the NSPO to collect remote sensing data. One of them,

TABLE 12.1
FORMOSAT-2 Remote Sensing Instrument Specifications

Payload	Remote Sensing Instrument (RSI) Imager of Sprites and Upper Atmospheric Lightning (ISUAL)
Weight	742 kg (with payload and fuel)
Shape	Hexahedron, height 2.4 m, outer diameter about 1.6 m
Orbit	891 km, sun-synchronous, passes over Taiwan Strait each day
Panchromatic (PAN)	0.45–0.90 μm
Multispectrum (MS)	0.45–0.52 μm blue; 0.52–0.60 μm green; 0.63–0.69 μm red; 0.76–0.90 μm NIR
Ground sampling distance (GSD)	2 m for black and white images; 8 m for color images
Swath	24 km
Mission life	5 years
Launch date	May 21, 2004 (Taipei Time)

FORMOSAT-2, takes land images for fulfilling Taiwan civilians' needs by providing black and white images with the resolution of 2 m and color images of 8 m resolution (Table 12.1). Also, an imager on board FORMOSAT-2 observes upper atmospheric lightning for the first time from a satellite. FORMOSAT-2 was successfully launched on May 21, 2004 (Taipei Time), and acquired the first image at 9:39 a.m. on June 4 [25]. To provide timely and low-cost image data, FORMOSAT-2 can daily monitor the environment and resources throughout the Taiwan main island, the offshore remote islands, Taiwan Strait, the surrounding ocean, and other regions under international cooperation. FORMOSAT-2 is expected to have many useful applications, such as agriculture and forestry predictions, natural disaster evaluations, land usage analyses, environmental monitoring, coastal search and rescue, and academic research and extension education. Especially for natural disaster evaluations, FORMOSAT-2 images are valuable for assessing damages through applying for tasking orders possibly within 1 day if weather allows. The conditions of flood-submerged fields, landslides, debris flows, and the structural integrity of hillsides also may be assessed. A couple of significant contributions by FORMOSAT-2 in monitoring and evaluating natural disasters in Taiwan include the flooding and debris flows brought by Typhoons Midully and Aere in summer 2004 [23,26].

12.4 FORMOSAT-2 IMAGES OF SOUTH ASIA TSUNAMI

On December 26, 2004, a 9.0-Mw earthquake was caused by the release of stresses accumulated as the Burma tectonic plate overrode the India tectonic plate [27]. Data from 60 GPSs in Southeast Asia were collected to identify the crustal deformation caused by the Sumatra earthquake [28]. This great earthquake generated a rupture expending at a speed of about 2.5 km/s toward the north-northwest and extending 1200–1300 km along the Andaman trough [29]. Subsequently, the significant movement of the seafloor resulted in a tsunami, or seismic sea wave, which struck the

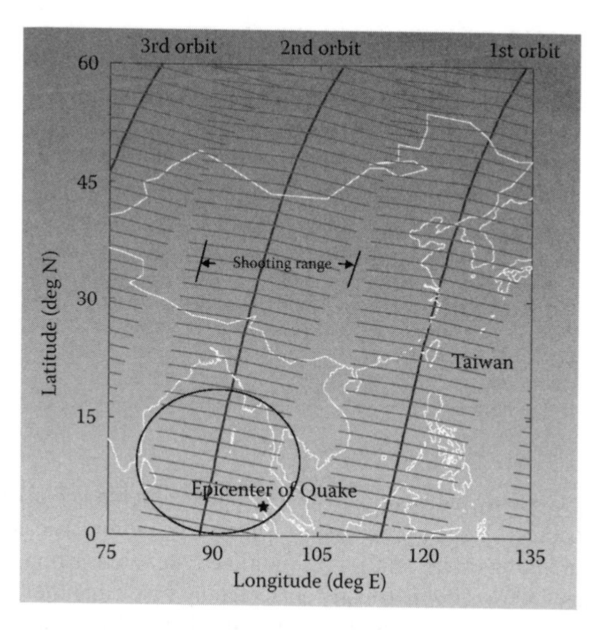

FIGURE 12.1 FORMOSAT-2 orbits covering the Asia area.

coastal regions of Sumatra, Thailand, Sri Lanka, southern India, and most South Asia areas. Over 170,000 people were reported dead or missing due to the tsunami unleashed by the quake.

For an instant response to the natural disaster, catastrophic images were acquired in 2 days by an emergency orbit scheduling. Passing over South Asia on its second orbit satellite as shown in Figure 12.1, FORMOSAT-2 was one of the first remote sensing satellites to successfully take images of Phuket Island and Banda Aceh, two heavily hit areas, within 2 days after the disaster, as listed in Table 12.2. Those images were processed and provided to aid rescuers working in those areas through channels of the Department of Foreign Affairs and research centers of Taiwan. Meanwhile, FORMOSAT-2 continued imaging up to an additional month for those seriously hit areas, including the west coast of Thailand, Sumatra of Indonesia, the eastern coast of India, Sri Lanka, and Andaman and Nicobar Islands. Highly-efficient data propagation increases the benefits from remote sensing data sharing and dissemination; therefore, the NSPO established a window through the Internet for the public access to FORMOSAT-2 image database. Several images of devastated regions were analyzed in this chapter as Figure 12.2. A standard level-II product of FORMOSAT-2 has been through image system geometric and radiometric correction [30,31]. For a better visual interpretation, fusion images were produced by using a multiplicative algorithm on 2 m panchromatic images of high spatial resolution and 8 m multispectral images of high spectral resolution. The fusion images provide rich information necessary for hazard assessment and mitigation.

Furthermore, for hypertemporal image analysis, three requirements must be met: (1) univariate at each temporal image; (2) precise coregistration; and (3) radiometric

TABLE 12.2
Process Timetable of FORMOSAT-2 Imagery

Sequence	Time Needed
Order	2 h
Scheduling	2 h
Imaging	2–24 h
Acquisition	2 h
Processing	2 h
Distribution	2 h
Total	12–34 h

consistency [3]. Ground control points (GCPs) and local reference maps could facilitate advanced georectification if necessary. Supremely, radiometric correction is the major process affecting image analysis of multiple-date satellite images. These multitemporal images were radiometrically corrected by first using the pseudoinvariant feature (PIF) method to locate possible dark and bright objects (the same buildings, roads, and water pixels on both images) and then processing the radiometric control set (RCS) method on the selected pixels to find out the correlation coefficients between images for each band of the multispectral images [32,33]. Under the assumption that at least some pixels have the same average surface reflectance among images acquired at different dates, the RCS method examines the band-to-band scattergrams

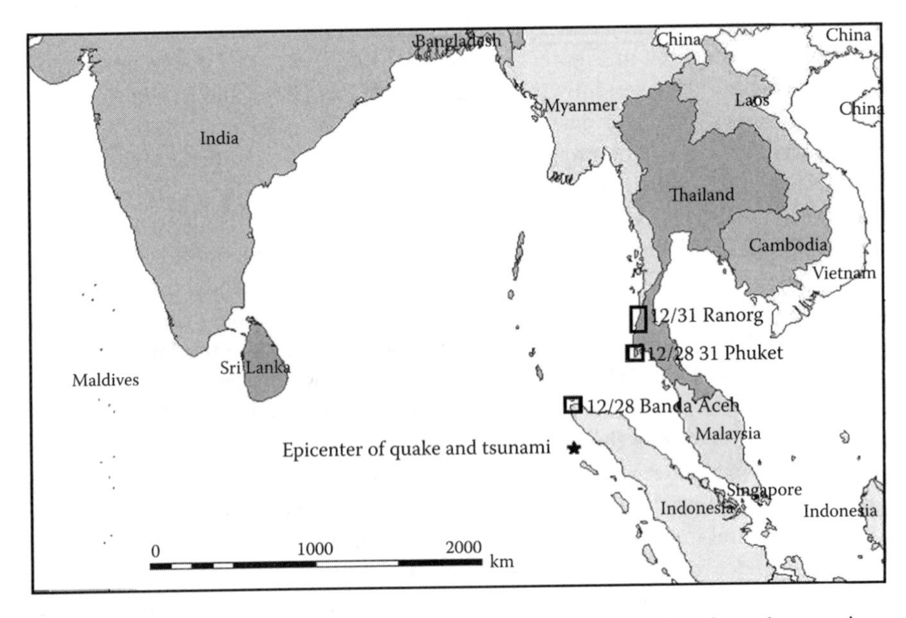

FIGURE 12.2 Illustrated study areas attacked by the Sumatra earthquake and tsunami.

in which the pixels have little or no variation between the dates of image acquisition. The radiometric methods were employed to correct the digital numbers (DNs) on the December 31 image based on the December 28 image for Phuket Island in this research. The radiometric rectification equations of the December 31 image were acquired through linear regression as follows:

$$DN_{band1'} = 1.2116\ DN_{band1} - 26.873; \quad R^2 = 0.917, \tag{12.1}$$

$$DN_{band2'} = 1.2165\ DN_{band2} - 18.395; \quad R^2 = 0.958, \tag{12.2}$$

$$DN_{band3'} = 1.1664\ DN_{band3} - 8.5458; \quad R^2 = 0.964, \tag{12.3}$$

$$DN_{band4'} = 1.2462\ DN_{band4} - 17.781; \quad R^2 = 0.972. \tag{12.4}$$

The high correlation coefficients of regression results show a significant confidence in the radiometric correction.

12.5 IMAGE PROCESS AND INTERPRETATION

After being corrected for geometry and postprocessed, the FORMOSAT-2 image is suitable for constructing about 1:24,000 scale maps in true or false color that can be analyzed for monitoring the disaster damages. The extent and localization of the tsunami damages were analyzed through either visual interpretation of the fusion images or digital analysis on multispectral FORMOSAT-2 images, including haze reduction, feature extraction, image classification, contrast enhancement, and band ratio analysis. To provide the promised information to help cope with the tsunami catastrophe, several important discovery and analysis results were addressed in the subset images of interest, which are useful for disaster region positioning, damage level assessment, rescue planning, and environmental impact evaluation (Figures 12.3 through 12.5).

The extent and localization of the damage were estimated by digital analysis of the differences between the struck and nonstruck areas. The VI (vegetation index) and NDVI were analyzed for revealing the extent of the intruded area by the tsunami. Near-infrared (NIR) reflectance intensity is related to mostly vegetation conditions and inversely to water content. Destroyed vegetation and high water content caused by the tsunami reduced NIR reflectance intensity in the inundated terrains. Immersed areas after the tsunami reflect weaker NIR than non-struck areas. Moreover, the scooped terrain by the tsunami flood wave increases red reflectance intensity. Thus, the VI or NDVI analysis is helpful for interpreting the boundary between the struck and nonstruck areas.

Figure 12.3 shows the destruction of Ranong, in the western part of Thailand along the Andaman Sea. Figure 12.3a1 is the false-color NIR fusion imagery of Khao Lak revealing the destructions by the tsunami, where there is a portion of Khao Lak—Ram Lu National Park. The long-term impact upon the ecological environment must be noticed in the future. Figure 12.3a2 is the NDVI analysis of the original multispectral images. A distinguishable difference of NDVI values between the struck and nonstruck areas along the coastline illustrates the extent of the tsunami

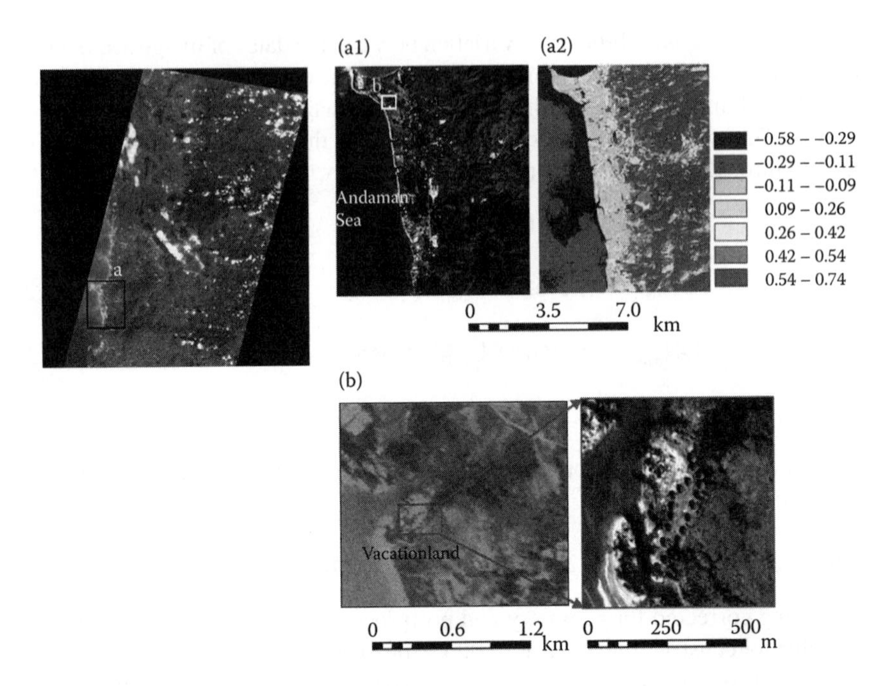

FIGURE 12.3 Destruction of Ranong, in the western part of Thailand along the Andaman Sea. (a1) Devastated area of Khao Lak, (a2) NDVI analysis helpful for distinguishing the extent of the tsunami destruction, and (b) overwhelmed wooden houses in the vacationland in Thap Lamu, Ranong.

inundation, resulting in a negative NDVI value. Also, the indentation eroded by the tsunami in the coastline becomes more evident in the NDVI map. Figure 12.3b shows the overwhelmed wooden houses in the vacationland in Thap Lamu, Ranong. Mud and clay brought by the flood wave covered most of the vacationland so that the traffic network was interrupted.

With many hotels and golf courses built in the neighborhood of the beach area, Thailand's Phuket is a world-famous resort island and were severely devastated in this disaster. Traffic facilities, such as roads, bridges, and airports, affect the efficiency of disaster relief. Important traffic infrastructures, such as the Sarasin Bridge connecting Phuket Island and mainland Thailand in Figure 12.4a and the Phuket International Airport for international aid in Figure 12.4b, were undamaged and usable for emergency rescue and recovery. An international airlift was still allowed to ferry critical aid and medicine to Phuket and also to take surviving travelers home. Figure 12.4c1 shows that Bang Thao, one of the coastal towns in the western Phuket Island, had only minor destruction after the tsunami. The clinic and clubs were safe to provide medical aids and disaster shelters. The main road, Scenic Coastal Drive, was not destroyed by the floodwaters. Moreover, to improve the visual appearance of waves, contrast enhancement was conducted on the extracted water body from the panchromatic imagery. Through histogram analysis and linear stretch, the waterfront was magnified. In this recreation beach area, the wave condition usually is alm inside the fiord and the wave front is parallel to and the wave direction perpendicular to the

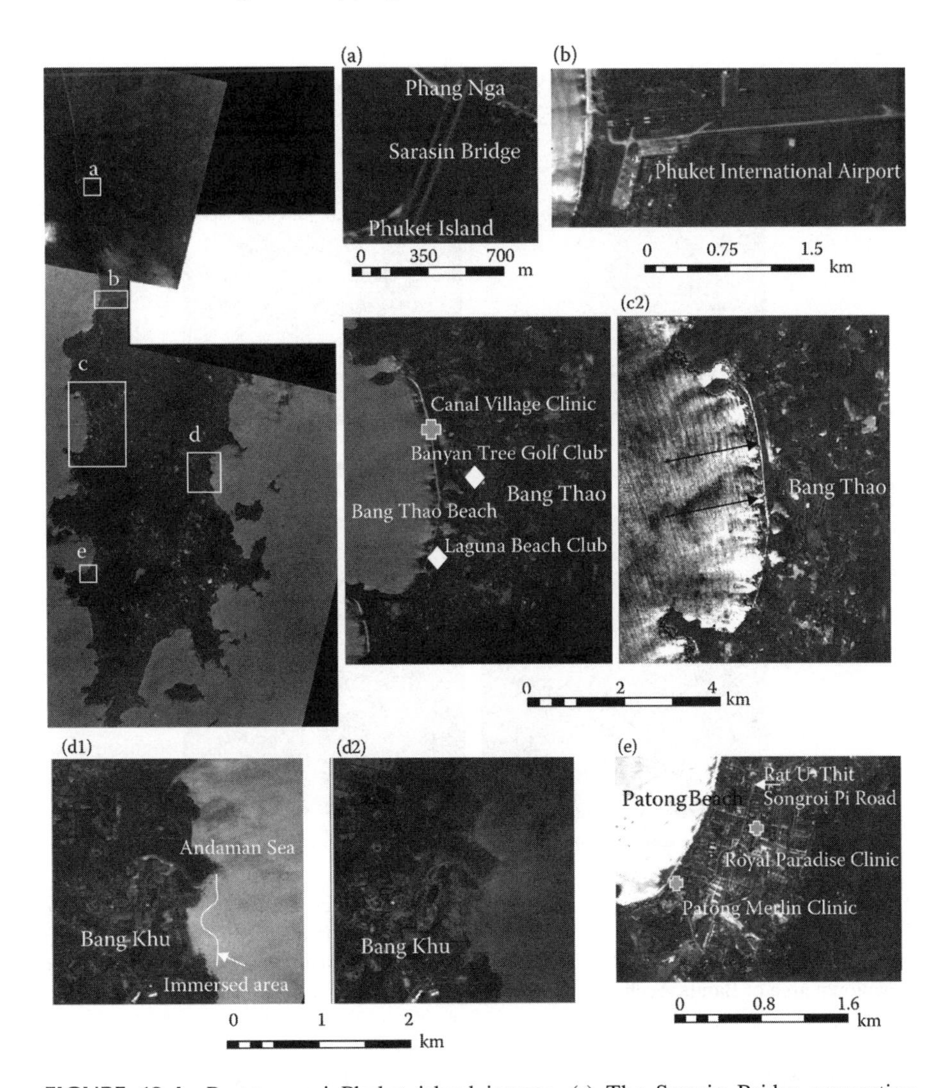

FIGURE 12.4 Post-tsunami Phuket island images. (a) The Sarasin Bridge connecting Phuket Island and mainland Thailand. (b) The Phuket International Airport used for receiving international aid. (c1) Bang Thao resort. (c2) The enhanced image with magnified waterfront areas; the attacking direction of the tsunami is shown by arrows. (d1) Infrared (IR) fusion image of Bang Khu on December 28. (d2) IR fusion image of Bang Khu on December 31. (e) The fusion image of Paton.

shoreline due to the dragging effect on the seabed and coastal terrain. The unusual wave height and pattern in the Ban Thao beach indicates the attacking direction of the tsunami, shown by red arrows in Figure 12.4c2. Figure 12.4d1 and d2 are the false-color NIR fusion images of Bang Khu in the eastern coast of Phuket which were taken on December 28 and 31, respectively. Comparing these two radiometrically corrected images, we discovered that flooding vanished in the coast region and the submerged land reappeared on December 31. Figure 12.4e is the fusion images

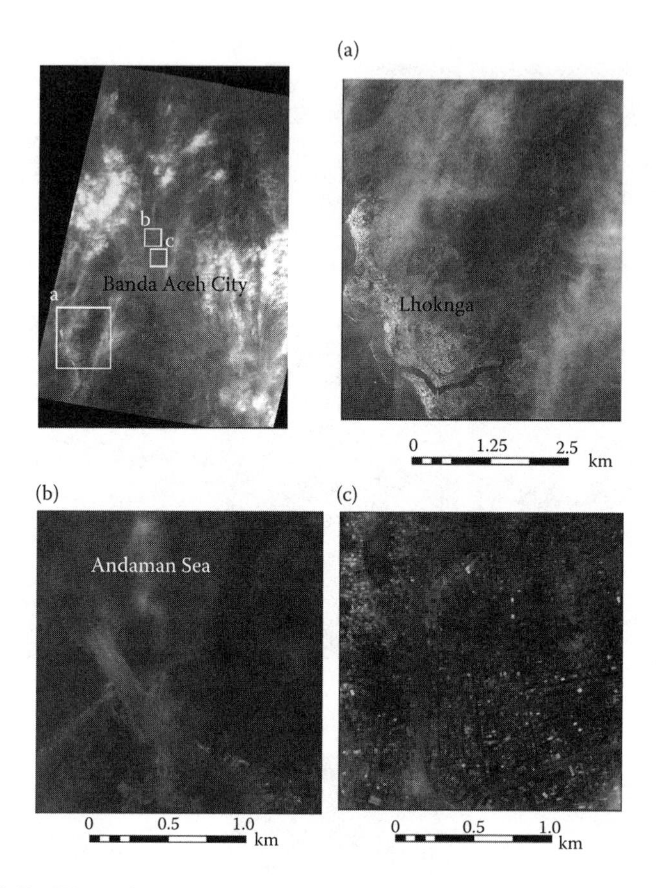

FIGURE 12.5 Worst-hit region—Aceh province. (a) Struck and flooded Lhoknga. (b) Polluted water of one tributary of the Aceh River with high suspended solids and wastes toward the Andaman Sea. (c) Badly destroyed traffic infrastructures and buildings in the downtown area of Banda Aceh city.

of Patong, which is a famous international resort town. The west side of Patong by Rat U-Thit Songroi Pi Road, where many scenic spots and quality hotels were located, was drastically slashed by the tsunami. There were two clinics providing medical aid in Patong were; Patong Merlin Clinic was struck by the tsunami and the Royal Paradise Clinic was safe to take care of the wounded people in the catastrophe.

Through haze reduction (by using a 3×3 high-frequency kernel) and Gaussian contrast adjustment on the FORMOSAT-2 image of December 28, the authorities and public were able to get a better glimpse of the devastated west coast of Indonesia's Sumatra Island—nearest the epicenter of the massive quake and tsunami. Many coastal areas of the northwestern Sumatra were badly hit by the quake-driven walls of massive water. Aceh province in particular suffered extensive damage and loss of life in this catastrophe. Figure 12.5a shows that the Lhoknga area was severely struck and flooded by the tsunami. The tsunami swashed up to about 3–4 km inland and destroyed the forests and farmland in the coast regions. More analyses of the tsunami effects in Banda Aceh, including inundation distance, shoreline erosion, and coseismic subsidence can also be found in Borrero's investigation report [34]. Figure 12.5b shows that one of the Aceh River tributaries crossing the downtown area toward the Andaman Sea was polluted by salty seawater, mud flow, and wastes. Thus, the invaded city was expected to be severely afflicted with the problems of water resources and environmental pollution. Pure water shortages and overwhelmed aquatic ecosystems need to be paid attention to during hazard recovery. Figure 12.5c is zoom-in subset showing the badly destroyed traffic infrastructures and buildings in the downtown area of Banda Aceh city. The city was flooded and blanketed with mud and rubble after the quake and tsunami, which increased the difficulty of emergency rescue and recovery.

12.6 CONCLUSIONS

Although satellite remote sensing has been advocated for monitoring terrain changes for a long time, satellite images with a high temporal resolution represent indeed a major progress in coastal disaster monitoring. FORMOSAT-2 can efficiently contribute timely information for disaster response and recovery due to its high spatial and temporal resolutions. Especially for an instant response to a natural disaster, FORMOSAT-2 images can be acquired in at most 2 days through an emergency order. In this application, FORMOSAT-2 proved to be a useful source in the monitoring of the severe Sumatra earthquake and tsunami disaster. Through digital image processes, some results of image analysis and interpretation were useful for instant response and rescue and were helpful for disaster recovery. The FORMOSAT-2 fusion images provide rich information necessary for disaster assessment and mitigation. FORMOSAT-2 imagery, especially fusion imagery, of December 28 and 29, 2004, clearly revealed the extension of flooding and widespread devastation to the coastal cities and surrounding areas. Evidently, the contribution of FORMOSAT-2 in providing the most up-to-date images available on hand can be appreciated. More advanced image processes can be executed based on the acquisition of local geographic and environmental information in order to prepare a sound and comprehensive recovery plan and indemnities distribution for those disaster areas. Remote sensing has been playing a key role in disaster mitigation and damage assessment nowadays and will flourish in disaster management as more advanced remote sensing satellites begin operations in the near future.

REFERENCES

1. Rydberg, A. and Borgefors, G., Integrated method for boundary delineation of agricultural fields in multispectral satellite images, *IEEE Geosci. Remote Sens.*, 39, 2514, 2001.
2. Murthy, C.S., Raju, P.V., and Badrinath, K.V.S., Classification of wheat crop with multitemporal images: Performance of maximum likelihood and artificial neural networks, *Int. J. Remote Sens.*, 24, 4871, 2003.
3. Piwowar, J.M., Pdeele, D.R., and LeDrew, E.F., Temporal mixture analysis of Artic sea ice imagery: A new approach for monitoring environmental change, *Remote Sens. Environ.*, 63, 195, 1998.

4. Yang, M.D., Merry, C.J., and Sykes, R.M., Integration of water quality modeling, remote sensing, and GIS, *J. Am. Water Resour. Assoc.*, 35, 253, 1999.
5. Chen, L., A study of applying genetic programming to reservoir trophic state evaluation using remote sensor data, *Int. J. Remote Sens.*, 24, 2265, 2003.
6. Cheng, K.S. and Lei, T.C., Reservoir trophic state evaluation using Landsat TM images, *J. Am. Water Resour. Assoc.*, 37, 1321, 2001.
7. Ganas, A., Papadopoulos, G., and Pavlides, S.B., The 7 September 1999 Athens 5.9 Ms earthquake: Remote sensing and digital elevation model inputs towards identifying the seismic fault, *Int. J. Remote Sens.*, 22, 191, 2001.
8. Gupta, B.P. and Joshi, B.C., Landslide hazard zoning using the GIS approach—a case study from Ramganga catchment, Himalayas, *Eng. Geol.*, 28, 119, 1990.
9. Lin, P.S., Hung, J.C., Lin, J.Y., and Yang, M.D., *Risk Assessment of Potential Debris-Flows Using GIS, Debris-Flow Hazards Mitigation: Mechanics, Prediction, and Assessment*, p. 608, Balkema, Rotterdam, 2000.
10. Lin, P.S., Lin, J.Y., Hung, J.C., and Yang, M.D., Assessing debris-flow hazard in a watershed in Taiwan, *Eng. Geol.*, 66, 295, 2002.
11. Kogan, F.N., Global drought watch from space, *Bull. Am. Meteorol. Soc.*, 78, 621, 1997.
12. Lin, C.Y., Lo, H.M., Chou, W.C., and Lin, W.T., Vegetation recovery assessment at the Jou-Jou mountain landslide area caused by the 921 Earthquake in Central Taiwan, *Ecol. Model.*, 179, 75, 2004.
13. Yang, M.D., Yang, Y.F., and Hsu, S.C., Application of remotely sensed data to the assessment of terrain factors affecting Tsao-Ling landside, *Can. J. Remote Sens.*, 30, 593, 2004.
14. Yang, M.D. and Yang, Y.F., Genetic algorithm for unsupervised classification of remote sensing imagery, In: *Proc. SPIE-IS&T Electronic Imaging, SPIE*, E.R. Dougherty, J.T. Astola, and K.O. Egiazarian (Eds), IS&T/SPIE, San Jose, 2004, p. 395.
15. Yang, M.D., Hsu, C.H., and Su, D.C., DBI and SI as fitness in GA classification for remote sensing imagery, In: *Proc. 2005 IEEE International Geoscience and Remote Sensing Symposium*, J. A. Benediktsson, W. Emery, and A. Camps (Eds), IEEE, Seoul, 2005.
16. Montoya, L., Geo-data acquisition through mobile GIS and digital video: An urban disaster management perspective, *Environ. Model. Softw.*, 18, 869, 2003.
17. Tralli, D.M., Blom, R.G., Zlotnicki, V., Donnellan, A., and Evans, D.L., Satellite remote sensing of earthquake, volcano, flood, landslide and coastal inundation hazards, *ISPRS J. Photogramm.*, 59, 185, 2005.
18. Ganas, A., Aerts, J., Astaras, T., Vente, D., Frogoudakis, E., Lambrinos, N., Riskakis, C., Oikonomidis, D., Filippidis, A., and Kassoli-Fournaraki, A., The use of Earth observation and decision support systems in the restoration of open-cast nickel mines in Evia, central Greece, *Int. J. Remote Sens.*, 25, 3261, 2004.
19. Van Der Sande, C.J., DeJong, S.M., and DeRoo, A.P.J., A segmentation and classification approach of IKONOS-2 imagery for land cover mapping to assist flood risk and flood damage assessment, *Int. J. Appl. Earth Obs.*, 4, 217, 2003.
20. Chen, C.M., Hepner, G.F., and Forster, R.R., Fusion of hyperspectral and radar data using the HIS transformation to enhance urban surface features, *ISPRS J. Photogramm.*, 58, 19, 2003.
21. Tramutoli, V., Cuomo, V., Filizzola, C., Pergola, N., and Pietrapertosa, C., Assessing the potential of thermal infrared satellite surveys for monitoring seismically active areas: The case of Kocaeli (Izmit) earthquake, August 17, 1999, *Remote Sens. Environ.*, 96, 409, 2005.
22. Rivereau, J.C., Spot data applied to disaster prevention and damage assessment, *Acta Astronaut.*, 35, 467, 1995.

23. Lin, C.W., Liu, C.C., Lee, S.Y., Tsang, Y.C., Lie, S.H., Chiu, Y.C., Wu, A.M., Wu, F.F., and Chen, H.Y., Assessment of the typhoon Mindulle induced hazard using ROCSAT-2 imagery—Chenyulan River and Tachia River watershed, In: *Proc. 23rd Conference of Surveying Theorem and Applications*, V.J.D. Tsai, S.P. Kao, and M.D.Yang (Eds), National Chung Hsing University, Taiwan, 2004, p. 367 (in Chinese).
24. NSPO, 2005, http://www.nspo.org.tw/e60/menu0401.html, accessed on July 21, 2005.
25. Wu, A.M. and Wu, F., Applications of ROCSAT-2 images, In: *Proc. 23rd Conference of Surveying Theorem and Applications*, V.J.D. Tsai, S.P. Kao, and M.D. Yang (Eds), National Chung Hsing University, Taiwan, 2004, p. 359 (in Chinese).
26. Chang, K.C., Wu, F., Chen, H., Chen, C.K., and Duan, C.R., Applying ROCSAT-2 image in 0702 hazard evaluation, In: *Proc. 23rd Conference of Surveying Theorem and Applications*, V.J.D. Tsai, S.P. Kao, and M.D. Yang (Eds), National Chung Hsing University, Taiwan, 2004, p. 339 (in Chinese).
27. USGS, 2005, http://earthquake.usgs.gov/eqinthenews/2004/usslav, accessed on July 21, 2005.
28. Vigny, C., et al., Insight into the 2004 Sumatra–Andaman earthquake from GPS measurements in southeast Asia, *Nature*, 436, 201, 2005.
29. Ammon, C.J., et al., Rupture process of the 2004 Sumatra–Andaman earthquake, *Science*, 308, 1133, 2005.
30. Wu, F., Wu, A.M., and Shieh, C.J., Development of image processing system for ROCSAT-2, In: *Proc. 5th International Symposium on Reducing the Cost of Spacecraft Ground Segments and Operations*, JPL, Pasadena, CA, 2003.
31. Chen, H., Wu, F., and Liu, C., Introduction to ROCSAT-2 terminal, In: *Proc. 23rd Conference of Surveying Theorem and Applications*, V.J.D. Tsai, S.P. Kao, and M.D. Yang (Eds), National Chung Hsing University, Taiwan, 2004, p. 333 (in Chinese).
32. Yang, X.Y. and Lo, C.P., Relative radiometric normalization performance for change detection from multi-date satellite images, *Photogramm. Eng. Remote Sens.*, 66, 967, 2000.
33. Ya'allah, S.M. and Saradjian, M.R., Automatic normalization of satellite images using unchanged pixels within urban areas, *Inform. Fusion*, 6, 235, 2005.
34. Borrero, J.C., Field data and satellite imagery of tsunami effects in Banda Aceh, *Science*, 308, 1596, 2005.

Section IV

Remote Sensing and In Situ *Measurements for Habitat Mapping*

13 Remote Sensing and *In Situ* Measurements for Delineation and Assessment of Coastal Marshes and Their Constituent Species

Martha S. Gilmore, Daniel L. Civco, Emily H. Wilson, Nels Barrett, Sandy Prisloe, James D. Hurd, and Cary Chadwick

CONTENTS

13.1 INTRODUCTION

A significant amount of the coastal wetlands along the Long Island Sound in the northeastern United States has been lost over the past century due to development, filling and dredging, or damaged due to human disturbance and modification. Global sea level rise is also likely to have a significant impact on the condition and health of coastal wetlands, particularly if the wetlands have no place to which to migrate. Beyond the physical loss of marshes, the species composition of marsh communities is changing. *Spartina alterniflora* (Loisel.) (salt cordgrass) and *Spartina patens* [(Aiton) Muhl.] (salt marsh hay), once the dominant species of New England salt marshes, are being replaced by monocultures of the nonnative genotype of *Phragmites australis* [(Cav.) Trin. ex Steud.] (common reed) in Connecticut marshes [1]. With the mounting pressures on coastal wetland areas, it is becoming increasingly important to identify and inventory the current extent and condition of coastal marshes located on the Long Island Sound estuary, implement a cost-effective way by which to track changes in wetlands over time, and monitor the effects of habitat restoration and management.

The identification of the distribution and health of individual marsh plant species like *Phragmites australis* using remote sensing is challenging because vegetation spectra are generally similar to one another throughout the visible to near-infrared (VNIR) spectrum and the reflectance of a single species may vary throughout the growing season due to variations in the amount and ratios of plant pigments, leaf moisture content, plant height, canopy effects, leaf angle distribution, and other structural characteristics. While vegetation phenology has long been recognized to be useful in discriminating species for vegetation mapping (e.g., [2,3]), most classification methodologies applied to wetlands have been based on image and/or ground reference data measured on a single date, which limits their applicability to images taken at other times. Previous work on the classification of marsh vegetation using multitemporal images (e.g., [4–6]) and light detection and ranging (LiDAR) (e.g., [7]) data have typically relied on the identification of endmembers, often derived from extensive field measurements. Such field measurements may be impractical if a goal is to inventory vegetation in even a small number of marshes. In this work, we test a new method that seeks to relate field measurements at a limited number of sites to image classification of the entire marsh.

Here we report on a novel approach whereby VNIR reflectance spectra of marsh vegetation measured in the field over the growing season were used to guide the classification of high spatial resolution multitemporal QuickBird data. Single-date LiDAR data also contributed to the classification. We sought to (1) determine the optimal times during the growing season for the discrimination of individual marsh plant species based on spectral reflectance and structure measured in the field, and (2) assess the utility of these field data to direct the classification of multitemporal images of the entire marsh, with particular attention to the mapping of the invasive species *Phragmites australis*. Our overall goal was to provide protocols that can be used to classify multiple- or single-date multispectral data that are available to wetland managers.

13.2 METHODS

13.2.1 Study Site

Ragged Rock Creek marsh is a 142-ha brackish tidal marsh located on the western shore of the Connecticut River, approximately 2.5 km north of its confluence with Long Island Sound (Figure 13.1). The vegetation at Ragged Rock Creek marsh is typical of Connecticut's estuarine tidal marshes, where the pattern of growth is generally controlled by salinity, a function of duration and magnitude of tidal inundation, and therefore elevation. The vegetation at Ragged Rock Creek marsh is a mosaic, ranging from patches of monospecific dominants with discrete boundaries, to mixed-species patches with more diffuse transitions. While our field survey identified 115 plant species at Ragged Rock Creek marsh, the vegetation was dominated by three communities: (1) *Spartina patens* meadows [*Spartina patens* (salt meadow grass) ± *Eleocharis* spp. (spike rushes), hereafter referred to as *Spartina patens*], (2) *Typha* spp., which include narrow-leaved cattail (*Typha angustifolia*) and hybrid cattail (*Typha* × glauca), and (3) nonnative common reed (*Phragmites australis*; Figure 13.2). *Typha* spp. and *Phragmites australis* typically formed dense monotypic stands and occupied the mid- to high-marsh areas and upper border, respectively. *Spartina alterniflora*, a common species in the salt marshes of Connecticut, is not abundant in the study area and was not inventoried.

13.2.2 Spectral Data Collection and Processing

Reflectance spectra were obtained using an ASD Fieldspec FR® spectroradiometer (Analytical Spectral Devices, Boulder, CO) with a wavelength range of 0.35–2.50 μm. Individual spectral measurements were an average of 5–10 scans and each canopy was generally sampled 10 or more times. These samples were then averaged to provide a single reflectance profile. Reflectance spectra were normalized to a white Spectralon® (Labsphere, Inc., North Sutton, NH) panel.

Reflectance spectra were collected for the three dominant vegetation communities at two sites in Ragged Rock Creek marsh (Figure 13.3): *Spartina patens* meadows, *Typha* spp., and *Phragmites australis*. To document between- and within-plant phenology, field spectra were collected from specific stands of each of the three plant communities monthly in the summer of 2004 at Site 1 (Figure 13.3c), approximately biweekly at Site 2 (Figure 13.3b) in the summer of 2005, and on a single date in the fall of 2006 (Table 13.1). The selected stands were dense monocultures so as to approximate best species' endmember characteristics.

To correlate better the field spectra to QuickBird satellite data, individual field reflectance spectra were averaged equally over the four QuickBird band intervals [Band 1 (blue): 0.45–0.52 μm; Band 2 (green): 0.52–0.60 μm; Band 3 (red): 0.63–0.69 μm; and Band 4 (NIR): 0.76–0.90 μm] to produce a simulated QuickBird band value. From these data, four spectral indices were found to be most useful for discrimination of species: (1) green/blue, (2) NIR/red, (3) red/green, and (4) the Normalized Difference Vegetation Index [NDVI = (NIR − red)/(NIR + red)]. The

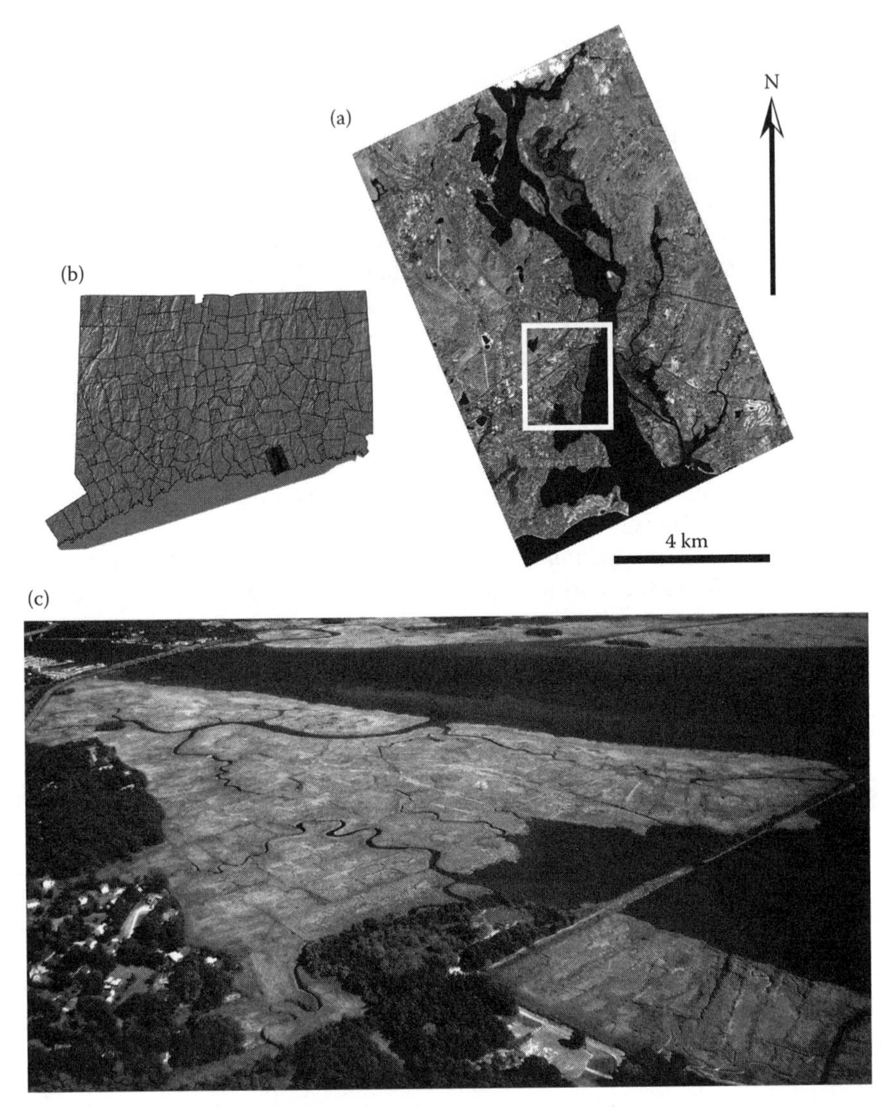

FIGURE 13.1 Location of Ragged Rock Creek marsh, Old Saybrook, CT: (a) July 20, 2004 QuickBird image (panchromatic band) of the mouth of the Connecticut River (the box indicates location of marsh); (b) location of (a). (c) Oblique aerial photograph of marsh taken in Fall 2003 looking to the northeast. (Photo credit: Connecticut Department of Environmental Protection.)

values of these indices over the growing season comprise a set of "radiometry rules" that were used to guide image segmentation and classification. Vegetation indices were used for classification instead of raw bands because they reduce data volume, provide information not available in a single band [8], and normalize differences in reflectance when using multiple images [9] and when comparing field with satellite data.

FIGURE 13.2 Three dominant salt marsh vegetation types at Ragged Rock Creek marsh: *Spartina patens* (foreground); *Typha* species (middle ground); and *Phragmites australis* (background). This chapter's fifth author (pictured) is approximately 6 ft tall.

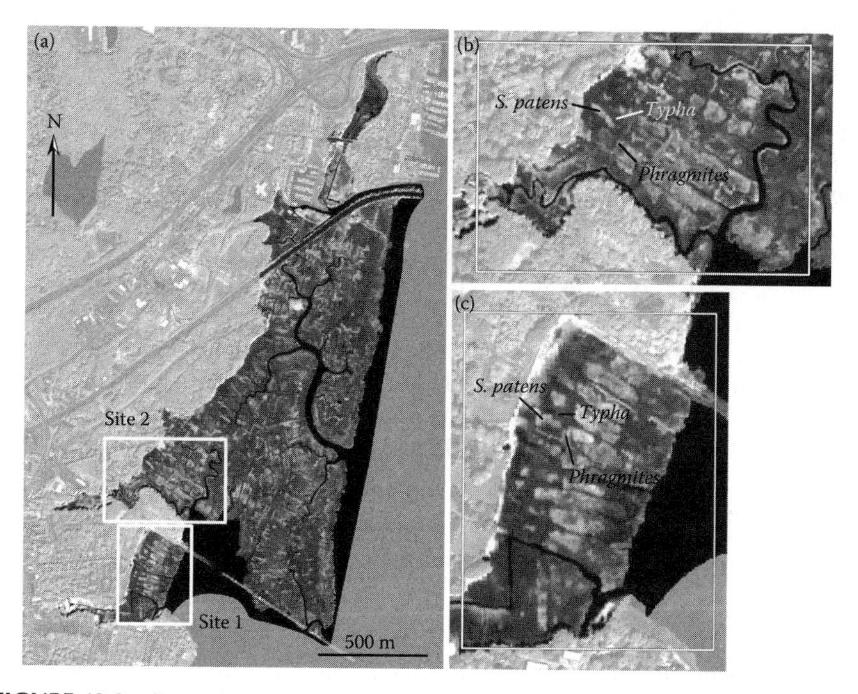

FIGURE 13.3 September 12, 2004 QuickBird image of Ragged Rock Creek marsh where Bands 4, 2, and 1 are converted to grayscale: (a) overview of marsh; (b) Site 2; and (c) Site 1. Arrows indicate approximate areas where reflectance spectra were measured in the field throughout the growing season. Spectra were measured at Site 1 in 2004 and Site 2 in 2005. (Adapted from Gilmore, M.S. et al., *Remote Sensing of Environment*, 112, 4050, 2008. With permission from Elsevier.)

TABLE 13.1
Dates of Data Acquisition at Ragged Rock Creek Marsh

Year	May	June	July	August	September	October
2004	**14**	**15**	**14**	**19**	**21**	
	27		2 20		12	*8*
2005	**27**	**9 27**	**13**	**1 12 26**	**9**	**1**
		2 17	23			
2006			31	13	**8**	
	←		Vegetation Survey			→

Source: Reprinted from Gilmore, M.S. et al., *Remote Sensing of Environment*, 112, 4051, 2008. With permission.

Notes: Regular font indicates QuickBird image collection, bold font indicates field spectra collection, and italics indicate airborne LiDAR data collection. Underlined dates indicate the QuickBird images used in classification.

13.2.3 Validation Data

A floristic inventory of the marsh was conducted throughout the summer of 2006, in large part to establish sets of training and validation data for image classification work. A set of 1000 randomly distributed point locations within the marsh was generated. At each location, 4 m^2 quadrats were placed and plant community composition and species abundance were recorded. Field GPS coordinates for each sampling site were recorded using Trimble GeoExplorer3 GPS units and were postprocessed to improve accuracy using Trimble Global Positioning System (GPS) Pathfinder Office, or coordinates were recorded using Garmin Map 76CS× GPS coupled with a CSI-Wireless MBX-3S Differential Beacon Receiver for enhanced accuracy using real-time differential correction. Data collected at each sampling site included species present, estimated percent cover for each species, height of dominant species, and sociability rank for each species based on its distribution (e.g., tightly clustered in one area versus distributed throughout) across the plot. Digital photographs were also taken at each site to document field conditions at the time the inventory was conducted. In total, 923 vegetation plots were recorded at Ragged Rock Creek; 877 were random plot locations and 46 were subjective plot locations opportunistically selected by the field teams as being unusual, rare, or monotypic plant communities. Some plots were revisited during the growing season; only one data point was maintained for each plot location resulting in a grand total of 917 field points. For the purposes of image classification accuracy assessment, all points were assigned to one of five dominant classes: *Phragmites*, *Typha* spp., *Spartina patens*, Water, and Other/Mixed to correspond with the image classification.

13.2.4 Image and LiDAR Processing

QuickBird satellite image data (2.44 m at nadir) were acquired for a 100-km^2 area at the mouth of the Connecticut River (Figure 13.1b) on nine dates from July 2003 to

November 2006 (Table 13.1). Each of the images was coregistered to the July 20, 2004 QuickBird image (UTM coordinate system, WGS84 datum, Zone 18 N) to assure consistent alignment. Of the nine scenes, five were selected for the final classification based on image quality (lack of clouds and haze), acquisition month and day, and importance as determined by field spectra data analysis (Figure 13.4). Although the images spanned several growing seasons, the month and day of acquisition were considered more important than the year of acquisition, because the monthly spectral variability was observed to be much greater than interannual variability in the dataset.

Multiple return LiDAR data were collected on October 8, 2004, at an altitude of 3000 ft using a Leica ALS50 airborne laser scanner. The LiDAR data points had a nominal ground sampling distance of 0.9 m and a reported horizontal accuracy of 0.5 m. Based on 22 points, the average vertical error between the bare earth LiDAR coverage and the control was 0.002 m with an root-mean-square error (RMSE) of 0.057 m. LiDAR nonground data points for the project area were reprojected to the UTM coordinate system (WGS84, Zone 18 N) and were converted to a 2.4-m resolution elevation grid using the Natural Neighbors interpolator in the 3D Analyst Extension to ArcGIS. The elevation grid was converted to an 8-bit unsigned integer format for use in the eCognition™ (Definiens AG, München, Germany) software. Because some of the LiDAR pulse penetrates the vegetation canopy, the elevation grid is an approximation of the top of the plant canopy within the study area (Figure 13.5).

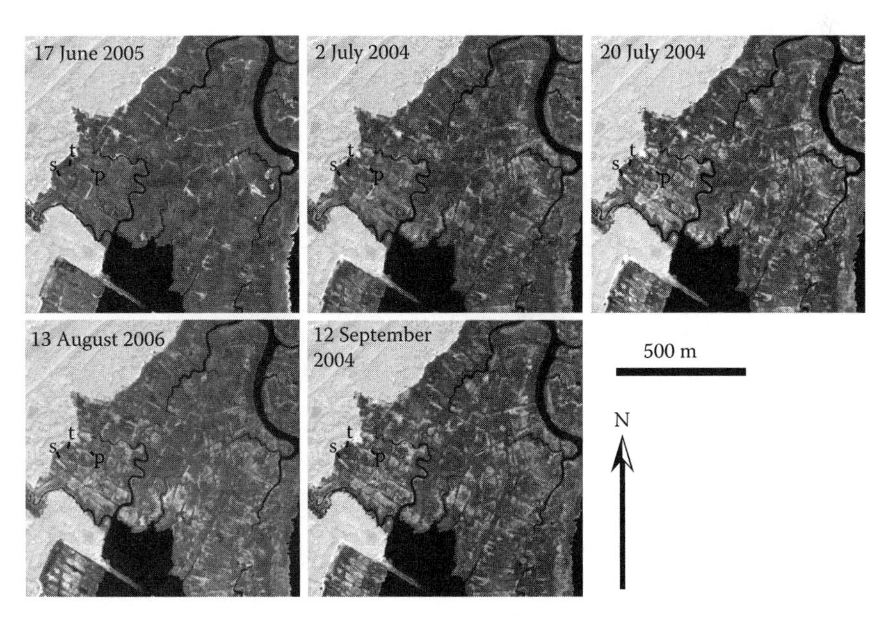

FIGURE 13.4 QuickBird multispectral imagery used in classification for a portion of Ragged Rock Creek marsh. Each image is a grayscale rendition of Bands 4, 2, and 1. Letters indicate position of example stands of the three primary vegetative species: p = *Phragmites australis*, t = *Typha* spp., and s = *Spartina patens* meadows.

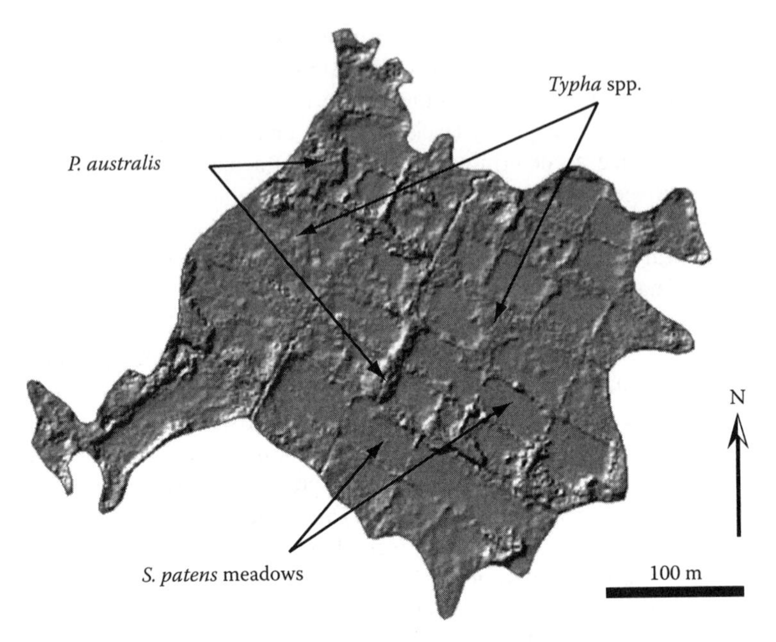

FIGURE 13.5 Hillshade representation of LiDAR top-of-canopy height data for a portion of the Ragged Rock Creek marsh (Site 2, Figure 13.3c). The elevation data range from ~0.4 to 5 m (Figure 13.8).

13.2.5 Classification

Classification was conducted using the object-oriented image analysis software eCognition™. Input data consisted of images of individual QuickBird bands from a single date, band ratio images from multiple dates, and LiDAR top-of-canopy information (Table 13.2). A polygon of the Ragged Rock Creek marsh was created and used to eliminate nonmarsh features such as houses, trees, and lawns from the input data. The input data were segmented into image objects, which are contiguous pixels that are grouped together into homogeneous polygon features. Spectral and spatial parameters (smoothness and compactness) were empirically adjusted to contribute to the segment (i.e., object) definition. The relative size of each image object is determined by a scale parameter, which sets the maximum allowed heterogeneity among image objects. A scale parameter of 20, and spectral and spatial contributions of 70% and 30%, respectively, was found to be the optimum values to best identify the plant communities within the Ragged Rock Creek marsh using QuickBird data. Each image object was treated as an individual entity in the classification.

13.2.6 Accuracy Assessment

Two-thirds ($n = 613$) of the field points from the floristic inventory were used for accuracy assessment, but only those that were at a distance greater than 2 m ($n = 379$) from each class boundary were used to ensure that the points were located within the class being evaluated and to minimize GPS inaccuracies and image registration

TABLE 13.2
Input Data [QuickBird (QB) and LiDAR] Used for Image Segmentation and Classification

	Image Type				
Image Date	**QB Band 2/ Band 1**	**QB Band 3/ Band 2**	**QB Band 4/ Band 3**	**QB Raw Bands 1, 2, 3 and 4**	**LiDAR**
June 17, 2005	—	—	X	—	—
July 2, 2004	X	X	—	—	—
July 20, 2004	—	—	—	X	—
August 13, 2006	X	X	X	—	—
September 12, 2004	X	X	X	—	—
October 8, 2004	—	—	—	—	X

errors. The accuracy assessment used a fuzzy set approach [10] producing measures of the frequency, magnitude, source, and nature of mapping error. Fuzzy sets acknowledge the inherent variation in vegetation communities by allowing for sites to exhibit some grade of membership among map classes. For example, sites may reasonably be members of multiple classes, or alternately, some sites may only show a poor resemblance to any of the map classes.

At each site, the Euclidean distances in ordination space to all vegetation class centroids [11,12] were inversely translated into ordinal ranks from 1 to 5, corresponding to the five linguistic levels of Gopal and Woodcock's [10] rating system of increasing correctness, labeled: (1) Absolutely Wrong, (2) Understandable but Wrong, (3) Reasonable or Acceptable Answer, (4) Good Answer, and (5) Absolutely Correct. Additionally, "expert" judgment was made, without the analyst's knowledge of the map labels, to make adjustments in linguistic values for the conspicuous dominance or absence of major species. The result was a matrix where each field point had a value between 1 and 5 assigned to each of the four potential vegetation classes. The accuracy assessment was completed by comparing the vegetation class rating of each sample site to the QuickBird classification result.

13.3 RESULTS

13.3.1 RADIOMETRY

The canopy reflectance spectra of each plant community were broadly similar, including absorptions typical of healthy photosynthesizing vascular plants (Figure 13.6). At the beginning of the growing season, each spectrum showed expected increases in the strength of the absorptions at approximately 0.45 and 0.68 μm and an increase in NIR reflectance. This trend continued in each species until the onset of senescence (Figure 13.6a). Comparison of the spectra of individual species showed that the magnitude and shape of the spectra can differ qualitatively on individual dates (Figure 13.6d).

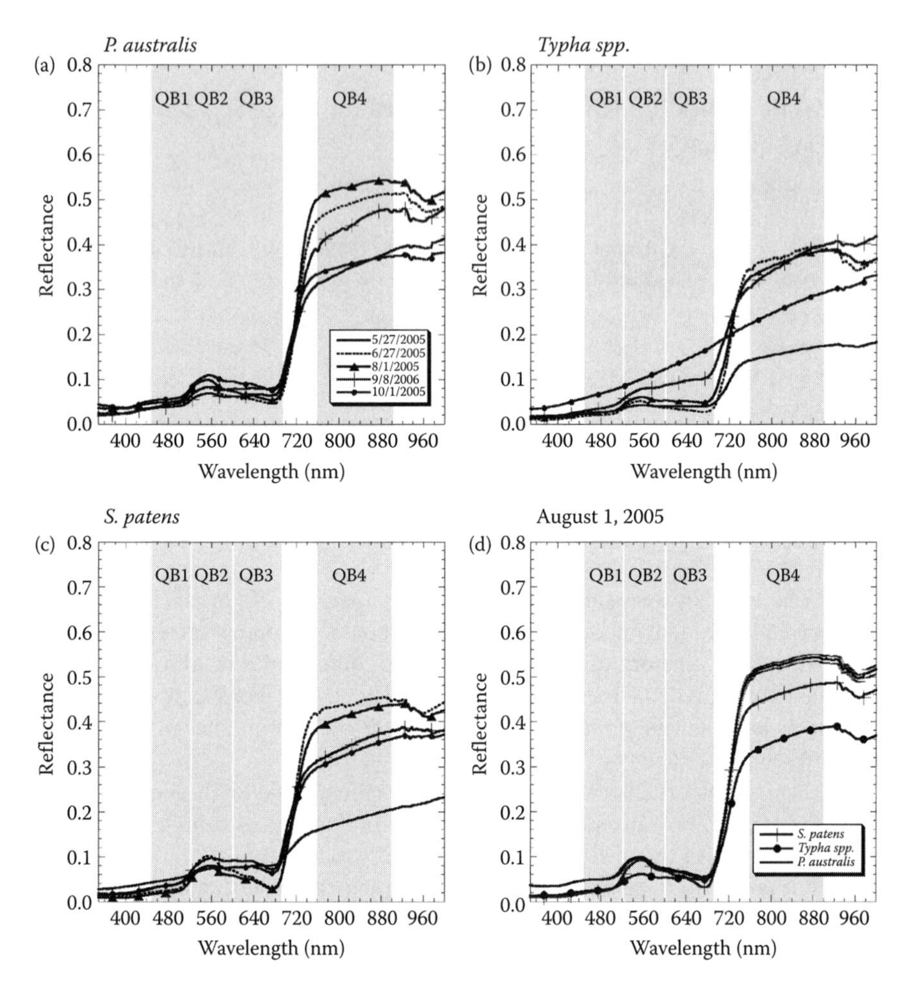

FIGURE 13.6 Example reflectance spectra of major plant species in Ragged Rock Creek marsh. Each spectrum is an average of at least 10 spectra. QuickBird band positions are indicated. Collection sites are shown in Figure 13.3b and c. Spectra of (a) *Phragmites*, (b) *Typha* spp., (c) *Spartina patens* throughout the growing season, and (d) Spectra of major marsh plant communities on August 1, 2005. One standard deviation of the averaged spectra is plotted for Phragmites in (c) and is typical of all spectra. (Adapted from Gilmore, M.S. et al., *Remote Sensing of Environment*, 112, 4052, 2008. With permission.)

To relate the spectral variability of the field data to QuickBird images, the reflectance spectra of each species were reduced to QuickBird band ratios and plotted as a function of average accumulated growing degree days [GDD_{50} = (average daily temperature − 50°F)] using temperature data records for the Groton, CT airport acquired by the National Climatic Data Center (Figure 13.7). The spectral behavior of the three marsh plant communities at the two field sites was broadly consistent over the growing seasons. Trends of NDVI and the NIR/red ratios were similar to each other,

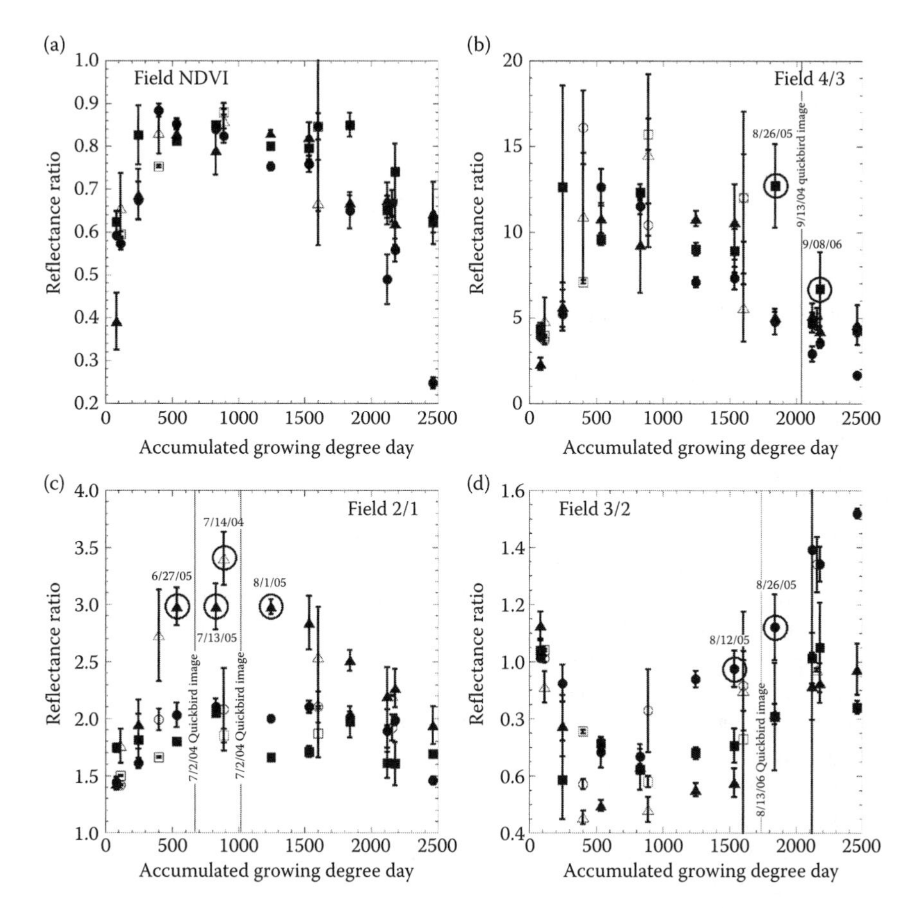

FIGURE 13.7 Field VNIR reflectance of *Phragmites* (squares), *Typha* spp. (circles), and *Spartina patens* (triangles) over the 2004–2006 growing seasons. Average accumulated growing degree days = (average daily temperature – 50°F). Field reflectance data recalculated as QuickBird (QB) bands: (a) NDVI, (b) Bands 4/3, (c) Bands 2/1, and (d) Bands 3/2. Each point is an average of at least 10 field measurements; error bars are one standard deviation. Dates of spectra acquisition are indicated in Table 13.1 where open symbols correspond to 2004 and closed symbols to 2005/2006. Circled data points refer to dates when spectral differences between the species were utilized to create classification rules for each species in available QuickBird images (indicated). (Adapted from Gilmore, M.S. et al., *Remote Sensing of Environment*, 112, 4054–4055, 2008. With permission.)

where the NIR/red ratio provides greater separability among the individual points (Figure 13.7a and b). All species demonstrated a rise in these indices at approximately GDD_{50} 400 (early June) corresponding with the green-up phase of plant growth and a subsequent decline in the indices corresponding to senescence. The behavior of the seasonal variation in each index varied with plant species: *S. patens* and *Phragmites* reached peak values at ~GDD_{50} 900 and *Typha* spp. at ~GDD_{50} 500. NDVI and NIR/red values for *Typha* spp. were generally higher than the other

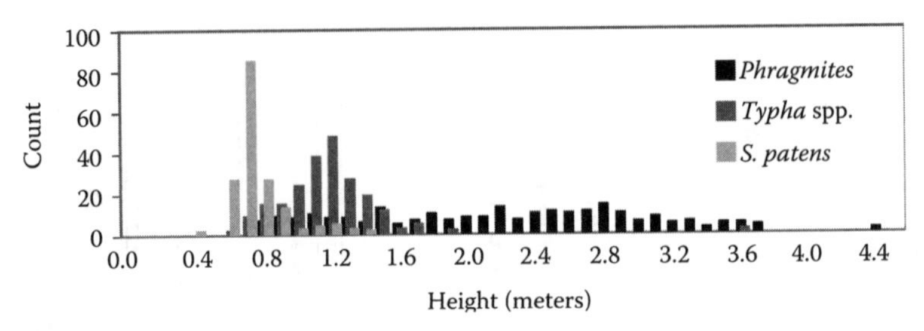

FIGURE 13.8 Histograms showing LiDAR height data for GPS field inventory plots where each plant class is dominant. (Reprinted from Gilmore, M.S. et al., *Remote Sensing of Environment*, 112, 4055, 2008. With permission.)

species near the time of their peak (mid to late June), while *Phragmites* values exceeded the other species in mid-August to early September.

The seasonal pattern of the QuickBird green/blue ratio of *S. patens* was distinct from the other two species (Figure 13.7c). Values of this ratio were similar for each species at the beginning of the season, but *Spartina patens* rose to a peak value near GDD_{50} 900 (mid-July). The absolute value for *Spartina patens* was twice that of *Typha* spp. and *Phragmites* from early June to early August.

The general seasonal pattern of the QuickBird red/green ratio for all species showed an initial decline from approximately GDD_{50} 200 to GDD_{50} 800 (late May to late June) and then an increase (Figure 13.7d). *Spartina patens* values in this index were lower than the other two species over GDD_{50} 400 to GDD_{50} 1500 (mid-June to early August). *Typha* spp. values exceeded those of the other species from mid-July onward.

Three simple band ratios were determined to be most useful in identifying at least one major plant community: the NIR/red ratio in late summer for *Phragmites*, the green/blue ratio in midsummer for *S. patens*, and the red/green ratio in late summer for *Typha* spp. (circled points, Figure 13.7). These periods showed both the greatest spectral separability among species and the best correspondence with the dates (month, day) of the QuickBird images available for classification (Table 13.1 and Figure 13.4).

On the Ragged Rock Creek marsh, the average height of *Phragmites* determined from the field survey points was 2.1 m ($n = 213$), that of *Typha* spp. was 1.1 m ($n = 206$) and that of *Spartina patens* was 0.7 m ($n = 160$; Figure 13.8).

13.3.2 QuickBird Classification

Figure 13.9 graphically illustrates the rules that were identified from the field spectra and implemented in the classification of the QuickBird images (Figure 13.10). Qualitatively, the classification identified contiguous areas of *Phragmites*, *Typha* spp., and *S. patens*. The distribution of these classes was broadly consistent with field observations, for example, the correspondence of *Phragmites* and negative correspondence of *Spartina patens* with creeks and ditches. The classification confirmed that the portion of Ragged Rock Creek marsh surveyed was dominated by the three species under study, where areas classified as *Typha* spp. and *Phragmites*

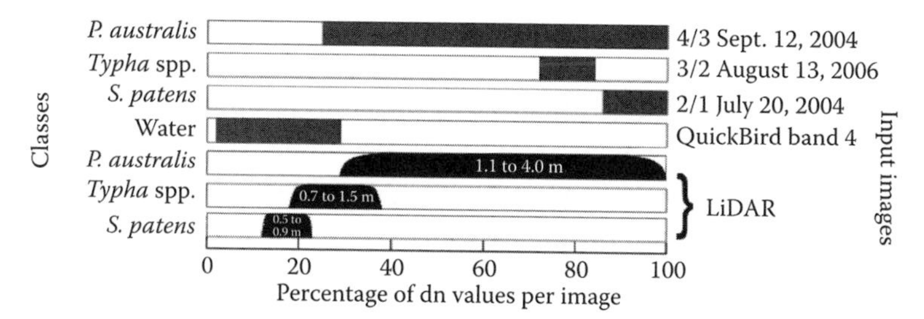

FIGURE 13.9 Thresholds used for knowledge-based rules implemented in eCognition for the classification of image objects. Each bar corresponds to one rule that was applied to an image used to define each class. (Reprinted from Gilmore, M.S. et al., *Remote Sensing of Environment*, 112, 4056, 2008. With permission.)

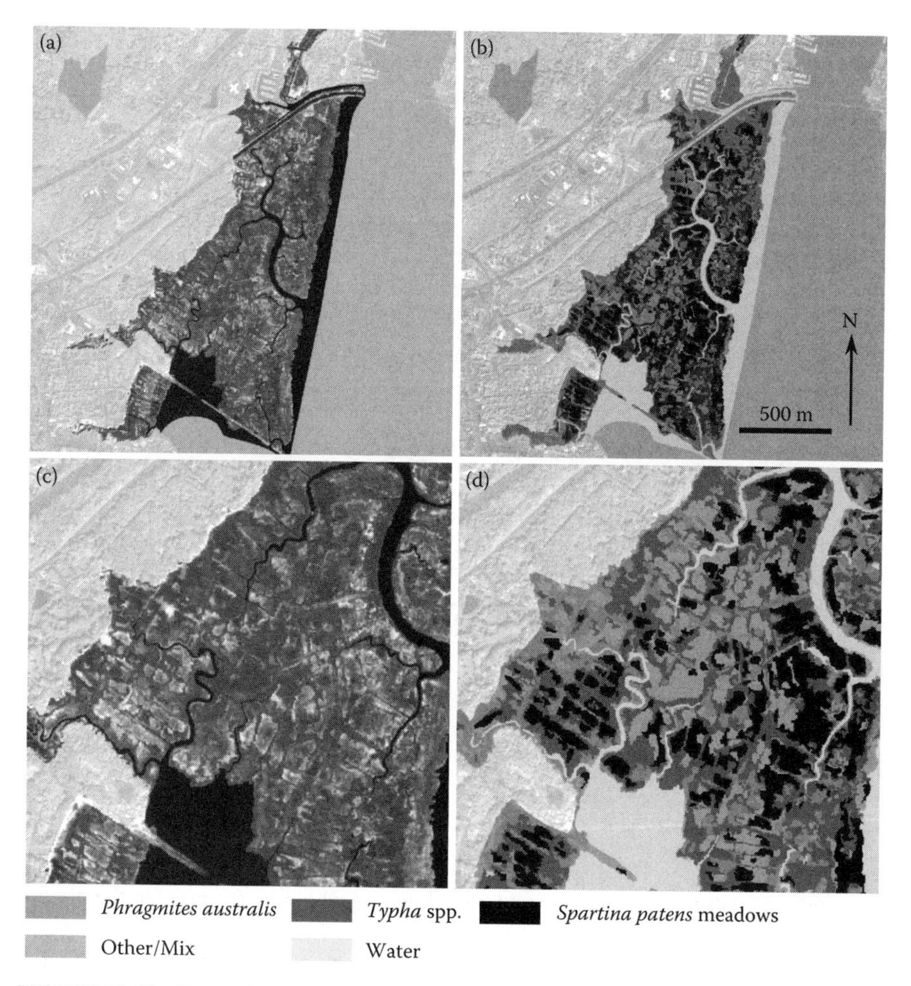

FIGURE 13.10 Ragged Rock Creek marsh classification results. (a) July 20, 2004 QuickBird panchromatic image; (b) Classification result; enlargement of a portion of the (c) QuickBird image; and (d) the classification.

TABLE 13.3
Fuzzy Accuracy Assessment Summary of Image Classification Using Two [10] Linguistics (Fuzzy Operators): (1) Absolutely Wrong; (2) Understandable but Wrong; (3) Reasonable or Acceptable; (4) Good Answer; and (5) Absolutely Correct

		Max (Linguistic = 5)		Right (Linguistic ≥ 3)	
Class	**Sites**	*N*	%	*N*	%
Phragmites australis	68	66	97.1	66	97.1
Typha spp.	157	99	63.1	119	75.8
Spartina patens	90	72	80.0	83	92.2
Other/mix	64	46	71.9	64	100
Total	379	283	74.7	332	87.6

comprised 36% (45.2 ha) and 24% (30.5 ha), respectively, *Spartina patens* covered 22% (27.8 ha), and other species covered 18% (22.9 ha) of the marsh.

A fuzzy accuracy measure for each map class and the overall (total) map were determined by the frequency of correct matches (and mismatches) using two fuzzy accuracy measures, MAX and RIGHT (Table 13.3). The MAX column indicates the strictest measure of accuracy (i.e., absolutely correct) where only the highest membership level was acceptable and is the closest measure to traditional accuracy assessment [13]. The RIGHT column indicates an acceptable map label choice (a rating level of ≥3). *Phragmites* had the highest accuracy with both MAX and RIGHT at 97%. *Typha* spp. was the least accurate with 63% MAX and 76% RIGHT. The *S. patens* meadows type was labeled correctly in most cases with a MAX of 80% and a RIGHT of 92%. The category Other/Mix had a MAX value of 72% but a RIGHT value of 100%, showing the greatest improvement of 28% of the RIGHT metric over the MAX. In general, the total weighted accuracy of all map labels adjusted for a real proportion of each map class had a MAX rating of nearly 77% and an acceptable RIGHT rating of nearly 89% [14].

13.4 DISCUSSION

13.4.1 Spectral Phenology

The spectral characteristics of vegetation are due to leaf pigments, plant structure (biomass and canopy architecture and cover), and plant health throughout the phenological cycle. Much of the spectral variability in the field data can be attributed to expected increases in plant pigments and biomass during the green-up phase of plant growth, the decline of these variables, and the increased contribution of background materials during senescence. *Phragmites* and *Typha* spp., which often occur as monocultures ≥2 m high, had higher NDVI and NIR/red values throughout the growing season than the low-growing *S. patens*. NIR index values peaked for *Typha* spp. in mid- to late June, corresponding to field observations of peak plant heights of ≥2 m, full development of flowers and wholly green leaves. *Phragmites* displayed peak NIR index values in mid-August to early September corresponding to peak plant heights of ≥3 m and the development of flowers. Moreover, by late August, *Typha* spp. had senesced resulting in a reduction in green biomass and likely an increase in the contribution of background materials (wrack, soil) to its spectrum.

The green/blue ratio of *S. patens* was dramatically higher than that of *Phragmites* or *Typha* spp. from mid-June to late August, due perhaps to inherent differences in the amount of chlorophyll *b* and carotenoids in these species, both of which absorb in the blue portion of the spectrum. The peak in the green/blue index for *S. patens* in mid-July corresponds to maximum pigment concentration at this time of year.

The field data generated the following set of spectral rules that may be applicable to the discrimination of *Phragmites*, *Typha* spp., and *Spartina patens* communities: (1) *Phragmites* is best distinguished by its high NIR response late in the growing season likely due to its high biomass, especially with respect to the other species, (2) *Typha* spp. is best distinguished by high red/green response in August due to senescence, and (3) *S. patens* is best distinguished by a unique green/blue ratio throughout the growing season attributed to pigments. The field data predict that seasonal spectral behavior should be evident within actual QuickBird data. To test this, we identified the image segments in eight QuickBird images containing the sites where the field point spectra were measured in 2005 and 2006. Average band values for all pixels within each segment were ratioed and plotted (Figure 13.11). In the QuickBird data, *Phragmites* uniquely displayed high NDVI and NIR/red values late in the growing season (Figure 13.11a and b), and *S. patens* had a relatively high green/blue values throughout the growing season (Figure 13.11c). The red/green values for *Typha* (Figure 13.11d) are greater than and separable from *Phragmites* throughout the growing season, but overlapped with *S. patens*. Thus, simple indices in four-band data such as QuickBird were adequate to measure and distinguish remotely the spectral characteristics of these species over the growing season.

The values of the QuickBird data spectral ratios (Figure 13.11) for the field targets over the growing season follow the patterns established in the field data (Figure 13.7). As the QuickBird data were collected at various times of day and tidal stage, this suggests that the canopy reflectance at the satellite level was dominated by vegetation, and the contribution of variable background elements like water and shade was minimal. The field targets were selected because they are monocultures with high areal density, so we expected little to no contribution of understory species in these data. Understory species may become an important part of the canopy spectrum in less dense stands, which may reduce classification accuracy. The agreement between field and satellite data supports the use of field spectra for the training of image classifiers.

The phenological trends seen here can vary as a function of plant vigor, which may depend on changes in salinity, climate, predation, or disturbance. Corrections for interannual and regional changes in these factors must therefore be performed when attempting to interpret classification results or select dates to maximize interspecies spectral discrimination. However, the spectral trends noted here may

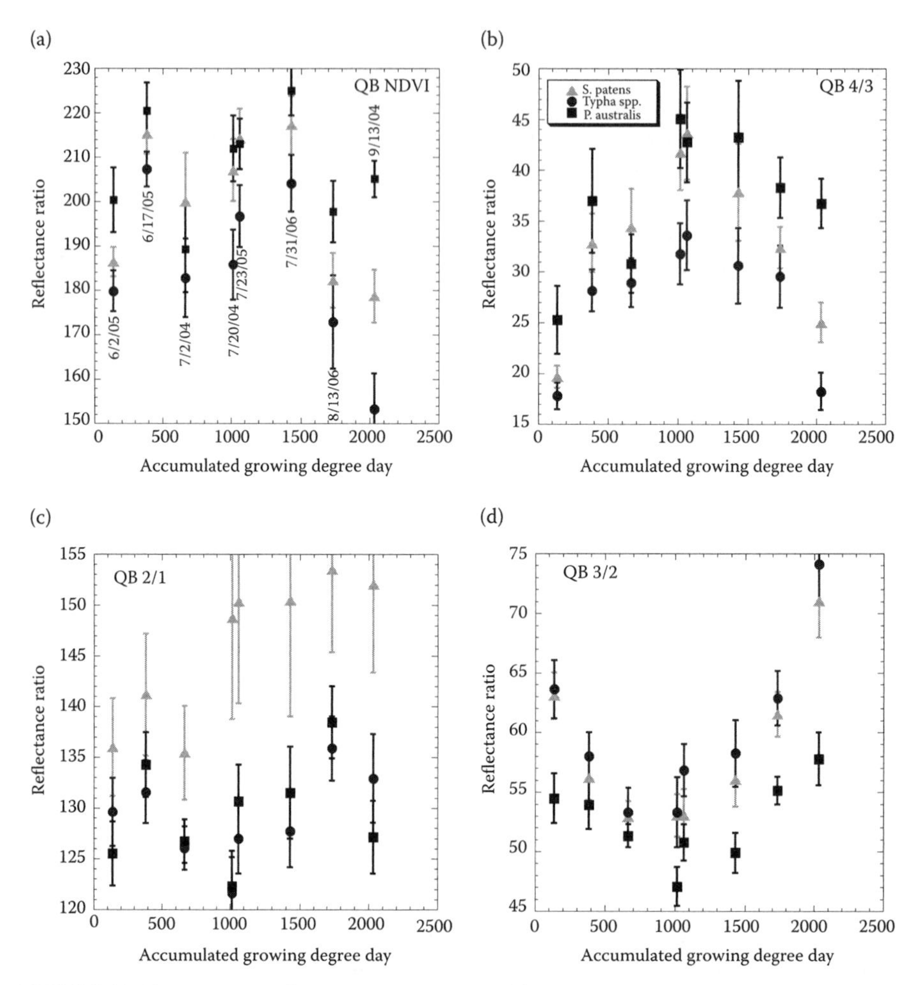

FIGURE 13.11 Average reflectance values ($\pm 1\sigma$) of image segments in QuickBird data containing the field targets (see Figure 13.3b and c): (a) NDVI; (b) Bands 4/3; (c) Bands 2/1; and (d) Bands 3/2. Average accumulated growing degree days = (average daily temperature −50°F). Image acquisition dates are indicated in panel (a). (Adapted from Gilmore, M.S. et al., *Remote Sensing of Environment*, 112, 4054–4055, 2008. With permission.)

have broader spatial and temporal application. We emphasize that in this study, the phenological behavior of the spectral indices is consistent over 3 years at two separate areas in both the field and satellite data of Ragged Rock Creek marsh. Furthermore, other studies show similar spectral behavior of wetland plant species. The late season peak in NIR reflectance we observed in our data for *Phragmites* is also noted for this species in VNIR field spectra measured in the New Jersey Meadowlands [15,16] and in the Yangtze Estuary [17]. *Phragmites* is also found to be most spectrally separable from *Typha* spp. in field VNIR spectra measured in the Hudson River Estuary in late August [18]. These studies suggest that the spectral variability utilized here may be applied to other areas, particularly if guided by any ecological assessment of local growth habits of marsh species.

13.4.2 Classification

The object-oriented classification rules yielded the highest MAX accuracies for points consisting of a single species. This approach was the most useful for discriminating *Phragmites*, which occurred most often as monotypic stands. Yet, the marsh mosaic also included another and perhaps more typical scenario, where several species are admixed in varying amounts. Treating the validation points in this study with linguistic values, fuzzy accuracy assessment, and the RIGHT accuracy measure enabled an arguably more realistic interpretation of vegetation stands with codominant or transitional species typical of many marshes. Similar improvements in accuracy between MAX and RIGHT measures have been noted for the mapping of marsh vegetation using QuickBird imagery with eCognition™ [19].

Using the RIGHT accuracy measure improved results for *S. patens*, which we suggest resulted from its distribution as understory in many of the validation quadrats. In these situations, the spectral signature of *S. patens* contributes to the signal even if taller species dominate the LiDAR first return (canopy) data. Also, *S. patens* was most often confused for bare ground/flotsam/wrack, which are all low-lying and spectrally bright and for other low growing species such as blackgrass (*Juncus gerardii*), bentgrass (*Agrostis stolonifera*), and switchgrass (*Panicum virgatum*).

Typha spp., the most overmapped class, was confused with *Phragmites* and a number of other species. Examination of the validation data shows that the classification misidentified many points where *Typha* spp. was recorded in low abundance. These points contain *Typha* spp., which occasionally has a diffuse clonal growth habit, within a mix of species of various heights, spectral signatures, and densities. Many of these species, like *Phragmites*, are observed to occur at similar heights (~1–2 m) to *Typha* spp. during the middle part of the growing season, including sedges (*Schenoplectus* species) and bulrushes (*Bolboschoenus* species). The wide range of species that are confused with *Typha* spp. is likely exacerbated by its lack of an extreme, defining spectral or height rule. In contrast, both *Phragmites* and *S. patens* have extremes in the LiDAR data (tallest and shortest, respectively) and have one spectral band ratio that is quite distinct (late season NIR/red and mid-season green/blue, respectively). The fact that *Typha* spp. covers the most area of the marsh may also increase the potential for commission error.

Elevation information, such as that derived from LiDAR, is an important feature of natural vegetation classification. *Phragmites* was observed to be approximately 1 m taller than the next tallest species in late summer. This characteristic distinguishes *Phragmites* from *Typha* spp. and *Spartina patens* in the LiDAR data, likely contributing to the success of its classification. While it is possible that the use of LiDAR as a sole discriminant could identify a large proportion of *Phragmites* occurrences, the spectral characteristics are likely necessary to distinguish between *Phragmites* and *Typha* spp. because there is considerable overlap in the height data of these species (Figures 13.7 and 13.8).

13.4.3 Implications

The canopy spectral and structural characteristics of *Phragmites*, *Typha* spp., and *S. patens* vary as a function of plant phenology in both field-collected spectra and QuickBird images. The plant communities were found to be separable spectrally in four-band data and thus we predict that other satellite remote sensing systems, such as GeoEye-1 (1.65 m, 4-band multispectral) or WorldView-2 (1.84 m, 4-band multispectral), could be used for marsh plant classification, obviating the need for more expensive hyperspectral airborne data. The spatial resolution of QuickBird images approximates the patch size of the mapped plant species in Ragged Rock Creek marsh, which likely contributes to the success of the classification. We would expect classification accuracy to degrade with lower spatial resolution data (e.g., Landsat or ASTER) as has been noted in other attempts to classify salt marsh vegetation [5].

The seasonal variations in spectral behavior make clear the necessity of multitemporal imagery for mapping multiple species on a complex tidal marsh. Our study demonstrates that field observations at a limited number of reference sites were sufficient for classification. The spectral and structural data predicted times during the growing season when each species was best discriminated. With this knowledge, a single date of imagery could be used to map adequately a single species. For example, in Ragged Rock Creek marsh, the high NIR response and height of *Phragmites* in the autumn distinguishes it uniquely among a mosaic of plant species typical of a brackish marsh. Natural resource land managers seeking to monitor *Phragmites* in this region could utilize single date color IR images, four-band satellite data, and/or LiDAR during late August to early September for classification. Classification could be accomplished at other marshes with field observations at a limited number of reference sites.

13.5 CONCLUSION

This study presents a new technique for the classification of major marsh plant species within a complex, heterogeneous tidal marsh using multitemporal QuickBird images, field reflectance spectra, and LiDAR height information. Analyses of the phenological variation of spectral and structural characteristics of marsh species measured in the field throughout the growing season were necessary to select the best dates to discriminate *Phragmites*, *Typha* spp., and *S. patens*. These three species are spectrally distinguishable at particular times of the year likely due to differences in biomass and pigments and the rate at which these change throughout the growing season. These differences are recognizable in high-resolution field spectra, in the spectra when degraded to the spectral resolution typical of four-band multispectral sensors, and in QuickBird images of the field sites.

Rules based on the spectral variability of species throughout the growing season were used to direct the classification of multitemporal QuickBird images. Classification accuracies for *Phragmites* were high due to the uniquely high NIR reflectance and height of this dense monoculture in the early fall. Mapping accuracies were best represented for all species by using a fuzzy accuracy assessment, which accommodates the complex mosaic of dominant and mixed plant communities of Ragged Rock Creek marsh. For classification, the spectral and LiDAR data are complimentary.

The results of this study demonstrate the utility of remote sensing to map certain types of marsh vegetation. Although multitemporal image data were utilized for the classification in this study, the phenological variations recognized here can be utilized for judicious selection and analysis of single date, four-band satellite or aerial image data having the same wavelength bands of coastal marshes along Long Island Sound. Such data are likely to be useful to coastal managers and may greatly facilitate the identification and inventory of marsh species.

ACKNOWLEDGMENTS

This study was funded by the U.S. EPA Long Island Sound Research Program grant no. LI97100901 ("Application of Remote Sensing Technologies for the Delineation and Assessment of Coastal Marshes and their Constituent Species") and, in part, by NASA grant no. NNL05AA14G ("Incorporating NASA's Applied Sciences Data and Technologies into Local Government Decision Support in the National Application Areas of Coastal Management, Water Management, Ecologic Forecasting and Invasive Species"). A portion of the data was provided by the NOAA Coastal Services Center Coastal Remote Sensing Project and the Institute for the Application of Geospatial Technology. The 2006 field work was funded by the Connecticut DEP Long Island Sound License Plate Program. We very much appreciate the contributions of Bill Moorhead and Joel Labella in the field.

REFERENCES

1. Warren, R.S., Fell, P.E., Grimsby, J.L., Buck, E.L., Rilling, G.C., and Fertik, R.A., Rates, patterns, and impacts of *Phragmites australis* expansion and effects of experimental *Phragmites* control on vegetation, macroinvertebrates, and fish within tidelands of the lower Connecticut River, *Estuaries*, 24, 90–107, 2001.
2. Reed, B.C., Brown, J.F., VanderZee, D., Loveland, T.R., Merchant, J.W., and Ohlen, D.O., Measuring phenological variability from satellite imagery, *Journal of Vegetation Science*, 5, 703–714, 1994.
3. Key, T., Warner, T.A., McGraw, J.B., and Fajvan, M.A., A comparison of multispectral and multitemporal information in high spatial resolution imagery for classification of individual tree species in a temperate hardwood forest, *Remote Sensing of Environment*, 75, 100–112, 2001.
4. Dennison, P.E. and Roberts, D.A., The effects of vegetation phenology on endmember selection and species mapping in southern California chaparral, *Remote Sensing of Environment*, 87, 295–309, 2003.
5. Belluco, E., Camuffo, M., Ferrari, S., Modenese, L., Silvestri, S., Marani, A., and Marani, M., Mapping salt-marsh vegetation by multispectral and hyperspectral remote sensing, *Remote Sensing of Environment*, 105, 54–67, 2006.
6. Judd, C., Stenberg, S., Shaughnessy, F., and Crawford, G., Mapping salt marsh vegetation using aerial hyperspectral imagery and linear unmixing in Humboldt Bay, California, *Wetlands*, 27, 1144–1152, 2007.
7. Rosso, P.H., Ustin, S.L., and Hastings, A., Use of LiDAR to study changes associated with *Spartina* invasion in San Francisco Bay marshes, *Remote Sensing of Environment*, 100, 295–306, 2006.

8. Coppin, P.R. and Bauer, M.E., Digital change detection in forest ecosystems with remote sensing imagery, *Remote Sensing Reviews*, 13, 207–234, 1996.
9. Singh, A., Digital change detection techniques using remotely-sensed data, *International Journal of Remote Sensing*, 10, 989–1003, 1989.
10. Gopal, S. and Woodcock, C., Theory and methods for accuracy assessment of thematic maps using fuzzy sets, *Photogrammetric Engineering & Remote Sensing*, 60, 181–188, 1994.
11. Podani, J., *Introduction to the exploration of Multivariate Biological Data*, p. 407, Backhuys Publishers, Leiden, The Netherlands, 2000.
12. Ter Braak, C.J.F. and Šmilauer, P. *CANOCO Reference Manual and User's Guide to Canoco for Windows: Software for Canonical Community Ordination*, p. 35, Centre for Biometry, Wageningen, CPRO-DLO, Wageningen, The Netherlands, 1998.
13. Congalton, R.G. and Green, K., *Assessing the Accuracy of Remotely Sensed Data: Principles and Practices*, p. 137, Lewis Publishers, Boca Raton, FL, 1999.
14. Gilmore, M.S. et al., Integrating multi-temporal spectral and structural information to map wetland vegetation in a lower Connecticut River tidal marsh, *Remote Sensing of Environment*, 112, 4050–4056, 2008.
15. Artigas, F.J. and Yang, J.S. Hyperspectral remote sensing of marsh species and plant vigour gradient in the New Jersey Meadowlands, *International Journal of Remote Sensing*, 26, 5209–5220, 2005.
16. Artigas, F.J. and Yang, J.S., Spectral discrimination of marsh vegetation types in the New Jersey Meadowlands, USA, *Wetlands*, 26, 271–277, 2006.
17. Gao, Z.G. and Zhang, L.Q., Multi-seasonal spectral characteristics analysis of coastal salt marsh vegetation in Shanghai, China, *Estuarine, Coastal and Shelf Science*, 69, 217–224, 2006.
18. Laba, M., Tsai, F., Ogurcak, D., Smith, S., and Richmond, M.E., Field determination of optimal dates for the discrimination of invasive wetland plant species using derivative spectral analysis, *Photogrammetric Engineering and Remote Sensing*, 71, 603–611, 2005.
19. Yu, Q., Gong, P., Clinton, N., Biging, G., Kelly, M., and Schirokauer, D., Object-based detailed vegetation classification with airborne high spatial resolution remote sensing imagery, *Photogrammetric Engineering and Remote Sensing*, 72, 799–811, 2006.

14 Quantifying Biophysical Conditions of Herbaceous Wetland Vegetation in Poyang Lake of Coastal China via Multitemporal SAR Imagery and *In Situ* Measurements

Limin Yang, Huiyong Sang, Hui Lin, and Jinsong Chen

CONTENTS

14.1 INTRODUCTION

Wetland ecosystems, known as the "kidneys of the earth," are an important habitat for aquatic flora and fauna and provide valuable services and goods for the human beings. The wetlands in Poyang Lake of the Southeast China coastal region are one of the first national natural reserves listed in the Ramsar convention in 1992. Poyang Lake is the largest freshwater lake in China and its natural wetland area covers over 4000 km^2 with diverse species of plants and vertebrate. Every year several million migratory birds live through the winter in this region (Liu and Ye, 2000). The Lake also plays an important role in flood control along the Yangtze River watershed. In the last several decades, however, overexploitation of the wetlands in Poyang Lake has altered seriously the ecosystem and reduced biodiversity. The area of wetlands in the Poyang Lake region has decreased by over 1000 km^2 and total water storage decreased by 6000 million m^3 due to reclamation (Wang et al., 2004).

Recognizing the urgency of this problem, a series of programs have been initiated by the local and provincial agencies since the late 1980s aiming at reversing the trends of wetland loss. One of the most important programs has been the implementation of a strategy for monitoring the status and trends of the wetlands by using remote sensing technology complemented with ground surveying. A challenging task for implementing such a capability for wetland restoration in the Poyang Lake region is how to cost-effectively monitor the wetland biophysical conditions so that an objective assessment on wetland status and trends and the effectiveness of all restoration efforts can be carried out.

14.2 RESEARCH OBJECTIVES

The purpose of this research was to evaluate the feasibility of quantifying wetland biophysical conditions and seasonal dynamics of wetland vegetation in the Poyang Lake region by use of multitemporal, multipolarization synthetic aperture radar (SAR) imagery and *in situ* measurements. The specific objectives of the research were (1) to understand seasonal changes of biophysical properties of wetlands by conducting field work over a growing season; (2) to understand radar backscattering mechanisms and quantify temporal variability of radar backscattering associated with different vegetation types within the region; and (3) to quantify the relationships between radar backscatters in HH and HV polarizations and the wetland vegetation biophysical parameters (e.g., plant height, above-ground/water fresh, and dry biomass) over a growing season.

14.3 PREVIOUS STUDY OF WETLANDS VIA REMOTE SENSING

Remote sensing techniques have proved useful in wetland studies at regional and global scales (e.g., a review by Ozesmi and Bauer, 2002). Much of the pioneering work on remote sensing of wetlands was accomplished by using optical remote sensing images. For instance, with a spatial resolution of about 80 m, multispectral scanner (MSS) data on the Earth Resources Technology Satellite (ERST-1) were used to map the extent of flooded areas in the United States (Rango and Salomonson, 1973, 1974; Deutsh and Ruggles, 1975). In all the studies, MSS band 7 (0.8–1.1 μm) was the most useful band for discriminating water from dry soil or vegetated surfaces, owing to a strong absorption of water in the near-infrared spectral range. Other space-borne optical sensors launched in later years with higher spatial resolution or wider spectral range (e.g., Landsat TM/ETM+) have been used to study wetlands. The most useful band of Landsat images for wetland identification is band 5 (1.55–1.75 μm), because of its ability to discriminate vegetation and soil moisture levels and improve separability between wetland types (Jensen et al., 1993).

Compared to the sensor onboard Landsat that acquires data with relatively high spatial resolution, other sensors such as the NOAA AVHRR and EOS MODIS have the ability to acquire data with high temporal resolution for monitoring the terrestrial ecosystems. The time series of satellite data have been widely used to map vegetation and its seasonal dynamics taking place in both natural and artificial wetlands (e.g., Hope et al., 2003; Pelkey et al., 2003; Boken et al., 2004; Ferreira and Huete, 2004).

In addition to optical sensors, systems operating in the microwave spectrum onboard satellites, space shuttles, and aircraft have also been used to study natural wetlands and rice paddies. Among many space-borne SAR systems, instruments operating at C-band (e.g., ERS-1, ERS-2, RADARSAT-1, and ENVISAT/ASAR) can collect data in either single or alternate polarizations and at different incidence angles. Such data have been used to map and monitor flood extent (Grings et al., 2006; Henry et al., 2006), wetland habitats (Bartsch et al., 2006, 2007a,b), paddy rice conditions (Le Toan, 1997; Shao et al., 2001; Chen et al., 2006), soil moisture and biophysical parameters (Townsend, 2001, 2002; Moreau and Le Toan, 2003; Grings et al., 2005), and seasonal hydrological patterns (Kasischke et al., 2003; Bourgeau-Chavez et al., 2005).

The relationship between radar backscattering in C-band and wetland vegetation condition is complex and varies both spatially and temporally. The radar signals are affected not only by characteristics of the wetlands, such as vegetation type and timing of flood events, but also by the parameters of the SAR instruments, such as wavelength, incidence angle, and polarization mode. For instance, it was found that the backscattering in C-band was sensitive to plant biomass and plant height in herbaceous wetland and paddy fields before a saturation point was reached (Le Toan, 1997; Moreau and Le Toan, 2003). However, studies also found that change in C-band backscattering was not sensitive to change of biomass in a wetland with short and sparse vegetation (Kasischke et al., 2003).

14.4 STUDY AREA

Poyang Lake is located in Jiangxi Province of China and is to the south of the Yangtze River (115°40′–117°E, 28°17′–29°50′N) (Figure 14.1a). Five tributary rivers (Gangjiang, Xiushui, Raohe, Xinjiang, and Fuhe) flow into the Poyang Lake. As Poyang Lake is connected to the Yangtze River, its flood condition is influenced not only by local precipitation and hydrological condition from the nearby rivers, but also by the backflows from the Yangtze River. Thus, the flooding pattern in the lake area varies both spatially and temporally.

A distinct alternation from a wet to a dry season driven by the subtropical monsoon atmospheric circulation characterizes the climate of the Poyang Lake region. The rainy season begins in early April, when the southeast monsoon starts to influence this region. On average, the annual precipitation is about 1482 m. The maximum monthly precipitation occurs in June, accounting for over 17% of the annual precipitation, whereas the minimum precipitation occurs in December.

Major land-cover types in the study area are herbaceous wetland vegetation around open water bodies, trees, and shrubs in uplands, bog, sandy areas, some residential areas, and crop lands. Natural wetlands are colonized by different types of vegetation. From the lake center toward the lake peripheral, different types of wetland vegetation are distributed that have adapted to different microenvironmental conditions. The type of vegetation includes floating vegetation, submerged vegetation

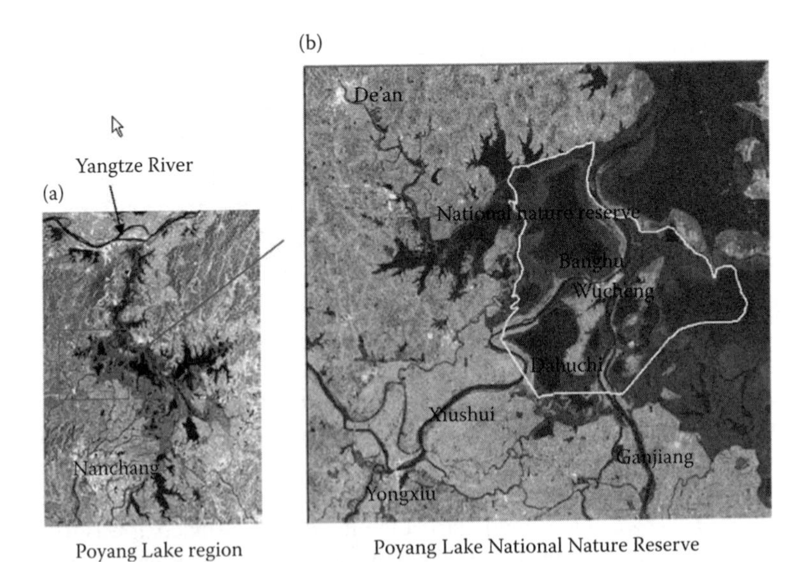

FIGURE 14.1 Study area. (a) The Poyang Lake region in Jiangxi Province of China to the south of the Yangtze River (115°40′–117°E, 28°17′–29°50′N). (b) The Poyang Lake Nature Reserve where field work was conducted in 2004–2005.

(e.g., *Potamogeton malainus* and *Vallisneria spiralis*), emergent aquatic vegetation (*Carex* spp.), and semiaquatic emergent tall vegetation, such as *Phragmites communis* Trin., *Miscanthus sacchariflorus*, and *Zizania caduciflora*.

Field study of this research was carried out within the Poyang Lake Nature Reserve situated in the northwest portion of the Poyang Lake (29°05′–29°17′N and 115°54′–116°08′E) (Figure 14.1b). This area covers about 80 km^2, and includes nine lakes: Bang Lake, Dahuchi Lake, Sha Lake, and six other smaller lakes. These lakes are often flooded and become a part of the Poyang Lake during the high water season in summer. The amount of water in the two rivers Gangjiang and Xiushui joining at Wucheng village influences variation of water level in the area.

14.5 SATELLITE DATA COLLECTION AND PREPROCESSING

The ENVISAT/ASAR, operating at C-band with a frequency of 5.33 GHz, is an advanced version of the SAR system and a successor of the ERS-1 and ERS-2 sensors. There are seven swaths (IS1–IS7), which can be selected at different incidence angles ranging from 15° to 45° for fine resolution imagery (12.5 m pixel size). For this study, 11 images collected between December 2004 and November 2005 were obtained. Most images are in alternating polarization precision (APP) mode (HH/HV or HH/VV) from swath IS4 to IS6 (Table 14.1). These time series ASAR images provide data over a growing season and cover a hydrological cycle of the Poyang Lake.

Orthorectification, speckle filtering, and radiometric calibration of all ASAR images are preprocessed before radar backscattering coefficients can be derived.

TABLE 14.1
Field Measurements, ENVISAT/ASAR Data, and Weather Conditions When Acquiring Satellite Images

Field Data		ASAR Data		Weather Conditions		
Dates	Samples	Dates	Swath and Polarization	Temperature (°C)	Precipitation (mm)	Relative Humidity (%)
N/A	N/A	December 12	IS4 HH/HV	10.6	0	69.8
March 27	15	March 27	IS4 HH/HV	11.8	21	92.3
April 14/15	25	April 14/15	IS5 HH/HV	16.4	0	52.7
April 20	22	April 18	IS7 HH/HV	18.2	6.7	76.5
May 24/25	24	May 25	IS6 HH/VV	21.0	0	81.1
N/A	N/A	June 24	IS5 HH/VV	28.3	0	76.8
N/A	N/A	August 14	IS4	28.1	6.4	80.5
September 19	11	September 18	IS4 HH/HV	28.9	0	73.8
October 18	21	N/A	N/A	17.4	0	71.9
October 26/27	26	October 26	IS6 HH/HV	17.4	0	92.1
November 18	9	November 19	IS4 HH/HV	9.8	3	73.7
November 28	25	November 27	IS4 HH/HV	13.6	0	92.3
December 29	24	N/A	N/A	N/A	N/A	N/A

The orthorectification for the ASAR images was conducted using the Ortho-Engine module of the PCI Geomatics. The Doppler Orbitography and Radiopositioning Integrated by Satellite (DORIS) microwave tracking system carried on ENVISAT provides accurate orbital information about the satellite. The PCI Ortho-Engine module utilizes this information together with a DEM dataset to conduct an orthorectification for all images, and an RMS position error within one pixel can be achieved using this method. For this study, a DEM dataset was created by digitizing from a 1:50,000 topographic map of the Poyang Lake region and was resampled to 12.5 m resolution (same as the pixel size of ASAR images). In addition, a Gamma map filtering was applied to all images to reduce speckles using a 5 × 5 pixel window.

14.6 DERIVATION OF BACKSCATTERING COEFFICIENTS FROM THE ASAR IMAGES

To analyze multitemporal SAR image, raw data from the ASAR level 1b precision image products were calibrated to backscattering coefficients using the equation

$$\sigma^0(\theta_i) = \frac{\mathrm{DN}_i^2}{K} \times \sin(\theta_i), \tag{14.1}$$

where σ^0 is the radar backscattering coefficient in a linear unit retrieved from the corresponding digital number (DN), θ_i is the local incidence angle considering the ellipsoid of the Earth and terrain influence at the ith pixel, and K is the absolute calibration constant that is dependent on the processor and ASAR product type.

Incidence angle has proved to have a significant effect on radar backscattering from different ground surfaces (Ulaby et al., 1986). In order to fully explore the relationships between vegetation parameters and the multitemporal ENVISAT/ASAR data, the influence of incidence angle on ASAR backscattering needs to be normalized because the images were acquired at different times and with different incidence angles (Shi et al., 1994; Baghdadi et al., 2001). Toward this end, the derived backscattering coefficient from Equation 14.1 can be normalized by

$$\sigma^0(\theta_i) = \sigma_{\mathrm{n}}^0 f(\theta_i), \tag{14.2}$$

where σ^0 is the backscattering coefficient at local incidence θ_i, σ_{n}^0 is a normalized backscattering coefficient, and $f(\theta_i)$ is a mean angular dependence function for a given scene of the ASAR image. The angular influence can be reduced or excluded after the backscattering coefficient is normalized by function $f(\theta_i)$. In a linear unit, $f(\theta_i)$ takes the form of cos $\alpha(\theta_i)$, where α is dependent on target texture and sensor condition (Shi et al., 1994; Baghdadi et al., 2001). For this study, the mean and standard deviation calculated over a 9 × 9 pixel moving window were used to quantify surface texture and to derive a reference value of α. The normalized ASAR backscattering coefficient was derived from Equation 14.2 after the α value was obtained.

For a distributed target (for instance, vegetation), the backscattering coefficients from an area of interest (AOI) should be further averaged to reduce the radiometric resolution errors due to speckles. This can be made by using an equation

$$\sigma^0 = \frac{1}{N}\left(\sum \sigma_i^0\right), \tag{14.3}$$

where σ^0 is a measurement of the backscattering coefficient corresponding to all N pixels within the AOI, and σ_i^0 is the backscattering coefficient at pixel i. Error related to radiometric resolution at a given level of confidence is a function of number N and the Equivalent Number of Look (ENL). For this study, it was estimated that a minimum number of 100 pixels needs to be averaged to ensure that the error of estimate is within ±1 dB at the 95% confidence interval (Laur et al., 2004).

In order to compute the ASAR backscattering for pixels corresponding to all sampling field plots, the geographic coordinates of each individual plot acquired by a GPS system were converted to a UTM projection, and then colocated onto the normalized ASAR backscatter images. The ASAR backscattering coefficient of each plot was then calculated using Equation 14.3, which represents different wetland vegetation types.

14.7 METHOD FOR FIELD DATA COLLECTION

Field survey was conducted from March to December 2005, corresponding as close as possible to the ENVISAT image acquisition dates (Table 14.1). A survey of plant species focused on the emergent aquatic and semiaquatic vegetation types, including *Carex* spp., *M. sacchariflorus*, and *Phragmites communis* at Bang Lake and Dahuchi Lake, all within the Poyang Lake Nature Reserve. Twelve field plots of *Phragmites communis* Trin. and more than 50 of *Carex* spp. were visited. Each field plot size was 0.5 × 0.5 m if the plant cover was homogeneous, otherwise it was 1 × 1 m. The fresh weight of soil samples, vegetation species, vegetation fresh weight, and vegetation height and coverage were measured and recorded at each plot. In addition, site photographs and geographic coordinates were recorded.

A linear sampling method was adopted for the field survey considering that different plants were concentrically distributed around the center of the lakes. Field sample plots covered all wetland vegetation zones and to capture different species. Multiple samples were collected for each plant species to account for its spatial variability. The sample plots were separated from each other by at least 100 m in distance.

For estimation of biomass, plant samples were first weighed for fresh weight and then dried in an oven to a constant weight for dry weight measurement. Soil samples were also dried in an oven to constant weight to get an estimate of dry weight. Fresh/dry biomass (g/m^2) is defined by the fresh/dry weight (g) divided by the plot size (m^2). Vegetation gravimetric moisture content (M_g) was used to estimate vegetation water content and was calculated by

$$M_g = \frac{M_w - M_d}{M_w}, \tag{14.4}$$

where M_w is the wet (fresh) weight of the vegetation sample and M_d is vegetation dry weight. Soil volumetric moisture content M_v (cm^3/cm^3) was evaluated by

$$M_v = \frac{M_w - M_d}{\rho_w / V_s}, \tag{14.5}$$

where M_w and M_d are wet weight and dry weight of the sampled soil, respectively, ρ_w is the density of water, and V_s is the volume of the soil sample.

14.8 RESULTS

14.8.1 Relationships between Radar Backscattering and Vegetation Biophysical Parameters of *Phragmites communis* Trin.

Based on a qualitative assessment of the field and radar data, the dependency of ASAR backscattering on vegetation parameters of *Phragmites communis* Trin., was believed to approximate a power relationship expressed as

$$\sigma^0 = a \times x^b, \tag{14.6}$$

where σ^0 is the backscattering coefficient, a and b are constant, and x is a vegetation parameter of interest (e.g., height, fresh/dry biomass, and vegetation water content). The scatterplots of radar backscatters against height, fresh and dry biomass and plant water content, and the fitted curves are presented in Figure 14.2. Table 14.2 summarizes the regression equations of ASAR backscatters in HH and HV polarizations as a function of vegetation height, fresh/dry biomass, and plant water content.

The height of *Phragmites communis* Trin. is one of the most important parameters affecting the incident ASAR signals, and the intensity of the radar backscattering in both HH and HV polarizations increased significantly as the height of the plant increased to above 2 m, with leaves developed in the vertical and horizontal directions. The high correlation between plant height and backscattering in HH and HV polarizations indicates that increase in height and the heterogeneity of the canopy enhanced the canopy volume backscattering. No saturation point of backscattering was evident within the observed range of height (0–2.8 m). Note that the correlation between the height of *Phragmites communis* Trin. and backscattering is positive for both HH and HV, yet the correlation is weaker in HV polarization ($R^2 = 0.59$) than in HH polarization ($R^2 = 0.90$).

Dry biomass of a plant is a function of its height and the number of stem and leaves. The sensitivity of radar backscattering to the amount of dry biomass of *Phragmites communis* Trin. was noted during its early growing stage, especially in HH polarization ($R^2 = 0.76$). However, the backscattering did not vary significantly after the dry biomass reached 1300 g/m^2. Compared to dry biomass, the relationship between radar backscattering and fresh biomass was less pronounced. A saturation level was observed in HH polarization as the fresh biomass reached 2500 g/m^2. In addition, the backscattering in HV was not correlated with fresh biomass ($R^2 = 0.07$).

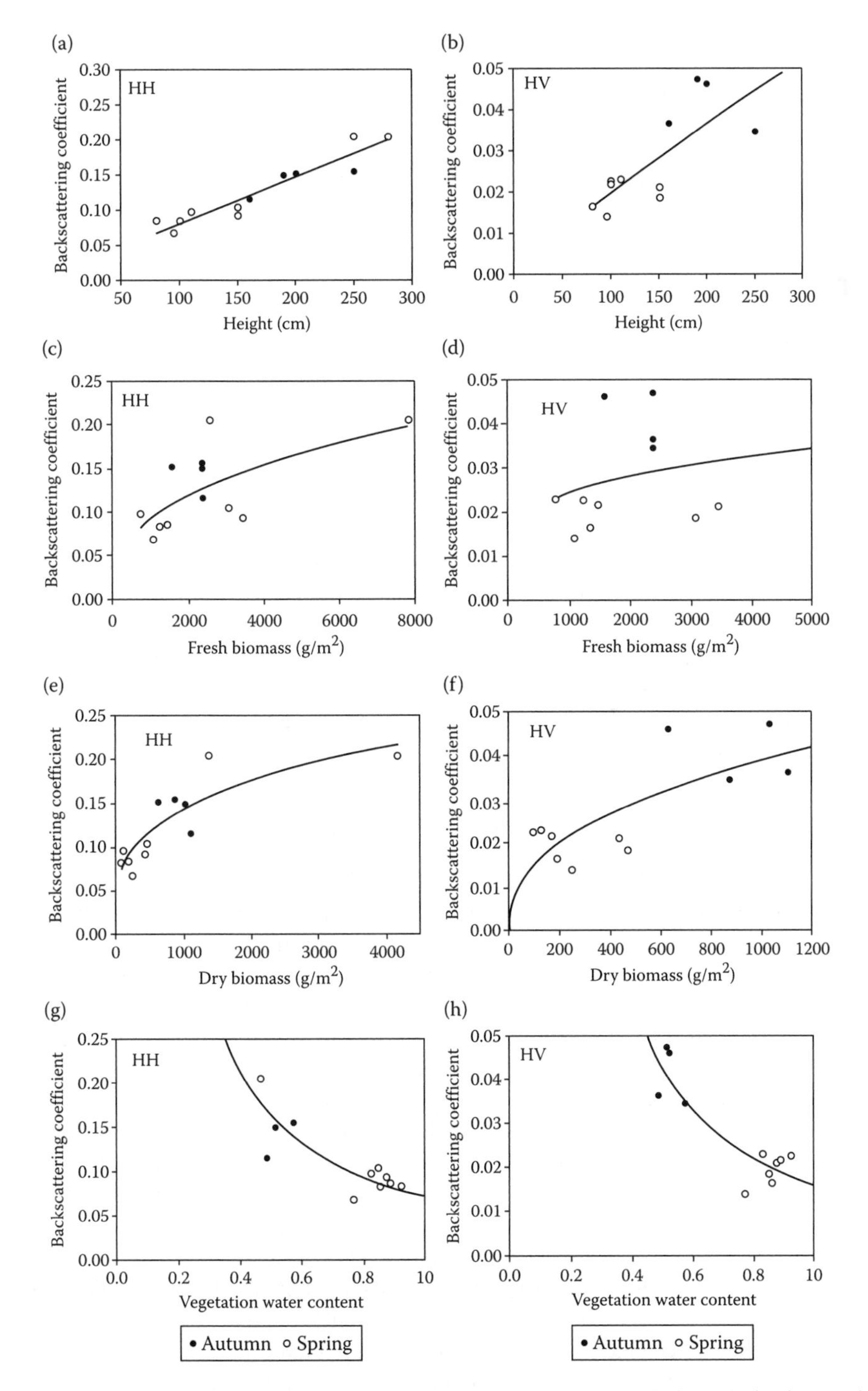

FIGURE 14.2 Relationships between ASAR backscatters in HH and HV polarizations and biophysical parameters of *Phragmites communis* Trin.

TABLE 14.2
α Value for Various Polarizations of ENVISAT/ASAR Images at Different Acquisition Dates

Vegetation Parameters	Polarization	Model Coefficients	R^2
Height	HH	$a = 0.0014$, $b = 0.8790$	0.9
	HV	$a = 0.0004$, $b = 0.8595$	0.59
Fresh biomass	HH	$a = 0.0072$, $b = 0.3695$	0.42
	HV	$a = 0.0054$, $b = 0.2174$	0.07
Dry biomass	HH	$a = 0.0202$, $b = 0.2852$	0.76
	HV	$a = 0.0024$, $b = 0.4042$	0.56
Water content	HH	$a = 0.0727$, $b = -1.1703$	0.75
	HV	$a = 0.0160$, $b = -1.4189$	0.8

In general, radar backscattering is positively correlated with the plant moisture content in both polarizations due to a higher dielectric constant of the wetter plants. Results from this study, however, show that the radar backscattering was strongly but negatively correlated with measured water content of *Phragmites communis* Trin. ($R^2 = 0.75$ for HH polarization, $R^2 = 0.8$ for HV polarization, Figure 14.2g and h), indicating a strong impact on ASAR backscattering caused by factors other than moisture content of vegetation.

14.8.2 Relationships between Radar Backscattering and Vegetation Biophysical Parameters of *Carex* spp.

As one of the most important grazing pastures for local livestock raising, the biomass of *Carex* spp. is a vital indicator for estimating livestock production in the Poyang Lake region. Figure 14.3 shows the measured height, fresh and dry biomass, and plant water content of *Carex* spp. plotted against the ASAR backscattering coefficient in HH and HV polarizations. Unlike the relationship illustrated in Figure 14.2 for *Phragmites communis* Trin., the correlation between the radar backscattering and all biophysical parameters of *Carex* spp. was rather weak.

When the height of *Carex* spp. was below 30 cm, the backscattering in HH and HV polarizations increased as the height of *Carex* spp. increased (Figure 14.3a and b). As the plant continued to grow above and beyond 30 cm; however, the backscattering coefficients in HH and HV decreased, which is likely due to a strengthened canopy attenuation associated with an increased canopy density.

The sensitivity of backscattering to water content of *Carex* spp. was evident only when the water content was very low (<0.6 cm^3/cm^3). Similarly, the sensitivity of backscattering to change of biomass of *Carex* spp. could be observed only at the lower biomass (<1500 g/m^3 for fresh biomass and <250 g/m^3 for dry biomass). This result implies that variables other than biomass of the *Carex* spp. have a major impact on radar backscattering, especially at the high biomass level.

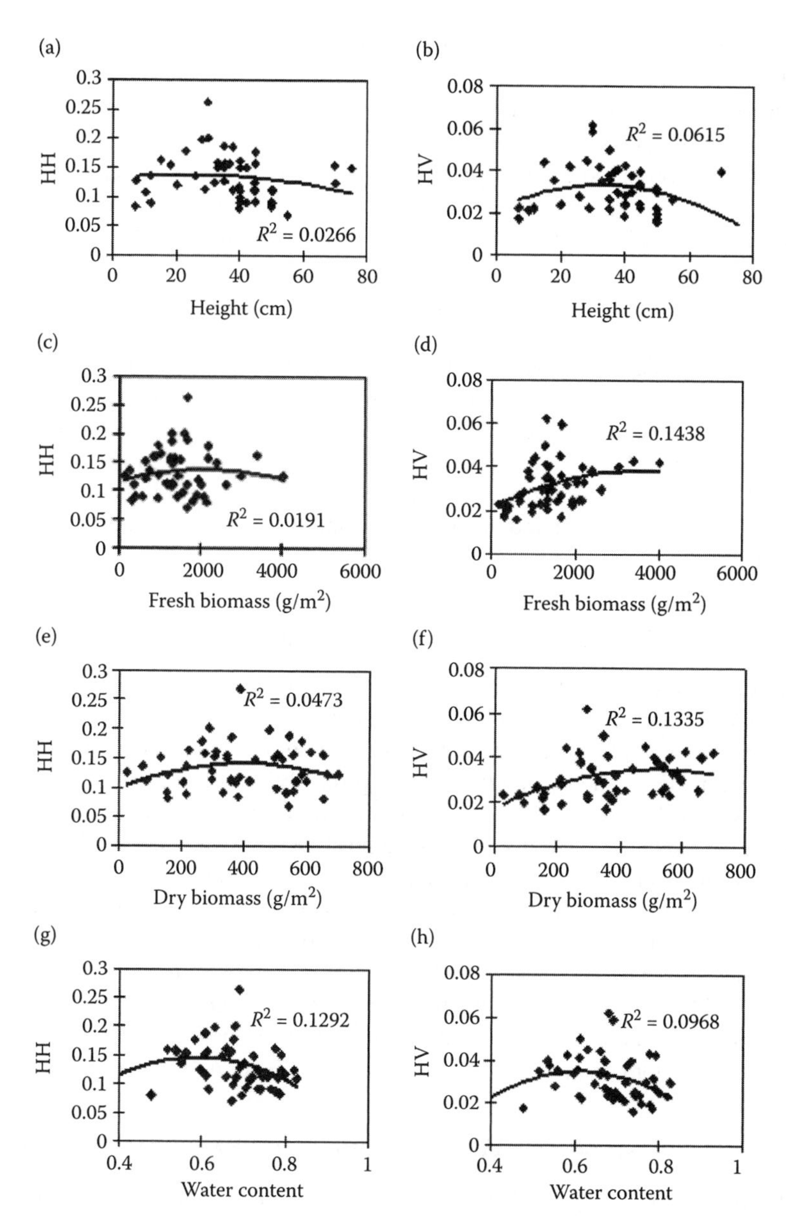

FIGURE 14.3 Scatterplots of ASAR backscatters in HH and HV polarizations against plant height, fresh biomass, dry biomass, and water content of *Carex* spp.

14.9 DISCUSSIONS

14.9.1 Plant Height and Radar Backscattering

At this study site, change in plant height of *Phragmites communis* Trin. affected significantly the magnitude of radar backscattering, especially in HH polarization.

This agrees with others who reported that the backscattering in C-band HH (RADARSAT) and VV (ERS) was positively correlated with plant height in the natural herbaceous wetlands and paddy fields (Le Toan, 1997; Costa et al., 2002; Kasischke et al., 2003). However, in this study, no saturation point in ASAR backscattering was observed for *Phragmites communis* Trin. when its height increased from 0.5 to 2.8 m. This result differs from other studies that reported that C-band backscattering reached saturation quickly as plant height increased (Le Toan, 1997; Costa et al., 2002).

Unlike the aforementioned characteristics of *Phragmites communis* Trin., the plant height of *Carex* spp. in this study site was correlated with radar backscattering only when the height was below 30 cm. The reason is that at the vegetative growth stage, the number and dimension of leaves of *Carex* spp. increased, which enhanced the volume backscattering from the canopy, while suppressing ground surface scattering. However, as plants continued to grow above and beyond 30 cm, HH and HV backscattering coefficients decreased, because of the strengthened canopy attenuation associated with higher canopy density and increased spatial coverage of the canopy.

14.9.2 Plant Biomass and Radar Backscattering

ASAR backscattering in HH polarization increased appreciably when dry biomass of *Phragmites communis* Trin. was below 1300 g/m^2; the backscattering was then saturated after the biomass increased further. The saturation level in HV polarization was lower than in HH at about 630 g/m^2. This is probably due to an increased extinction of radar signals by plant canopy as that continued to develop. Similar to the relation between backscattering and dry biomass, a positive but weaker correlation between backscattering and fresh biomass was also observed. A saturation point of backscattering in response to change of fresh biomass was reached at 2200 g/m^2, which is much higher than those reported in other studies conducted in some paddy fields and herbaceous wetlands (Le Toan, 1997; Costa et al., 2002; Kasischke et al., 2003; Moreau and Le Toan, 2003). Unlike the *Phragmites communis* Trin., the biomass of *Carex* spp. shows little correlation with radar backscattering, indicating that the ability to estimate biomass of *Carex* spp. using the ENVISAT ASAR is rather limited.

14.9.3 Plant and Surface Water Condition and Radar Backscattering

As mentioned previously, the height of *M. sacchariflorus* plays an important role in the radar backscattering mechanism under nonflooded conditions. Once flooded, however, the effects of plant height on radar backscattering in both polarizations were not as pronounced. Under such conditions, the backscattering process was complicated by changes of groundwater depth. Thus, a specular reflection from the water surface could dominate the backscattering process.

Unlike findings reported by others that the increase of plant water content often enhances radar backscattering in HH polarization at C-band for an agricultural field (Ferrazzoli et al., 1997), the moisture content of *M. sacchariflorus* at this study site showed a strong but negative effect on ASAR backscattering in HH and HV polarizations. There is little correlation between its water content and the intensity of radar backscattering observed for *Carex* spp., in either HH or HV polarization.

14.9.4 Vegetation Structure and Radar Backscattering

For *Phragmites communis* Trin., the surface roughness of the air–vegetation boundary is random; thus its surface scattering can be ignored. However, the leafy canopy of *Carex* spp. started to droop down as its growing stage progressed, and was influenced by wind condition and high water content. As the dense leaves stretched in a direction in accordance to local wind, the vegetation surface roughness is not random. Therefore, the air–vegetation surface scattering from the *Carex* spp. can make a contribution to the total radar backscattering.

Generally speaking, change in HV polarization as a "depolarization" phenomenon associated with leaf structure is more sensitive to change in vegetation parameters than it is in HH polarization. However, in this study site, the impact of vegetation parameters on radar signals was very complicated due to influence by other environmental conditions (wind, dews, rain drops, and surface water). As is shown in Figure 14.3, polarized backscattering in HH and HV did not increase notably after the height of *Carex* spp. reached 30 cm and its fresh biomass was above 1300 g/m^2. The saturation level of radar backscattering as is related to plant height and fresh biomass of *Carex* spp. was much lower than that of *Phragmites communis* Trin.

14.10 SUMMARY AND CONCLUSIONS

The study quantifies the relationship between radar backscattering and the biophysical conditions (plant height, fresh and dry biomass, and vegetation water content) and seasonal dynamics in the Poyang Lake region through field survey and by use of multitemporal and multipolarization ENVISAT ASAR imagery. Temporal variability of radar backscattering in HH and HV polarizations in response to biophysical characteristics and seasonal change of two dominant wetland species (*Phragmites communis* Trin. and *Carex* spp.) was analyzed.

Among all vegetation biophysical parameters, the height of *Phragmites communis* Trin. was the most significant parameter affecting variation of radar backscatters, especially in HH polarization under nonflooded condition. Dry biomass is another parameter influencing the radar backscattering. The backscattering intensity is more sensitive to change of dry biomass than it is to fresh biomass. The water content of *Phragmites communis* Trin. was negatively correlated with backscattering due to an impact on radar signals by canopy structures and ground surface water change. Overall, the dependence of backscattering on plant biophysical parameters was more pronounced in HH polarization than it is in HV polarization.

Compared with *Phragmites communis* Trin., the sensitivity of ASAR backscattering to seasonal changes of plant biophysical parameters of *Carex* spp. with a leafy canopy was much less pronounced, due to large variations of its geometric structure caused by local wind condition, and its bimodule growth cycle. The intensity of backscattering was more sensitive to vegetation parameters in HV polarization than it is in HH polarization. The backscattering from *Carex* spp. in both polarizations reached saturation in the early growing season. Overall, the plant canopy structures of *Carex* spp. and its seasonal variation played a more important role in affecting ASAR backscattering than did other plant parameters such as water content.

The results from this study present both an opportunity and challenges in using ENVISAT ASAR data for monitoring wetland biophysical conditions in Poyang Lake region. Because of the complex nature of the natural wetlands and their biophysical conditions, it is necessary to conduct research by using both optical and microwave remote sensing data synergistically. For microwave remote sensing data, use of L-band data should help as it can penetrate plant canopy and provide more information about the ground condition of the wetlands during a growing season. In addition, Fully-Polarimetric SAR and InSAR data should be explored for their potential for quantification of biophysical conditions of the wetlands. A more frequent field survey collecting wetland vegetation and biophysical conditions concurrent with date and time of satellite data acquisition is necessary to better understand the mechanisms of remote sensing signals and the wetland vegetation and biophysical conditions over a season, so that a robust and efficient monitoring system of the Poyang Lake region can be implemented.

ACKNOWLEDGMENTS

We are grateful to Chongjun Tang, Ying Liu, Shuming Bao, and the staff at the Jiangxi Normal University of China for their efforts in field campaigns and geospatial data collection. Thanks are due to Xiangming Xiao of the Institute of Earth, Ocean and Space, the University of New Hampshire, USA, for his insightful comments on the research. This research is partially supported by the Research Grant Council of Hong Kong SAR, P.R. China, and a Research Grant from the Jiangxi Normal University of China. We thank an anonymous referee for detailed and constructive comments.

REFERENCES

Baghdadi, N., Bernier, M., Gauthier, R., and Neeson, I. 2001. Evaluation of C-band SAR data for wetlands mapping. *International Journal of Remote Sensing* 22: 71–88.

Bartsch, A., Wagner, W., Scipal, K., Pathe, C., Sabel, D., and Wolski, P. 2006. ENVISAT ASAR global mode capabilities for global monitoring wetlands. In: *Globewetland Symposium*, Frascati.

Bartsch, A., Kidd, R.A., Pathe, C., Scipal, K., and Wagner, W. 2007a. Satellite radar imagery for monitoring inland wetlands in boreal and sub-arctic environments. *Aquatic Conservation—Marine and Freshwater Ecosystems* 17: 305–317.

Bartsch, A., Pathe, C., and Wagner, W. 2007b. Wetland mapping in the West Siberian Lowlands with ENVISAT ASAR global mode, In: *ENVISAT Symposium 2007*, Montreux, Switzerland.

Boken, V.K., Hoogenboom, G., Kogan, F.N., Hook, J.E., Thomas, D.L., and Harrison, K.A. 2004. Potential of using NOAA-AVHRR data for estimating 150 irrigated area to help solve an inter-state water dispute. *International Journal of Remote Sensing* 25: 2277–2286.

Bourgeau-Chavez, L.L., Smith, K.B., Brunzell, S.M., Kasischke, E.S., Romanowicz, E.A., and Richardson, C.J. 2005. Remote monitoring of regional inundation patterns and hydroperiod in the greater everglades using synthetic aperture radar. *Wetlands* 25: 176–191.

Chen, J., Lin, H., Liu, A., Shao, Y., and Yang, L. 2006. A semi-empirical backscattering model for estimation of leaf area index (LAI) of rice in southern China. *International Journal of Remote Sensing* 27: 5417–5425.

Costa, M.P.F., Niemann, O., Novo, E., and Ahern, F. 2002. Biophysical properties and mapping of aquatic vegetation during the hydrological cycle of the Amazon floodplain using JERS-1 and Radarsat. *International Journal of Remote Sensing* 23: 1401–1426.

Deutsh, M. and Ruggles, F. 1975. Optical data processing and projected applications of ERTS-1 imagery covering the 1973 Mississippi River Valley floods. *Water Resources Bulletin* 10: 1023–1039.

Ferrazzoli, P., Paloscia, S., Pampaloni, P., Schiavon, G., Sigismondi, S., and Solimini. D. 1997. The potential of multifrequency polarimetric SAR in assessing agricultural and arboreous biomass. *IEEE Transactions on Geoscience and Remote Sensing* 35: 5–17.

Ferreira, L.G. and Huete, A.R. 2004. Assessing the seasonal dynamics of the Brazilian Cerrado vegetation through the use of spectral vegetation indices. *International Journal of Remote Sensing* 25: 1837–1860.

Grings, F., Ferrazzoli, P., Karszenbaum, H., Tiffenberg, J., Kandus, P., Guerriero, L., and Jacobo-Berrles, J.C. 2005. Modeling temporal evolution of Junco marshes radar signatures. *IEEE Transactions on Geoscience and Remote Sensing* 43: 2238–2245.

Grings, F.M., Ferrazzoli, P., Jacobo-Beriles, J.C., Karszenbaum, H., Tiffenberg, J., Pratolongo, P., and Kandus, P. 2006. Monitoring flood condition in marshes using EM models and ENVISAT ASAR observations. *IEEE Transactions on Geoscience and Remote Sensing* 44: 936–942.

Henry, J.B., Chastanet, P., Fellah, K., and Desnos, Y.L. 2006. Envisat multi-polarized ASAR data for flood mapping. *International Journal of Remote Sensing* 27: 1921–1929.

Hope, A.S., Boynton, W.L., Stow, D.A., and Douglas, D.C. 2003. Interannual growth dynamics of vegetation in the Kuparuk River watershed, Alaska based on the normalized difference vegetation index. *International Journal of Remote Sensing* 24: 3413–3425.

Jensen, J.R., Cowen, D.J., Althausen, J.D., Narumalani, S., and Weatherbee, O. 1993. An evaluation of the Coastwatch Change Detection Protocol in South Carolina. *Photogrammetric Engineering and Remote Sensing* 59: 1039–1046.

Kasischke, E.S., Smith, K.B., Bourgeau-Chavez, L.L., Romanowicz, E.A., Brunzell, S., and Richardson, C.J. 2003. Effects of seasonal hydrologic patterns in south Florida wetlands on radar backscatter measured from ERS-2 SAR imagery. *Remote Sensing of Environment* 88: 423–441.

Le Toan, T., Ribbes, F., Wang, L.F., Floury, N., Ding, K.H., Kong, J.A., Fujita, M., and Kurosu, T. 1997. Rice crop mapping and monitoring using ERS-1 data based on experiment and modeling results. *IEEE Transactions on Geoscience and Remote Sensing* 35: 41–56.

Liu, X. and Ye, J. 2000. *Jiangxi Wetlands*, 1st edition. Chinese Forestry Publishing Company, Beijing.

Moreau, S. and Le Toan, T. 2003. Biomass quantification of Andean wetland forages using ERS satellite SAR data for optimizing livestock management. *Remote Sensing of Environment* 84: 477–492.

Ozesmi, S.L. and Bauer, M.E. 2002. Satellite remote sensing of wetlands, *Wetlands Ecology and Management* 10 (5): 381–402.

Pelkey, N.W., Stoner, C.J., and Caro, T.M. 2003. Assessing habitat protection regimes in Tanzania using AVHRR NDVI composites: Comparisons at different spatial and temporal scales. *International Journal of Remote Sensing* 24: 2533–2558.

Rango, A. and Anderson, A. 1974. Flood hazard studies in the Mississippi River basin using remote sensing. *Water Resources Bulletin* 10: 1060–1081.

Rango, A. and Salomonson, V.V. 1973. Repetitive ERTS-1 observations of surface water variability along rivers and low-lying areas, *Amer. Wat. Resour. Assoc. Proc. no. 17*, Remote Sensing and Water Resources Management, pp. 191–199.

Shao, Y., Fan, X.T., Liu, H., Xiao, J.H., Ross, S., Brisco, B., Brown, R., and Staples, G. 2001. Rice monitoring and production estimation using multitemporal RADARSAT. *Remote Sensing of Environment* 76: 310–325.

Shi, J.C., Dozier, J., and Rott. H. 1994. Snow mapping in Alpine regions with synthetic-aperture radar. *IEEE Transactions on Geoscience and Remote Sensing* 32: 152–158.

Townsend, P.A. 2001. Relationships between vegetation patterns and hydroperiod on the Roanoke River floodplain, North Carolina. *Plant Ecology* 156: 43–58.

Townsend, P.A. 2002. Estimating forest structure in wetlands using multitemporal SAR. *Remote Sensing of Environment* 79: 288–304.

Ulaby, F.T., Moore, R.K., and Fung, A.K.1986. *Microwave Remote Sensing Active and Passive, Vol. II: Radar Remote Sensing and Surface Scattering and Emission Theory*. Artech House, Dedham, MA.

Wang, X., Fan, Z., Cui, L., Yan, B., and Tan, H. 2004. *Wetland Ecosystem Assessment of Poyang Lake*, 1st edition. Science Publishing Company, Beijing.

15 EO-1 Advanced Land Imager Data in Submerged Aquatic Vegetation Mapping

Eric R. Akins, Yeqiao Wang, and Yuyu Zhou

CONTENTS

15.1 INTRODUCTION

Submerged aquatic vegetation (SAV) habitats such as seagrass communities are among the most productive coastal habitats vital to estuarine ecosystems. Seagrasses serve a vital function in nutrient cycling, sediment stabilization, and maintaining water quality in the coastal zone (Newell and Koch, 2004; Larkum et al., 2006) and provide food and habitat for fish and other fauna (Heck et al., 1989, 1995; Hughes et al., 2002; Lazzari and Stone, 2006). The most prominent seagrass in the Northeast Atlantic region of the United States is eelgrass (*Zostera marina* L.), which grows in quiet embayments along the coast. Eelgrass has been deemed a critical marine resource and is currently protected by both Federal (Clean Water Act; 33 U.S.C. 26 Section 1251 *et seq.*) and state legislations (e.g., Long Island South Shore Estuary Reserve Comprehensive Management Plan). Even with this protection, eelgrass is declining in some areas. Seagrass communities are adversely affected by direct human impacts, such as eutrophication and physical damage from boating and fishing, and indirect human impacts stemming from climate change, such as sea level rise and increased ultraviolet irradiance (Duarte, 2002). Increases in human population and changes in land use in the coastal zone contribute to increased nitrogen loading in estuaries and the effects of nitrogen loading on eelgrass have been extensively quantified (Burkholder et al., 1992; Short and Burdick, 1996; Hauxwell et al., 2003). Additionally, the threat of climate change on seagrasses and coastal ecosystems through rising sea levels, increased temperatures, changes in water chemistry, and increased storm activity is expected to be severe (Short and Neckles, 1999; Orth et al., 2006). Inventory and monitoring are critical components in seagrass ecosystem management, restoration, and protection and provide key insights into the overall health of estuarine ecosystems.

Traditional mapping of seagrass is accomplished by manual delineation of aerial photography combined with *in situ* sampling. Satellite imagery such as Landsat and SPOT data has been used to map seagrass coverage (Ferguson and Korfmacher, 1997; Pasqualini et al., 2005; Schweizer et al., 2005). Aerial hyperspectral scanners have also been used to map seagrasses (Dierssen et al., 2003). Recently, high-resolution satellite imagery such as QuickBird has been used in seagrass mapping (Wolter et al., 2005; Mishra et al., 2006).

In support of the goal of involving high spatial resolution remote sensing in coastal resources inventory and monitoring, the Northeast Coastal and Barrier Network of the National Park Service (NPS)'s Inventory and Monitoring (I&M) Program sponsored a study that employed high spatial resolution QuickBird satellite data for terrestrial vegetation and SAV mapping within the boundary of Fire Island National Seashore (FINS) (Wang et al., 2007). The study also compared and validated agreement between satellite-derived vegetation maps and the delineation result from conventional aerial photograph interpretation so that repeated monitoring of vegetation change can be achieved using satellite remote sensing. In this study, SAV habitats were mapped into three categories of high-density seagrass (HDS), low-density seagrass (LDS), and unvegetated bottom. Underwater videography, global positioning system (GPS)-guided field reference photography, and bathymetric data were used to support remote sensing image classification and information extraction. This study achieved approximately 75% overall classification accuracy for SAV mapping.

Although high spatial resolution of the QuickBird-2 satellite data helped identify the spatial patterns of the SAV communities, the lack of information on shortwave spectral coverage limited the data capacity and quality of SAV classification and mapping. As SAV habitats are mostly in shallow coastal waters, short wavelengths in light blue portion of the spectrum possess better capacity for penetration of water column and therefore could provide the information for improved SAV mapping.

The Advance Land Imager (ALI) is a sensor system that provides shortwave spectral coverage in the blue band. The overall goal of this study was to employ ALI sensor data on temperate coastal SAV identification and mapping. Through this relationship we hope to streamline the mapping methods and techniques and make them easily transferable and sustainable for coastal resource and environmental management.

15.2 MATERIALS AND METHODS

15.2.1 Study Area

The Long Island South Shore in Suffolk County, New York, is characterized by relatively calm embayment protected from the incoming surf and high energy of the ocean by the Long Island barrier island system. The FINS is a member of the Long Island barrier island system. Designated as a national park in 1964, Fire Island extends about 52 km from Fire Island Inlet to Moriches Inlet. The administrative boundary of FINS extends 305 m into the Atlantic Ocean to the south and 1219 m north into the Great South Bay. Great South Bay, the largest shallow estuarine bay in the state of New York, comprises back barrier and tidal creek marsh, intertidal flats, and eelgrass beds. The Bay varies in width from approximately 300 m to 11 km with an average depth of approximately 1 m at mean low water and stretches from South Oyster Bay in the west to Moriches Bay in the east (Figure 15.1). Moriches and Shinnecock Bays are shallow bays characterized by salt marsh and strong tidal exchange through inlets from the ocean. This study focused on the bay side of the barrier islands from Great South Bay to Shinnecock Bay. The primary species of seagrass found within the study area are widgeon grass (*Ruppia maritma*) and eelgrass (*Zostera marina*).

15.2.2 Imagery and Reference Data

The EO-1 satellite was launched in November 21, 2000, on a technology validation/demonstration mission as part of the NASA's New Millennium Program. Currently, the EO-1 satellite is operating in an extended mission status managed jointly by

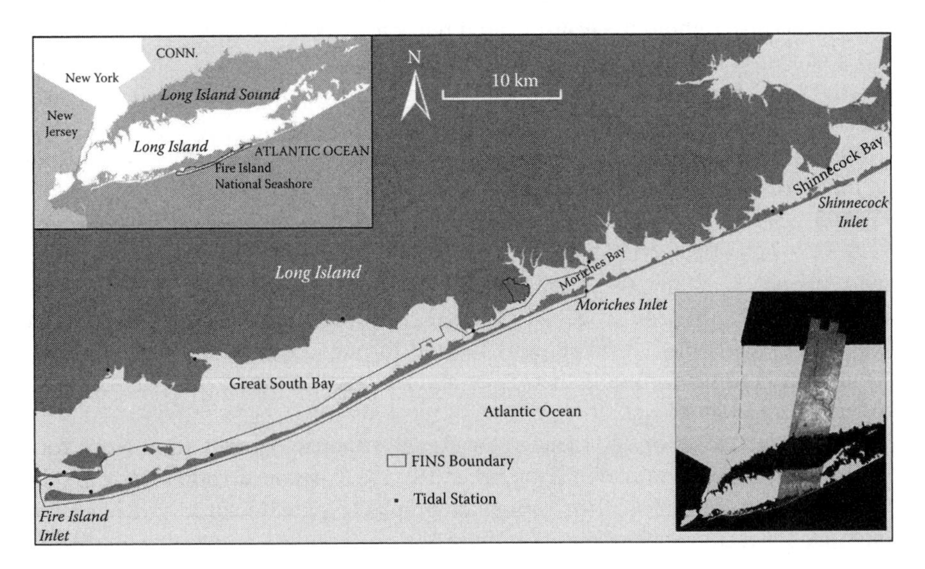

FIGURE 15.1 Study area and location of tidal stations. The inset shows the swath of ALI imagery that covered a portion of the study area.

TABLE 15.1
ALI Spectral Designation and Scaling Factor

ALI Band	ALI Spectral Range (μm)	Scaling Factor	Offset
1	0.048–0.69	0.024	−2.2
2	0.433–0.453	0.045	−3.4
3	0.45–0.515	0.043	−4.4
4	0.525–0.605	0.028	−1.9
5	0.63–0.69	0.018	−1.3
6	0.775–0.805	0.011	−0.85
7	0.845–0.89	0.0091	−0.65
8	1.2–1.3	0.0083	−1.3
9	1.55–1.75	0.0028	−0.6
10	2.08–2.35	0.00091	−0.21

L = (DN × scaling factor) + offset.
L = radiance (mW cm^{-2} sr^{-1} μm^{-1}).

NASA and the U.S. Geological Survey (USGS). The EO-1 satellite orbits in a sun-synchronous, near-polar orbit, traveling north to south on the descending orbital node crossing the equator at approximately 10:00 a.m. and revisiting every 16 days.

Among sensors on board the EO-1 satellite included the multispectral ALI and the hyperspectral Hyperion. The primary purpose of the ALI sensor was to validate and demonstrate technology for the Landsat Data Continuity Mission. ALI operates in a pushbroom fashion and has a 37-km swath by 42 or 185 km scene length (Figure 15.1). ALI data possess 30 m spatial resolution for nine spectral bands covering the visible light to the near-infrared portion of the electromagnetic spectrum, and a panchromatic band with 10 m spatial resolution (Table 15.1). The noticeable spectral band that covers the short blue wavelength between 0.433 and 0.453 μm could provide better penetration for SAV identification.

The original imagery data consisted of three separate image scenes acquired on April 6, 2004 (41°14.8′–72°46.817′ NW, 41°7.917′–72°21.7′ NE, 40°28.233′–73°2.05′ SW, 40°21.417′–72°37.2′ SE); March 7, 2006 (42°14.942′–72°40.596′ NW, 42°8.372′–72°14.083′ NE, 40°29.29′–73°11.521 SW, 40°22.798′–72°45.689′ SE); and August 8, 2006 (42°18.583′–72°27.917′ NW, 42°11.733′–72°2.633′ NE, 40°34.6′–73°2.5′ SW, 40°27.833′–72°37.833′ SE). We checked the NOAA water level tidal predictions* from stations located throughout the Long Island Sound (Figure 15.1) to determine if there was a correlation between time of ALI data collection and water level.

We obtained a set of 2004 true-color digital orthophotographs from New York State Geographic Information Systems (GIS) Clearinghouse (http://www.nysgis.state.ny.us/). We mosaicked the orthophotographs (Figure 15.2a) to provide reference data for rectification of the ALI images.

* http://co-ops.nos.noaa.gov (December 2008 checking).

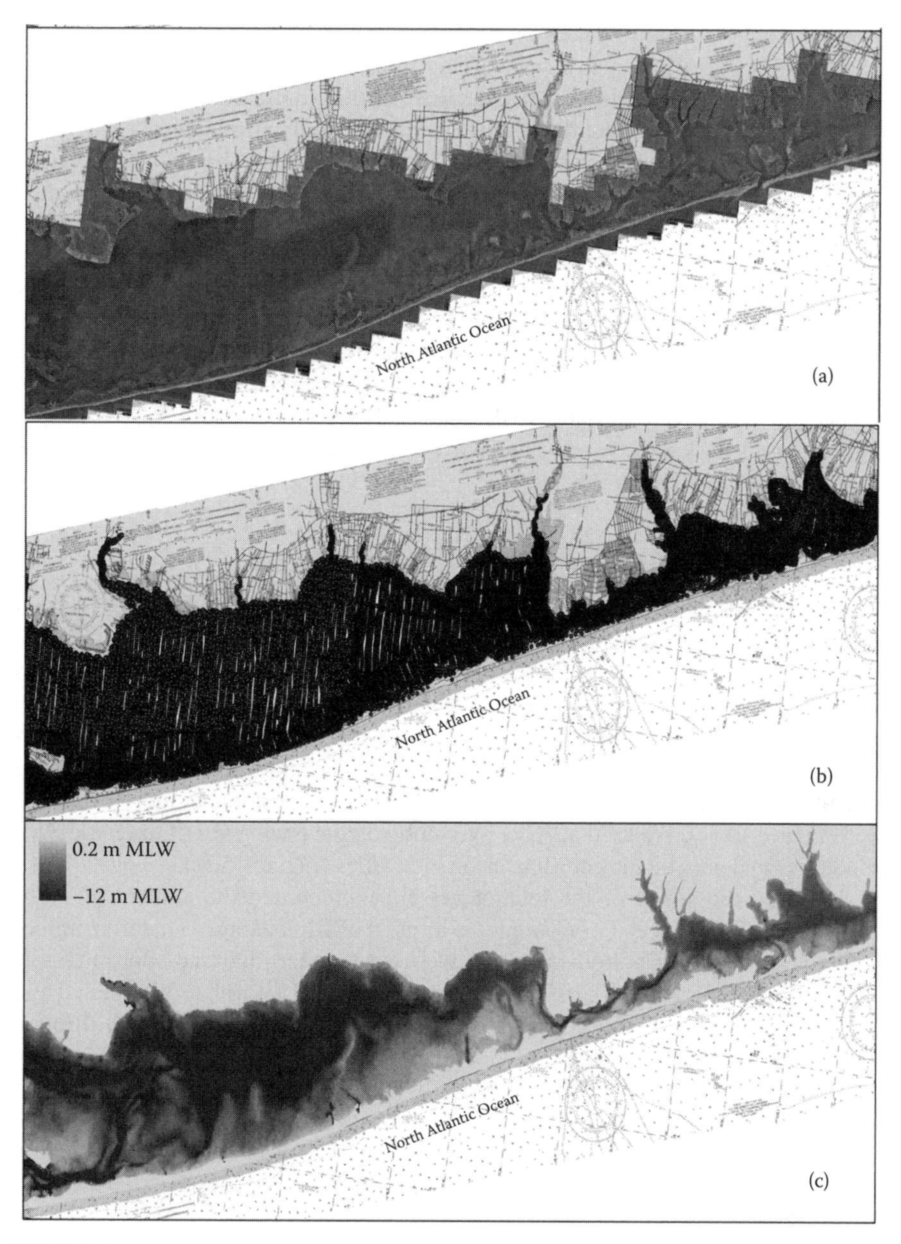

FIGURE 15.2 Examples of reference data including (a) the true-color digital orthophotograph, (b) bathymetric sounding points, and (c) the dredged channels and deep water areas derived from the bathymetric sounding point data.

The March 2006 dataset coincided with water level roughly 0.15–0.60 m above mean lower low water. The April 2004 and August 2006 dataset coincided with water level roughly at mean lower low water to 0.15 m above. The average tidal range at all the stations across the study area is 0.15–0.90 m. Following NOAA CCAP protocols for remote sensing data collection for SAV mapping, data collection should occur within 2 h of mean low water. This is often not the case due to the sun-synchronous orbit orientation of the satellite. In this study, water level was determined not to adversely affect the quality of datasets used in this study.

Since the albedo of deep water in visible color is similar to that of dense eelgrass beds, areas of deep water can be a considerable source of error in digital image classification of eelgrass habitats. Bathymetry is highly correlated with SAV distributions. The depth range in which SAV grow is variable and dependent on the local physical and environmental characteristic as well as the light limitations of the plant. In general, SAV beds do not exist in any location that is deeper than 2 m mean low water in the Great South Bay. We used approximately 85,000 sounding points from the surveys collected by the NOAA National Geophysical Data System (GEODAS)* to generate the bathymetric data as a depth mask (Figure 15.2b and c). With the depth data we could mask out all areas known to be too deep for eelgrass or widgeon grass growth in the Great South Bay.

15.2.3 Field Reference Database

To support SAV mapping, we employed an underwater video transect database that was created by a separate study (Wang et al., 2007) as field reference of benthic habitats. For that effort, we attached an underwater video camera to a sled and towed the camera sled behind a vessel (Figure 15.3a). The position, heading, and speed of the vessel were output from a GPS receiver located on the vessel and overlaid on the live video image. To increase the spatial accuracy of the coordinates we placed the Trimble ProXR GPS receiver in a kayak and positioned directly over the sled (Figure 15.3b). We used a VCR/TV combo on the towing vessel to record SAV conditions and geographic coordinates on video files (Figure 15.3c).

Based on the speed of the towing vessel, we recorded the amount of video equaling 30 m transects. For example, at a speed of 1 knot approximately 1 min of video covers 30 m of bottom traveled. We used the start and end coordinates of each segment to create linear feature coverage in GIS format. We related the recorded video files to the coordinates captured on the towed GPS unit through time stamp and viewed each 30 m video and observed the percent of SAV within each transect. We estimated the percent by dividing the number of seconds in which SAV was observed in the camera field of view by the total seconds in each clip (Figure 15.3d). This yielded a percent cover of SAV over each 30 m long by 0.25 m wide transect. We noted the dominant species of SAV in each video transect. We conducted direct observations on top of the water surface and recorded SAV conditions in association with observed patches of SAV (Figure 15.3e and f). During the fieldwork we documented about 400 field photographs of water-surface SAV patches, and about 7 h of underwater video that covered 206 of 30 m transects for SAV beds.

We re-examined the site in October 2008 as a confirmation of the video database. We linked the digital video file corresponding to each line segment to the ALI image

* http://www.ngdc.noaa.gov/mgg/gcodas/geodas.html (December 2008 checking).

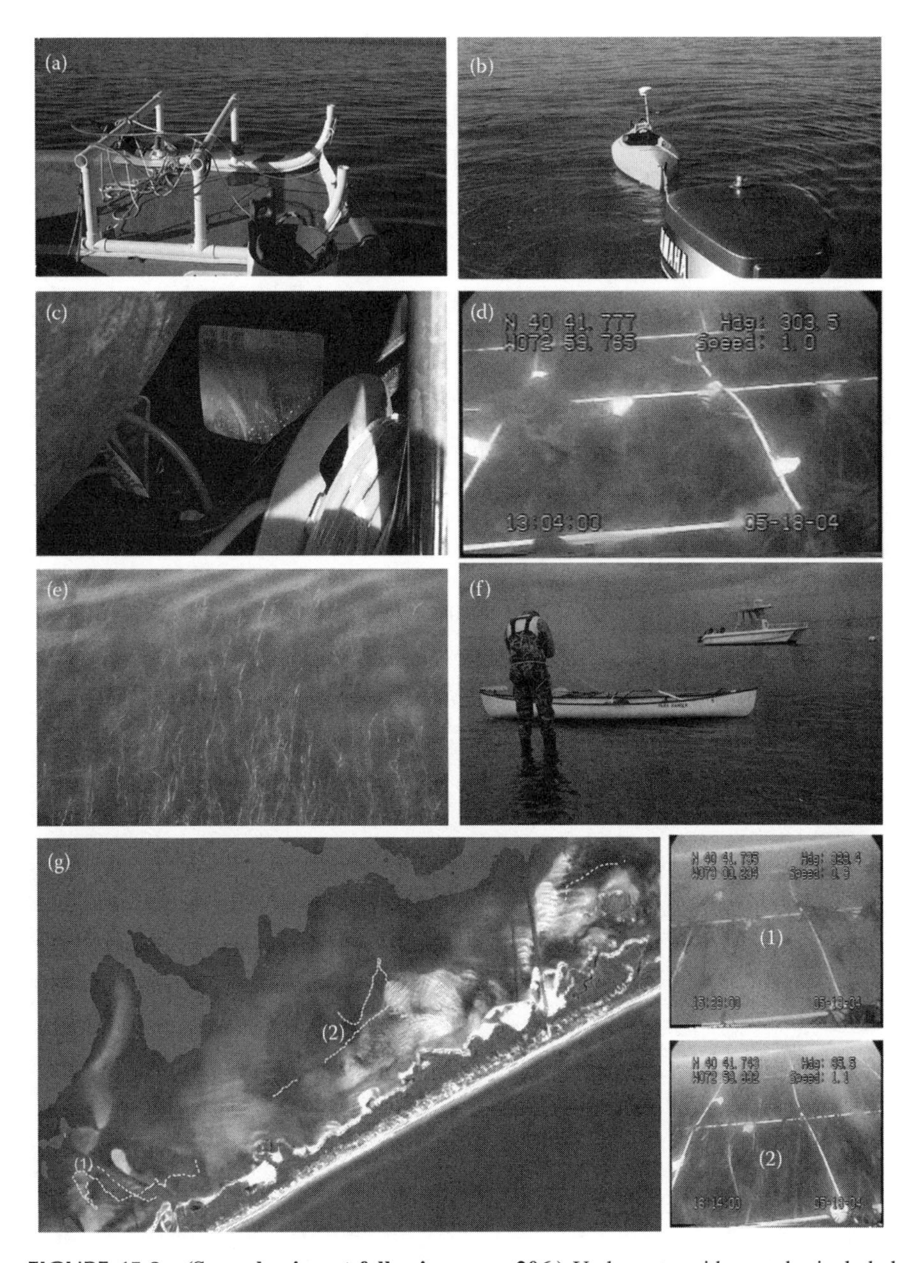

FIGURE 15.3 **(See color insert following page 206.)** Underwater videography included (a) a video camera mounted on a slide, (b) GPS tracking of underwater camera, (c) VCR/TV monitoring equipment, (d) underwater footage, (e) and (f) direct field observation, and (g) example locations of underwater video transects.

(Figure 15.3g). The fieldwork and observations provided essential independent reference data for identifying vegetation types to be mapped.

The New York Department of State and the NOAA Coastal Services Center conducted benthic habitat mapping in 2002, which used 1:1200 color orthophotography to map SAV by visual interpretation and manual delineation. The produced SAV reference maps of Great South Bay were used in this study for seagrass habitat identification and accuracy assessment.

15.2.4 Initial Image Processing

We first converted the ALI images to units of radiance (W m^{-2} sr^{-1} μm^{-1}) (Table 15.1) and then to at-sensor reflectance using

$$\rho_P = \frac{\pi L_\lambda d^2}{\mathrm{ESUN}_\lambda \cos\theta_s}, \tag{15.1}$$

where ρ_p is the planetary reflectance (unitless), L_λ is spectral radiance at the sensor's aperture, d is the Earth-to-Sun distance in astronomic units at the acquisition date, ESUN_λ is the mean solar exoatmospheric irradiance, and θ_s is the solar zenith angle in degrees (90° minus the solar elevation angle). We then conducted a resolution merge by integration of ALI multispectral (30 m) and panchromatic (10 m) bands. We selected the principal component spectral sharpening with the cubic convolution resampling method. The resolution merge improved the interpretability of the ALI imagery data by having finer spatial resolution with multispectral information (Figure 15.4a).

FIGURE 15.4 Resolution merged and georectified ALI image data acquired on March 7, 2006 (a) and the masked subset image used in classification (b).

We registered the ALI images by image-to-image geometric rectification using the 2004 true-color aerial orthophotograph as a reference base. The reported accuracy of the orthophotograph was 1.2 m. We selected from 35 ground control points between each ALI image and the reference true-color digital image. We registered each scene to an accuracy of 0.5 pixels, determined from the root-mean-square error. This yielded a combined accuracy of 6.7 m between our registration and the reported accuracy of the base image.

To eliminate the spectral confusion between inland vegetation and submerged habitats, we masked out all the land areas so that the classification process can be focused on coastal waters only. We created a land mask for each image by conducting a 2-class unsupervised classification using Band 7 (0.845–0.89 μm) of the ALI data. The infrared spectrum of the ALI data separated the water and land areas successfully. The land masks were compared to the orthophotograph for accuracy examination and to remove areas of inland water. The resulting data was used to mask out pixels of land area from the ALI images. In addition, we used the depth mask generated from the bathymetry data to mask out water areas deeper than 2 m from the ALI images. The final subset and masked image data were ready for the classification process (Figure 15.4a and b).

15.2.5 Digital Image Classification

We conducted an initial image classification using unsupervised ISODATA algorithm for a data screening. The results concluded that the April 6, 2004 and August 8, 2006 ALI data could not generate a high-quality SAV map due to weather and water conditions at the time of image acquisitions. We then set aside the two images and kept the March 7, 2006 image as the working data.

We utilized the transect data from underwater videography to identify SAV signatures for a supervised classification. To compensate for environmental differences and reduce the complexity of image qualities, we split the March image into two sections. We defined three categories: HDS with 50–100% coverage, LDS with 5–50% coverage, and unvegetated bottom or sand (SND) with less than 5% SAV coverage. This classification scheme was the same as the reference study by Wang et al. (2007). Upon finishing the individual classifications we mosaicked results into a final classification. We created a 5-m buffer for the transect segments, overlaid the vector polygons on the masked ALI images, and copied to AOIs to create signatures for supervised classification.

15.3 RESULTS

The total subset from ALI image in the final data covered about 163 km^2, of which approximately 18%, or 2940 ha, was classified as SAV. Among those 65% (1909 ha) was classified as HDS, and 35% (1031 ha) was classified as LDS. Underwater transect data confirmed the classification for the covered areas (Figure 15.5a and b). The final classification of SAV is displayed in Figure 15.5c.

We conducted an accuracy assessment for SAV using a total of 300 randomly selected pixel locations, 100 from each category. Overall classification accuracy was

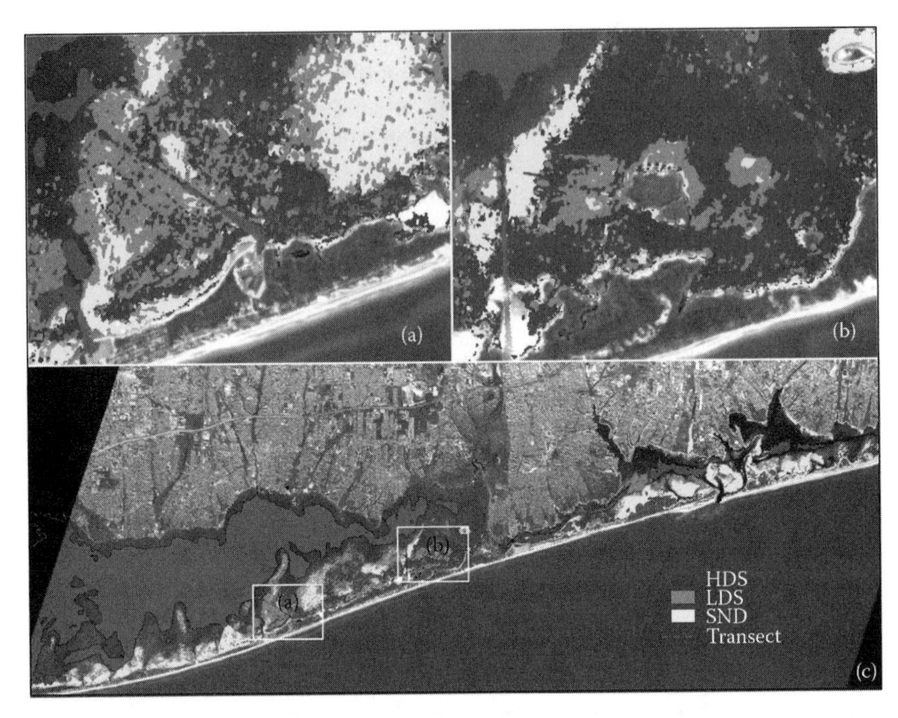

FIGURE 15.5 **(See color insert following page 206.)** Zoomed display for selected sections of (a) and (b) and the SAV classification map displayed on top of the ALI image (c).

approximately 76%. The producer and user accuracy for HDS were 87% and 73%, respectively. The producer and user accuracy for LDS were 84% and 55%, respectively (Table 15.2). The confusion Matrix for the classification is shown in Table 15.3. Difficulty in discriminating between LDS and unvegetated bottom existed at multiple sites.

A comparison of the overlapping region within the FINS boundary of the SAV maps obtained by classification of QuickBird-2 and ALI data generated comparable

TABLE 15.2
Accuracy Assessment for SAV Classification

				Accuracy		
Class	Reference Totals	Class Totals	Number Correct	Producers (%)	Users (%)	κ
HDS	63	75	55	87.3	73.3	0.63
LDS	49	75	41	83.7	54.7	0.42
SND	113	75	72	63.7	96.0	0.92

Overall classification accuracy = 75.7%.
Overall κ statistics = 0.64.

TABLE 15.3
Confusion Matrix for SAV Classification

Class	HDS	LDS	SND	Total
HDS	71	10	19	100
LDS	7	60	33	100
SND	2	2	96	100
Total	80	72	148	

TABLE 15.4
Comparison of Areas (ha) of SAV Classification Using ALI and QuickBird Data for the Same Study Area

	ALI	QuickBird
HDV	765	713
LDV	525	501
SND	970	783
Other	888	1151

results for the categories of HDS and LDS (Table 15.4). The "other" category represents shallow open water not classified as HDS, LDS, or SND. The disparity between the QuickBird-2 and the ALI classification of unvegetated bottom and "other" was due to differences in data quality and capacity. Apparently, classification of ALI data generated less unidentified "other" area than the same process by using the QuickBird-2 data.

15.4 CONCLUSIONS AND DISCUSSIONS

The integration of *in situ* and remote sensing data is critical for success in classifying and mapping SAV. Traditional mapping of seagrass habitats has been accomplished using a combination of aerial photographs and *in situ* sampling by divers using scuba. Satellite imagery offers advantages over aerial photography, including larger individual scenes with spectral uniformity and increased reimaging capability. However, satellite imagery has disadvantages such as inflexibility with fixed data acquisition times and difficulty in meeting the environmental requirements for seagrass mapping.

In this study, we used *in situ* observation in data preparation, classification, and accuracy assessment. SAV habitats are dependent on environmental conditions and light limitations. We used sounding data to create a depth mask to eliminate a possible source of error due to spectral confusion between SAV and unvegetated bottom. We used tidal data to determine water level at the time of ALI data acquisition to reveal any possible influence of water depth on SAV mapping. During classification, we employed underwater videography and transect data from the field reference database as well as direct observations to help in selection of training signatures. This is very helpful primarily for identification of areas with HDS and LDS coverage. We employed transect data and direct water surface observations as the reference in accuracy assessment of the SAV classification. We reference high spatial resolution digital orthophotography at every stage of this study. The integration of *in situ* data proved invaluable in ALI image processing and SAV mapping.

Monitoring SAV provides key insights into estuarine ecosystem conditions because of SAV dependency on environmental parameters such as water quality, temperature, light, and nutrients. Understanding the role of SAV in the ecosystem is invaluable to resource managers involved in mapping and monitoring the coastal environment. SAV habitats have been linked to water-quality standards to assess ecosystem health in Chesapeake Bay (Dennison et al., 1993) and have been used as a nutrient pollution indicator in New England estuaries (Lee et al., 2004). SAV act as early indicators of change in the coastal environment ever more important as resource managers face the challenges of climate change.

In this study, we tested the capabilities of the sensor's shortwave spectral coverage previously untested in SAV classification and mapping. The results show promise in future applications of SAV classification using ALI data or other sensors with similar spectral coverage. Integration of bathymetry data, orthophotography, and GIS information is the key to establishing accurate classification of SAV. With ALI's swath width of 37 km, it is possible to capture a larger area than higher resolution sensors to eliminate spectral differences in scenes due to differences in environmental factors. Satellite remote sensing data provide a cost-effective means of creating and updating SAV coverage maps necessary in monitoring and identifying change in the dynamic environment of coastal estuaries and barrier islands.

We also identified limitations in this study that are common to remote sensing applications in coastal environment. SAV data acquisition should coincide with peak SAV biomass to accurately determine the extent of SAV distribution. The growing season of eelgrass peaks in early spring and peak biomass is achieved in August and September. Of the three datasets obtained, only the March 2006 dataset at the start of the growing season was appropriate to use in this study. Successful data acquisition for SAV mapping is dependent on sunlight, atmospheric conditions, wind and wave action, but most importantly, water clarity. Seagrass minimal light requirements for survival range between 4% and 29% of incident light measured just below the water surface much higher than phytoplankton and algae (Dennison et al., 1993). Light attenuation characteristics of Great South Bay adversely affected by instances of "brown tide" caused by the microalga *Aureococcus anophagefferens* have been documented (Dennison et al., 1989). Eelgrass dieback concurrent with "brown tides" were observed in Great South Bay and eelgrass distribution declined in postbloom years. In Moriches and Shinnecock Bays, eelgrass was not affected by the "brown tide" events possibly due to higher rates of tidal flushing and eelgrass distribution inexplicably increased in postbloom years. Seagrass detection and classification rely entirely on the ability for light penetration of the water column and the degraded water clarity due to "brown tide" or other suspended elements in the water column is a possibility of the spectral confusion of the April and August ALI datasets.

Spectral confusion between areas of LDS and unvegetated bottom occurred in several areas of the classification. This could be due to the relatively low spatial resolution (10 m) of the data inability to resolve patchiness of the seagrass in areas of low density. Confusion between floating and attached macroalgae and both HDS and LDS is probable and may not have been identified in the accuracy assessment. The selection of Hyperion hyperspectral bands integrated with ALI and high spatial resolution datasets could reveal new insights into SAV mapping.

In an attempt to assess the contribution of the shortwave blue spectral band (0.433–0.453 μm) of the ALI sensor in the identification and classification of SAV, we conducted a principal component analysis on the dataset used in the classification process (Jensen, 2005). The results indicated that all ALI spectral bands are highly correlated for the subset dataset that focus on water surface only. The first principal component accounted for 99.45% of the variance for the subset data and all ALI bands displayed a high degree of correlation with the first principal component. By subsetting and masking out land and bay water areas deeper than 2 m, distinct spectral variation primarily from terrestrial features was removed. We were unable to determine a unique contribution of the shortwave blue band in classification of SAV using principal component analysis. Further testing is recommended for gaining insights into the effectiveness of ALI short blue wavelength in SAV mapping.

ACKNOWLEDGMENTS

Dr. James Irons of the NASA GSFC and Dr. Lawrence Ong of the JPL provided the ALI imagery data. We appreciate the logistic support and field assistance by Michael Bilecki, Jordan Raphael, Paul Czachor, and Jason Flynn of the FINS and Sara Stevens of the Northeast Coastal and Barrier Network of the NPS. This study was funded by the NASA Rhode Island Space Grant Consortium grant no. NNG05GG71H ("Integration of NASA EO-1 Advanced Land Imager and Hyperion Data for temperate coastal submerged aquatic vegetation mapping"). The referenced field data and SAV map were from a study funded by the NPS grant no. H4525037056, under the title "Remote sensing of terrestrial and submerged aquatic vegetation in Fire Island National Seashore: Towards long-term resource management and monitoring," in which Michael Traber performed underwater videography exercise.

REFERENCES

Burkholder, J.M., Mason, K.M., and Glasgow, H.B., Jr., 1992, Water-column nitrate enrichment promotes decline of eelgrass *Zostera marina:* Evidence from seasonal mesocosm experiments, *Marine Ecology Progress Series*, 81: 163–178.

Dennison, W.C., Marshall, G.J., and Wigland, C., 1989, Effect of "Brown Tide" shading on eelgrass (*Zostera marina* L.) distributions, in: E.M. Cosper, V.M. Bricelji, and E.J. Carpenter (Eds), *Novel Phytoplankton Blooms: Causes and Impacts of Recurrent Brown Tides and Other Unusual Blooms, Coastal and Estuarine Studies 35*, pp. 675–692, Springer, Berlin.

Dennison, W.C., Orth, R.J., Moore, K.A., Stevenson, C., Carter, V., Kollar, S., Bergstrom, P.W., and Batiuk, R.A., 1993, Assessing water quality with submersed aquatic vegetation, *BioScience*, 43: 86–94.

Dierssen, H.M., Zimmerman, R.C., Leathers, R.A., Downes, V., and Davis, C.O., 2003, Ocean color remote sensing of seagrass and bathymetry in the Bahamas banks by high resolution airborne imagery, *Limnology and Oceanography*, 48: 444–455.

Duarte, C.M., 2002, The future of seagrass meadows, *Environmental Conservation*, 29: 192–206.

Ferguson, R.L. and Korfmacher, K., 1997, Remote sensing and GIS analysis of seagrass meadows in North Carolina, USA, *Aquatic Botany*, 58: 241–258.

Hauxwell, J., Cebrian, J., and Valiela, I., 2003, Eelgrass Zostera marina loss in temperate estuaries: Relationship to land-derived nitrogen loads and effect of light limitation imposed by algae, *Marine Ecology Progress Series*, 247: 59–73.

Heck, K.L., Jr., Able, K.W., Fahay, M.P., and Roman, C.T., 1989, Fishes and decapod crustaceans of Cape Cod eelgrass meadows: Species composition, seasonal abundance patterns and comparison with unvegetated substrates, *Estuaries*, 12: 59–65.

Heck, K.L., Jr., Able, K.W., Roman, C.T., and Fahay, M.P., 1995, Composition, abundance, biomass, and production of macrofauna in a New England Estuary: Comparisons among eelgrass meadows and other nursery habitats, *Estuaries*, 18: 379–389.

Hughes, J.E., Deegan, L.A., Wyda, J.C., Weaver, M.J., and Wright, A., 2002, The effects of eelgrass habitat loss on estuarine fish communities of Southern New England, *Estuaries*, 25: 235–249.

Jensen, J.R., 2005, *Introductory Digital Image Processing: A Remote Sensing Perspective*, 3rd edition, Prentice-Hall, Upper Saddle River, NJ, 1996.

Larkum, A.W.D., Orth, R.J., and Duarte, C.M. (Eds), 2006, *Seagrasses: Biology, Ecology and Conservation*, Springer, The Netherlands.

Lazzari, M.A. and Stone, B.Z., 2006, Use of submerged aquatic vegetation as habitat by young-of-the-year epibenthic fishes in shallow Maine nearshore waters, *Estuarine, Coastal and Shelf Science*, 69: 591–606.

Lee, K.S., Short, F.T., and Burdick, D.M., 2004, Development of a nutrient pollution indicator using the seagrass, *Zostera marina*, along nutrient gradients in three New England estuaries, *Aquatic Botany*, 78: 197–216.

Mishra, D., Narumaiani, S., Rundquist, D., and Lawson, M., 2006, Benthic habitat mapping in tropical marine environments using QuickBird Multispectral data, *Photogrammetric Engineering and Remote Sensing*, 72: 1037–1048.

Newell, R.I.E. and Koch, E.W., 2004, Modeling seagrass density and distribution in response to changes in turbidity stemming from bivalve filtration and seagrass sediment stabilization, *Estuaries*, 27: 793–806.

Orth, R.J., Carruthers, T.J.B., Dennison, W.C., Duarte, C.M., Fourqurean, J.W., Heck, K.L., Jr., Hughes, A.R., Kendrick, G.A., Kenworthy, W.J., Olyarnik, S., Short, F.T., Waycott, M., and Williams, S.L., 2006, A global crisis for seagrass ecosystems, *Bioscience*, 56: 987–996.

Pasqualini, V., Pergent-Martini, C., Pergent, G., Agreil, M., Skoufas, G. Soures, L., and Tsirika, A., 2005, Use of spot 5 for mapping seagrasses: An application to *Posidonia oceanica*, *Remote Sensing of the Environment*, 94: 39–45.

Schweizer, D., Armstrong, R.A., and Posada, J., 2005, Remote sensing characterization of benthic habitats and submerged vegetation biomass in Los Roques Archipelago National Park, Venezuela, *International Journal of Remote Sensing*, 26: 2657–2667.

Short, F.T. and Burdick, D.M., 1996, Quantifying eelgrass habitat loss in relation to housing development and nitrogen loading in Waquoit Bay, Massachusetts, *Estuaries*, 19: 730–739.

Short, F.T. and Neckles, H.A., 1999, The effects of global climate change on seagrasses, *Aquatic Botany*, 63: 169–196.

Wang, Y., Traber, M., Milstead, B., and Stevens, S., 2007, Terrestrial and submerged aquatic vegetation mapping in Fire Island National Seashore using high spatial resolution remote sensing data, *Marine Geodesy*, 30: 77–95.

Wolter, P.T., Johnston, C.A., and Niemi, G.J., 2005, Mapping submergent aquatic vegetation in the US Great Lakes using Quickbird satellite data, *International Journal of Remote Sensing*, 26: 5255–5274.

16 Remote Sensing Applications Used to Inventory and Monitor Natural Resources in North Atlantic Coastal National Parks

Sara Stevens and Courtney Schupp

CONTENTS

16.1 INTRODUCTION

16.1.1 National Park Service Inventory and Monitoring Program

National Park Service (NPS) managers across the country are confronted with increasingly complex and challenging issues that require a broad-based understanding of the status and trends of each park's natural resources as a basis for making decisions, working with other agencies, and communicating with the public to protect park natural systems and native species (Natural Resource Challenge). Mandated to preserve and protect park resources for and through future generations, park managers must know the current condition of park natural resources and understand how they are changing over time in order to be successful. To address this need for reliable, scientifically based information for resource protection, the NPS has implemented a strategy to inventory and monitor key natural resources. This strategy is called the Inventory and Monitoring (I&M) Program and includes three major components: (1) completion of basic natural resource inventories upon which monitoring efforts can be based; (2) development of experimental prototype monitoring programs in a selected number of parks to evaluate alternative monitoring designs and strategies; and (3) implementation of ecological monitoring of critical parameters (i.e., "vital signs") in all parks with significant natural resources.

As part of the NPS's effort to "improve park management through greater reliance on scientific knowledge," a primary role of the I&M Program is to collect, organize, and make available natural resource data and to contribute to the Service's institutional knowledge by transforming data into information through analysis, synthesis, and modeling. This role includes finding, explaining, and applying the most effective mapping technologies, which increasingly include remote sensing technologies for inventorying park resources.

To facilitate collaboration, information sharing, and economies of scale in inventory and monitoring, the NPS organized more than 270 parks with significant natural resources into 32 "I&M Networks" linked by geography and shared ecosystem characteristics (Fancy et al., 2009). The following goals of the program are shared by all 32 Networks:

- Inventorying the natural resources under NPS stewardship to determine their nature and status.
- Monitoring park ecosystems to better understand their dynamic nature and condition and to provide reference points for comparisons with other, altered environments.
- Establishing natural resource inventory and monitoring as a standard practice throughout the National Park System that transcends traditional program, activity, and funding boundaries.
- Integrating natural resource inventory and monitoring information into NPS planning, management, and decision-making.
- Sharing NPS accomplishments and information with other natural resource organizations and forming partnerships for attaining common goals and objectives.

16.1.2 NPS Northeast Coastal and Barrier Network

As part of the I&M Program, the Northeast Coastal and Barrier Network (NCBN) was established by a charter in 1999. This Network is made up of eight parks found along the northeastern Atlantic seaboard (see Figure 16.1 for a map of NCBN parks), and all but one of them is within the coastal zone. The NCBN includes Cape Cod National Seashore (CACO) in Massachusetts, Fire Island National Seashore (FIIS) and Sagamore Hill National Historic Site (SAHI) located in New York, Gateway National Recreation Area (GATE) with units in both New York and New Jersey, Assateague Island National Seashore (ASIS) located in Maryland and Virginia, Thomas Stone National Historic Site (THST) also located in Maryland, and Colonial National Historical Park (COLO) and George Washington Birthplace National Monument (GEWA) both located in Virginia. Together, these eight parks contain approximately 59,220 ha (146,300 acres) of park lands, representing some of the most ecologically similar collections of lands within the NPS. They consist of critical coastal habitat for many rare and endangered species, and serve as migratory corridors for birds, sea turtles, and marine mammals. They also protect vital coastal wetlands, essential to water quality, fisheries, and the biological diversity of coastal, nearshore, and terrestrial environments.

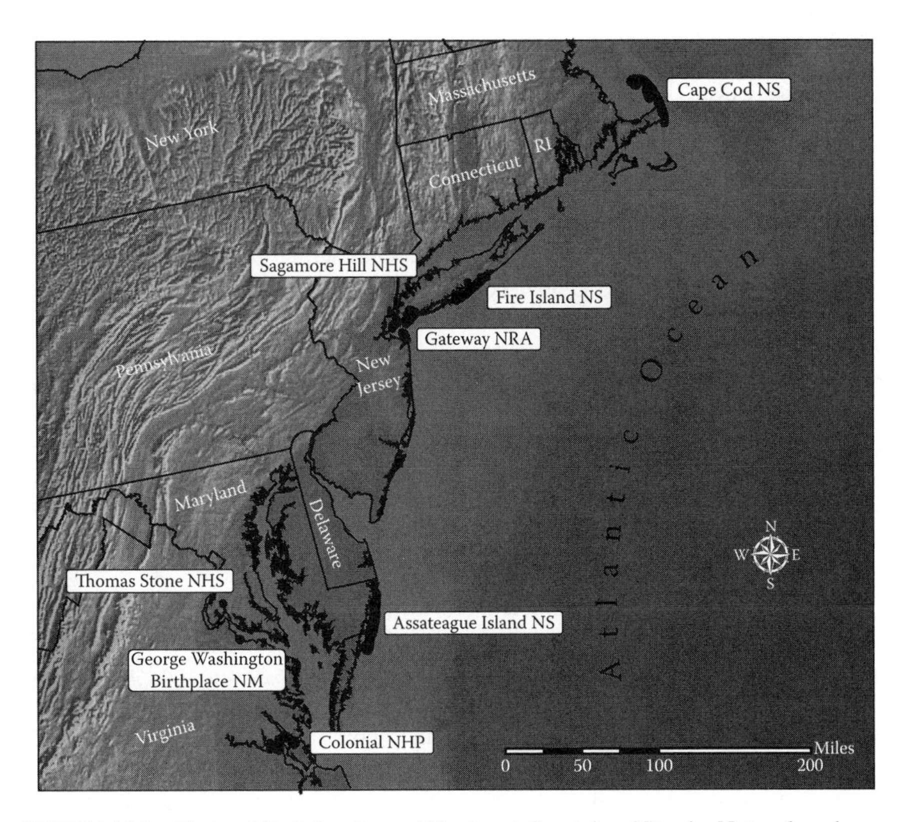

FIGURE 16.1 National Park Service and Northeast Coastal and Barrier Network parks.

In the development of the NCBN's long-term ecological monitoring plan, key natural resource issues and associated ecological indicators or "vital signs" were identified for each of the NCBN parks (Stevens and Entsminger, 2004; Stevens et al., 2005). The issues identified as highest priority were those that park managers needed additional information on, in order to appropriately address management, as well as those affecting multiple parks in the network.

Because all but one NCBN park is within the coastal zone, common issues are related to estuarine water quality, seagrass condition and distribution, salt marsh condition, and habitat change, landscape change, and shoreline change. Managers of the Network's coastal parks must address unique management issues because of the rapidly changing and dynamic nature of the coastal environment. Unlike inland parks, these parks experience the constant shifting of their biotic communities and more dramatically accelerated geomorphologic change. Barrier island parks such as FIIS and ASIS are constantly being shaped and reshaped, eroded and accreted. Their natural existence being dependent upon the shifting of sand, wave energy, and tidal fluctuations makes them challenging to manage in concert with adjacent land uses. Large storm events, such as tropical storms, hurricanes, and extratropical storms, nor easters, result in unique and dynamic ecosystems and can contribute to significant alteration of these parks. These storms generally cause rapid beach erosion, dune displacement, and coastal flooding and can create even more dramatic change when they coincide with high astronomical or spring tides.

The NCBN monitoring program was designed to assist parks in acquiring long-term datasets using systematically gathered and scientifically rigorous approaches (Oakley et al., 2003). Working with scientists from multiple agencies and academic institutions, the Network is using a variety of tools and methodologies to monitor change and condition of salt marshes, park estuaries, shorelines, beaches and dunes, and vegetation, fish, and bird communities. The Network continues to develop standardized protocols for monitoring, recently completing protocols for shoreline change using ground-based Global Positioning System (GPS), and estuarine water quality and seagrass condition and distribution monitoring (Kopp and Neckles, 2009). Other protocols under development will measure salt marsh elevation, nekton communities, salt marsh vegetation, nitrogen inputs to park estuaries, and coastal topography. Although as much as 75% of some park's resources are underwater, submerged resources were not selected for monitoring during the initial planning phase of the NCBN program. Historically these resources have gone unmanaged, leading to a lack of existing data and information available to park managers. Scientists and park managers have emphasized the need to inventory, map, and monitor these resources.

A variety of remote sensing and analysis techniques are being explored for use in coastal monitoring by both the Network monitoring program and park staff in order to increase knowledge about natural and cultural resources within the Parks, and to enhance on-the-ground field efforts by NPS staff and scientists to inventory and map these resources. Data acquisition includes incorporating remote sensing and GPS technologies, such as vessel-based bathymetry, Light Detection And Ranging (LiDAR) topography, aerial photography, and satellite imagery. LiDAR data are being collected to monitor geomorphologic change, satellite imagery data are being tested to monitor vegetation community and landscape change, and a variety of vessel-mounted technologies are being tested to inventory and map submerged marine resources including cultural relicts, benthic habitats, bathymetry, and seafloor characteristics. Initial work using remote sensing data for submerged vegetation mapping has also been tested in one NCBN park (Wang et al., 2007). Although these remotely sensed datasets are being collected for specific trend analyses as part of the NCBN program, NCBN parks use these data to answer many park-specific management questions, such as measuring impacts of natural events and human actions, triaging habitat protection in the face of climate change, and siting infrastructure to increase longevity. LiDAR especially has been used to answer a variety of management questions related to changes in elevation and topographic control on habitat productivity. Examples from ASIS include the relationship between elevation, feeding habitat, and nest site selection by piping plovers, a federally listed Threatened and Endangered bird species found along the Atlantic coast; and the effects of grazing by feral horses on vegetation and dune topography (De Stoppelaire et al., 2004). Resource managers at FIIS, ASIS, and GATE use LiDAR to address a number of specific management questions that pertain to human-induced shoreline change. ASIS, for example, needed to develop a cost-effective and focused strategy to mitigate the effects of a beach engineering project that was preventing overwash processes along part of the island. LiDAR data enabled park managers to identify low-elevation areas that, with minimal manipulation, could serve as overwash pathways and thereby restore morphodynamic processes and piping plover habitat. Several examples illustrating the use of remote sensing applications in ecological monitoring and in park management efforts are described below.

16.2 REMOTE SENSING APPLICATIONS USED IN THE NCBN MONITORING PROGRAM

16.2.1 Coastal Topography Monitoring Using EAARL LiDAR

From the onset of the NCBN program, the Network has collaborated with ASIS (Brock et al., 2002), the U.S. Geological Survey (USGS), and the National Aeronautics and Space Administration (NASA) to collect Experimental Advanced Airborne Research Light Detection and Ranging (EAARL LiDAR) elevation data for all of the NCBN parks. This is a pulsed, green-wavelength, waveform-resolving laser-ranging system that has the capability to collect coastal land elevations and canopy height, and, in areas with clear and calm water, it can simultaneously collect adjacent submerged topography. Its sensor suite includes raster scanning, water-penetrating full waveform adaptive LiDAR, a 20 cm per pixel color infrared, down-looking color digital camera, an 80 cm per pixel RGB digital camera, a hyperspectral scanner, an array of precision kinematic GPS receivers that provide for submeter georeferencing of each laser and multispectral sample, and a precision Digital Miniature Attitude Reference System (DMARS) inertial measurement units (IMU) for precision attitude determination (Nayegandhi et al., 2006). EAARL has the real-time capability to detect, capture, and automatically adapt to each laser return backscatter over a large range, detecting considerable vertical variations of the surface being measured.

These features enable the collection of dramatically different surface types in a single EAARL overflight, so it is able to map a variety of landform types including coral reefs, bright sandy beaches, coastal vegetation, and forests.

As part of the long-term monitoring program, the Network is developing a protocol in collaboration with scientists from Rutgers University for monitoring coastal beach and dune topography using EAARL data. The goal is to contribute to an assessment of condition and rate of change for park managers so that they can identify and determine potential threats to existing park natural and cultural resources, as well as park infrastructure, and manage accordingly. This protocol will be implemented in four of the Network parks: CACO, FIIS, ASIS, and the Sandy Hook Unit of GATE. The protocol will provide guidelines to monitor the rates of spatial and temporal change in these continuously evolving barrier island and coastal dune systems. Topographic change and dune movement will be quantified, and areas of elevation loss and horizontal migration identified for park managers. Specific measures of existing conditions provided to park management will include the height and width of dunes and berms, the slope and perturbations of the foreshore, the height of cliffs, the identification of overwash fans, and the edge of vegetation along the beach. Subsequent sampling will illustrate how these conditions are changing over time. Temporal changes in geomorphology, as well as area and volumetric change of the cross-shore and along-shore sediments, will also be analyzed.

Traditionally, beach and dune topographic surveys entailed solely collecting elevation data along transects seaward to a benchmark on land. Covering a large area of beach using these traditional methods can be very labor and time intensive. Recently, LiDAR data have provided a more efficient means of collecting topographic information system-wide in parks (Brock et al., 2004). Although LiDAR can also be cost prohibitive, the dataset collected is extremely dense, allowing for three-dimensional analysis of change along the entire beach and dune system rather than an analysis along individual two-dimensional transect lines. The utilization of EAARL allows for over 100 km of coastline to be surveyed during a 3–4-h mission.

The ability to sample large areas rapidly and accurately is especially useful in mapping morphologically dynamic regions such as barrier island parks such as Assateague Island, Fire Island, and Sandy Hook, as well as the dynamic dune systems found at CACO. Spatially complete assessments of topographic change can also be made after storm events by comparing the poststorm measurements against baseline data for these parks. The NCBN coastal topography monitoring protocol will include a combination of methods to measure the rate of beach and dune topographic change through the use of LiDAR-derived digital terrain models (DTMs), as well as transect data obtained with ground-based Real-Time Kinetic Global Positioning System (RTK GPS). The intent is for park managers to be better able to anticipate changes related to future climate variability, park use and management, and other human-induced alteration. Currently, the protocol is to collect EAARL LiDAR on a biennial basis in each park and supplement these surveys with ground-based GPS profile surveys of the beach and dune systems. The LiDAR data will provide the necessary three-dimensional spatial extent needed to fully document change on a system-wide scale. The ground survey measurements will be collected to assess the accuracy of the remotely sensed LiDAR data, as well as to provide two-dimensional profiles of the beach and dune systems in areas of concern as defined by individual park management needs. The two-dimensional surveys may be conducted at times outside of the window of LiDAR data collection such as directly following storm events or on a more intensive survey schedule to address within-year variation in specific park areas. Over the long term, this monitoring program provides park managers with a picture and analysis of how these coastal systems have changed and are changing over time.

In addition to the coastal topography monitoring program, the Network is working with scientists from the University of Rhode Island and the USGS to develop historical DTMs for CACO, FIIS, and ASIS, as well as the Sandy Hook unit of GATE. These historical DTMs are being derived from aerial photographs of the parks from as far back as the 1950s. These historical models will then be compared to current LiDAR-based DTMs to synthesize a record of geomorphologic change in these coastal parks that will span 30–40 years. This information will provide park managers with insight into not only what has already occurred, but what they might expect in the future. Figure 16.2 is an example of a historical DTM extracted from

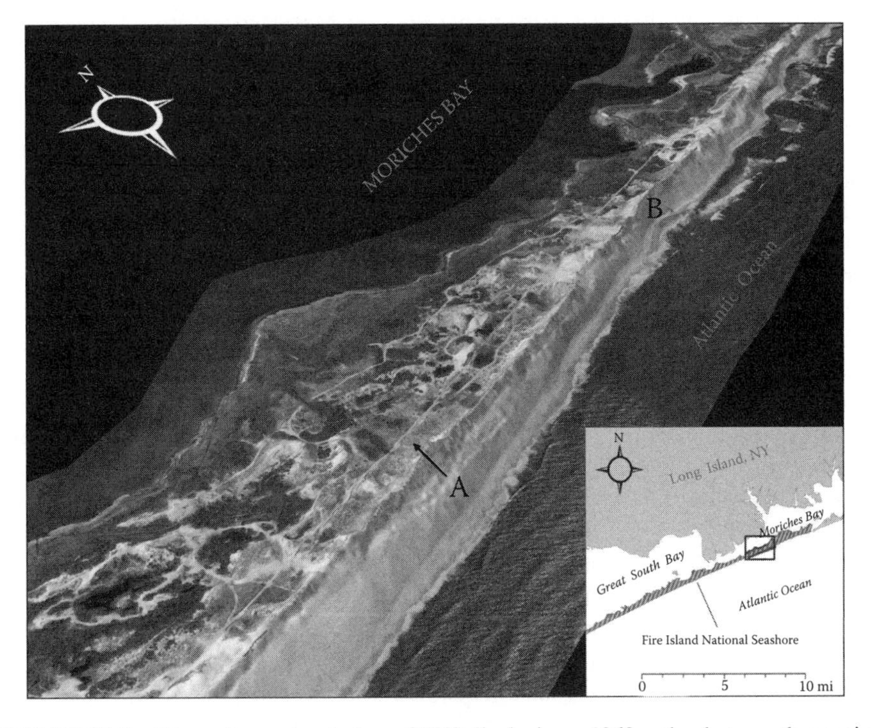

FIGURE 16.2 Three-dimensional view of FIIS displaying a 1969 orthophotograph mosaic draped over 1969 topography. Note the well-defined dune crest (A) and the relatively narrow beach (B) along this portion of Fire Island. Historical topography and orthophotography are the product of photogrammetric processing. (Rachel Hehre, University of Rhode Island, Kingston, RI.)

1967 aerial photography of FIIS draped with a DTM derived from EAARL LiDAR data collected in 2007. These models will help scientists make recommendations to park management concerning shoreline vulnerability, infrastructure development, and protection and restoration efforts based on which areas of the park have accreted or eroded dramatically over time.

16.2.2 LiDAR and Coastal Park Management Scenarios

LiDAR has proven to have many applications in all of the NCBN parks and is rapidly becoming a highly valuable management tool. These data have since been used for a number of applications including habitat and vegetation mapping, storm-event assessment, and coastal geomorphologic change evaluation in areas of concern within the parks.

16.2.2.1 Using LiDAR Elevation Data to Evaluate Changes in Island Morphology and Volume on ASIS

ASIS has been using LiDAR elevation data to guide and evaluate several mitigation projects in an area impacted by both a major storm and the subsequent human responses to that storm. In 1933, following a major hurricane, what was once a continuous barrier spit along the coastline of Delaware, Maryland and Virginia breached, creating what is now called, Ocean City Inlet, and now functions as the northern boundary of ASIS. By 1935, jetties were built along both sides of the newly formed inlet to stabilize the navigable waterway.

For over 70 years, the jetties have altered sand transport, leading to unnatural erosion and accelerated shoreline migration on the northern portion of Assateague Island, including a lower elevation, reduced sand volume, and a westward shift of more than 350 m, as measured via ground-based surveys such as shoreline position and remote sensing data including LiDAR and aerial photographs. These changes resulted in significant geomorphologic, habitat, and biotic changes that would not have occurred under natural circumstances. Coastal geologists studying these changes predicted that without mitigation the northern portion of Assateague Island would continue to destabilize and that a new inlet or inlets would be created by storm events in the near future. A new inlet or inlets would also have serious implications for the adjacent mainland communities and region as a whole, causing increased rates of shore erosion, loss of estuarine habitats, and decreased flows through the existing Ocean City and Sinepuxent channels, resulting in increased maintenance needs (U.S. Army Corps of Engineers, 1998).

As a result of the continued rate of accelerated shoreline migration in the mid-1990s, the U.S. Army Corps of Engineers (USACE) and ASIS initiated a regional feasibility study for mitigating inlet impacts and restoring the Assateague Island to as natural a condition as possible, in addition to addressing other priority shoreline erosion issues in the Ocean City area. The study resulted in a plan called the North End Restoration Project, which included partnerships with the State of Maryland, Maryland Department of Natural Resources, Worcester County, Maryland, and the Town of Ocean City, Maryland, and which also gained the support of the local community. The first, or "short-term," phase of the restoration project was designed to provide a one-time infusion of sand to replace a portion of the sediment lost over the past 60 years due to the effects of the jetties. The second phase, the "long-term sand management" component, addresses the ongoing and future effects of the jetties by re-establishing, via mechanical sand bypassing, a "natural" sediment supply for northern Assateague Island that reflects historic, preinlet rates. To evaluate the success and progress of mitigation efforts, the plan also included measurable criteria, such as overwash frequency.

On January 28 and February 4, 1998, prior to the start of the restoration plan, extratropical cyclones occurred over Assateague Island. During each storm, wave heights were recorded at 7 m off the coast of Ocean City, Maryland, and one of the most severely impacted areas included the north end of Assateague Island, where an overwash area extended several miles, and where scouring extended down to the peat layer, with large overwash fans deposited in the estuary. LiDAR surveys collected before and soon after the storm enabled quantification of volumetric and elevational losses. After 65 years of the Ocean City Inlet jetty system acting as a barrier to longshore transport, a major breach on the north end of Assateague Island had become a distinct possibility. In an effort to prevent this possibility from becoming a reality, an "emergency storm berm" of 1.4 million m^3 was constructed at an elevation of 3.3 m (NGVD29) along the most severely impacted portion of the island.

In September 2002, the first stage of the North End Restoration Project began with 1.4 million m^3 of sand being placed seaward of the mean high waterline to replace a portion of the sand lost to the inlet over the past decades (Figure 16.3).

FIGURE 16.3 The first stage of the North End Restoration Project took place during September 2002 as sediment is deposited on the beach of northern ASIS through pipes and then spread out by bulldozers, extending the shoreline into the ocean. (USACE photograph by Paul J. Smith.)

The second stage of the restoration project, to restore sediment transport to Assateague Island, was then initiated in January 2004, and continues with sediment dredged from the inlet and tidal delta complex and then mechanically placed in the nearshore waters of Assateague Island twice a year.

In an effort to monitor the success of the North End Restoration Project, as well as the condition and effects of the emergency storm berm, NPS and USACE researchers have been collecting a variety of ecological and geomorphological monitoring data in the project area. The use of LiDAR has dramatically improved the abilities of Assateague managers to characterize the changes resulting from the restoration project on a much broader scale than once possible with ground-based GPS technology. LiDAR data have been particularly important in calculating the changes in island shape, sand volume, and beach and berm heights along the entire north end resulting from inlet stabilization, storm impacts and recovery, cross-shore and alongshore movement of mechanically bypassed sand between the nearshore placement site and the beach, and evolution of embryo dunes and shrub communities in the sheltered area west of the emergency berm. These LiDAR data also provide context for smaller-scale projects and analyses, including monthly shoreline position mapping to evaluate accelerated short-term shoreline change following jetty tightening, and small areas surveyed by all terrain vehicle (ATV)-mounted RTK GPS to guide and quantify manipulation of the emergency berm topography to allow overwash processes and plover habitat creation. Although the LiDAR accuracy is too low to use for quantifying small-scale changes, such as the centimeter-scale elevation accuracies required for achieving desired overwash frequencies, the LiDAR and aerial photographs provide context that is extremely useful in interpreting data that are collected in smaller areas on a more frequent monitoring basis, such as shoreline position and widely spaced topographic profiles of the berm and dunes.

Using LiDAR and historic DTMs in conjunction with other remote sensing technologies including offshore wave gauges and current meters that map hydrodynamic pathways, North End Restoration Project partners are working to refine a sediment budget model for the Ocean City–Assateague region in order to understand how sediment moves between the nearshore, the inlet, the beaches, and into the islands' interiors. Rather than attempting to interpolate topography between widely spaced topographic profiles along the beaches, particularly where profile locations have varied over the decades, the volume changes calculated using historic DTMs will provide estimates of volume change for the entire north end over several decades, and will be used to quantify sediment budgets and chronological changes in those budgets. With a combination of historic datasets, remotely sensed data, and smaller-scale ground-based monitoring and surveys, the park and its partners are better able to assess the success and progress of the North End Restoration Project to restore natural ecosystem processes and sediment supply to the island.

16.2.2.2 CACO Historical Topography Surveys

EAARL LiDAR data collected in 2005 will be used to quantify a century of change along CACO beaches when compared to baseline data collected from 1887 to 1889 by Henry L. Marindin, one of the first coastal surveyors of the U.S. Coast and Geodetic Survey, conducted a study of shoreline change of the outer Cape. In

Marindin's study, transects were surveyed from onshore points across the dunes, bluffs, beaches, and continuing offshore for 1–2 km. Two hundred and twenty-nine profiles were collected, spaced at approximately 300 m intervals. Marindin reported latitude and longitude for the origin of each profile, the azimuth of each line, and the distance and elevation of each observation point. In order to look at change over a century, scientists from CACO, in collaboration with scientists from the Provincetown Center for Coastal Studies, have begun to resurvey Marindin's profiles using RTK GPS receivers for navigation and measurement (Giese and Adams, 2007). In addition to resurveying the offshore portion of the profiles, a complete set of elevations for the onshore portion of all profiles will be extracted from EAARL LiDAR data collected in 2005 (Figure 16.4, three-dimensional model of Marindin profiles).

Like the historical DTM study mentioned above, CACO staff will use this information to gain a better understanding of the effects of coastal change, a high priority of Cape Cod management because of immediate threats to costly facilities and park recreational uses. Many of the park's coastal sites are highly valued community resources, such as public beach facilities. Maintenance of CACO's public facilities and infrastructure are increasingly affected by coastal change at a time when the pressures from users and the demand for coastal access and facilities magnify the value of these facilities, as demonstrated by the increase in park visitation over the last 20 years. Since coastal change is almost unpredictable and construction requires a long planning horizon, expense to the government can be significant and disruption of public services is likely to occur. Uncertainty and short-term unpredictability of change often lead to emergency actions with higher costs to the park and visitors.

Coastal parks with ocean beaches suffer annual losses to storms, and extensive repairs, cleanup, and repaving of parking lots are often required. In 1978, the Eastham

FIGURE 16.4 Topographic profiles of CACO, Massachusetts, collected between 1887 and 1889 by Henry L. Marindin, draped over a modern DTM extracted from LiDAR collected in 2005. (Roland Duhaime, NPS FTSC, University of Rhode Island; Imagery source: MassGIS 2001.)

Coast Guard Beach facility within CACO was inundated by a severe coastal storm and resulting overwash. As a result, the park lost the use of the parking lot and facilities. This led to the construction of a new shuttle parking lot estimated at about $750,000 in 1986, so that the public use area would remain accessible. In 1996 and 1998, two lighthouses that are important cultural resources in the park, the Nauset Light in Eastham, Massachusetts and the Highlands Lighthouse in North Truro, Massachusetts, were moved away from the receding shoreline at a cost of $1.5 million and $300,000, respectively.

The pace of coastal erosion, and the cost of maintaining landscapes and facilities under siege by coastal change, requires a deeper understanding of, and ability to predict, how coastal landforms will change. Creating predictive models or improving capability to anticipate changes should be an integral part of planning for new, cost-efficient, sustainable solutions. Technologies such as GPS, geographic information system (GIS), and LiDAR are invaluable tools in providing new data and analyses to describe the status and evolving morphology of coastal systems. When these new technologies and data are coupled with historical data to analyze trends, models can be developed to describe coastal systems for future park facilities planning and management.

16.2.3 Monitoring Park Resources Using High Spatial Resolution Remote Sensing Data

16.2.3.1 Monitoring Terrestrial and Submerged Aquatic Vegetation on FIIS Using High Spatial Resolution Remote Sensing Data

As part of the NPS Inventory Program, vegetation maps based on aerial photography and ground surveys are being developed for more than 270 I&M Program parks including the eight NCBN parks. These maps have already proved to be extremely useful for park managers and scientists. When developing the NCBN monitoring program, park staff consistently voiced the need to have updated vegetation maps to ensure that monitoring efforts captured the dynamic nature of the vegetation communities in these coastal and barrier island settings. Mapping park-wide vegetation via the existing methodologies is costly, time consuming, and labor intensive, making it impractical for repeated measurement on a regular basis. Because of this, exploration of new data sources and mapping methodologies that could efficiently update the existing vegetation maps quickly and accurately was necessary. In addition, submerged aquatic vegetation (SAV) such as seagrass species and beds were not mapped as part of the original NPS vegetation mapping project. SAV is known to grow within the marine administrative boundaries of CACO, FIIS, and ASIS. Therefore, the ability to map this component was tested during an effort at FIIS.

In collaboration with remote sensing specialists from the University of Rhode Island, the NCBN conducted a pilot vegetation community and submerged aquatic habitat mapping project using high spatial resolution QuickBird-2 satellite imagery (Wang et al., 2007). The maps were created by using both a stratified classification and an unsupervised classification for mapping terrestrial vegetation communities that were first identified during the initial aerial photograph-based vegetation mapping effort for each park.

FIIS was used as the 2008 pilot park for protocol development because completed map products were available from 2002, and analysis of vegetation community change could be conducted over a 7-year time span. The vegetation map that was completed for Fire Island in 2002 as part of the NPS Vegetation Mapping Program (Klopfer et al., 2002) has been used extensively for management and planning purposes, despite being obsolete due to the dynamic nature of the barrier island system, especially within the rapidly changing areas of the island. The 2002 map was based on 1997 hardcopy color infrared and true-color aerial photographs (1:1200 scale) that were digitally scanned (600 dpi) and georeferenced for heads-up digitizing; by the time the map was delivered, it represented vegetation conditions from 5 years prior to its completion. The final classification system included five broadly defined vegetation groups including salt marsh, dune grassland, dune shrubland, interdunal swale, and forest/shrubland, further separated into 39 subcategories (FIIS Map products are available at http://biology.usgs.gov/npsveg/fiis/index.html).

In order to update the 2002 Fire Island vegetation map in a cost-efficient manner, high spatial resolution satellite remote sensing data were chosen as the data of choice. Although aerial photographs have better spatial resolution than imagery from satellite sensors, the greater expense in both data acquisition and processing time make photographs much less timely and cost-effective for repeated measures as part of a long-term monitoring program. Satellite sensors, on the other hand, offer advantages over aerial photograph interpretation in terms of cost and labor for repeated vegetation mapping. QuickBird-2 satellite imagery was used for this project. The QuickBird-2 satellite data possesses 0.61 m spatial resolution for the panchromatic band and 2.5 m spatial resolution for the multispectral bands (visible to near-infrared). The capability of repeated data acquisition from this satellite could easily facilitate processing for long-term monitoring of vegetation change.

Between April and May 2004, high spatial resolution QuickBird-2 satellite imagery was collected and consisted of separate image scenes to cover the entire FIIS study area. Due to water clarity issues during April and May 2004, a follow-up scene for the SAV mapping was ordered for May 2005, a timeframe chosen to increase the chances of optimal water clarity conditions. Biomass of SAV along the Great South Bay side of Fire Island begins to increase dramatically during the early spring and peaks in mid-to-late June. However, the period of peak biomass in June coincides with the time when plankton blooms create poor water clarity conditions. For this reason, the timing of image acquisition for submerged mapping was early-to-late spring when plant biomass was rising and water clarity less of an issue. Chapter 9 of this book describes more details about this SAV mapping effort.

Through classification of the 2004 satellite images, and comparison with the original 2002 terrestrial vegetation map for Fire Island, change in vegetation types was identified for a 7-year period. The overall map accuracy of the 2004 classification was approximately 82%. The overall map accuracy for the SAV map was approximately 69%. The results of this pilot study are promising and will be used by the NCBN to develop a standardized protocol for monitoring vegetation change using high spatial resolution satellite imagery. The next step in development of this protocol will be to determine the temporal sampling frame that will be used for monitoring for each park. Barrier island parks will need updating on a more regular basis in comparison to inland parks. Change analysis metrics will be identified that will best assist park managers in determining the current and long-term trends in condition of park vegetation communities.

16.2.3.2 Mapping Salt Marsh Habitats in the GATE Using High Spatial Resolution Remote Sensing Data

Productivity of salt marshes is a primary indicator of ecosystem health (Leibowitz and Brown, 1990). Nationwide, emergent salt marshes declined by an estimated 14,450 acres in 10 years from 1986 to 1997 (Dahl, 2000). In November 2000, the New York Department of Environmental Conservation (NYDEC) released a study that showed a sustained trend of significant salt marsh loss in Jamaica Bay over the past 100 years. The Jamaica Bay Wildlife Refuge encompasses 2500 acres within the boroughs of Brooklyn and Queens in New York City. This refuge, the only one in the National Park System, provides a variety of habitats for more than 300 kinds of waterfowl and shorebirds. It is a critical stopover area along the Eastern Flyway migration route and is one of the best bird-watching locations in the Western Hemisphere (GATE, 2003). Interpretation of historical aerial photographs shows that 51% of salt marshes in the Bay in 1929 had been lost by 1999. Although the salt marshes have been protected since 1972 as part of GATE, a recent study shows that 38% of salt marshes in Jamaica Bay in 1974 had been lost by 2002 (Hartig et al., 2002). The salt marsh change in Jamaica Bay is similar to the trends found in other salt marshes elsewhere in the northeastern United States. Therefore, Jamaica Bay is ideal for developing a protocol for salt marsh mapping and change detection for monitoring future extents of salt marshes in the northeast United States.

Given that salt marsh monitoring requires repeatable and reliable updates of land-cover maps, it is necessary to explore new data and approaches, including high spatial resolution satellite imagery that could efficiently update the salt marsh maps. In this project, scientists at the University of Rhode Island developed a protocol using QuickBird-2 high spatial resolution satellite imagery as the primary data source. There were two goals of this project. The first was to test the reliability of satellite remote sensing data for long-term change detection of salt marshes. The second was to examine the utility of data generated with high spatial resolution satellite imagery in comparison with historical data generated from aerial photography. This study employed an unsupervised classification scheme to extract information about salt marsh coverage from QuickBird-2 images. Three categories of overall salt marsh composition in Jamaica Bay, including (1) *Spartina* area coverage in excess of 50%, (2) *Spartina* area coverage of 10–50%, and (3) areas consisting of mudflats, were identified for mapping purposes. For specific islands, a High Marsh category was added and could include a variety of different plants, such as *Spartina patens*, *Phragmites*, *Distichlis spicata*, and others. These classes allowed comparisons between the new data and historical salt marsh data. The final salt marsh map provides the updated status and coverage of the salt marshes in the Jamaica Bay area. Details about this study are described in Chapter 9 of this book.

16.3 CONCLUSIONS AND DISCUSSIONS

For inventorying park resources, remote sensing technologies have several main advantages over other traditional survey techniques, such as using a GPS. One major benefit is the ability to collect data over large areas in a short period of time, such as a 1-day LiDAR survey that can map the topography for an entire large park, whereas an ATV-mounted or hand-carried GPS survey would take a much longer time to accomplish. A related benefit is the ability to collect a snapshot of conditions, such as those following a particular storm or within a season, whereas the extended period of time required for on-the-ground surveys could instead capture a range of conditions created by several weather events or seasonal transitions during the months-long survey period. Remote sensing technologies can also help in mapping inaccessible coastal areas that lack access roads, submerged lands and water, or parks that have experienced a natural disaster and therefore have few available staff or infrastructure to support ground-based surveys.

The main drawbacks to remote sensing are that they often require more lead time to allow for logistical planning, and therefore may miss the optimal survey window; data are sometimes more costly to acquire and process; surveys often require external or additional funding sources because they can rarely be collected by existing park equipment and park staff; and grant applications could justify the funding needs only on the anticipation of possible needs rather than on a response to known needs.

Remote sensing is best used for acquiring data over large areas of parks, whether as a baseline inventory or to measure changes over time. Ground-based surveys are best used for acquiring data within small areas, particularly when surveying these areas repeatedly as part of a monitoring study. Ground-based survey has advantages when a high level of accuracy is required, such as centimeter-scale overwash elevation data or ground-truthing of older or remotely sensed datasets; when it is important to respond quickly to events, such as surveying topographic response immediately following a coastal storm; and when funding and logistics-planning for advance contracting of remote surveys is unavailable. Because conventional and remote sensing survey techniques both have advantages and drawbacks, the two approaches are best seen as complementary, with datasets providing information and context to enhance the value of each other. The best survey choice for any given event or project will depend on the balance of constraints within park staff expertise, time-sensitivity, funding availability, severity of the event or importance of the project, existing datasets, and the scales of both coverage area and acceptable error.

Whether collecting baseline information on park resources and conditions, working to identify long-term trends in park conditions, monitoring the status of park resources on a regular basis, identifying event-related changes, evaluating the success of projects, or responding to changes or proposals outside of park boundaries, park managers need to have reasonably accurate, reliable data in order to justify the best resource management decisions. With reliable scientific data, parks can respond responsibly and effectively to regional development proposals, can identify areas with the highest protection needs and focus efforts on those resources, can protect and restore ecosystems and resources, and can make the most efficient use of time and funding. Without credible scientific information, parks must take their best guesses at existing conditions, existing resources, and potential impacts of any actions taken at a regional or park level. NPS is charged to "preserve unimpaired the natural and cultural resources" of the National Park System, and must choose effective, efficient, accessible methods to accomplish its mission. Remote sensing capabilities demonstrate and offer a tremendous promise for effective and rapid collection of significant information for parks.

REFERENCES

Brock, J.C., Krabill, W.B., and Sallenger, A.H., 2004, Barrier Island morphodynamic classification based on LiDAR metrics for North Assateague Island, Maryland, *Journal of Coastal Research*, 20(2): 498–509.

Brock, J.C., Wright, C.W., Sallenger, A.H., Krabill, W.B., and Swift, R.N., 2002, Basis and methods of NASA airborne topographic mapper LiDAR surveys for coastal studies, *Journal of Coastal Research*, 18: 1–13.

Dahl, T.E., 2000, *Status and Trends of Wetlands in the Conterminous United States 1986 to 1997*, U.S. Department of the Interior, Fish and Wildlife Service, Washington, DC, 82pp.

De Stoppelaire, G.H., Gillespie, T.W., Brock, J.C., and Tobin, G.A., 2004, Use of remote sensing techniques to determine the effects of grazing on vegetation cover and dune elevation at Assateague Island National Seashore, *Impact of Horses Environmental Management*, 34: 642–649.

Fancy, S.G., Gross, J.E., and Carter, S.L., 2009, Monitoring the condition of natural resources in US National Parks, *Environmental Monitoring and Assessment*, 151: 161–174.

Gateway National Recreation Area (GATE), 2003, A Report on Jamaica Bay, February 2003, National Park Service, U.S. Department of the Interior.

Giese, G.S. and Adams, M., 2007, Cape Cod National Seashore Cape Cod Shoreline Change and Resource Protection, Unpublished Project Management Proposal, Cape Cod National Seashore.

Hartig, E.K., Gornitz, V., Kolker, A., Mushacke, F., and Fallon, D., 2002, Anthropogenic and climate-change impacts on salt marshes of Jamaica Bay, New York, *Wetlands*, 22: 71–89.

Klopfer, S., Olivero, A., Sneddon, L., and Lundgren, J., 2002, Final Report of the NPS Vegetation Mapping Project at Fire Island National Seashore, Unpublished Technical Report, Conservation Management Institute, GIS and Remote Sensing Division, College of Natural Resources, Virginia Tech, VA, CMI-GRS–02-03.

Kopp, B.S. and Neckles, H.A., 2008, A protocol for monitoring estuarine nutrient enrichment in coastal parks of the National Park Service Northeast Region, Natural Resource Report NPS/NCBN/NRR—2009/110. National Park Service, Fort Collins, Colorado.

Leibowitz, N.C. and Brown, M.T., 1990, Indicator strategy for wetlands. In: C.T. Hunsaker and D.E. Carpenter (Eds.). Ecological Indicators for the Environmental Monitoring and Assessment Program. US EPA, Office of Research and Development. EPA 600/3-90/060.

Nayegandhi, A., Brock, J.C., Wright, C.W., and O'Connell, M.J., 2006, Evaluating a small footprint waveform-resolving LiDAR over coastal vegetation communities, *Photogrammetric Engineering and Remote Sensing*, 72(12): 1407–1417.

Oakley, K.L., Thomas, L.P., and Fancy, S.G., 2003, Guidelines for long-term monitoring protocols, *Wildlife Society Bulletin*, 31: 1000–1002.

Stevens, S., Milstead, B., Albert, M., and Entsminger, G., 2005, Northeast Coastal and Barrier Network Vital Signs Monitoring Plan, Technical Report NPS/NER/NRTR–2005/025, National Park Service, Boston, MA.

Stevens, S.M. and Entsminger, G., 2004, Northeast Coastal and Barrier Network Information Management Plan, Unpublished Technical Report, National Park Service, Northeast Coastal and Barrier Network.

U.S. Army Corps of Engineers, 1998, Ocean City, Maryland, and Vicinity Water Resources Study Final Integrated Feasibility Report and Environmental Impact Statement, Appendix D, Restoration of Assateague Island, Baltimore, MD.

Wang, Y., Traber, M., Milstead, B., and Stevens, S., 2007, Terrestrial and submerged aquatic vegetation mapping in Fire Island National Seashore using high spatial resolution remote sensing data, *Marine Geodesy*, 30: 77–95.

Section V

Effects of Land-Use/Land-Cover Change in Coastal Areas

17 Coastal Area Land-Cover Change Analysis for Connecticut

James D. Hurd, Daniel L. Civco, Emily H. Wilson, and Chester L. Arnold

CONTENTS

17.1 INTRODUCTION

Concern about the economic, environmental, and cultural toll of urbanization continues to grow, particularly in our rapidly developing coastal areas [1,2]. Of the estimated 3,502,309 (2007 U.S. Census estimate) people living in the state of Connecticut, almost 40% reside in the 36 coastal towns of the state. These towns encompass about 897 square miles, or only about 18% of the state's total land area, but offer some unique anthropogenic and natural landscapes. For example, Connecticut's two largest populated cities, Bridgeport and New Haven, are located within this region. Once thriving industrial centers, these cities represent some of the most intensively developed areas of the state. Several towns along the western coast comprise the "Gold Coast" of Connecticut, wealthy suburbs of New York City, which provide their own unique form of housing, commercial development, and open space patterns. Also, the two largest casinos in the United States, Foxwoods and Mohegan Sun, in addition to Pfizer Pharmaceutical's new Global Research and Development headquarters are located in the eastern region, all recently contributing to significant housing growth and related impacts on the coastal landscape of Connecticut.

In juxtaposition, the coastal region of Connecticut also contains numerous ecologically important natural areas. These include regions like the lower Connecticut River, home to internationally recognized tidal wetland communities; the Cockaponsett Triangle, a largely undeveloped forested region near the central Connecticut coast; and numerous state parks, state forests, and tidal wetlands that serve as protected areas and habitat. Additionally, the entire Connecticut coastal upland borders and drains into Long Island Sound. The Sound has been designated as an estuary of national significance* by the U.S. Environmental Protection Agency (EPA), and serves as feeding, breeding, and nursery areas for many aquatic and migratory bird species.

The coastal area of Connecticut is diverse, but is under significant pressure from both anthropogenic and natural forces. These include continued economic and development pressures, expansion of invasive species, loss of important and vital habitat, and potential impacts caused by climate change and severe weather events, to name a few. Land use is an underlying factor that can be assessed and managed with proper information provided to local communities and managers.

Land-cover data provide a critical source of spatial information that enables managers to assess conditions at a given point in time. This information can be used to better understand the cumulative effects of development, including impacts of land cover on water quality and indicators that link land-cover change with ecosystem health. In addition, land cover from multiple years can be used to document land-cover change, providing the ability for managers to analyze the potential impacts of these changes on the coastal ecosystem.

This chapter describes the development of the Connecticut's Changing Landscape (CCL) land-cover dataset and application in two projects to assess the land-cover condition of coastal Connecticut: the Coastal Area Land Cover Analysis Project (CALCAP) and the Coastal Riparian Buffers Analysis Project.

17.2 CCL LAND COVER

The CCL dataset, developed by the Center for Landuse Education and Research (CLEAR) at the University of Connecticut, comprises four dates of consistently created land cover for the entire state of Connecticut, plus intersecting watersheds and nine towns in south central Massachusetts that comprise the Quinebaug and Shetucket Rivers Valley National Heritage Corridor (along with 26 towns in eastern Connecticut) (Figure 17.1). The primary purpose for the development of the CCL land cover was to serve as source data for various landscape characterization models, such as forest fragmentation and urban growth modeling. It has, however, proven extremely valuable as a

* Information about the U.S. Environmental Protection Agency's National Estuary Program can be found at http://www.epa.gov/nep/

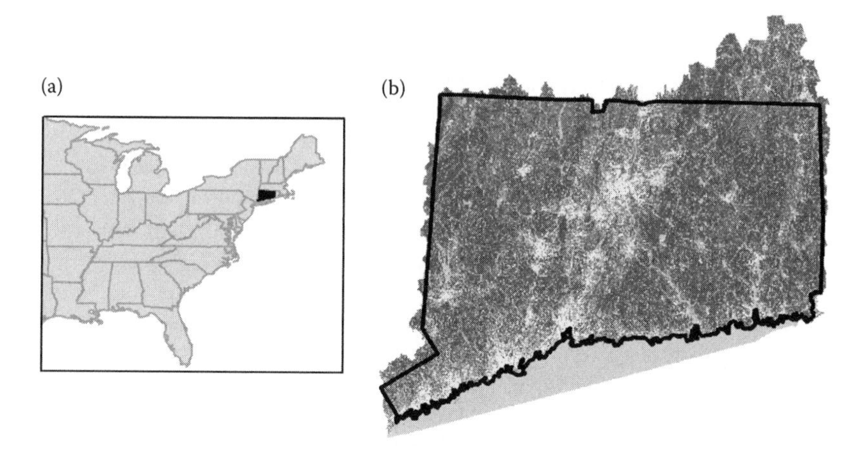

FIGURE 17.1 (a) Location of Connecticut in the northeastern United States. (b) Grays showing the extent of the CCL land cover as it relates to the state boundary of Connecticut (black outline).

means to visually depict general landscape change in Connecticut. Overall, the CCL study has become a major resource for researchers, state agencies, regional and local planners, nonprofit organizations, the public and the press, with over 600 different organizations having downloaded the actual data from the CLEAR web page.

17.2.1 Data

Land cover for the state of Connecticut was derived based on the classification of Landsat Thematic Mapper (TM) or Enhanced Thematic Mapper (ETM) satellite imagery covering four temporal periods from 1985, 1990, 1995, and 2002. The primary Landsat scene, covering most of the study area of Connecticut, came from the World Referencing System (WRS) Path 13/Row 31. The secondary scene, covering the extreme southeastern portion of the state, was acquired from a portion of the Landsat scene from WRS Path 12/Row 31 (Figure 17.2). All data were reprojected into Connecticut state plane NAD83 feet using a nearest neighbor resampling

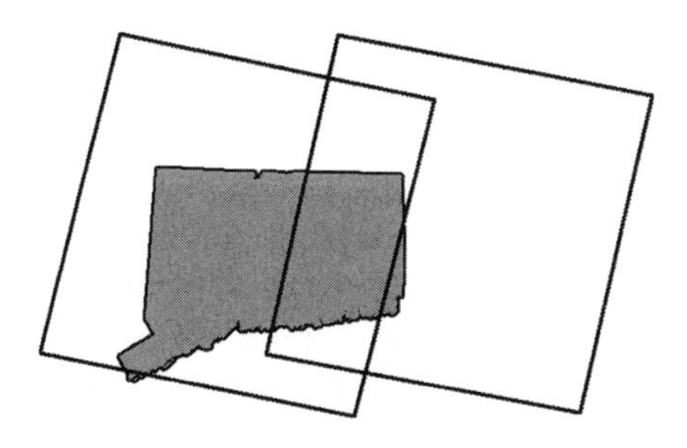

FIGURE 17.2 Primary scene (left, Path 13/Row 31) and secondary scene (right, Path 12/Row 31) boundaries used in land-cover classification.

TABLE 17.1
Landsat Dates Used for the Classification of Each Time Period for the Primary and Secondary Landsat Scenes

Classification Year	Primary Scene WRS Path 13/Row 31	Secondary Scene WRS Path 12/Row 31
1985	April 26, 1985	August 9, 1985
1990	August 30, 1990	September 8, 1990
1995	August 28, 1995	September 19, 1994
2002	September 8, 2002	July 31, 2002

method. The spatial resolution of the Landsat pixels was rounded up to 100 ft. Additionally, all the Landsat images were aligned to eliminate the identification of changed areas resulting from misalignment issues.

The initial classification was performed on the April 26, 1985 Landsat TM image (primary scene). The secondary Landsat TM image was from August 9, 1985. Cloud and cloud shadow regions covering some of the northwest portion of the primary image were extracted and substituted with a May 4, 1988 Landsat TM scene. Most of this area consisted of forest cover and was not impacted by potential change in land cover between these two time periods. These Landsat image datasets served as the basis for the 1985 classification. Classifications for subsequent periods utilized the image dates identified in Table 17.1.

17.2.2 Initial Classification Process

A variety of image processing techniques turned the raw Landsat TM imagery into quantifiable land-cover datasets. Three main steps are described, including (1) linear road features, (2) unsupervised classification and cluster identification for nonroad areas, and (3) cleanup.

Because one of the original uses of this data was forest fragmentation modeling, it was important to identify the road network as a fragmenting feature. The first step in classifying linear road features in a pixel-based classification was to identify a corridor that contained the road network and any adjacent developed areas. All improved (paved) roads were selected from the vector road coverage and rasterized to the pixel size of the source imagery (100 ft). This layer was buffered five pixels (500 ft) on either side to account for areas of misalignment or errors between the road coverage and Landsat TM image data. All Landsat image pixels contained within the 5 pixel buffer area were extracted for analysis. The intent was to create an image data layer from which the classification of roads could be focused by minimizing nonroad pixels.

ERDAS Imagine SubPixel Classifier™ (SPC) engineered by Applied Analysis Inc. (AAI) was used to classify road corridor pixels. The SPC is a supervised classifier that enables the detection of materials of interest (MOIs) as whole or fractional pixel composition, with a minimum detectable threshold of 20% and in increments of 10% (i.e., 20–30%, 30–40%,…,90–100%). Because of tonal variations in the built landscape, MOIs representing different brightness classes of road and paved surfaces (i.e., dark, medium, and bright surfaces) were selected to be mapped. Any pixel identified by the SPC, regardless of its percent composition, was considered a developed pixel.

Final results of the SPC did not fully extract the road network. To enhance further the results of the SPC, knowledge-based (KB) classification was employed. Those pixels not identified as developed through the SPC technique were extracted for further evaluation. Landsat TM bands 4 (near-infrared) and 5 (middle-infrared) provided the most contrast between developed pixels and other pixels. In the ERDAS Imagine Knowledge Engineer, a rule was created that used value ranges (129–143 for band 4; 128–193 for band 5) to identify developed pixels that met the criteria for both Landsat TM bands. In addition, a pixel also had to be contained within the rasterized road layer. The result of this procedure was the identification of additional developed pixels not identified during the subpixel classification process.

Unfortunately, the SPC and KB classification processes were still not able to extract the full extent of the road network. To correct for this problem, the rasterized road layer was ultimately embedded into the final classification and onscreen digitizing was conducted to remove areas of misalignment or error. While this process did not extract the full extent of the road network, it was able to identify a significant number of developed pixels within the buffered road area, which proved beneficial in determining the true road alignment. A similar process was performed on the August 9, 1985 Landsat TM image covering the southeast portion of the analysis area.

The remaining image pixels not assessed above in the road identification were then classified. To begin, those pixels identified as developed in the previous step were eliminated from the full Landsat TM image. An area of approximately 180 square miles along the central coast of Connecticut was then subset from the overall analysis area for use in deriving classification signature statistics. This area was selected because of its largely heterogeneous landscape patterns, which contained a significant amount of those categories identified in the classification scheme (Table 17.2). Unsupervised Iterative Self-Organizing Data Analysis Technique (ISODATA) classification was performed generating 100 signature clusters. These clusters were identified and labeled into the appropriate land-cover category.

Maximum likelihood classification was applied to the entire analysis area using selected signatures derived through the ISODATA process. Classification was performed one class at a time specifying a distance image as the output. The distance file produces an image where pixel values represent the spectral distance from the

TABLE 17.2
Classification Scheme Used for the CCL Land Cover

1. Developed	5. Coniferous forest	9. Tidal wetland
2. Turf and grass	6. Water	10. Barren land
3. Other grasses and agriculture	7. Nonforested wetland	11. Utility corridor
4. Deciduous forest	8. Forested wetland	

class signature. The lower the value, the more similar a pixel is to a specific class signature. This procedure was repeated for each class. Visual examination of the distance image with the Landsat TM image resulted in the identification of thresholds that were used with the Knowledge Engineer to derive a land-cover image.

Pixels remaining as unclassified were further extracted from the Landsat TM image. ISODATA classification was performed on these remaining pixels. The clusters were identified and labeled to the appropriate category. The resulting classifications were then merged to create a single classified image with all pixels belonging to a single category.

Next, several steps were taken to cleanup the classification. First, a digital elevation model was used to eliminate areas misclassified as wetlands due predominately to steep northwest-facing slopes. Using the Knowledge Engineer, any pixel identified as nonforested or forested wetlands that fell on a slope of 12° or more was reassigned to deciduous forest. Several majority filters were used to eliminate specific isolated pixels resulting in a more uniform classification. Lastly, detailed "heads-up" digitizing was used to remove any remaining apparent errors and to also include utility corridors, which can be considered as significant fragmenting features to the forest landscape. Utility corridors were digitized out of the deciduous and coniferous forest classes only. The overall intent in developing a land-cover image using a hierarchical approach was to continually eliminate those pixels that were easily classified and narrow down those pixels that were more difficult. Remaining errors would potentially be cleaned during the "heads-up" digitizing phase of the classification to produce a highly accurate 1985 land-cover map that would be used to derive subsequent land-cover maps for 1990, 1995, and 2002.

17.2.3 Subsequent Classification Process

Cross-correlation analysis (CCA) was chosen as the method for determining subsequent land cover because it overcomes many of the limitations of conventional change detection methods and is able to produce a consistent set of land cover. Cross-correlation works by using the land-cover categories identified in the base land-cover image to derive an "expected" class average spectral response [3,4]. This information is used to derive a Z-statistic for each pixel falling within a given land-cover type. The Z-statistic describes how close a pixel's response is to the "expected" spectral response of its corresponding class value in the land-cover image. Pixels that have undergone change between the date of the land-cover image and the multispectral image will produce high Z-statistic values, whereas pixels that have not changed will produce low Z-statistic values. The benefit of this technique is that it eliminates the problems associated with radiometric and phenological differences that are so readily experienced when performing change detection.

For the CCL, CCA was applied to five groups of land-cover categories. These groups include (1) water; (2) deciduous, coniferous, and forested wetlands; (3) turf and grass and agriculture; (4) barren; and (5) nonforested and tidal wetlands. Using the 1985 land cover, pixels belonging to each group were extracted from an August 30, 1990 Landsat TM image (i.e., for the deciduous, coniferous, and forested wetlands group, pixels classified as these in 1985 were extracted from the August 30,

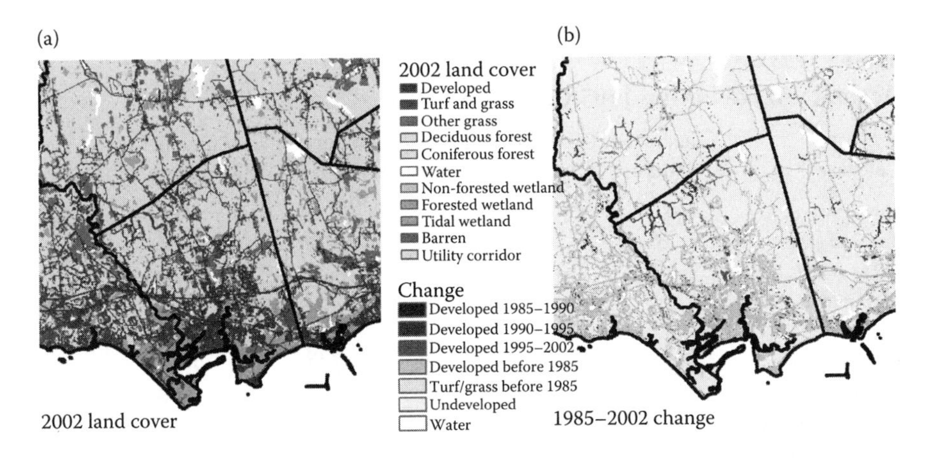

FIGURE 17.3 A black and white example of the 2002 land cover (a) and the 1985–2002 change map (b). Normally, these data are shown in color.

1990 TM image). The CCA procedure was applied to the extracted pixels and the results were visually examined with the recent image data to determine the threshold between probable change pixels and nonchanged pixels. Those pixels identified as changed were extracted from the August 30, 1990 TM image. ISODATA classification was performed to identify the category that each pixel now belonged. These steps were repeated for each class group. Once completed, each group of classifications was combined into a single image and edited to remove apparent errors. These pixels were then fused with the 1985 land cover to produce an updated 1990 land-cover image. This updated 1990 land cover was then applied to the 1995 Landsat TM images using the CCA method and then the resulting 1995 land cover applied to the 2002 Landsat ETM image. An example of the land cover and land-cover change maps are provided in Figure 17.3.

17.2.4 Accuracy Assessment

Accuracy was determined by selecting an independent stratified random sample of pixels for each classification. The number of sample pixels varied per date and is provided in Tables 17.3 through 17.6. Each of these pixels was examined using the available source Landsat imagery in conjunction with available orthophotograph quarter quadrangles for 1995 and 2004 for the entire state of Connecticut to identify the true land cover. Accuracy analysis provides statistics that describe overall classification accuracy and κ coefficient of agreement (Tables 17.3 through 17.6). The overall classification accuracy is a percentage expressed as the number of correctly classified sample pixels over the total number of sample pixels. This percentage indicates how accurate the classification is with respect to the reference data [5]. The κ coefficient of agreement is a measure of the actual agreement minus chance agreement. This value provides a better measure of how the classification performs compared to the reference data because it takes into account those pixels committed to other categories, not just those pixels correctly classified [6,7].

TABLE 17.3
Accuracy Table for the 1985 CCL Land Cover

Category	Reference Totals	Classified Totals	Number Correct	Producer's Accuracy (%)	Consumer's Accuracy (%)	κ Statistic
Developed	83	70	63	75.90	90.00	0.8825
Turf and grass	29	38	27	93.10	71.05	0.6945
Other grasses and agriculture	42	48	36	85.71	75.00	0.7294
Deciduous forest	117	130	103	88.03	79.23	0.7364
Coniferous forest	51	46	42	82.35	91.30	0.9042
Water	51	51	49	96.08	96.08	0.9568
Nonforested wetland	33	30	29	87.88	96.67	0.9645
Forested wetland	42	34	33	78.57	97.06	0.9682
Tidal wetland	39	37	36	92.31	97.30	0.9709
Barren land	32	34	25	78.13	73.53	0.7190
Utility corridor	32	34	29	90.63	85.29	0.8439

Notes: Overall accuracy = 85.51%, overall κ statistic = 0.8400, and number of sample points = 552

TABLE 17.4
Accuracy Table for the 1990 CCL Land Cover

Category	Reference Totals	Classified Totals	Number Correct	Producer's Accuracy (%)	Consumer's Accuracy (%)	κ Statistic
Developed	104	84	78	75.00	92.86	0.9124
Turf and grass	26	35	25	96.15	71.43	0.7004
Other grasses and agriculture	47	51	42	89.36	82.35	0.8074
Deciduous forest	122	119	104	85.25	87.39	0.8390
Coniferous forest	42	46	34	80.95	73.91	0.7181
Water	51	52	49	96.08	94.23	0.9365
Nonforested wetland	37	33	29	78.38	87.88	0.8702
Forested wetland	46	40	36	78.26	90.00	0.8911
Tidal wetland	38	35	33	86.84	94.29	0.9387
Barren land	19	35	18	94.74	51.43	0.4973
Utility corridor	30	32	29	96.67	90.63	0.9010

Notes: Overall accuracy = 84.88%, overall κ statistic = 0.8286, and number of sample points = 562.

TABLE 17.5
Accuracy Table for the 1995 CCL Land Cover

Category	Reference Totals	Classified Totals	Number Correct	Producer's Accuracy (%)	Consumer's Accuracy (%)	κ Statistic
Developed	90	71	66	73.33	92.96	0.9173
Turf and grass	26	34	23	88.46	67.65	0.6620
Other grasses and agriculture	48	51	42	87.50	82.35	0.8084
Deciduous forest	175	184	157	89.71	85.33	0.7940
Coniferous forest	47	44	41	87.23	93.18	0.9261
Water	60	59	59	98.33	100.0	1.000
Nonforested wetland	35	32	31	88.57	96.88	0.9668
Forested wetland	33	32	28	84.85	87.50	0.8678
Tidal wetland	32	33	32	100.0	96.97	0.9680
Barren land	27	37	25	92.59	67.57	0.6606
Utility corridor	33	31	30	90.91	96.77	0.9659

Notes: Overall accuracy = 87.83%, overall κ statistic = 0.8580, and number of sample points = 608.

TABLE 17.6
Accuracy Table for the 2002 CCL Land Cover

Category	Reference Totals	Classified Totals	Number Correct	Producer's Accuracy (%)	Consumer's Accuracy (%)	κ Statistic
Developed	92	58	55	59.78	94.83	0.9366
Turf and grass	22	36	20	90.91	55.56	0.5351
Other grasses and agriculture	45	48	33	73.33	68.75	0.6566
Deciduous forest	105	104	87	82.86	83.65	0.7931
Coniferous forest	40	45	34	85.00	75.56	0.7343
Water	53	52	51	96.23	98.08	0.9785
Nonforested wetland	31	30	28	90.32	93.33	0.9289
Forested wetland	39	35	32	82.05	91.43	0.9070
Tidal wetland	27	30	27	100.0	90.00	0.8943
Barren land	19	32	18	94.74	56.25	0.5452
Utility corridor	27	30	27	100.0	90.00	0.8943

Notes: Overall accuracy = 82.40%, overall κ statistic = 0.8015, and number of sample points = 500.

17.3 COASTAL AREA LAND COVER ANALYSIS PROJECT

The release of the statewide CCL land-cover data sparked a lot of interest. As a result, the Office of Long Island Sound Programs at the Connecticut Department of Environmental Protection contacted CLEAR to do the CALCAP. The CALCAP used the CCL land cover extensively to examine and provide an improved understanding of how and where development within Connecticut's coastal area and lower Connecticut River towns may be affecting significant ecological and coastal recreation areas. Examples of such areas range from the ecologically important coves of the lower Connecticut River that contain rare species habitat, to major coastal recreation destinations. The information provided by this project was used, or is being used, to develop new and updated coastal area protection and management plans such as Connecticut's Coastal and Estuarine Land Conservation Program (CELCP plan), Comprehensive Wildlife Management Plan, the Statewide Comprehensive Outdoor Recreation Plan (SCORP), municipal plans of conservation and development, state parks/wildlife management area plans, private land conservation organization protection plans and Long Island Sound Stewardship Initiative plans. The objectives of the study that utilized the CCL land-cover data were as follows:

1. Quantify 2002 land cover and 1985–2002 land-cover change within the Connecticut coastal area using the CCL land-cover data.
2. For each local watershed intersecting the coastal area, estimate the percent of impervious surface and quantify the amount and percent of development.

The coastal area for the CALCAP was defined as the 36 towns that constitute the Connecticut coastal area and an additional six towns that border the lower Connecticut River (Figure 17.4). For the first objective, the 2002 land cover and 1985–2002

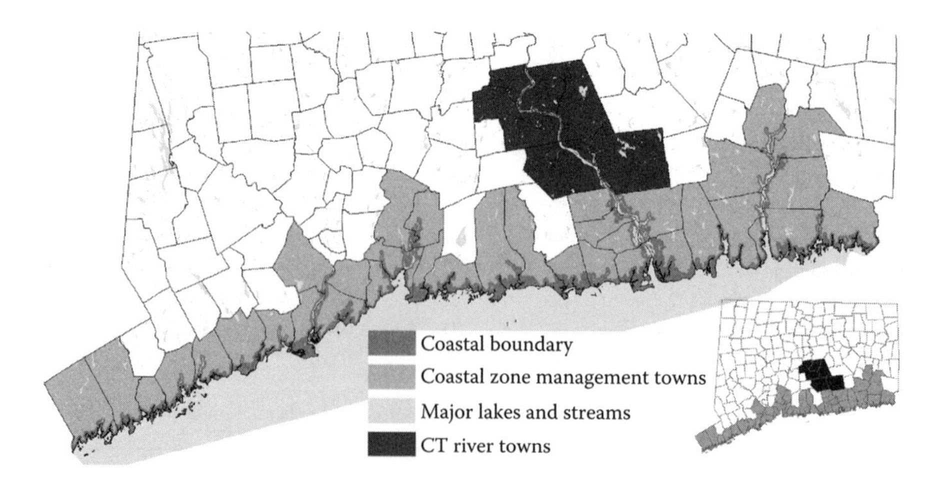

FIGURE 17.4 Study area for the CALCAP consisted of the 36 coastal zone management towns and an additional six towns along the Connecticut River.

land-cover change maps for the state of Connecticut were used as the input data. The raster land-cover data were converted to vector format, maintaining the grid structure as boundaries to the vector polygons for proper alignment and analysis of the various datasets. Polygon layers representing the coastal area towns and coastal boundary area were used to summarize the land cover and land-cover change data for each of the areas. Additionally, land cover and land-cover change statistics were summarized based on each coastal municipality.

The findings from this basic analysis show that the degree of development is much larger (51% total development) for the immediate coastal boundary than for the coastal towns (34% total development). Both these areas exceed the state average of 23% developed, highlighting the heightened degree of development along Connecticut's immediate coast and the potential impact our development patterns are having on the coastal resources of the state (Figure 17.5). Furthermore, 2.6% of the coastal land area was changed to development during the 17-year time period, which is slightly higher than the statewide percentage of 2.4%.

The second objective examined the land-cover characteristics and impervious surfaces within the local watershed delineations that intersect the coastal towns. These basins represent the smallest watershed units available from the Connecticut Department of Environmental Protection. For each watershed, the percent of developed land from the 2002 land cover was calculated as well as an estimate of the area of impervious surfaces. Impervious surfaces such as asphalt, concrete, and roofing materials prevent infiltration of rainfall into the soil, disrupting the water cycle and affecting both the quantity and quality of our water resources. Research has shown the amount of impervious surface in a watershed to be a reliable indicator of the impacts of development on water resources [8–12].

To determine estimates of impervious surfaces, the Impervious Surface Analysis Tool (ISAT) was used. This GIS-based tool uses land-cover information and related impervious coefficients to derive estimates of imperviousness for a given analysis unit. These coefficients (Table 17.7) represent the average impervious surface area in a typical pixel of each land-cover category. ISAT was used on each of the four dates

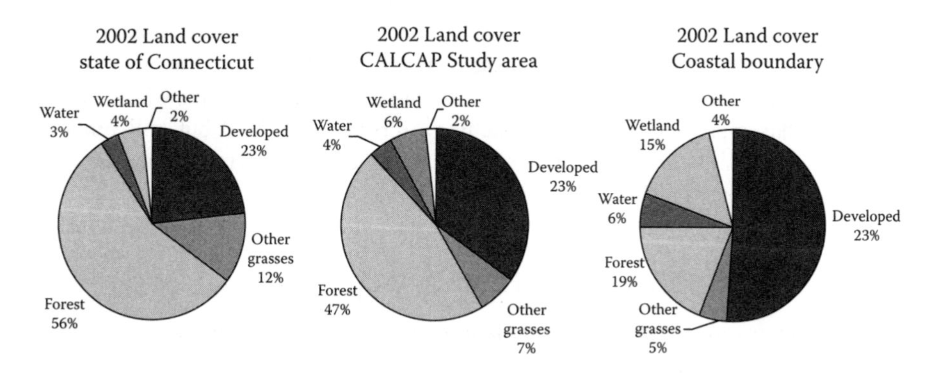

FIGURE 17.5 Summary of the land-cover breakdown comparing the state of Connecticut (left), the coastal study area (center), and the coastal boundary (right). The developed land-cover class shown in the pies is a combination of the developed and turf/grass classes.

TABLE 17.7
ISAT Coefficients for CCL Land Cover

	Population Density		
Land-Cover Class	**High**	**Medium**	**Low**
Developed	42.26	26.07	22.67
Turf and grass	12.87	12.09	8.58
Other grasses and agriculture	11.56	6.25	2.97
Deciduous forest	5.08	2.91	1.37
Coniferous forest	14.98	3.17	1
Water	4.25	0.77	0.46
Nonforested wetland	5.98	2.29	0.48
Forested wetland	1.2	1.03	0.46
Tidal wetland	1.02	1.63	3.11
Barren land	19.92	12.29	8.18
Utility corridor	5.52	0.8	1.2

of land cover (1985, 1990, 1995, and 2002) and therefore revealed changes in impervious area over time. From this information, watersheds crossing critical thresholds were identified. Research on imperviousness and water quality has determined that when a watershed exceeds 10% imperviousness, stream health typically begins to deteriorate [8,12]. When a basin exceeds 25% imperviousness, stream health typically becomes severely compromised, requiring extensive restoration.

The findings from this portion of the analysis show that the western half of coastal Connecticut is more developed than the eastern region and therefore contains more impervious surface (Figure 17.6a and b). In terms of increases in imperviousness over the 17-year period (1985–2002), the greatest increases are found in the extreme eastern region of the coastal area and also along the central part of the coastal area (Figure 17.6c). In contrast, the area along the lower Connecticut River remained relatively unchanged. Of the 1763 basins analyzed, 80 basins crossed from under 10% imperviousness to over 10% imperviousness and 18 basins crossed from under 25% imperviousness to over 25% imperviousness (Figure 17.6d). Perhaps the most significant result from the impervious estimates is the identification of basins where imperviousness has increased and is approaching a critical threshold but still remains below. These basins should receive extra attention before they cross either the 10% or 25% impervious threshold.

17.4 COASTAL RIPARIAN BUFFERS ANALYSIS PROJECT

This project examined land cover and land-cover change within the watersheds and riparian corridors of coastal Connecticut based on the CCL land-cover information. Riparian, or streamside corridors, are known to be environmentally important areas critical to stream stability, pollutant removal, and both aquatic and terrestrial wildlife habitat. These areas are also sometimes referred to as “buffer” areas. The

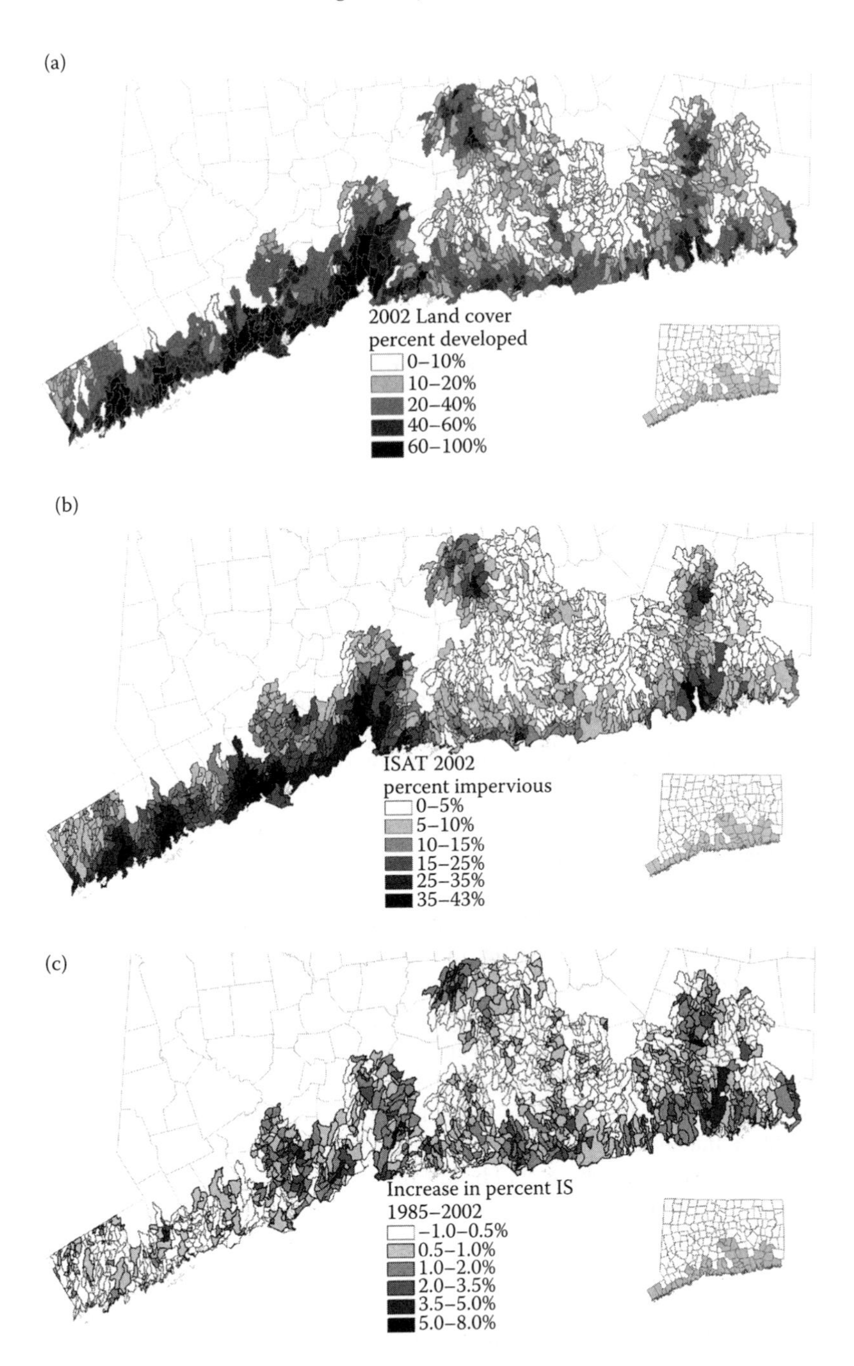

FIGURE 17.6 Watershed statistics generated as part of the CALCAP. (a) According to the 2002 land cover, percent of each basin in the developed category. (b) Imperviousness by basin as a result of the ISAT. (c) Percent increase in imperviousness (measured by the ISAT tool) between 1985 and 2002. (d) Those basins crossing critical impervious thresholds between 1985 and 2002.

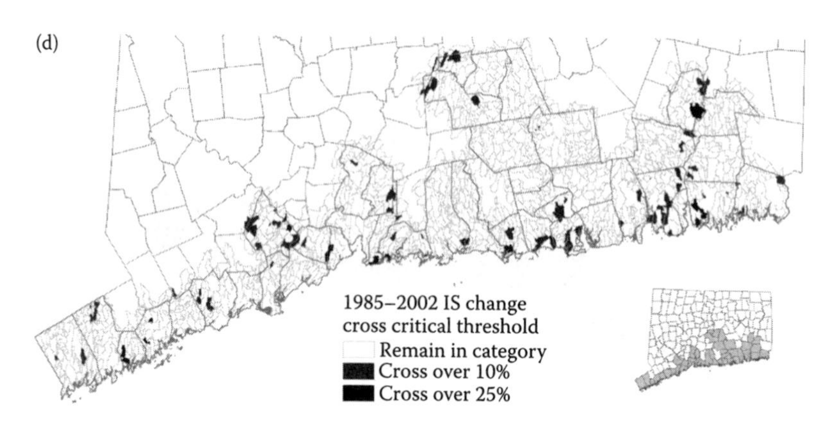

FIGURE 17.6 Continued.

purpose of this study, funded by the Long Island Sound Study National Estuary Program, was to provide local officials, researchers, landowners, and other interested parties an overview of the status of riparian corridors draining to the Long Island Sound, and to provide understanding of land-cover trends within these areas. The objectives of this project that utilized the CCL land-cover data were to

1. Provide an overall picture of the state of riparian areas in Long Island Sound's immediate drainage basin.
2. Provide an assessment of the condition of the riparian areas at three buffer widths: 100, 200, and 300 ft.

The CCL land cover from 2002 and land-cover change from 1985 to 2002 were analyzed to derive a variety of statistics for both entire basins and buffer zones within basins that cover the coastal region of Connecticut. Basins were characterized for land cover and land-cover change at the subregional level of organization, resulting in 167 study units. In addition, within these basins data were compiled for riparian corridors of three different widths: 100, 200, and 300 ft (to either side of the stream).

To provide an overall picture of the state of riparian areas in the Long Island Sound's immediate drainage basin, the first step of the analysis used 2002 CCL land cover to characterize the subregional basins in the study area. The result is a tabulation containing the area of each land-cover class for each subregional basin. Next, the 1985 and 2002 land-cover datasets were used to analyze land-cover change. For every basin, the area in acres was calculated for each land-cover change class. These include developed before 1985, turf and grass before 1985, water, undeveloped changed to developed between 1985 and 2002, and undeveloped changed to turf and grass between 1985 and 2002. Turf and grass are included on the change maps as it is usually associated with developed areas. The percent increase in developed land (change to developed between 1985 and 2002/area that was developed before 1985) was also calculated. Additionally, statistics for forest cover of the subregional basins were developed from the overall land-cover statistics. Forest cover is becoming recognized in the literature as a land-cover statistic that is a good indicator of watershed health [13].

The second objective was to assess the condition of land cover within riparian buffer areas of three sizes: 100, 200, and 300 ft. Before creating buffer zones, the data layer "to be buffered" needed to be created. This data layer had to be a seamless combination of streams, water body shorelines, and tidal wetland edges. Various GIS layers were necessary including stream networks, water body polygons, and tidal wetland polygons. When combined, inconsistencies were found that resulted in missing areas, or holes, that would have been incorrectly treated as islands in the analysis and, when buffered, would be incorrectly included in the buffer zone. All cases of these holes were removed via "heads-up" digitizing, resulting in a data layer that was a continuous line of streams, water body shores, and tidal wetland edges. Buffering this continuous line created a data layer consisting of seamless buffer zones at 100, 200, and 300 ft from the "to be buffered" line (Figure 17.7).

The basin polygons and 100, 200, and 300 ft buffer zones were used to create a number of statistics for each basin. Statistics include total area (acres) of the basin, total water area (acres) for the basin, and for each of the 100, 200, and 300 ft buffer zones: acres outside buffers and water, buffer area (acres), percent of basin in buffer only, and percent of basin in buffer and water.

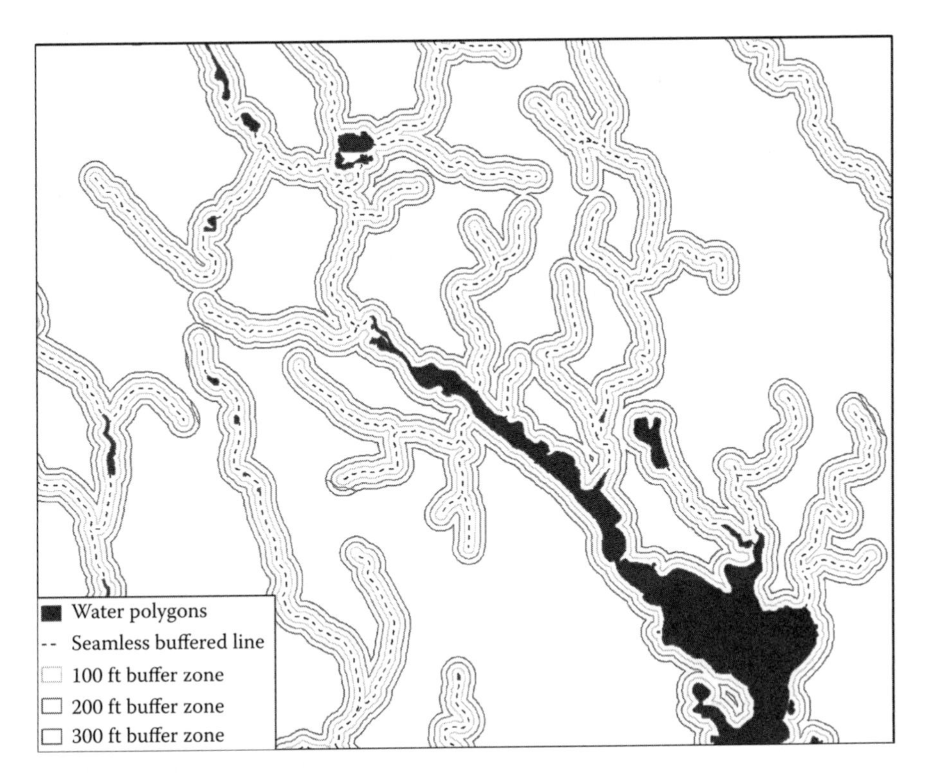

FIGURE 17.7 Example of buffer zones.

The land-cover data were clipped so that a GIS file consisted of land cover within the 100 ft buffer zone, land cover within the 200 ft buffer zone, and land cover within the 300 ft buffer zone. To simplify the results for the purposes of characterizing the riparian areas, the CCL land-cover data were condensed into three categories (other than water). This "simplified land cover" was as follows:

- *Natural vegetation*: deciduous forest, coniferous forest, forested wetlands, nonforested wetlands, and tidal wetlands classes. This class was seen as the most environmentally desirable condition of a riparian area.
- *Other vegetation*: turf and grass and "other grasses and agriculture" classes. This class was seen as of intermediate environmental value, since it was vegetated but likely associated with developed land, and possibly producing pollutant problems if the lawn or field was treated with chemicals and fertilizer.
- *Non-vegetation*: developed and barren classes.

The land cover and land-cover change data for the 100 and 300 ft buffer zones were summarized by basin to characterize the status of coastal riparian areas. The 200 ft buffer zone was excluded due simply to the unwieldy amounts of data and information. To better compare subregional basins, the investigators measured land-cover change not only in acres, but in "percentage of 1985 natural vegetation lost by 2002," a metric that normalized for differences in basin size. This simple metric allowed comparison of relative difference not only in land-cover change between basins, but between the 100 and 300 ft buffer zones. The analysis revealed the basins that had experienced the greatest loss of natural vegetative cover in the riparian zone (Figure 17.8).

In addition, all 167 subregional basins were analyzed using a combined basin-wide/buffer zone metric that was developed by Goetz et al. [14] for the Chesapeake Bay region. Goetz found that the best predictor of stream health, as determined by an intensive county field sampling program, was an index that combined basin-wide impervious cover and tree cover within the 100 ft buffer. This index was applied using data of percent impervious cover derived from the ISAT and percent of "natural vegetation" as described previously (Figure 17.9). The index was calculated using the following values, from the Goetz study (Table 17.8).

17.5 RESULTS AND DISCUSSIONS

The CCL land-cover data reinforce the reality that, in general, Connecticut is urbanizing, and the coastal region is already largely an urban one. Connecticut's coastal towns are about 50% more developed than the state average and the area within Connecticut's coastal boundary (defined by a line 1000 ft inland of coastal waters, tidal wetlands, or the inland extent of the 100-year frequency coastal flood, whichever if further inland) is more than twice as developed as the state average. Much of this development is pre-1985, and reflects the "urban sea" era of Long Island Sound during the first half of the twentieth century, when much of the coastline west of New Haven was used for commercial and industrial purposes.

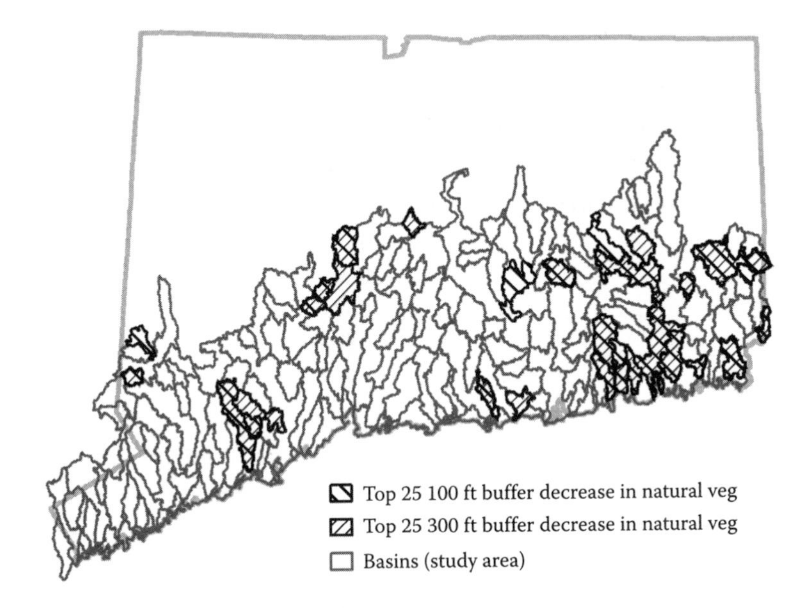

FIGURE 17.8 Riparian buffers analysis identified which basins had the greatest loss of natural vegetation in the 100 ft buffer zone and the 300 ft buffer zone.

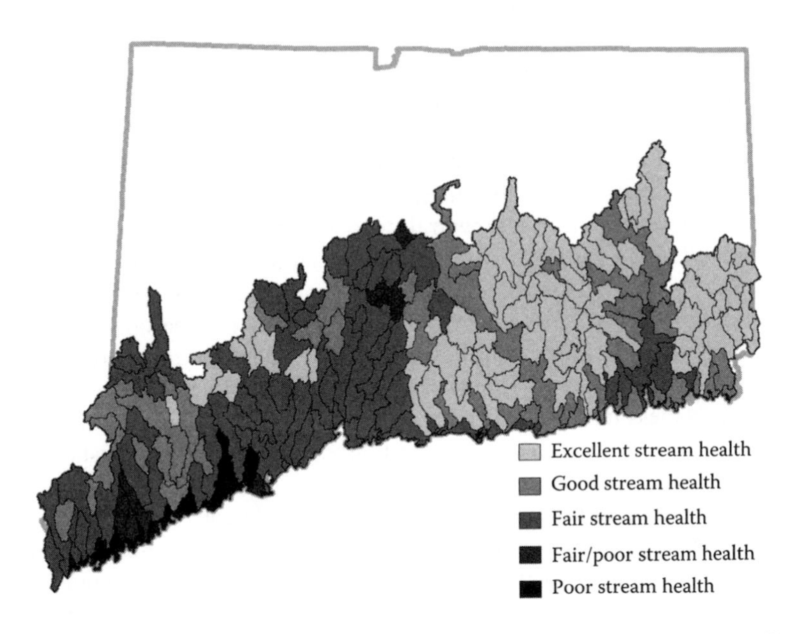

FIGURE 17.9 Results when applying basin impervious surface area and 100 ft buffer zone vegetative area to identify stream health following Goetz et al. [14].

TABLE 17.8
Ranges of Percent Imperviousness in the Basin and Percent of Natural Vegetation in the 100 ft Riparian Buffer Zone and How, Together, These Values Reveal Stream Health

Percent Impervious Surface	Percent Natural Vegetation	Stream "Health"
≤6	≥0.65	Excellent
≤10	≥0.60	Good
≤25	≥0.40	Fair
>25	<0.40	Poor
>10	<0.60	Fair/poor
		Other

Source: Goetz, S.J., et al., 2003, *Remote Sensing of Environment*, 88: 195–208.

The land-cover change data show where development pressure has occurred in recent decades. Although the absolute amount of new developed land is not overwhelming (2.6% of the total project area), the percent change in development identifies the eastern shoreline as a "hot spot" of development pressure. Coastal towns east of New Haven (including six additional Connecticut River towns within Connecticut's Coastal and Estuarine Land Conservation Program project area), experienced an average 15% increase in development between 1985 and 2002. This increase is mostly due to residential development.

The impervious cover modeling shows that while few changes have occurred along the highly developed western shoreline, coastal watersheds farther east are experiencing increases in impervious cover that the national research base indicates will initiate water quantity and quality impacts. The major increases in impervious cover naturally correlate with the increases in developed land, with the biggest increases in the towns from the central coast (Madison, CT) east to the Rhode Island border. During the 17-year study period, a large number of local watersheds crossed the 10% threshold that studies suggest signals the onset of stream degradation. There are very few local coastal watersheds within the coastal boundary that are still under the 10% impervious cover threshold, and all of these are located on the central coast (Guilford, CT) to the east. Fewer than 10 watersheds have impervious cover levels less than 5%, and most of these are located along the easternmost coast (Stonington, CT) with a few scattered throughout the central coast (Guilford, CT, and Old Lyme, CT).

Management implications suggest a distinctly different "East–West" coastal resource protection strategy built upon these landscape data and trends. Along the western shoreline, management must focus on remediation and restoration, seizing opportunities during redevelopment to enact strategies such as "low-impact development," urban stream corridor restoration, and connecting existing open space parcels to the extent possible. The land-cover maps of the greater New Haven area (mid-coast) highlight the challenges of working in this urban environment. East of New Haven, more opportunities exist for proactive strategies such as land acquisition, local land-use regulation protective of coastal resources, and most importantly, natural resource-based local land-use planning that maximizes protected natural lands and reduces the impacts of new development. The maps illustrate that such landscapes still exist and that where such opportunities still exist, they should be acted upon. Information of the type generated by this study informs state agencies concerned with managing the coast.

CLEAR's land-use education programs, which typically conduct more than 125 workshops a year at the community level, indicate that when the data presented here are shared with coastal community decision-makers and other groups, it can become a powerful stimulant to a wide range of local actions that can help to better protect Connecticut's coastal resources.

This Riparian study constitutes the first statewide assessment of the land-cover status of coastal riparian areas. As far as the authors know, it is the first study of its kind in the nation. The objective of the study was to develop a "triage"-type overview that would allow federal, state, and local land managers and researchers to comparatively assess the status of the riparian areas, and the changes that have been occurring over the last 15–20 years.

The study has accomplished that goal. The land-cover characterization of the 167 subregional basins was very similar to the statewide average; this was not a surprise, as the study area ended up covering about half of the entire state of Connecticut. Within the riparian buffer areas, increases in development and losses of natural vegetation increased with increasing buffer width, both in terms of acreage (expected) and in relative terms as depicted by percent changes of 2002 over 1985 levels (not expected). The relative loss of natural vegetation was smallest in the 100 ft zone (1.6% loss), increased in the 200 ft zone (2.2%), and increased again in the 300 ft zone (2.6%). This could partly be due to relatively larger errors incurred in the 100 ft width, but is more likely evidence that Connecticut's watercourse and wetland laws are having at least some effect in retarding deforestation along riverbanks and wetlands.

"Hot spot" areas where relatively forested riparian zones are quickly losing natural vegetation were identified, with the largest concentration of these being, as expected, in southeastern Connecticut (East Lyme, CT to Stonington, CT). This is consistent with the CCL study, which showed that area of the state to be experiencing some of the most rapid expansion of development in the state.

The literature on land-cover indicators is plentiful in the area of impervious cover and its relationship to stream health. It is less robust, but growing, in the area of forest cover and its relationship to watershed health, and there are only a few studies that relate riparian zone cover to watershed or stream health. One of the most thorough studies, conducted in the Chesapeake Bay region, shows that the best predictor of stream quality is a combined indicator of overall basin imperviousness and forest cover within the 100 ft buffer zone [14]. The 167 coastal basins of Connecticut were assessed using these criteria, and the resulting map provides a striking East/West dividing line in predicted stream quality, at approximately the westernmost extent of the greater Connecticut River drainage area.

Finally, in addition to meeting the original objectives, this project has produced data that should be useful to managers and researchers beyond the original scope. First, creating the seamless watercourse, water body, and wetlands "to be buffered" layer was a considerable technical feat, and this data layer may be of further use to other researchers and managers. Second, the land cover and land-cover change data contained in the many project spreadsheets comprise a large body of data that this study has only begun to mine. Last, the results of this study, mostly in map form, are already being incorporated into CLEAR outreach programs for municipal officials.

17.6 CONCLUSIONS

The CCL land cover has been used extensively in and of itself and is a critical dataset in many studies and statistics, such as the two described here. The studies and their results have been folded into many statewide and local planning efforts and are a key piece of information in many of CLEAR's outreach education programs, such as the Nonpoint Education for Municipal Officials (NEMO) municipal workshops* Linking Land Use to Water Quality, Conducting a Community Resource Inventory, and Open Space Planning and Planning for Habitats. However, time passes and the 2002 land cover has become dated. CLEAR analysts have recently completed a CCL Version 2 dataset. This dataset includes 2006 land cover, which extends the data four more years making a CCL span of 21 years of land-cover change for the entire state. It also includes improvements. The most significant is the addition of an Agricultural Field class. For the first time, land-cover change can be assessed in terms of change in farmland fields, not just the generic "other grasses" class. This has been a common request as municipalities try to assess the loss of local farms and protect those still in existence. All other classes have been further refined making the Version 2 CCL a completely new set of land cover for all dates.

The land cover and associated analyses have informed the heated debates on smart growth and urban sprawl. The objective, factual land-cover maps help bring all sides together and enable different organizations to work from the same base information with the ultimate goals of both economic prosperity and protecting the landscape of Connecticut.

ACKNOWLEDGMENTS

This material is based upon work supported by the National Aeronautics and Space Administration. In addition, the CALCAP was funded by NOAA's Office of Ocean and Coastal Resource Management through the Connecticut Department of Environmental Protection Office of Long Island Sound Programs (OLISP). The Coastal Riparian Buffers Analysis Project was funded by EPA's Office of Long Island Sound Programs.

* See http://nemo.uconn.edu for more information.

REFERENCES

1. Bank of America, 1995, *Beyond Sprawl: New Patterns of Growth to Fit the New California*, Bank of America Environmental Polices and Programs, San Francisco, CA, 11pp.
2. Sierra Club, 1997, *Sprawl Costs us All: A Guide to the Costs of Sprawl and How to Create Livable Communities in Maryland*, Sierra Club Foundation, Annapolis, MD, 16pp.
3. Koeln, G. and Bissonnette, J., 2000, Cross-correlation analysis: Mapping landcover change with a historic landcover database and a recent, single-date multispectral image, in *Proceedings of the 2000 ASPRS Annual Convention*, Washington, DC.
4. Smith, F.G.F., Herold, N., Jones, T., and Sheffield, C., 2002, Evaluation of cross-correlation analysis, a semi-automated change detection, for updating of an existing landcover classification, in *Proceedings of the 2002 ASPRS Annual Convention*, Washington, DC.
5. Story, M. and Congalton, R.G., 1986, Accuracy assessment: A user's perspective, *Photogrammetric Engineering and Remote Sensing*, 52(3): 397–399.
6. Congalton, R.G., Oderwald, R.G., and Mead, R.A., 1983, Assessing Landsat classification accuracy using discrete multivariate analysis statistical techniques, *Photogrammetric Engineering and Remote Sensing*, 49(12): 1671–1678.
7. Fung, T. and LeDrew, E., 1988, The determination of optimal threshold levels for change detection using various accuracy indices, *Photogrammetric Engineering and Remote Sensing*, 54(10): 1449–2454.
8. Schueler, T.R., 1994, The Importance of Imperviousness, *Watershed Protection Techniques*, 1(3): 100–111.
9. Arnold, C.L. and Gibbons, C.J., 1996, Impervious surface coverage: The emergence of a key environmental indicator, *Journal of the American Planning Association*, 62(2): 3243–258.
10. Brabec, E., Schulte, S., and Richards, P.L., 2002, Impervious surfaces and water quality: A review of current literature and its implications for watershed planning, *Journal of Planning Literature*, 16(4): 499–514.
11. Clausen, J.C., Warner, G., Civco, D., and Hood, M., 2003, Nonpoint education for municipal officials impervious surface research, Final Report, Connecticut DEP, 18pp.
12. Schueler, T.R., 2003, Impacts of impervious cover on aquatic systems, *Watershed Protection Research Monograph No. 1*, Center for Watershed Protection, Ellicott City, MD.
13. Booth, D.B., Hartley, D., and Jackson, R., 2002, Forest cover, impervious-surface area, and mitigation of stormwater impacts, *Journal of the American Resources Association*, 38(3): 835–846.
14. Goetz, S.J., Write, R.K., Smith, A.J., Zinecker, E., and Schaub, E., 2003, IKONOS imagery for resource management: Tree cover, impervious surfaces, and riparian buffer analyses in the mid-Atlantic region, *Remote Sensing of Environment*, 88: 195–208.

18 Effects of Increasing Urban Impervious Surface on Hydrology of Coastal Rhode Island Watersheds

Yuyu Zhou, Yeqiao Wang, Arthur J. Gold, and Peter V. August

CONTENTS

18.1 INTRODUCTION

Increasing the impervious surface area (ISA) from urban and suburban development is a particularly important component of human-induced land-use and land-cover change (LULCC). ISA is a critical factor affecting the cycling of terrestrial runoff and associated materials to and within ocean margin waters. Increasing the ISA causes a series of environmental problems for freshwater and estuarine ecosystems through the alteration of the frequency and magnitude of runoff. It impacts watershed hydrology in terms of influencing the runoff and base flow and causes problems such as soil erosion and nonpoint source pollution (Arnold and Gibbons, 1996). Runoff from urbanized watersheds frequently contains high concentrations of nutrients and pollutants, which can be harmful to freshwater and marine organisms, and therefore, it has significant impacts on sensitive tidal creeks and estuaries systems (Paul and Meyer, 2001; Schiff et al., 2002). The coastal state of Rhode Island experiences urban runoff problems resulting from urban development. The understanding of how increasing the ISA influences the environment, especially the rainfall–runoff relationship, can provide valuable information on land management activities. Wegehenkel et al. (2006) found that a precise estimation of settlement areas in a catchment together with an improved estimation of the degree of the actual imperviousness is required for precise calculations of surface runoff and the flood peaks even in a rural catchment with a relatively low amount of settlement areas.

Reported studies provided useful information about the impact of land-cover change and increasing impervious surface on infiltration and runoff and validated the importance of ISA in hydrology (Arnold and Gibbons, 1996; Paul and Meyer, 2001). Current studies are often limited to the general relationship between ISA derived from land-cover types and environmental problems due to a lack of precise ISA data. It is important to investigate how hydrological processes are affected by the spatial heterogeneity of ISA, and the influence of ISA on hydrological cycle in watersheds, in particular for those that drain into coastal waters.

Modeling with high spatial resolution data may provide useful information on planning and management activities such as low-impact development. Studies demonstrated that hydrological modeling is an effective method to evaluate the impact of ISA on watershed hydrology (Dunn and Mackay, 1995; Booth and Jackson, 1997; Brun and Band, 2000; Fohrer et al., 2001; Lahmer et al., 2001; Lee and Heaney, 2003; Ott and Uhlenbrook, 2004; Wegehenkel et al., 2006; Dougherty et al., 2007). In terms of temporal and spatial change of parameters in hydrological models, distributed models have advantages over lumped models because lumped models use aggregated empirical parameters that lack clear physical meanings. Also it is difficult to evaluate the change of model parameters on the hydrological process based on lumped models (Kuchmenta et al., 1996). Distributed hydrological models are based on physical, chemical, and biological theories and the results can be more reliable than lumped models if the geospatial input data are available at appropriate scales.

Because of a lack of high spatial resolution ISA data, hydrological models usually employ estimated ISA from land-cover data or assign all urban areas as having the same ISA coverage (Ott and Uhlenbrook, 2004). However, using estimated percentage of ISA instead of precise high spatial resolution ISA data may cause considerable errors in the rainfall–runoff modeling and introduce uncertainty in evaluating the impact of ISA on watershed hydrology. ISA data extracted from high spatial remote sensing imageries can provide more precise information for hydrological modeling.

Other remote sensing-derived information can improve hydrology modeling from several aspects. For example, remote-sense-derived leaf area index (LAI) can improve simulated hydrographs, a marked change in the relative proportions of actual evapotranspiration comprising canopy evaporation, soil evaporation, and transpiration (Andersen et al., 2002). As remote sensing data are from different sensor systems with variable spatial resolutions, a single grid-based model faces challenges when using data that have been derived from multiple remote sensing products. For example, problems exist such as information lost in scaling using coarser spatial resolution units. Therefore, an effective approach for data scaling needs to be considered.

We developed a *distributed object-oriented rainfall–runoff simulation* (DORS) model to study the rainfall–runoff relationship with the high spatial resolution ISA derived from remote-sensed data (Zhou and Wang, 2008a). The DORS model takes land-cover types as the spatial units. The advantages of using land-cover types as the spatial units include reducing the data volume, increasing computational efficiency, strengthening representation of watersheds, and being able to employ data products in different scales. Instead of using empirically estimated proportions of ISA from land-cover data, the DORS model incorporates high spatial resolution ISA. Those modeling features and precise data improve the performance of hydrological modeling and reduce the uncertainty in evaluating the impact of ISA.

Regression and correlation analyses have been useful tools to quantify the relationship between response and predicator variables. However, spatial dependence received less attention in the conventional regression methods for geospatial observations with an assumption that the observations were independent. Without consideration of spatial dependence, regression methods could have errors in estimating regression coefficients, coefficient of determination, and significance level. Spatial regression techniques with consideration of spatial dependence have been extensively used in analyses of spatially explicit data (Ji and Peters, 2004; Anselin et al., 2006; Aguiar et al., 2007; Luck, 2007). In studies of spatial-related problems, the consideration of spatial dependence improves the fitness of predictive models.

In this chapter, we analyzed the ISA-related watershed characteristics and simulated the hydrology pattern in selected watersheds with various level of urbanization in the state of Rhode Island. We checked the spatial dependence in regression analysis and compared the performance of ordinary least square (OLS), spatial error, and spatial lag regression models for the relationship between watershed characteristics and hydrology. We quantified the relationship between watershed hydrology using the discharge per area and ratio of runoff to base flow and predicator variables of percent ISA, distance from ISA to stream, and stream density by OLS and spatial lag regression models in watershed scale.

18.2 STUDY AREA AND DATA

18.2.1 Study Area

We delineated a watershed in Rhode Island that has long-term flow records from U.S. Geological Survey (USGS) gaging station for model validation. The outlet for the watershed coincided with the gaging stations of USGS Chipuxet River at

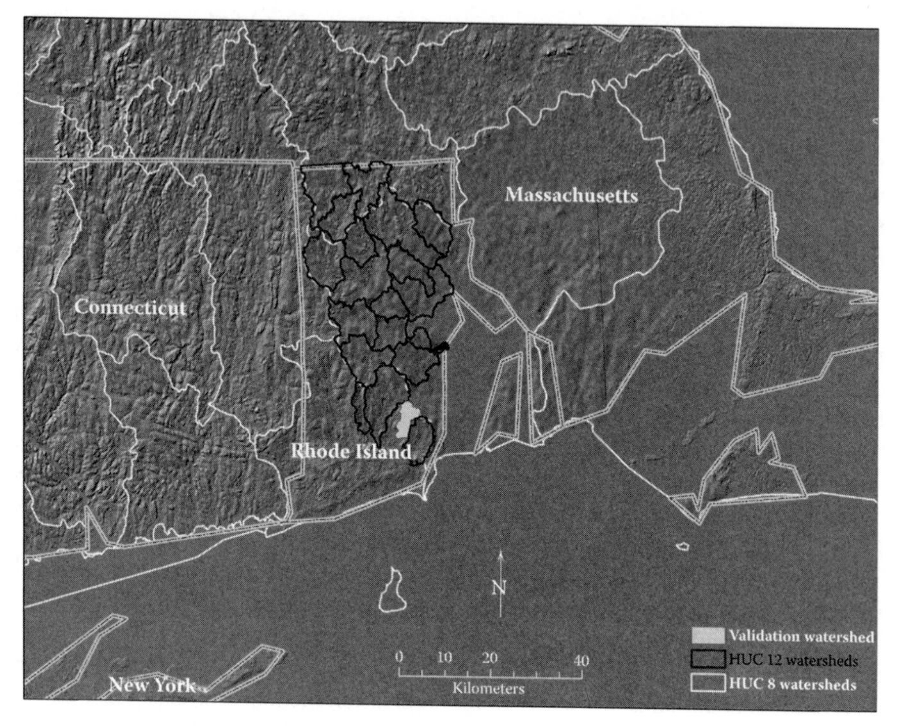

FIGURE 18.1 Study area and selected watersheds.

West Kingston, Rhode Island. The watershed with an area of 2515 ha is a fourth-order stream system that drains two small lakes and receives both through groundwater and surface water inputs. The watershed is centered at 41.5°N and 71.5°W and lies on the border of three towns: Exeter, South Kingstown, and North Kingstown (Figure 18.1). The elevation varies from 20 to 90 m. It is a coastal climate with a monthly temperature of 22°C in July and −1°C in January in 1997. The annual precipitation in 1997 was approximately 1200 mm. The year 1997 is the selected time period for which we tested the simulations in this study. The main soil types are silty loam and sandy clay loam that have relatively medial value of porosity and hydraulic conductivity. The watershed includes typical land-cover types in Rhode Island including agriculture, forest, and urban.

We selected 20 watersheds at Hydrologic Unit Code (HUC) 12 level representing a variety degrees of ISA cover and different degrees of urbanization in Rhode Island to perform the impact analyses. The watersheds lie within the Blackstone River watershed and the Pawcatuck-Wood watershed both at HUC 8 level between 41°25′ and 42°1′ N latitude and 71°20′ and 71°47′ W longitude (Figure 18.1). The areas of the study watersheds range from 1500 to 13,000 ha. The elevation varies from sea level to 240 m. The land-cover types are distributed heterogeneously in the study watersheds. According to our previous study, 10% of Rhode Island has been covered by ISA in 2004 (Zhou and Wang, 2007).

18.2.2 Data

The data used in this study include high spatial resolution ISA, digital elevation model (DEM), land cover, vegetation index, soil type, and meteorological data. ISA is one of the most important factors impacting watershed hydrology, and it is the focus of this study. We extracted ISA information from 1 m high spatial resolution orthophotograph using an object-oriented classification method (Zhou and Wang, 2008b). In the 20 study watersheds, percentages of ISA range from 3% to 29%. The dynamic range of ISA percentage permits an evaluation of the impact of ISA on hydrology. We developed DEM from 1:100,000 USGS Hypsography Dataset 10 m Elevation Contour Lines and converted the vector data to 10 m resolution grid. We obtained land-cover data from the Landsat TM image at 30 m spatial resolution. The major land-cover types include coniferous forest, deciduous forest, mixed forest, wetland, urban, grass, agriculture, and water (Novak and Wang, 2004). We delineated the stream networks using the DEM and the National Hydrography Dataset (NHD) data (USGS, 2007). We derived soil type data from 1996 USDA/NRCS 1:15,840 SSURGO soils and employed the lookup table between soil types and soil properties to extract the soil parameters required for our simulation model (Zhou and Wang, 2008a). We obtained LAI from vegetation index of simple ratio (SR) with the help of land-cover data. We adapted the relationship between SR and LAI (Zhou et al., 2007) and modified it for Landsat TM with the adjustment coefficient. We used meteorological data including hourly precipitation, daily maximum temperature, minimum temperature, mean temperature, humidity, and radiation from the weather station at Kingston, Rhode Island. We used meteorological data from March 20 to April 28, 1997, in all simulations studies. We selected early spring as the period of simulation to examine the rainfall–runoff relationships. In most years stream flow in early spring in the study area is characterized by both storm runoff and base flow due to an excess of precipitation compared to evapotranspiration. Figure 18.2 illustrates the data layer concept for a selected example watershed for the simulation modeling.

18.3 METHODS

18.3.1 Hydrological Modeling and Validation

We adopted and modified the algorithms from the distributed hydrological models by Chen et al. (2005) and Wigmosta et al. (1994). These models processed the regular grid as spatial units and simulated the groundwater and overland water movement. Similar to other common models, these models provide an effective system to consider the flow of water in watershed and use a modeling framework for water flow among neighboring spatial grid units. For our application of the rainfall–runoff study, we modified the algorithm of hydrological processes from these grid-based models to spatial units of objects and added important processes such as infiltration.

The horizontal boundary of the simulated area is a watershed delineated from a DEM where divides between neighboring watersheds are identified. Vertically, the simulation extends from the saturated zone in the soil to the top of vegetation canopy. The model divides the study watershed into basic spatial units based on land cover

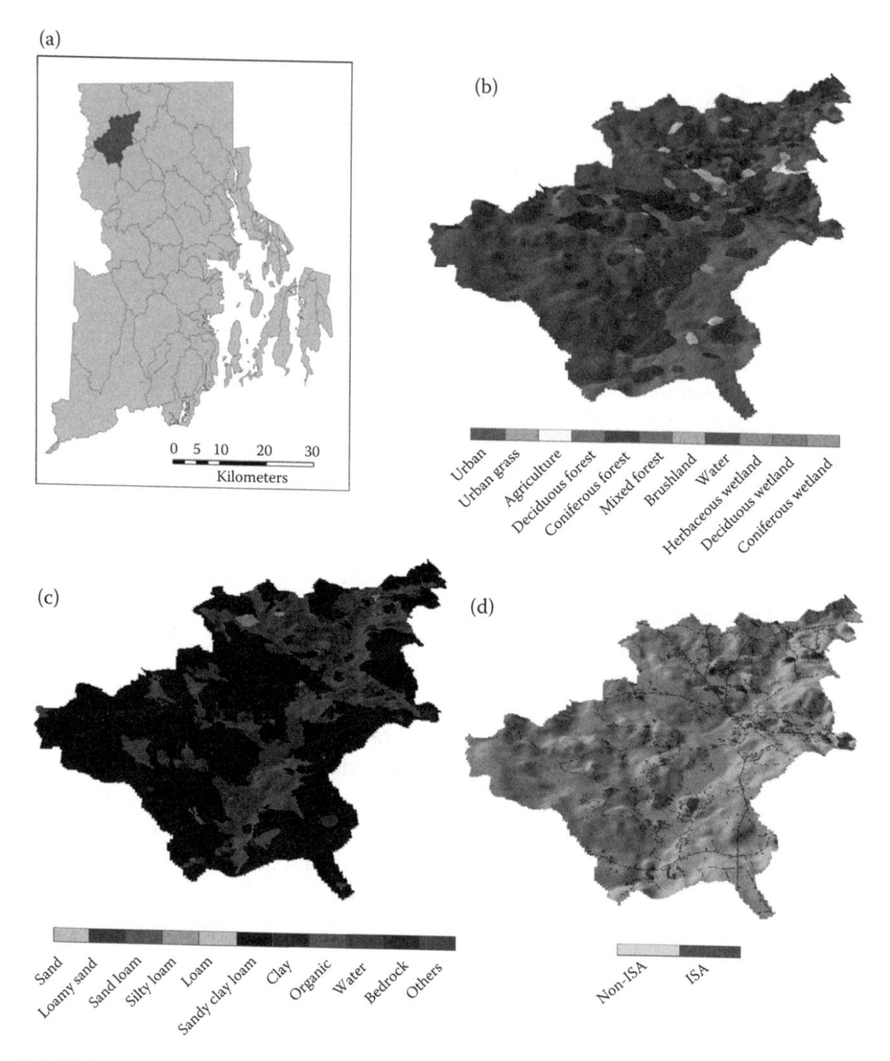

FIGURE 18.2 **(See color insert following page 206.)** Location of an example watershed (a) and some of the selected data layers involved in modeling, including land-cover types (b), soil types (c), and high spatial resolution (1 m) ISA (d).

and DEM data. The model treats each spatial unit, the object, as a unique vegetation–soil system. According to the need of hydrological processes simulation, the model divides an object vertically into five strata, that is, overstory canopy, understory, litter or moss layer, soil unsaturated zone, and saturated zone. We perform model simulations of the hydrological processes at the object level. The size of the object is controlled based on the user-determined area thresholds in segmentation. Therefore, the model splits the objects with large area into small ones to make the runoff and groundwater routing more reasonable.

The model includes major components of segmentation, parameterization, interception, infiltration, evapotranspiration, saturated flow, overland flow, and channel flow. The DORS model takes objects, group of pixels, as spatial units to reduce data volume, increase computational efficiency, strengthen representation of watersheds, and utilize the data in different scales. The model provides a framework to integrate remote sensing data and the derived products in different scales for the simulation of hydrological process in a watershed. The object-based algorithm also improves the simulation efficiency. Object-by-object routing produces three-dimensional representation of surface and saturated subsurface flow.

The parameters of precipitation, solar radiation, DEM, land cover and ISA, LAI, and soil properties parameters are major inputs of the DORS model. All input parameters are spatially converted to object level after the segmentation process. Other required parameters such as Manning roughness coefficient for the runoff routing are determined based on the land-cover types. The major output is the hourly discharge at the outlet of the watershed. Information such as spatial distribution of runoff, evapotranspiration, and ground flow can be retrieved at any time in the simulation period. The framework of the model is illustrated in Figure 18.3.

The simulated discharge at the outlet of the watershed was sensitive to the extent of generated runoff. We focus our evaluation on the model's abilities to capture the temporal discharge patterns through the runoff and base flow simulation with the incorporation of high spatial resolution ISA. We integrated the USGS hourly discharge at the outlet to a daily level to permit comparison and validation with the daily and integrated output from the simulation model.

Discharge change pattern is a critical index indicating the watershed hydrology, and is mainly affected by runoff that has an important impact on the materials

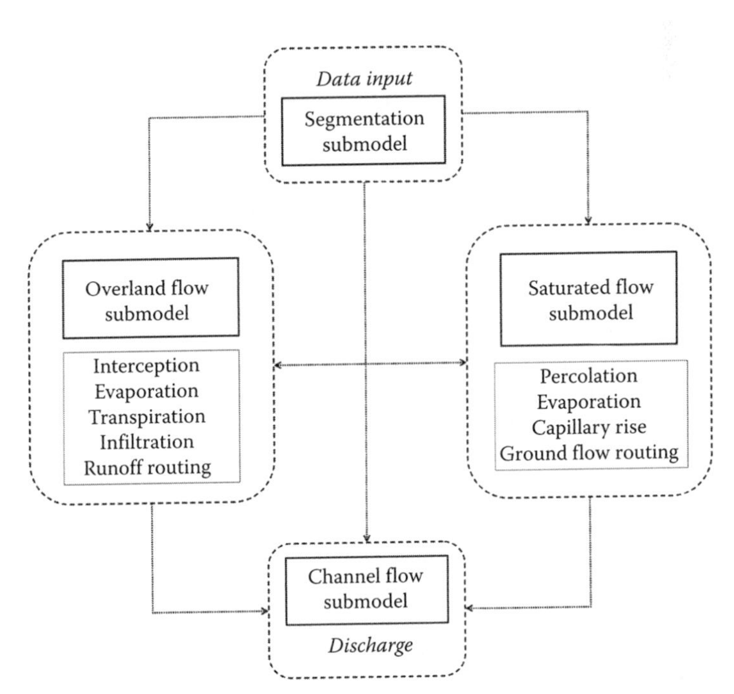

FIGURE 18.3 Framework of the DORS model.

movement and associated pollution in watershed. In order to check the model performance, we used two indicators to compare the simulated and measured daily discharge during the study period. The first is the ratio of absolute error to the mean R_a, which is used by the World Meteorological Organization. The second is the Nash coefficient R_2.

$$R_a = 100\% \times \frac{\sum_{i=1}^{n} \left| Q_{mi} - Q_{si} \right|}{n\bar{Q}_m}, \tag{18.1}$$

$$R_2 = 1 - \frac{\sum_{i=1}^{n} (Q_{mi} - Q_{si})^2}{\sum_{i=1}^{n} (Q_{mi} - \bar{Q}_m)^2}, \tag{18.2}$$

where R_a is the relative absolute error to the mean, R_2 is the Nash coefficient, Q_{mi} is the measured discharge, Q_{si} is the simulated discharge, $\bar{Q}_m$ is the mean value of observed daily discharge, and n is the number of days.

18.3.2 Spatial Regression

Discharge pattern is a critical index indicating the watershed hydrology. It is mainly affected by runoff that has important impacts on materials movement and associated pollution in a watershed (Arnold and Gibbons, 1996). The ratio of runoff to base flow is frequently used to represent the hydrological response to different combinations of land use, especially increasing ISA (Harbor, 1994; Fohrer et al., 2001; Lee and Heaney, 2003). The DORS model has the capacity for exporting discharge, runoff, and base flow. Therefore, this study used two indicators, discharge per area and ratio of runoff to base flow, to analyze the hydrological pattern in the 20 study watersheds. The ratio of runoff to base flow is unitless. The unit of discharge per area is normalized to mm/day using watershed area.

We built indicators including percentage of ISA and distance from ISA to stream to quantify the ISA and its spatial patterns in the 20 study watersheds. We derived these indicators from the 1 m high spatial resolution ISA and stream network data. This study used spatial analysis in GIS to derive distance from ISA to stream and used the mean value for each watershed in the analyses. In addition, this study included an indicator of stream density to further explore the variation of hydrology pattern in study watersheds. We calculated all indicators for each study watershed. Percentage of ISA is unitless, the unit of distance from ISA to stream is normalized to km/km^2 using watershed area, and the unit of stream density is normalized to km/km^2 using watershed area.

We used regression methods to analyze the hydrological response to spatial configuration and percentage of ISA (Zhou et al., 2009). As the spatial dependence in observations is not considered in OLS regression techniques, we used spatial lag and spatial error regression methods (Anselin et al., 2006) for their capacity in analyzing geospatial data and the possible spatial dependence in observations. We performed the diagnostics for multicollinearity and spatial dependence and test of Lagrange Multiplier (LM) and robust LM before the regression analyses. Analyses were performed with GeoDa (version 0.9.5-i) software (Anselin et al., 2006).

We checked three regression models using the response variables of discharge per area and ratio of runoff to base flow, and a set of predicator variables including percentage of ISA, distance from ISA to stream, and stream density from the 20 study watersheds. The equations for OLS (Equation 18.3), spatial lag (Equation 18.4), and spatial error (Equation 18.5) regression models were described as

$$R = \beta_0 + \beta_1(\text{PerISA}) + \beta_2(\text{Dist}) + \beta_3(\text{StreamDen}) + \varepsilon, \tag{18.3}$$

where R is discharge per area and ratio of runoff to base flow, β_0 is the intercept, β_1, β_2, and β_3 are regression coefficients of the predicator variables, ε is random error, PerISA is the percentage of ISA, Dist is the distance from ISA to stream, and StreamDen is the stream density.

$$R = \rho WR + \beta_0 + \beta_1(\text{PerISA}) + \beta_2(\text{Dist}) + \beta_3(\text{StreamDen}) + \varepsilon, \tag{18.4}$$

where WR is a spatially lagged dependent variable for weights matrix W and ρ is a parameter from regression. In this study, we obtained the W based on the method of K-nearest neighbors (Anselin et al., 2006).

$$R = \beta_0 + \beta_1(\text{PerISA}) + \beta_2(\text{Dist}) + \beta_3(\text{StreamDen}) + \upsilon, \quad \upsilon = \lambda W\upsilon + u, \tag{18.5}$$

where υ is a vector of spatially autocorrelated error terms, u is a vector of independent and identically distributed (i.i.d.) error, and λ is a parameter from regression.

The R^2 in the spatial regression is a pseudo R^2. It is not comparable with the measure in OLS regression. The pseudo R^2 serves as a rough estimate on the explanatory power of the models (Piha et al., 2007). Compared with R^2, the Akaike information criterion (AIC) is a more suitable measure of performance for spatially correlated data (Anselin et al., 2006). Therefore, the comparison of the model performance in this study was based on AIC. The model with the lowest AIC value is the best.

18.4 RESULTS

18.4.1 Model Validation

The modeled and the measured daily discharge at the outlet and daily precipitation from hourly data in the simulation period from March 20 to April 28, 1997, are illustrated in Figure 18.4. The result indicates that the model performs reasonably well in simulating discharge in the study watersheds. The modeled and the measured daily discharge showed similar temporal patterns. The modeled and the measured mean daily discharge at the outlet over the 40-day period are 1.30 and 1.25 m³/day

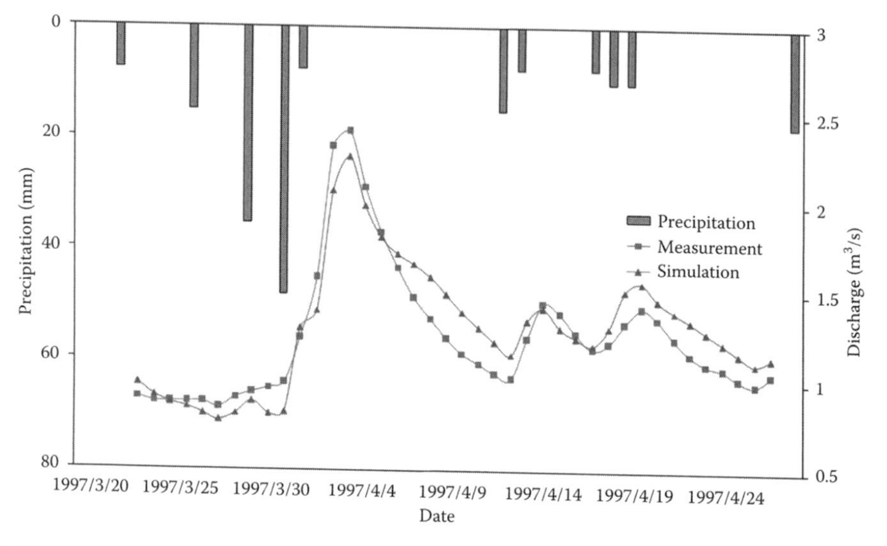

FIGURE 18.4 Temporal variations of measured and modeled discharge at outlet and scaled daily precipitation for the test watershed.

for the test watershed. The correlation coefficient between these two sets of discharge results was 0.98. The ratio of absolute error to the mean R_a and the Nash coefficient R_2 for the simulated and the measured daily discharge during the study period are 9.4% and 0.86 in the test watershed. The validation result indicated that the DORS model is capable of capturing the relationship between rainfall and runoff in the study watershed.

18.4.2 Spatial Regression

The regression results demonstrate that AIC are 32.68, 34.66, and 30.32, respectively, for the OLS, spatial lag, and spatial error regression models with three predicator variables for relationship between the discharge per area and percentage of ISA, distance from ISA to stream, and stream density. Moran's *I* score for the relationship between discharge per area and percentage of ISA and stream density is not significant, indicating no significant spatial autocorrelation in the residuals. The LM test of lag and error are not significant, also indicating the absence of spatial dependence. The results indicate that the coefficients for ISA in OLS models are significant at the $\alpha = 0.01$ level. ISA can explain 46% variation in the observations. The inclusion of stream density improves the explained variation. There is small improvement in explained variation after the inclusion of distance from ISA to stream because of the collinearity between stream density and distance from ISA to stream. The multicollinearity test using GeoDa indicates that the regression between discharge per area and percentage of ISA and stream density have a lower score than that between discharge per area and all three predicator variables. Therefore, we selected the OLS model with percentage of ISA and stream density to analyze the relationship between discharge per area and watershed characteristics. The result indicates that increases in percentage of ISA and/or stream density correspond to increases in discharge per area.

The results demonstrate that AIC are –5.27, –10.96, and –10.52, respectively, for OLS, spatial lag, and spatial error regression models with three predicator variables for relationship between ratio of runoff to base flow and percentage of ISA, distance from ISA to stream, and stream density. The analysis of spatial dependence shows that the Moran's I score of 0.18 ($p < 0.01$) for the relationship between ratio of runoff to base flow and percentage of ISA and distance from ISA to stream is highly significant, indicating spatial autocorrelation in the model residuals. The LM test of lag and error are both significant ($p < 0.05$), while the robust LM test of error is not significant. The result indicates that the spatial lag regression model is best for the regression analysis between ratio of runoff to base flow and watershed characteristics. Based on AIC comparisons, the spatial lag regression model was superior to OLS in regression with different independent variables. The results indicate that the coefficients for ISA in the OLS models are significant at the $\alpha = 0.1$ level and that ISA can explain 31% variation in the observations. The inclusion of distance from ISA to stream improves the explained variation. There is a small difference in explained variation after the inclusion of stream density because of the collinearity between stream density and distance from ISA to stream. The test of multicollinearity indicates that the regression between ratio of runoff to base flow and percentage of ISA and distance from ISA to stream have a much lower score than that between ratio of runoff to base flow and all three predicator variables. Therefore, we selected the spatial lag model with the percentage of ISA and distance from ISA to stream to evaluate the relationship between ratio of runoff to base flow and watershed characteristics. The determining factor coefficients in the spatial lag model indicate that increases in percentage of ISA and/or decreases in distance from ISA to stream correspond to increases in ratio of runoff to base flow.

18.5 DISCUSSION

18.5.1 Hydrological Modeling

We developed a DORS model to simulate hydrological processes at a watershed scale using spatial inputs of remote sensing data in varied scales. This model combined three major submodels that simulate surface runoff, groundwater, and channel flow, and it has dynamic linkage between these submodels. The model provides a framework to integrate remote sensing raw data and derived products in different scales for the hydrological process in a watershed, and the object-based algorithm also reduces the simulation time and data volume. Object-by-object routing produces three-dimensional representation of surface and saturated subsurface flow. The DORS model uses land-cover-based objects, rather than pixels as modeling units. Therefore, this model not only can utilize the data at varied scale, but also keeps a balance between computation speed and representation of spatial heterogeneity.

We validated the model in a complex watershed in coastal Rhode Island. We compared the modeled daily discharge at the watershed outlet with measurement in the

USGS observation station. The result indicates that the model performs well for the purpose of modeling the runoff and base flow in this complex watershed. Furthermore, the model has the capacity of hourly simulation for short-term applications.

18.5.2 Watershed Hydrology

Impervious surface changes the hydrological pattern in watersheds through increasing runoff production and reducing the time of concentration of surface runoff. The change of hydrology is likely to cause problems such as soil erosion, water pollution, and ecological degradation in streams. The analyses using OLS and spatial regression models demonstrate that the coefficient for the predicator variable of ISA percentage is always statistically significant in all relationships. In all regression models, ISA has a positive impact on the ratio of runoff to base flow and discharge per area. The comparison also indicates that the spatial lag regression model has the better performance for the relationship between ratio of runoff to base flow and predicator variables, while the OLS regression model provides the similar performance as spatial regression models for the relationship between discharge per area and predicator variables. We compared the strength of the percentage of ISA in relationship between ratio of runoff to base flow and watershed characteristics considering the OLS and spatial lag regression models. Using the spatial lag regression model, the magnitude of determining factor coefficient gets lower. It indicates that the impact of ISA on hydrology may be overestimated without consideration of spatial dependence.

The coefficient of determination (R^2) in all regression models indicates that ISA explains an important part of variations in hydrological pattern. This also implies that there are other factors contributing to the unexplained variations although percentage of ISA significantly impacts hydrology. In the regression analyses, results indicate that inclusion of distance from ISA to stream and stream density improves the explained variation in hydrological pattern. The regression analysis indicates that the distance from ISA to stream has negative impacts on ratio of runoff to base flow and that stream density has positive impacts on discharge per area.

18.6 CONCLUSIONS

We developed a remote-sensed-data-driven and -distributed object-oriented hydrological model to study the rainfall–runoff relationship with the high spatial resolution ISA. The modeling processes take objects based on land-cover data as the fundamental spatial units in order to reduce data volume, increase computational efficiency, strengthen representation of watersheds, and utilize the data in varied scales. With the hydrological simulation and extracted high spatial resolution ISA information, we analyzed the spatial dependence in regression analyses and compared three regression models in construction of relationships between hydrology pattern and predicator variables of percent ISA, distance from ISA to stream, and stream density. The incorporation of high spatial resolution ISA in the hydrological model improves hydrological simulation and reduces the uncertainty in regression relationship. By evaluating the indicators of ratio of runoff to base flow and discharge per area, we found that the percentage of ISA is closely related to the hydrology pattern although there are some unexplained variations in the regression models. Without consideration of spatial dependence in the OLS model, the magnitude of ISA impact on hydrology in terms of the ratio of runoff to base flow was estimated inaccurately. From the established relationship between hydrological pattern and watershed characteristics, the magnitude of change in hydrology with the change of ISA and other factors can be derived. The relationships between watershed characteristics and hydrology will be of help to decision-makers in watershed management. The method in this study provides flexible tools and information to evaluate the impact of complex practice in planning and management activities on watershed hydrology. This is important in particular for studying and managing coastal watersheds where the terrestrial and ocean waters are connected.

ACKNOWLEDGMENT

This study was funded by the Rhode Island Agricultural Experimental Station (project no. RI00H330).

REFERENCES

Aguiar, A.P.D., Câmara, G., and Escada, M.I.S., 2007, Spatial statistical analysis of land-use determinants in the Brazilian Amazonia: Exploring intra-regional heterogeneity, *Ecological Modelling*, 209(2–4): 169–188.

Andersen, J., Dybkjær, G., Jensen, K.H., Refsgaard, J.C., and Rasmussen, K., 2002, Use of remotely sensed precipitation and leaf area index in a distributed hydrological model, *Journal of Hydrology*, 264: 34–50.

Anselin, L., Syabri, I., and Kho, Y., 2006, GeoDa: An introduction to spatial data analysis, *Geographical Analysis*, 38(1): 5–22.

Arnold, C.A., Jr. and Gibbons, C.J., 1996, Impervious surface coverage: The emergence of a key urban environmental indicator, *Journal of the American Planning Association*, 62: 243–258.

Booth, D.B. and Jackson, C.R., 1997, Urbanization of aquatic systems: Degradation thresholds, stormwater detection, and the limits of mitigation, *Journal of the American Water Resources Association*, 33(5): 1077–1090.

Brun, S.E. and Band, L.E., 2000, Simulating runoff behavior in an urbanizing watershed, *Computers, Environment and Urban Systems*, 24: 5–22.

Chen, J.M., Chen, X., Ju, W., and Geng, X., 2005, Distributed hydrological model for mapping evapotranspiration using remote sensing inputs, *Journal of Hydrology*, 305(1–4): 15–39.

Dougherty, M., Dymond, R.L., Grizzard, T.J., Godrej, A.N., Zipper, C.E., and Randolph, J., 2007, Quantifying long-term hydrologic response in an urbanizing basin, *Journal of Hydrologic Engineering*, 12(1): 33–41.

Dunn, S.M. and Mackay, R., 1995, Spatial variation in evapotranspiration and the influence of land use on catchment hydrology, *Journal of Hydrology*, 171(1–2): 49–73.

Fohrer, N., Haverkamp, S., Eckhardt, K., and Frede, H., 2001, Hydrologic response to land use changes on the catchment scale, *Physics and Chemistry of the Earth* (*B*), 26(7–8): 577–582.

Harbor, J.M., 1994, A practical method for estimating the impact of land use change on surface runoff, groundwater recharge, and wetland hydrology, *Journal of the American Planning Association*, 60(1): 95–108.

Ji, L. and Peters, A.J., 2004, A spatial regression procedure for evaluating the relationship between AVHRR-NDVI and climate in the northern Great Plains, *International Journal of Remote Sensing*, 25(2): 297–311.

Kuchment, L.S., Demidiv, V.N., Naden, P.S., Cooper, D.M., and Broadhurst, P., 1996, Rainfall–runoff modelling of the Ouse basin, North Yorkshire: An application of a physically based distributed model, *Journal of Hydrology*, 181: 323–342.

Lahmer, W., Pfiitzner, B., and Becker, A., 2001, Assessment of land use and climate change impacts on the mesoscale, *Physics and Chemisty of the Earth (B)*, 26: 565–575.

Lee, J.G. and Heaney, J.P., 2003, Estimation of urban imperviousness and its impacts on storm water systems, *Journal of Water Resource Planning and Management*, 129(5): 419–426.

Luck, G.W., 2007, The relationships between net primary productivity, human population density and species conservation, *Journal of Biogeography*, 34(2): 201–212.

Novak, A. and Wang, Y., 2004, Effects of suburban sprawl on Rhode Island's forest: A Landsat view from 1972 to 1999, *Northeastern Naturalist*, 11: 67–74.

Ott, B. and Uhlenbrook, S., 2004, Quantifying the impact of land-use changes at the event and seasonal time scale using a process-orientated catchment model, *Hydrology and Earth System Sciences*, 8(1): 62–78.

Paul, M.J. and Meyer, J.L., 2001, Streams in the urban landscape, *Annual Review of Ecological Systems*, 32: 333–365.

Piha, M., Tiainen, J., Holopainen, J., and Vepsäläinen, V., 2007, Effects of land-use and landscape characteristics on avian diversity and abundance in a boreal agricultural landscape with organic and conventional farms, *Biological Conservation*, 140(1–2): 50–61.

Schiff, K., Bay, S., and Stransky, C., 2002, Characterization of stormwater toxicants from an urban watershed to freshwater and marine organisms, *Urban Water*, 4, 215–227.

U.S. Geological Survey, 2007, National Hydrography Dataset, http://nhd.usgs.gov/ (Last date accessed: July 25, 2007).

Wegehenkel, M., Heinrich, U., Uhlemann, S., Dunger, V., and Matschullat, J., 2006, The impact of different spatial land cover data sets on the outputs of a hydrological model: A modelling exercise in the Ucker catchment, North-East Germany, *Physics and Chemistry of the Earth*, 31(17): 1075–1088.

Wigmosta, M.S., Vail, L.W., and Lettenmaier, D.P., 1994, A distributed hydrology-vegetation model for complex terrain, *Water Resources Research*, 30: 1665–1679.

Zhou, Y. and Wang, Y., 2007, An Assessment of impervious surface areas in Rhode Island, *Northeastern Naturalist*, 14(4): 643–650.

Zhou, Y., Zhu, Q., Chen, J., Wang, Y., Liu, J., Sun, R., and Tang, S., 2007, Observation and estimation of net primary productivity in Qilian Mountain, western China, *Journal of Environmental Management*, 85(3): 574–584.

Zhou, Y. and Wang, Y., 2008a, A distributed and object-oriented rainfall–runoff simulation model with high spatial resolution impervious, *ASPRS 2008 Annual Conference and Technology Exhibition*, April 28–May 2, Portland, OR.

Zhou, Y. and Wang, Y., 2008b, Extraction of impervious surface areas from high spatial resolution imageries by multiple agent segmentation and classification, *Photogrammetric Engineering and Remote Sensing*, 74(7): 857–868.

Zhou, Y., Wang, Y., Gold, A., August, P., and Boving, T., 2009, Assessing impact of urban impervious surface on watershed hydrology using distributed object-oriented simulation and spatial regression, *GeoJournal*, accepted.

19 Contemporary Land-Use/Land-Cover Change in Coastal Pearl River Delta and Its Impact on Regional Climate

Limin Yang, Wenshi Lin, Lu Zhang, Hui Lin, and Dongsheng Du

CONTENTS

19.1 INTRODUCTION

Land use/land cover (LULC) is one of the most convincing aspects of the global change that has occurred in the terrestrial ecosystem (Meyer and Turner II, 1994; IPCC, 2001). Many changes in LULC reflect the impacts of human activities on global environment (e.g., Houghton et al., 1999). Change in LULC is also recognized as a main driver affecting the local, regional, and global climate (e.g., Charney et al., 1977; Chase et al., 1996; Stohlgren et al., 1998; Eastman et al., 2001; Foley et al., 2005). For instance, urbanization alters the urban–rural surface energy balance, affects the thermal stratification of the urban boundary layer, the local-scale atmospheric circulation, and the aerosol environment (Changnon and Huff, 1986; Shepherd 2005). Urbanization also affects precipitation through increases in hygroscopic nuclei, turbulence transfer, convection, rain-producing clouds, and the addition of water vapor from anthropogenic sources (Souch and Grimmond, 2006), all of which can lead to an altered pattern in urban precipitation frequency and intensity (e.g., Shepherd, 2006). The impact of LULC change on regional-scale climate has also been well documented (e.g., Dickinson, 1983; Sellers et al., 1996; Pielke et al., 1997; Xue et al., 2001).

The Pearl River Delta (PRD) region in southeast China is one of the fastest growing regions in the world as a result of a rapid urbanization since the 1980s. About 60 million people reside in the region, including 7 million in Hong Kong, half a million in Macao, and the rest in Mainland China. The economy of the region was traditionally based on a labor-intensive agricultural production. Since the late 1980s, however, the region has undergone an unprecedented LULC change as a result of policy change and a rapid economic development (Yeh and Li, 1997, 1998; Seto et al., 2002; Weng, 2002). Such an enormous development has brought about fundamental changes in LULC driven primarily by the conversion of the large-scale agricultural lands to urban land use. The LULC changes in turn could also affect the local and regional climate over time (Lo et al., 2007). For instance, a long-term climate record indicates a warming trend in the winter season in southeastern China by approximately 0.32° per decade from 1960 to 1996 (Liang and Wu, 1999), which is believed to be partially attributable to the LULC change of the region. Another study estimated an increase of mean surface temperature of 0.05°C per decade, attributable to urbanization from 1979 to 1998 in southeastern China (Zhou et al., 2004). A trend of decreasing winter season precipitation from 1988 to 1996, which coincides with the urban development of the PRD region, has been reported (Kaufmann et al., 2007).

The primary objective of the research reported here was to quantify the major LULC changes and their impact on regional climate of the coastal PRD region via satellite remote sensing by using a mesoscale climate model. The uniqueness of the PRD region makes it an ideal area for such a study because (1) although a number of studies on LULC change and related driving forces have been successfully conducted in some local areas (one city or a few counties or a single Landsat scene, e.g., the studies by Weng, 2002; Seto et al., 2002; and Seto and Kaufmann, 2003), an LULC change study at the regional scale is lacking; (2) research on the magnitude and impact of LULC change on regional climate over the last three decades has not been carried out due to lack of LULC change data over the period for the region.

19.2 LULC CHANGE IN PRD 1980–2000

19.2.1 Study Area

The coastal PRD region, by definition of this study, consists of East River, West River, and North River drainage basins located in southeastern China. It ranges from 111° to 116°E and from 21° to 24°N, occupying approximately 96,500 km^2. The region is characterized by a subtropical monsoon climate, distinguished by a moderate, cool, and dry season between October and April, and a long, warm to hot rainy season between May and September. The average annual temperature ranges from 21°C to and 23°C, and the average annual total precipitation varies from 1600 to 2600 mm. An evergreen forest is the predominant natural vegetation type in the region, mostly growing in the hilly and mountainous areas. This climate, together with a fertile soil, makes the region well suited for agriculture. Major crops growing in the area include the paddy rice, sugarcane, banana, and lychee, to name a few.

19.2.2 Data for LULC Change Study

Three datasets were used for LULC classification and LULC change detection, including a Landsat GeoCover image dataset, a digital elevation model (DEM) dataset derived from the Shuttle Radar Topography Mission (SRTM), and an existing LULC dataset developed by the Chinese Academy of Sciences in late 2000 based on both automated and visual interpretation of satellite images and aerial photographs as well as data from field surveys. The Landsat GeoCover was the main data source that was used for performing LULC classification and for LULC change detection. This dataset was complied from the archived images acquired by Landsat MSS/TM/ETM+ (Multispectral Scanner System/Thematic Mapper/Enhance Thematic Mapper Plus) sensors (Tucker et al., 2004). A total of 18 Landsat GeoCover scenes *ca.* 1980 (MSS), 1990 (TM), and 2000 (ETM+) covering the PRD region were used for this study. All these images were acquired during the dry seasons from September to December from 1980 to 2000, which minimizes the seasonal and interannual variability among the images. The nominal spatial resolutions for MSS and TM/ETM+ images are 57 and 28.5 m, respectively.

The second dataset is the digital elevation derived from the SRTM data. The SRTM project produced a high-resolution digital topographic data for 80% of the Earth's land surface by means of SAR interferometry technology (Rodriguez et al., 2005). The absolute vertical accuracy of the elevation data is about 16 m at 90% confidence. A mosaic of SRTM DEM data at 90 m resolution for the PRD was acquired and refined.

19.2.3 Methodology for LULC Change Study

Numerous methods have been developed for detecting LULC changes using remote sensing data. For this study, a postclassification approach was adopted. A digital classification method was used to generate three land-cover maps *ca.* 1980, 1990, and 2000, followed by a postclassification comparison among all three maps for detecting land-cover changes from the 1980s to 1990s and from the 1990s to 2000. The adoption of the postclassification approach is based on the fact that the Landsat MSS images and TM/ETM+ images have different spatial resolution and spectral bands, and there is little information on the atmospheric conditions at the time of image acquisition.

Change detection using the classification-based method is relatively straightforward and it does not require a rigorous radiometric correction among all archived satellite images. However, the effectiveness of the method depends on the quality of the land-cover classification produced for each epoch. For this study, great attention was paid to obtain a classification with high accuracy.

A decision-tree classification algorithm was used to generate three LULC maps for the 1980s, 1990s, and 2000 using an eight-category LULC classification system, including dryland cropland, irrigated cropland, forest, shrubland, grassland, barren land, urban/developed land, and water body. The decision-tree algorithm is a non-parametric statistic technique. It does not require the normality of input variables and can handle both categorical and continuous data. The algorithm provides a robust and efficient way to extract information from a large quantity of satellite and ancillary data, and it yields classification rules that can be readily interpreted and applied spatially (e.g., Hansen et al., 2002). The algorithms use in this study provides several advanced options for conducting training and rule-based modeling for classification. Major features include the boosting and cross-validation procedures.

The classification procedure is shown in Figure 19.1. Briefly, training sample data were first selected by using a stratified random sampling method from the Landsat GeoCover image. The stratification was made by using the existing LULC data. Quality control was conducted to ensure a correct LULC label assigned to each training sample unit. A decision tree algorithm was then used to establish the classification rules through a learning process using the training data. After the initial classification, the quality of the classification was evaluated through a cross-validation procedure. If the cross-validation result was unsatisfactory, an adjustment of training samples was made and used to produce another set of rules for a new classification. This process was repeated until a "best" set of rules was obtained, and the classification rules were applied to the entire scene to generate the final LULC map.

19.2.4 Spatial and Temporal Pattern of LULC Change in PRD

Figure 19.2 (left) shows LULC maps of the 1980s and 2000. The overall accuracy of the LULC classification achieved by using the decision-tree method, based on an independent data interpreted from the historical Landsat images and aerial photographs, is 75% for a land-cover map using the Landsat MSS images and around 85% for using TM/ETM+ images. Table 19.2 lists three confusion matrices pertinent

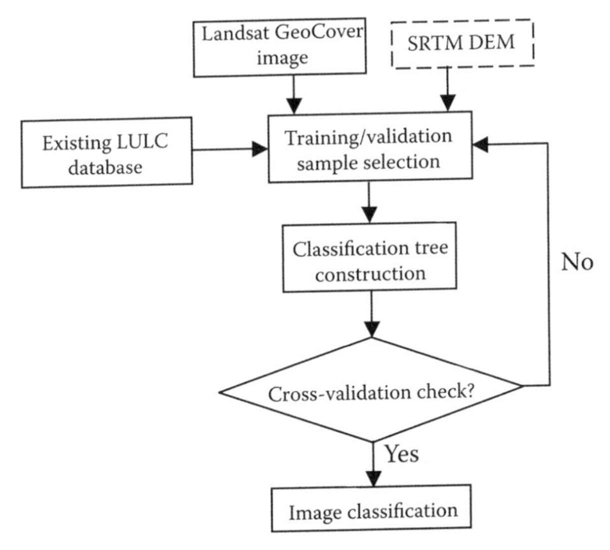

FIGURE 19.1 Workflow of LULC classification using the DTC method.

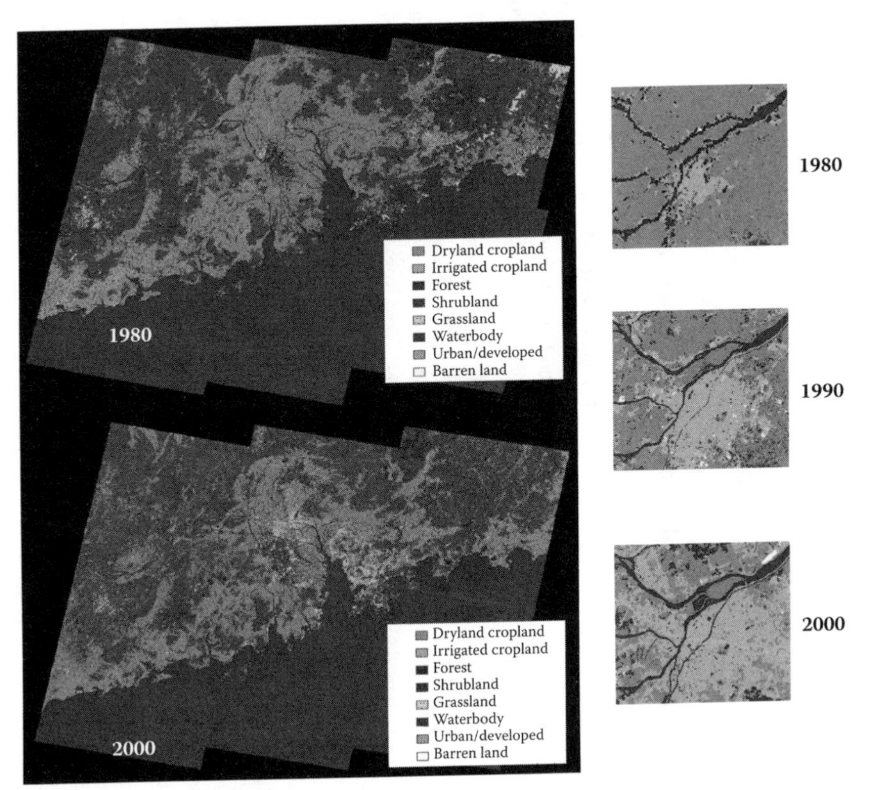

FIGURE 19.2 (Left) Land use and land cover of the PRD region, *ca*. 1980 and 2000; (Right) LULC map of Dongguan city, from left to right: the LULC map of 1980, 1990, and 2000 (light gray tone represents urban land use).

TABLE 19.1

Temporal and Areal Mean Values of Key Variables in the PBL from the Two Experiments Subject to Two Different Land Conditions

	Domain-Averaged Value		Land-Averaged Value		Urban-Averaged Value	
	USGS Land Use (LC1980)	Urban Expansion (LC2000)	USGS Land Use (LC1980)	Urban Expansion (LC2000)	USGS Land Use (LC1980)	Urban Expansion (LC2000)
Mean 2 m temperature (°C)	21.76	21.83	20.62	20.71	21.24	21.60
Mean 2 m daily maximum temperature (°C)	25.90	26.16	26.02	26.41	26.26	27.88
Mean 2 m minimum temperature (°C)	18.15	18.07	16.10	15.97	16.88	16.33
Mean 2 m mixing ratio (g/kg)	11.69	11.67	11.19	11.17	11.37	11.36
Mean relative humidity (%) at level $\sigma = 0.995$	60.0	59.6	60.1	59.4	59.2	57.8
Sensible heat flux (W/m^2)	24.9	28.9	16.6	21.5	28.5	86.0
Latent heat flux (W/m^2)	198.1	192.2	117.0	105.6	128.8	41.8
PBL height (m)	827	839	765	783	780	856

Notes: The time mean values are monthly averaged quantities for October 2004, whereas the areal mean values include three averaged quantities within the domain D03, that is, domain-averaged quantities over all grid points, land-averaged quantities over all land grid points, and urban mean quantities over all urban grid points. Note that urban grid points are indicated by the red grids in Figure 19.3.

to each of the three land-cover classification maps (the 1980s, 1990s, and 2000s). Inclusion of the SRTM DEM improved the land-cover classification accuracy by separating land-cover types in between mountainous and hilly areas and the coastal plans.

As expected, the urban development as a result of conversion from the agricultural lands was the predominant driver of the LULC changes occurring in the region from the 1980s and 2000. Much of the land-use change occurred through a conversion of the artificial wetlands, fishing ponds to urban land use, or by land reclamations along the coastal areas. As a typical example, Figure 19.2 (right) illustrates the LULC changes in the city of Dongguan from the 1980s to 2000. The urban expansion of the Dongguan is clearly seen from the maps, mostly driven by a fast increase in the manufacturing sector through the joint ventures and many other investments as a result of the policy change initiated in the late 1980s.

Besides changes due to urban expansion, there is also a distinct change in forest and woodlands from the 1980s to 2000. While a decrease of forest land is evident as a result of cutting and land-use change in many rural areas, there is also some increase of woodlands in many hilly areas of the region, mostly due to implementation of conservation policy and a large-scale campaign for planting of pine and other fast-growing tree species in the region after 1980. This has altered many areas that were formally occupied by grass and shrubland, and is one of the important LULC changes of the region. Overall, the result of the LULC change study indicates that conversion from cropland to urban development, that is, urbanization, is the prevailing LULC change occurring in the PRD region during the period from 1980 to 2000, followed by the conversion from agri- and aquacultral businesses to urban development induced by land conversion.

19.3 ASSESSMENT OF THE IMPACT OF LULCC CHANGE ON REGIONAL CLIMATE THROUGH MODELING

19.3.1 Model Framework

To investigate the impact of urbanization on the regional climate, field experiments are often neither well suited (Molders and Olson, 2004) nor feasible. This is also the case for the PRD region and the main reason why a numerical modeling approach was taken for this study. Specifically, a Fifth-Generation NCAR/PSU (National Center for Atmospheric Research, Pennsylvania State University) Mesoscale Model (MM5) model was employed to simulate the impact of urbanization on the monthly climate conditions. Two simulations were conducted; one was to model regional climate conditions of a dry autumn in October 2004, and another one for a hot summer of June 2005. Note that in the design of this study, urbanization meant a decrease in surface and canopy albedo and soil moisture, an increase in roughness length and stomatal resistance, and a change in empirical parameters used to calculate transpiration (Molders and Olson, 2004). Other impacts such as change in aerosol, moisture, and heat release on local and regional climate after urban expansion were neglected.

The MM5 model used for this study is a limited area, non-hydrostatic, terrain-following sigma-coordinate model that is capable of simulating or predicting region-

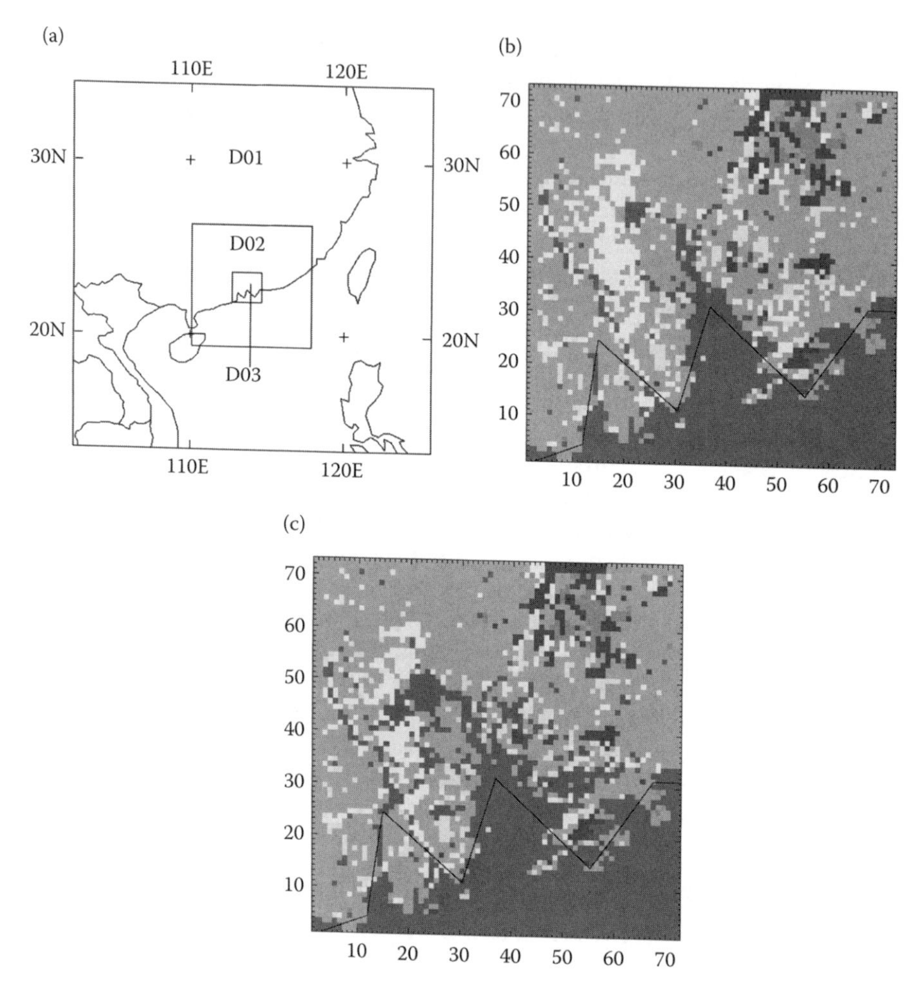

FIGURE 19.3 **(See color insert following page 206.)** Computation domains and land-cover datasets used for MM5 simulations: (a) Computed domains. D01: 27 km grid spacing; D02: 9 km grid spacing; D03: 3 km grid spacing. (b) Original LC1980 data in the domain D03. (c) LC2004 data in the domain D03. The only change between (b) and (c) is the urban expansion.

al-scale atmospheric circulation (Grell et al., 1994). Three nested computational domains with a grid spacing of 27, 9, and 3 km (hereafter referred to as D01, D02, and D03, respectively) on a Mercator conformal map projection were used in the experiments, as is shown in Figure 19.3. The simulations were run in a two-way interactive mode. The use of coarse domain captured the changes in large-scale forcing, while the fine domain was designed to resolve the urban areas. There are 31 sigma levels vertically from the top to surface level in the model. The first layer above ground was about 38 m in thickness, and the top level was at 50 hPa.

Two different land-cover datasets of the PRD region were used to investigate the potential impact of LULC change on climate. The land-cover types in and around the major urban areas representing the 1980s and 2000 is shown in Figures 19.3b and c, respectively. The land-cover data of the 1980s (hereafter referred to as "LC1980") was developed by the USGS in the mid-1990s, representing the urban condition of the PRD region in the 1980s, while another land-cover data of 2000 (hereafter referred to as "LC2000") represents a changed landscape in 2000 after the urban expansion took place. The LC2000 was generated using the Landsat GeoCover images by this study. Based on the two land-cover maps, 21 grid cells (cell size is 9 km^2) are identified as urban in the LC1980, whereas 329 grid cells belong to urban in the LC2000. Therefore, the urban expansion from the 1980s to 2000 is approximately 2772 km^2.

19.3.2 Model Parameterization

Parameterizations for model physics included a medium range forecast planetary boundary layer (PBL) scheme (Hong and Pan, 1996) and the NCAR Community Climate Model (CCM2) longwave and shortwave scheme (Hack et al., 1993) for all experiments. An advanced land surface model (LSM), namely the National Centers for Environmental Prediction, Oregon State University, Air Force, and Hydrologic Research Lab (NOAH) LSM, was selected (Chen and Dudhia, 2001), which has several enhancements on urban land-use treatments (Lo et al., 2004). The Dudhia's simple ice phase cloud microphysics scheme (Dudhia, 1989) was used in the outermost domain D01, whereas the Reisner's mixed phase cloud microphysics scheme (Reisner et al., 1998) was used in D02 and D03. For domains D01 and D02, the Kain–Fritsch cumulus parameterization scheme (Kain and Fritsch, 1993) was employed, whereas for domain D03, no cumulus parameterization scheme was introduced assuming that convection was reasonably well resolved by the explicit microphysics (Weisman et al., 1997). Finally, a four-dimensional data assimilation was carried out in the outermost domain D01.

19.3.3 Initial and Boundary Conditions

The initial and boundary conditions were interpolated from the $1° \times 1°$ resolution global reanalysis data produced by the National Center for Environmental Prediction (NCEP) (Kistler et al., 2001), and was vertically interpolated to 31 sigma levels. Initial conditions for the two finer domains D02 and D03 were obtained by interpolating from the coarsest domain. As the monthly mean soil temperature and moisture were used for initialization of the LSM, the simulation results for the first 17 days were excluded from the analysis in order to reduce the influence of the initial value spin-up (Yang et al., 1995; Gupta et al., 1999).

We used the five-layer soil model (Dudhia, 1996) in the MM5 model in which the soil temperature and moisture were always initialized by climatological soil temperature and wetness, which are a function of the land-use types but time-invariant. The model setup by reinitializing every 24 h was designed to test the first-order effect of urban expansion. In all of the atmospheric simulations, the 13–36 h atmospheric forecast was used for the analysis. This approach, similar to that of Westrick and Mass (2001), was adopted to reduce the total error that would have accumulated over the relatively long monthly forecast period, and to bypass the uncertainty in a regional climate model related to lateral boundary conditions (Wu et al., 2005).

19.4 IMPACT OF LULCC ON REGIONAL CLIMATE CONDITION IN A DRY AUTUMN IN OCTOBER 2004

19.4.1 Shelter-Level Temperature

As aforementioned, October 2004 was chosen for simulation, which was a dry autumn in the PRD. The observed monthly mean temperature and the simulated monthly mean temperature (2 m above the surface) are shown in Figures 19.4a and 19.5, respectively. The magnitude of the simulated results was similar to that observed in the northern part of the Guangdong province, but it was lower than the observed value in the area near the coast. The simulated temperature difference between using the LC2000 and LC1980 data is illustrated by several heat islands in the region (Figures 19.5c and 19.6c). In domain three (urban areas), the maximum difference in

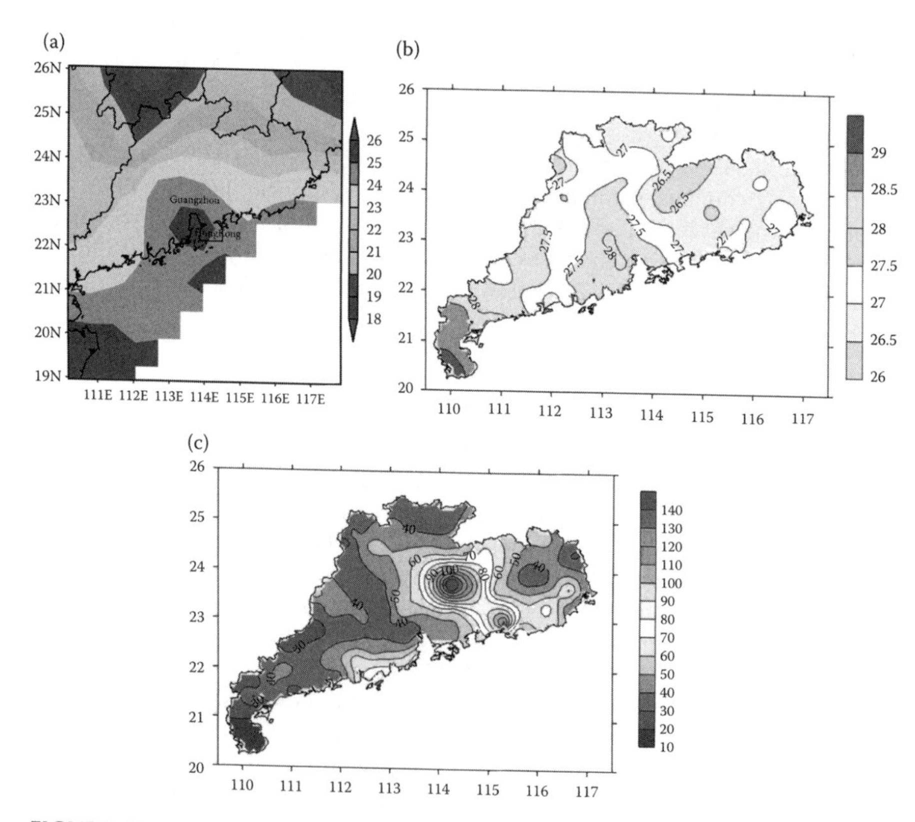

FIGURE 19.4 (See color insert following page 206.) (a) Observed monthly mean 2 m temperature (unit: °C) of October 2004. (b) Observed monthly average 2 m temperature (unit: °C) of June 2005. (c) Observed monthly total precipitation (unit: cm) of June 2005. All data are provided by the Beijing Climate Center.

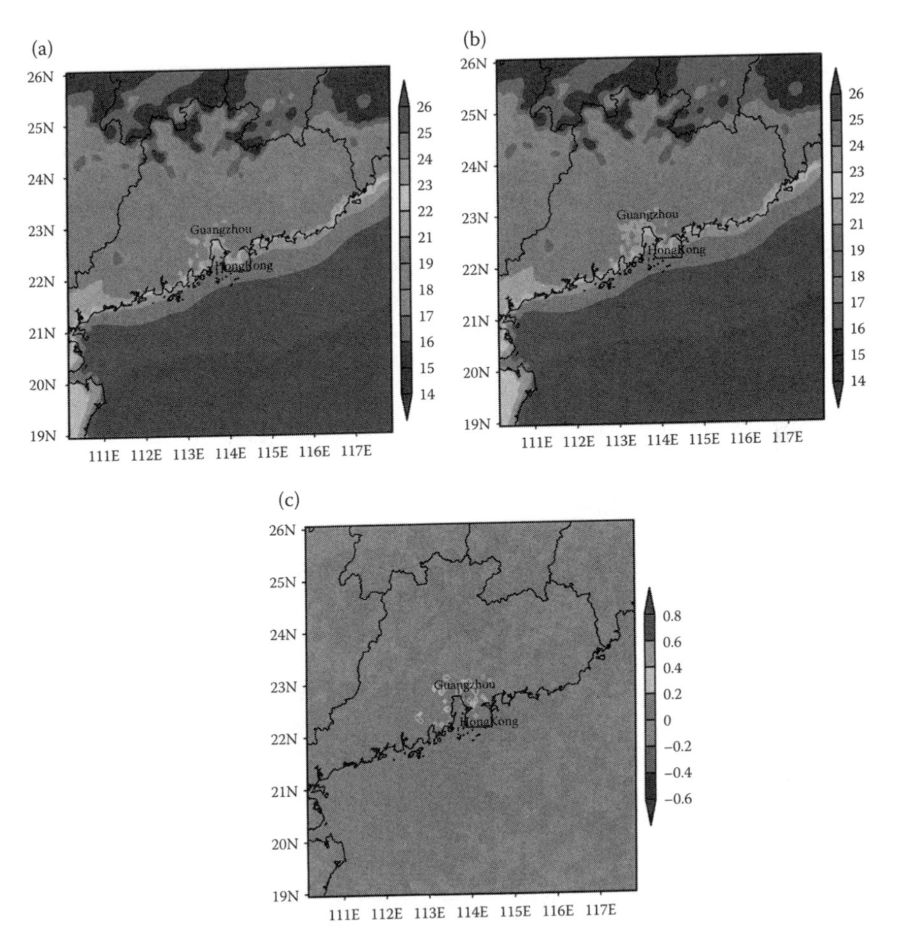

FIGURE 19.5 **(See color insert following page 206.)** Simulated monthly mean 2 m temperature (unit: °C) of October 2004 in the domain D02 (a) with LC1980 data, (b) with LC2000 urban expansion data, (c) and the difference between (b) and (a).

monthly mean temperature using LC1980 versus LC2000 reached 0.9°C over the urban areas in Guangzhou, Dongguan, Zhongshan, and Shenzhen. The urban heat islands are also clearly shown in simulated monthly maximum temperature between urban and rural areas (Figure 19.7).

The spatial-averaged values of climate variables can often reflect the extent and effect of land–atmosphere interactions over land on a larger spatial scale (Poveda and Mesa, 1997). For this study, the monthly mean temperature and other climate variables were averaged over several spatial units as is shown in Table 9.1. Note that the averaged value of temperature changed from 21.76°C to 21.83°C averaged over domain three, from 20.62°C to 20.71°C over land, and from 21.24°C to 21.60°C over the urban areas when LC1980 and LC2000 data were used, respectively. The monthly mean daily maximum temperature averaged within the domain, the land, and the urban areas increased by 0.26°C, 0.39°C, and 1.62°C, respectively.

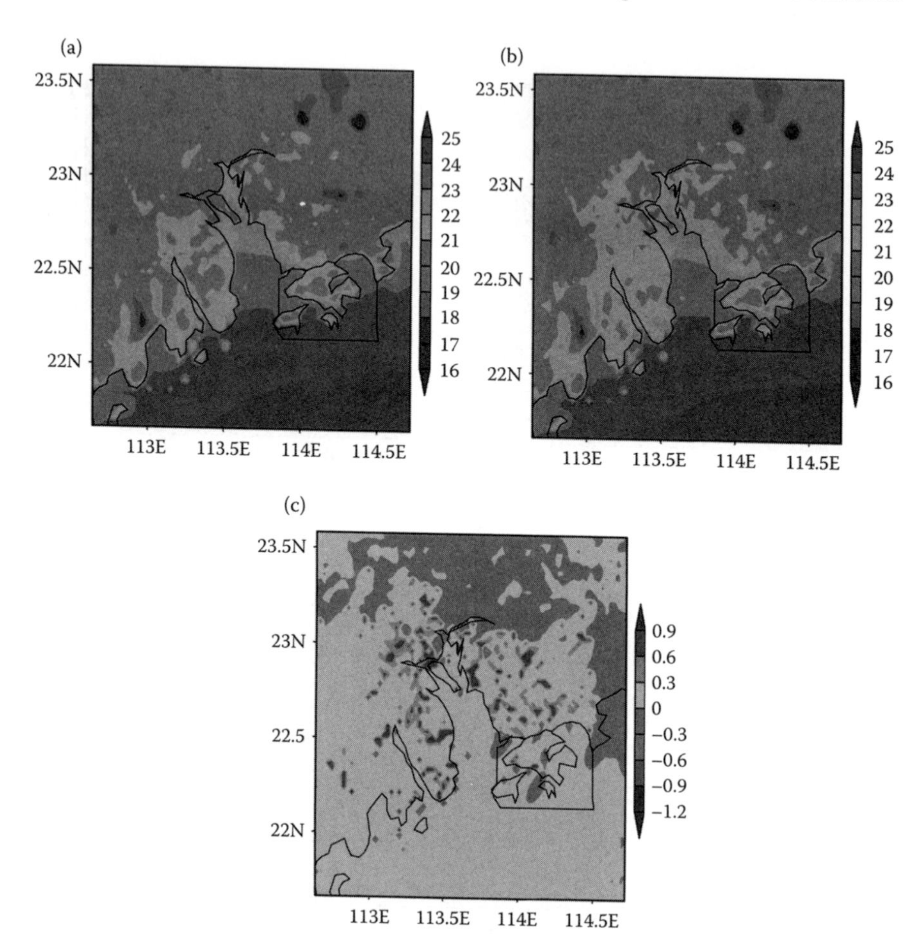

FIGURE 19.6 **(See color insert following page 206.)** Same as Figure 19.5, except for the domain D03 (unit: °C).

19.4.1.1 Moisture Condition

The effect of urban expansion on the atmospheric humidity was indicated by a decrease of the mixing ratio by 0.4 g/kg in the city of Guangzhou and Dongguan (Figure 19.8a and b), whereas the maximum decrease of the relative humidity in the urban areas was 2.4%. The monthly mean absolute humidity averaged over three spatial units (D03, land, and urban areas) decreased by 0.02, 0.02, and 0.01 g/kg, respectively, using LC1980–LC2000 data. The monthly mean relative humidity averaged over the three units changed from 60% to 59.6%, 60% to 59.4%, and 59.2% to 57.8%, respectively.

19.4.1.2 Surface Fluxes

The differences of monthly mean sensible and latent heat fluxes between LC2000 and LC1980 are shown in Figure 19.9. The sensible heat flux in eastern and northeastern PRD increased with the maximum increase reaching 90 W/m^2. In contrast,

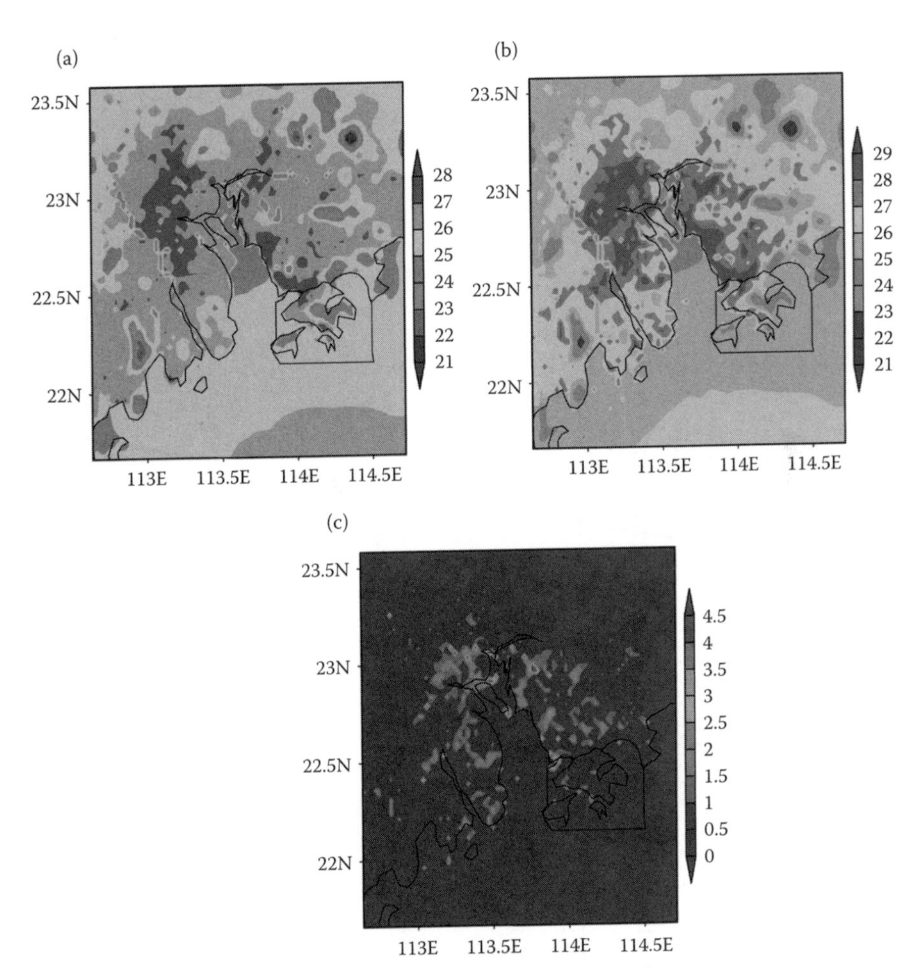

FIGURE 19.7 **(See color insert following page 206.)** Same as Figure 19.5, except for daily maximum 2 m temperature (unit: °C).

the latent heat flux decreased in all urbanized areas with the maximum decrease of 300 W/m^2. The monthly mean surface sensible heat fluxes averaged within the domain D03, land, and urban areas increased, respectively, from 24.8 to 28.9 W/m^2, 16.6 to 21.5 W/m^2, and 28.5 to 86.0 W/m^2 from LC1980 to LC2000 (Table 19.1).

19.4.1.3 PBL Height

The spatial distribution and difference in simulated monthly height of the PBL associated with LC1980 and LC2000 are shown in Figure 19.10. Geographically, the PBL was higher toward the coastal region and lower inland, and was relatively deep in the urban areas. Temporally, the height of the PBL over the urban areas increased significantly when LC2000 was used. The average increase was 20–80 m near the urban areas, and the maximum increase reached 180 m in Guangzhou. The monthly mean height of the PBL averaged within the domain D03, land, and urban areas also

TABLE 19.2a
Confusion Matrix of Land-Cover Classification of PRD Region Using Landsat MSS Imagery of the 1980s

	Classified Data									
Reference Data	**Ic**	**Dc**	**Fr**	**Sh**	**Gr**	**Wt**	**Ur**	**Ba**	**Total**	**Percent**
Irrigated cropland (Ic)	740	49	41	12	7	15	29	1	894	82.8
Dry cropland (Dc)	144	59	21	11	4	2	24	6	271	21.8
Forest (Fr)	44	15	1136	17	28	2	6	0	1248	91.0
Shrubland (Sh)	46	10	77	11	7	0	4	0	155	7.1
Grassland (Gr)	21	7	102	6	49	1	7	0	193	25.4
Water body (Wt)	21	0	5	0	0	259	13	1	299	86.6
Urban/developed (Ur)	52	18	5	2	3	14	162	8	264	61.4
Barren land (Ba)	9	4	0	0	0	1	5	96	115	83.5
Total	1077	162	1387	59	98	294	250	112	3439	
Percent	68.7	36.4	81.9	18.6	50.0	88.1	64.8	85.7		73.04

TABLE 19.2b
Confusion Matrix of Land-Cover Classification of PRD Region Using Landsat TM Imagery of the 1990s

	Classified Data									
Reference Data	**Ic**	**Dc**	**Fr**	**Sh**	**Gr**	**Wt**	**Ur**	**Ba**	**Total**	**Percent**
Irrigated cropland (Ic)	595	8	87	17	2	1	6	0	716	83.1
Dry cropland (Dc)	21	33	25	22	18	1	10	0	130	25.4
Forest (Fr)	75	10	1048	24	21	1	2	0	1181	88.7
Shrubland (Sh)	38	27	58	30	21	0	4	0	178	16.9
Grassland (Gr)	1	10	47	12	118	0	6	0	194	60.8
Water body (Wt)	2	2	2	0	0	390	4	0	400	97.5
Urban/developed (Ur)	2	6	2	2	0	1	432	7	452	95.6
Barren land (Ba)	0	0	0	0	0	1	5	114	120	95.0
Total	734	96	1269	107	180	395	469	121	3371	
Percent	81.1	34.4	82.6	28.0	65.6	98.7	92.1	94.2		81.87

TABLE 19.2c
Confusion Matrix of Land-Cover Classification of PRD Region Using Landsat ETM+ Imagery of the 2000s

	Classified Data									
Reference Data	**Ic**	**Dc**	**Fr**	**Sh**	**Gr**	**Wt**	**Ur**	**Ba**	**Total**	**Percent**
Irrigated cropland (Ic)	282	30	15	7	0	4	6	3	347	81.3
Dry cropland (Dc)	59	55	13	3	0	0	6	4	140	39.3
Forest (Fr)	11	10	685	2	6	3	4	1	722	94.9
Shrubland (Sh)	9	7	24	2	1	3	3	1	50	4.0
Grassland (Gr)	0	0	11	0	48	0	0	0	59	81.4
Water body (Wt)	0	0	1	0	0	348	2	1	352	98.9
Urban/developed (Ur)	9	4	8	0	1	3	116	5	146	79.5
Barren land (Ba)	6	2	1	0	0	1	2	54	66	81.8
Total	376	108	758	14	56	362	139	69	1882	
Percent	75.0	50.9	90.4	14.3	85.7	96.1	83.5	78.3		84.50

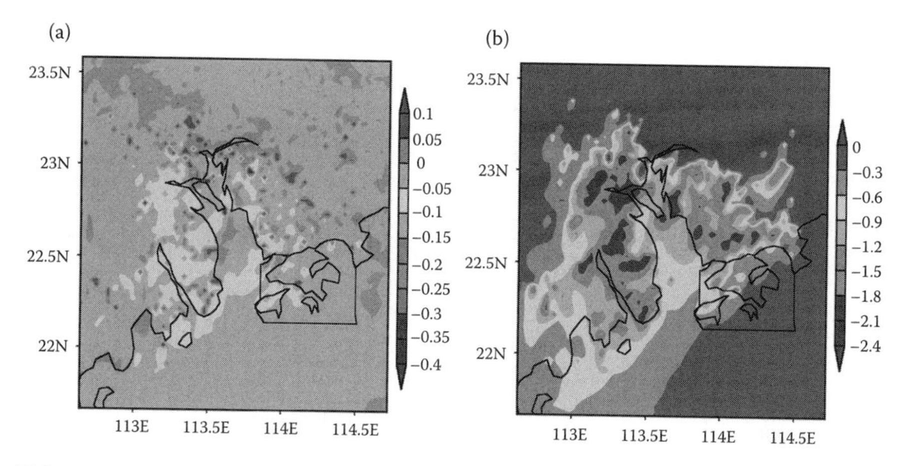

FIGURE 19.8 **(See color insert following page 206.)** Same as Figure 19.5, except for (a) 2 m mixing ratio (unit: g/kg) and (b) relative humidity at level $\sigma = 0.995$ (unit: %).

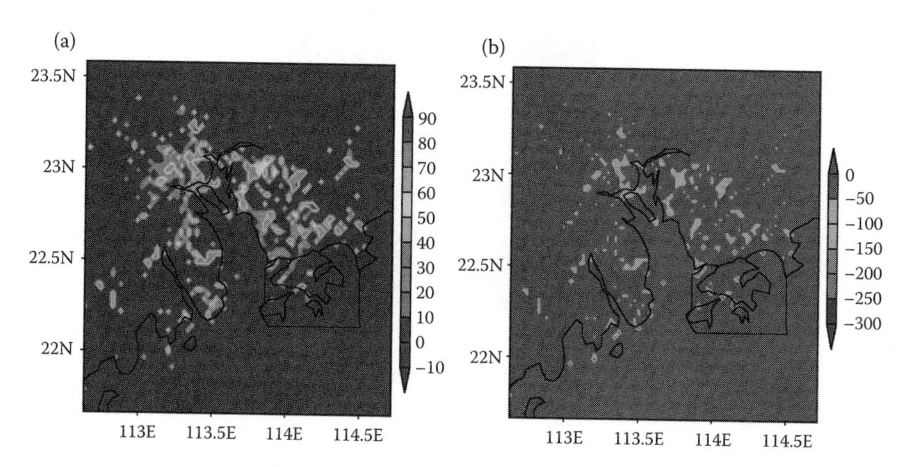

FIGURE 19.9 **(See color insert following page 206.)** Same as Figure 19.5, except for (a) sensible heat flux (unit: W/m^2) and (b) latent heat flux (unit: W/m^2).

increased from 827 to 839 m, 765 to 783 m, and 780 to 855 m using LC1980–LC2000 data, respectively. The daily mean values of the spatially averaged PBL height within the domain D03, land, and urban areas were all increased when LC2000 data were used.

19.5 IMPACT OF LULCC ON REGIONAL CLIMATE CONDITION IN SUMMER OF JUNE 2005

19.5.1 Shelter-Level Temperature

A summer month of June 2005 was chosen for the second model simulation. The simulated monthly mean air temperatures (Figure 19.11a and b) and difference

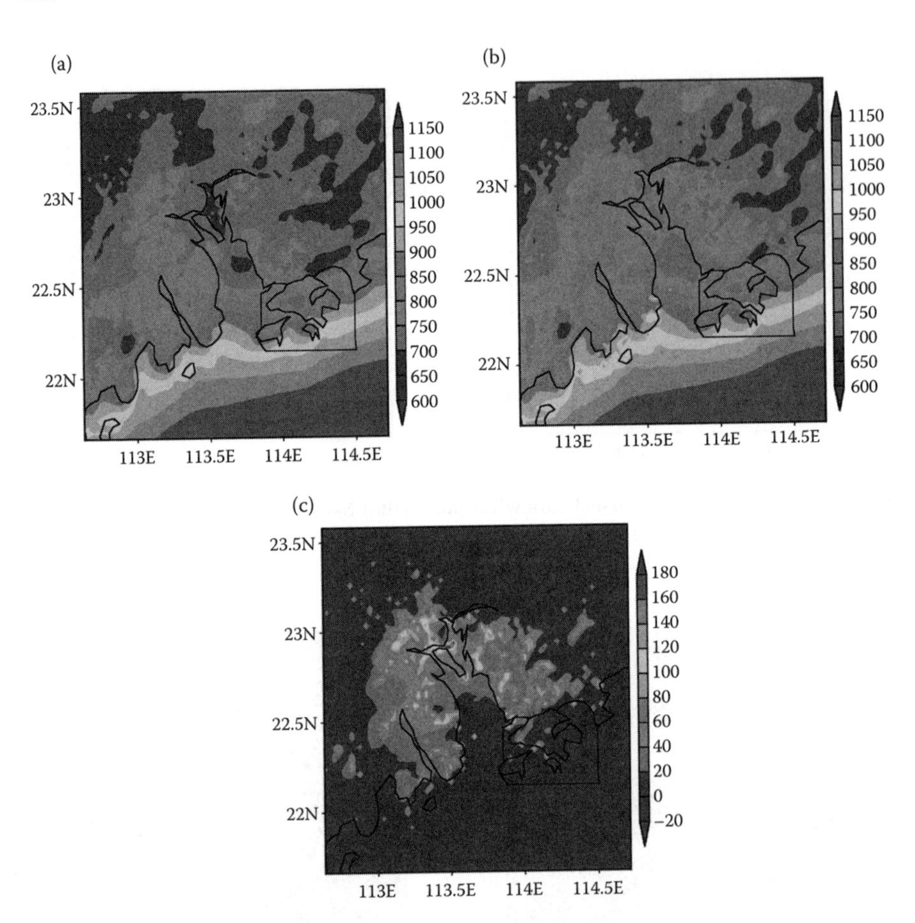

FIGURE 19.10 **(See color insert following page 206.)** Same as Figure 19.5, except for PBL height (Unit: m).

(Figure 19.11c) of domain D02 using LC1980 and LC2000 are shown in Figure 19.11. By and large, the spatial pattern of simulated monthly mean temperature is in good agreement with that of observed monthly mean temperature (Figure 19.4b). Figure 19.11c shows the increase in monthly mean temperature from 1980 to 2000 from 0.1°C to 0.3°C, and the maximum difference of 0.3°C was found in the city of Shenzhen, attributable to the rapid urbanization as a result of the establishment of the Shenzhen special economic zones in the 1980s. Figure 19.11d and e shows the monthly mean temperature and temperature difference in and around the urban areas of Guangzhou, Shenzhen, Dongguan, and Zhongshan.

19.5.1.1 Moisture and Precipitation

A persistent rainfall event occurred in Guangdong province during June 2005, with the maximum total rainfall of 160 cm received in Dongguan city. The observed total precipitation in June 2005 for Guangdong province is shown in Figure 19.4c. Figure 19.12a and b depicts the simulated monthly total precipitation for the same period using LC2000 and LC1980, respectively. The maximum difference between

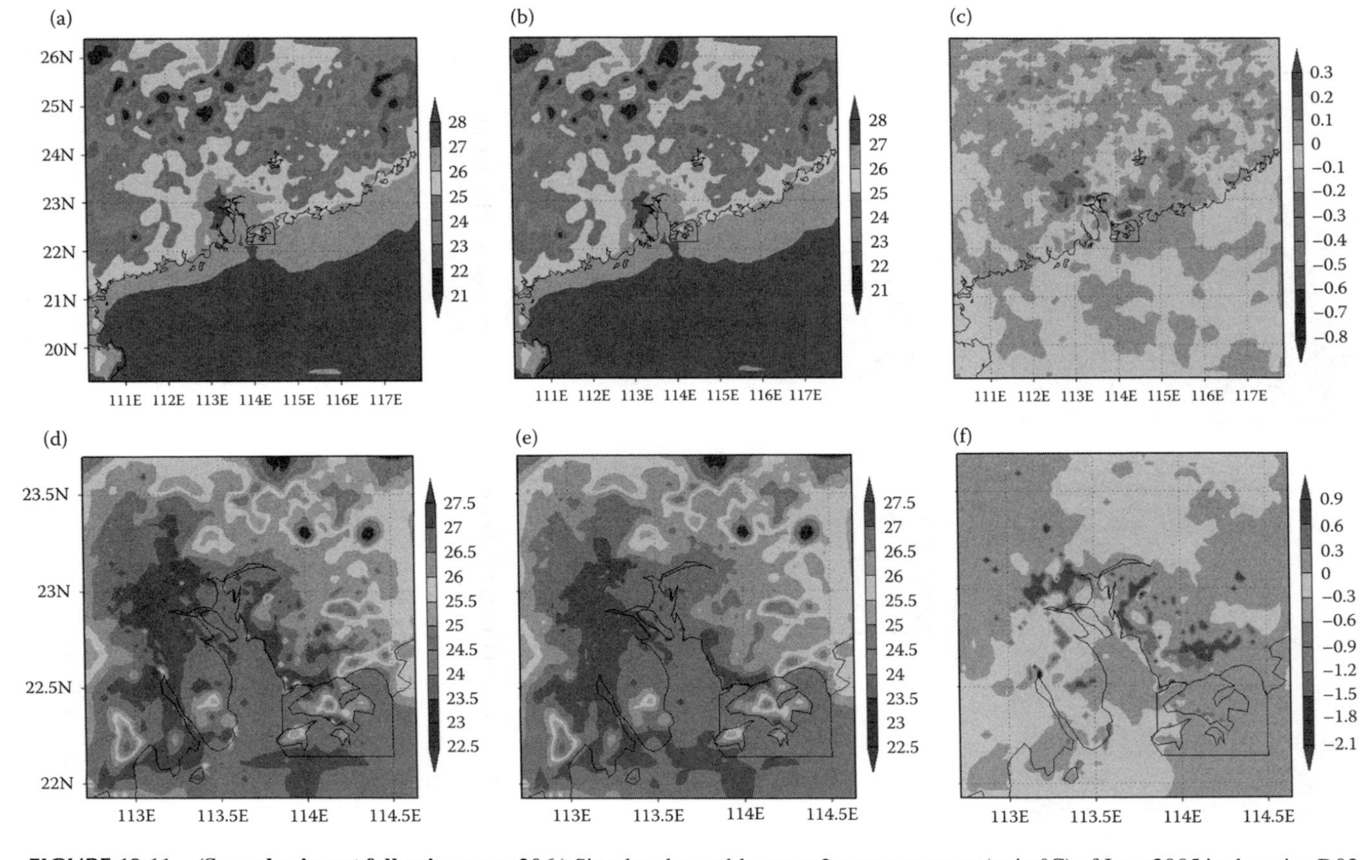

FIGURE 19.11 **(See color insert following page 206.)** Simulated monthly mean 2 m temperature (unit: °C) of June 2005 in domains D02 and D03: (a) D02 with LC2000; (b) D02 with LC1980; (c) the difference between (a) and (b); (d) D03 with LC2000; (e) D03 with LC1980; and (f) the difference between (d) and (e).

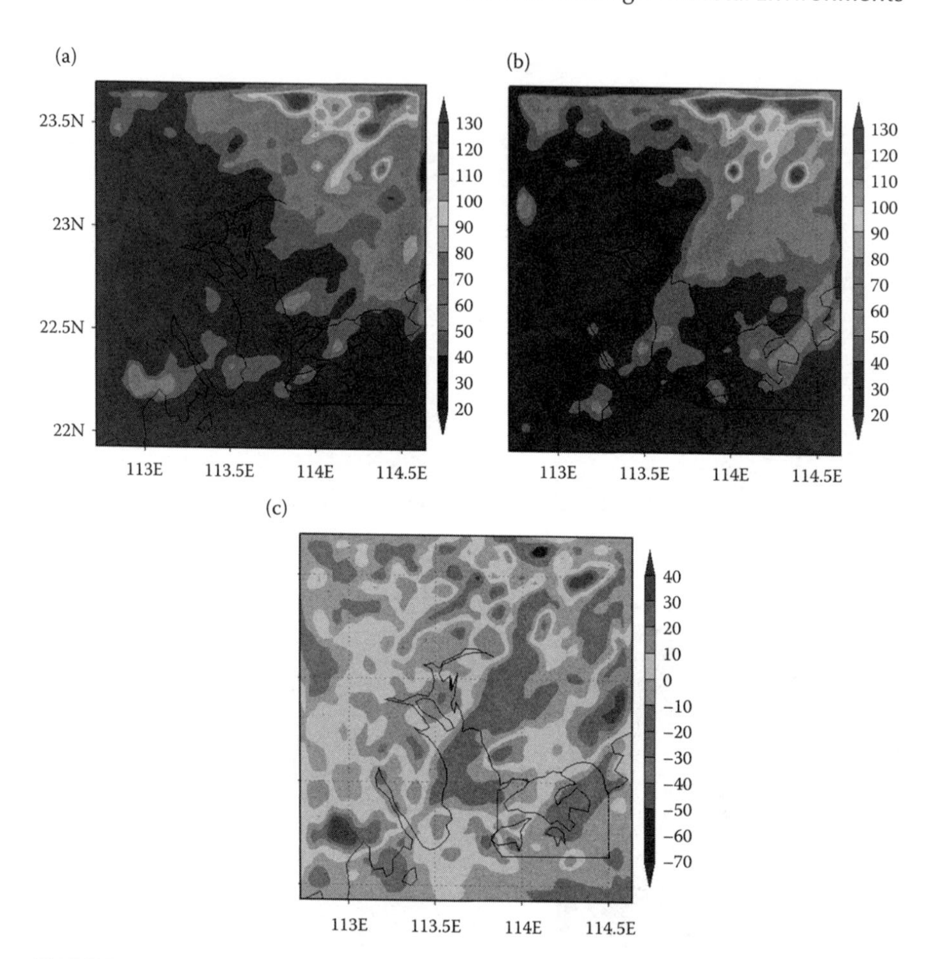

FIGURE 19.12 **(See color insert following page 206.)** Simulated monthly total precipitation (unit: cm) of June 2005 in domain D03: (a) with LC2000; (b) with LC1980; and (c) the difference between (a) and (b).

the two simulations in monthly accumulated precipitation was 10 cm in the urban areas of Guangzhou, Zhongshan, Zhuhai, and Shenzhen.

As the urban area possesses little moisture than does the rural land surface, both simulated mixing ratio and relative humidity in all urban areas were lower (Figure 19.13). The maximum decrease in mixing ratio was 1.4 g/kg in Guangzhou, Dongguan, Shenzhen, and Zhuhai. The largest decrease in simulated relative humidity was 1.8% in Guangzhou, Dongguan, and Shenzhen, and a smaller decrease was found in Zhongshan and Zhuhai.

19.5.1.2 Surface Fluxes

The differences in simulated monthly mean sensible and latent heat fluxes between LC2000 and LC1980 scenario are shown in Figure 19.14a and b. The maximum increase of 30 W/m^2 of sensible heat flux was found in urbanized areas. As expected,

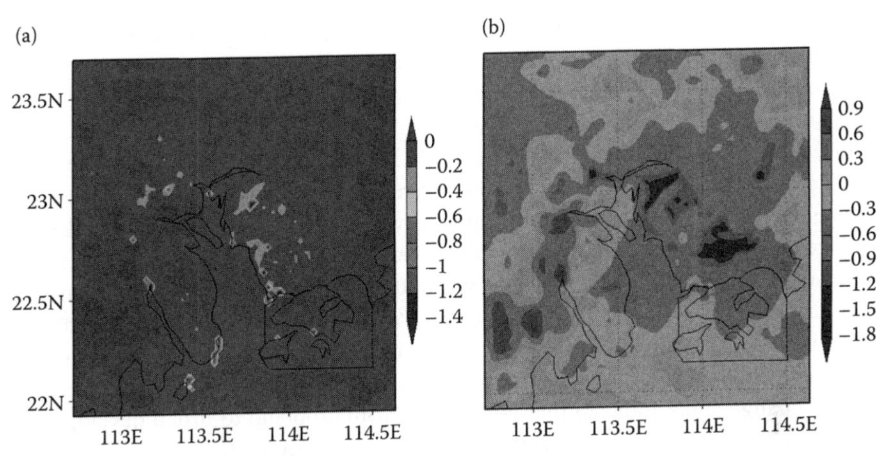

FIGURE 19.13 **(See color insert following page 206.)** Differences between LC2000 and LC1980 in (a) monthly mean 2 m mixing ratio (unit: g/kg); (b) monthly mean relative humidity (unit: %).

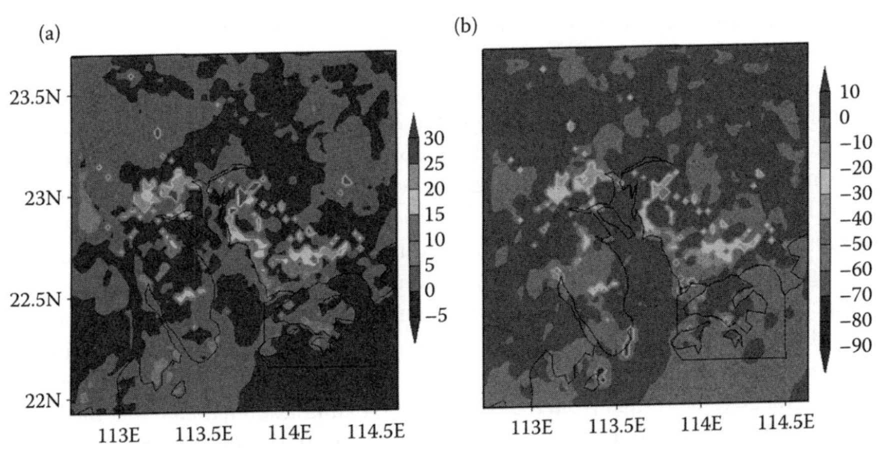

FIGURE 19.14 **(See color insert following page 206.)** Differences between LC2000 and LC1980 in (a) monthly mean sensible heat flux (unit: W/m^2); (b) monthly mean latent heat flux (unit: W/m^2).

areas with increased sensible heat flux correlated to the locations with high monthly mean temperature–areas that had undergone urban expansion. In contrast, latent heat flux decreased in all urbanized areas and the average decrease was about 30 W/m^2. Again the areas with decreased latent heat flux corresponded to areas where urban expansion occurred.

19.6 SUMMARY AND CONCLUSION

As one of the most rapidly developing regions in the world, the PRD region has undergone rapid socioeconomic transformations in the past three decades. As a result, significant changes in LULC have taken place and such changes have an important impact on the regional climate. This study quantified the regional land-cover changes by producing and analyzing three LULC maps *ca.* 1980, 1990, and 2000, using Landsat GeoCover imagery aided by other ancillary data. This new land-cover change dataset provides a basis for studying the LULC change of the region.

The development of the PRD region as a result of extensive conversion from the agricultural land use to urban development was a predominant process of LULC change that occurred in the region from the 1980s and 2000. Much of the conversion involved alteration from the artificial wetlands and fishing pond to the urban land use, or land reclamations along the coastal areas. This development was mostly driven by the fast increase in the manufacturing and exporting industry through the joint ventures and many other investments from the 1980s. Overall, the results of the LULC change study indicate that conversion from cropland to urban development, that is, urbanization, is the prevailing LULC change occurring in the PRD region during the period from 1980 to 2000, followed by the conversion from agri- and aquacultural businesses to urban development induced by land reclamations in the coastal areas. Besides urban expansion, the change of forest and woodland in the region is another important process of LULC change. The increased woodland and forest areas are the result of the policy change in conservation strategy and the reforestation efforts implemented after the 1980s.

The impacts of LULC changes, especially the urbanization on regional climate over the PRD, were investigated through numerical simulation by using a mesoscale climate model (MM5). Two different land-cover data of the PRD region were used for simulating impact of LULC change on weather and climate for a wet summer from June 1, 2005 to July 2, 2005, and a dry autumn of October 2004. The results suggest that the distinct heat islands formed in expanded urban areas in the city of Guangzhou, Shenzhen, and Dongguan are as a result of increased monthly mean air temperature in both summer and autumn. Other impacts due to the urban expansion include decreasing mixing ratio and relative humidity, and increasing (decreasing) of sensible (latent) heat flux.

Specifically, the results from the October 2004 simulation indicate that the maximum difference in monthly mean temperature was 0.9°C over urban areas. The monthly mean relative humidity decreased by 1.4% in urban areas. The maximum decrease of mixing ratio was 0.4 g/kg in the city of Guangzhou and Dongguan, whereas the maximum decrease of relative humidity was 2.4%. The increase of sensible heat flux in the urban land showed a maximum of 90 W/m^2, whereas the maximum decrease of latent heat flux was about 300 W/m^2. The monthly mean PBL height increased from 20 to 80 m, and the maximum positive change was 180 m over the urban land in Guangzhou.

The simulated monthly mean temperature of June 2005 agrees with that observed in the northern part of Guangdong province, but it was lower than that observed near the coastal areas. Impact of urban expansion on spatial distribution of precipitation is complex in that it increased in some urban areas, while it decreased in others. The domain-averaged and land-averaged values, which are used to indicate the land atmosphere interaction over a larger spatial extent, also show that the regional climate is sensitive to changes from urban expansion in the PRD region.

As one of the regional study, the results from this research provide useful information on major LULC changes from the 1980s to 2000 and their impact on regional climate by using satellite remote sensing data and a mesoscale climate model. However, the model used in this study does not explicitly account for the complex urban canopy structures and their impact on physical and radiative processes within the urban boundary layer. The recent development of a coupled modeling system, which encompasses an urban canopy layer model within a mesoscale climate model (e.g., Zhang et al., 2008; Otte et al., 2004), shows improved capability for simulating the impact of urbanization on local and regional climate. This new approach will lead to a better understanding of feedbacks between LULC change and regional climate in future.

ACKNOWLEDGMENTS

This study was supported by the Research Grant Council of Hong Kong SAR, P.R. China (ref. no. CUHK4642/05H), and the Direct Grant for Research 2005–2006 of the Chinese University of Hong Kong (project code 2020881). The Landsat images and DEM data were provided by USGS EROS Center. We thank the Resource and Environmental Scientific Data Center of the Chinese Academy of Science for supplying a China LULC dataset. The authors would like to thank two anonymous referees for valuable comments.

REFERENCES

Changnon, S.A. and Huff, F.A., 1986, The urban-related nocturnal rainfall anomaly at St. Louis, *Journal of Climate and Applied Meteorology*, 25: 1985–1995.

Charney, J., Quirk, W.J., Chow, S.H., and Kornfield, J., 1977, A comparative study of the effects of albedo change on drought in semi-arid regions. *Journal of the Atmospheric Sciences*, 34: 1366–1385.

Chase, T.N., Pielke, R.A., Kittel, T.G.F., Nemani, R., and Running, S.W., 1996, The sensitivity of a general circulation model to large-scale vegetation changes, *Journal of Geophysical Research*, 101: 7393–7408.

Chen, F. and Dudhia, J., 2001, Coupling an advanced land surface hydrology model with the Penn State-NCAR MM5 modeling system. Part I: Model implementation and sensitivity, *Monthly Weather Review*, 129: 569–585.

Dickinson, R.E., 1983, *Land Surface Processes and Climate-surface Albedo and Energy Balance: Advances in Geophysics*, Academic Press, New York, 48pp.

Dudhia, J., 1989, Numerical study of convection observed during the winter monsoon experiment using a mesoscale two-dimensional model, *Journal of the Atmospheric Sciences*, 46: 3077–3107.

Dudhia, J., 1996, A multi-layer soil temperature model for MM5. Preprint from the Sixth PSU-NCAR Mesoscale Model Users' Workshop, Boulder CO (http://mmm.ucar.edu/mm5/lsm/lsm-docs.html).

Eastman, J.L., Coughenour, M.B., and Pielke, R.A., 2001, Does grazing affect regional climate? *Journal of Hydrometeorology*, 2: 243–253.

Foley, J.A., DeFries, R., Asner, G.P., Barford, C., Bonan, G., Carpenter, S.R., Chapin, F.S., Coe, M.T., Daily, G.C., Gibbs, H.K., Helkowski, J.H., Holloway, T., Howard, E.A., Kucharik, C.J., Monfreda, C., Patz, J.A., Prentice, I.C., Ramankutty, N., and Snyder, P.K., 2005, Global consequences of land use: Review, *Science*, 309: 570–574 (DOI: 10.1126/science.1111772).

Grell, G.A., Dudhia, J., and Stauffer, D.R., 1994, A description of the fifth-generation Penn State/NCAR mesoscale model (MM5), NCAR Technical Note, NCAR/TN-398+STR, National Center for Atmospheric Research, Boulder, CO, 117pp.

Gupta, H.V., Bastidas, L.A., Sorooshian, S., Shuttleworth, W.J., and Yang, Z.L., 1999, Parameter estimation of a land surface scheme using multicriteria methods, *Journal of Geophysical Research*, 104(D16): 19491–19503.

Hack, J.J., Boville, B.A., Briegleb, B.P., Kiehl, J.T., Rasch, P.J., and Williamson, D.L., 1993, Description of the NCAR Community Climate Model (CCM2), NCAR Technical Note, NCAR/TN-382+STR, National Center for Atmospheric Research, Boulder, CO, 108pp.

Hansen, M.C., DeFries, R.S., Townshend, J.R.G., Sohlberg, R., DiMiceli, C., and Carroll, M., 2002, Towards an operational MODIS continuous field of percent tree cover algorithm: Examples using AVHRR and MODIS data, *Remote Sensing of Environment*, 83(1–2): 303–319.

Hong, S.Y. and Pan, H.L., 1996, Non-local boundary layer vertical diffusion in a medium-range forecast model, *Monthly Weather Review*, 124: 2322–2339.

Houghton, R.A., Hackler, J.L., and Lawrence, K.T., 1999, The U.S. carbon budget: Contributions from land-use change, *Science*, 285: 574–578.

IPCC 2001, *Climate Change 2001: Scientific Basis*, 881pp, Cambridge, United Kingdom and New York, NY, Cambridge University Press.

Kain, J.S. and Fritsch, J.M., 1993, Convective parameterization for mesoscale models: The Kain–Fritsch scheme. The Representation of Cumulus Convection in numerical Models, *Meteorological Monograph*, 46: 165–170.

Kaufmann, R.K., Seto, K.C., Schneider, A., Liu, Z., Zhou, L., and Wang, W., 2007, Climate response to rapid urban growth: Evidence of a human-induced precipitation deficit, *Journal of Climate*, 20(10): 2299.

Kistler, R., Kalnay, E., Collins,W., et al., 2001, The NCEPNCAR 50-Year Reanalysis: Monthly Means CD-ROM and Documentation. *Bulletin of the American Meteorological Society*, 82: 247–268.

Liang, J.Y. and Wu, S.S., 1999, Climatological diagnosis of winter temperature variations in Guangdong, *Journal of Tropical Meteorology*, 15(3): 221–229 (in Chinese).

Lo, C.F., Lau, A.K.H., and Fung, J.C.H., 2004, Impact of land surface process in MM5 over Hong Kong and Pearl River Delta, 14th MM5 User's Workshop, June 22–25, 2004, Boulder, CO.

Lo, C.F., Lau, A.K.H., Chen, F., Fung, J.C.H., and Leung, K.K.M., 2007, Urban modification in a mesoscale model and the effects on the local circulation in the Pearl River Delta region, *Journal of Applied Meteorology and Climatology*, 46: 457–476.

Otte, T.L., Lacser, A., Dupont, S., and Ching, J., 2004, Implementation of an urban canopy parameterization in a mesoscale meteorological model, *Journal of Applied Meteorology*, 43: 1648–1665.

Molders, N. and Olson, M.A., 2004, Impact of urban effects on precipitation in high latitudes, *Journal of Hydrometeorlogy*, 5: 409–429.

Pielke, R.A., Lee, T.J., Copeland, J.H., Eastman, J.L., Ziegler, C.L., and Finley, C.A., 1997, Use of USGS-provided data to improve weather and climate simulations, *Ecological Applications*, 7: 3–21.

Poveda, G. and Mesa, O.J., 1997, Feedbacks between hydrological processes in tropical South American and large-scale ocean-atmospheric phenomenon, *Journal of Climatology*, 10: 2690–2702.

Reisner, J., Rasmussen, R.M., and Bruintjes, R.T., 1998, Explicit forecasting of supercooled liquid water in winter storms using the MM5 mesoscale model, *The Quarterly Journal of the Royal Meteorological Society*, 124: 1071–1107.

Rodriguez, E., Morris, C.S., Belz, J.E., Chapin, E.C., Martin J.M., Daffer, W., and Hensley, S., 2005, An assessment of the SRTM topographic products, Technical Report JPL D-31639, Jet Propulsion Laboratory, Pasadena, CA.

Sellers, P.J., Randall, D.A., Collatz, G.J., Berry, J., Field, C., Dazlich, D.A., Zhang, C., and Bounouam L., 1996, A revised land-surface parameterization (SiB2) for atmospheric GCMs. Part 1: Model formulation, *Journal of Climatology*, 9: 676–705.

Seto, K.C. and Kaufmann, R.K., 2003, Modeling the drivers of urban land use change in the Pearl River Delta, China: Integrating remote sensing with socioeconomic data, *Land Economics*, 79: 106–121.

Seto, K.C., Woodcock, C.E., Song, C., Huang, X., Lu, J., and Kaufmann, R.K., 2002, Monitoring land-use change in the Pearl River Delta using Landsat TM, *International Journal of Remote Sensing*, 23: 1985–2004.

Shepherd, J.M., 2005, A review of current investigations of urban-induced rainfall and recommendations for the future, *Earth Interactions*, 9: 1–27.

Shepherd, J.M., 2006, Evidence of urban-induced precipitation variability in arid climate regimes, *Journal of Arid Environments.*, 67: 607–628.

Souch, C. and Grimmond, S., 2006, Applied climatology: Urban climate, *Progress in Physical Geography*, 30: 270–279.

Stohlgren, T.J., Chase, T.N., Pielke, R.A., Kittel, T.G.F., and Baron, J., 1998, Evidence that local land use practices influence regional climate and vegetation patterns in adjacent natural areas. *Global Change Boil*, 4: 495–504.

Tucker, C.J., Grant, D.M., and Dykstra, J.D., 2004, NASA's global orthorectified landsat data set, *Photogrammetric Engineering and Remote Sensing*, 70(3): 313–322.

Weisman, M.L., Skamarock, W.C., and Klemp, J.B., 1997, The resolution dependence of explicitly modeled convection, *Monthly Weather Review*, 125: 527–548.

Weng, Q., 2002, Land use change analysis in the ZLC2000jiang Delta of China using satellite remote sensing, GIS and stochastic modeling, *Journal of Environmental Management*, 64: 273–284.

Westrick, K.J. and Mass, C., 2001, An evaluation of a high resolution hydrometeorological modeling system for prediction of a cool-season flood event in a coastal mountainous watershed. *Journal of Meteorology*, 2: 161–180.

Wu, W.L., Lynch, A.H., and Rivers, A., 2005, Estimating the uncertainty in a regional climate model related to initial and lateral boundary conditions, *Journal of Climatology*, 18: 917–932.

Yang, Z.L., Dickinson, R.E., Henderson-Sellers, A., and Pitman, A.J., 1995, Preliminary study of spin-up processes in land-surface models with the first stage data of PILPS phase 1(a), *Journal of Geophysical Research*, 100: 553–578.

Xue, Y., Zeng, F.J., Mitchell, K., Janjic, Z., and Rogers, E., 2001, The impact of land surface processes on the Simulation of the U.S. hydrological cycle: A case study of 1993 US flood using the Eta/SSiB regional model, *Monthly Weather Review*, 129: 2833–2860.

Yeh, A., and Li, X., 1997, An integrated remote sensing and GIS approach in the monitoring and evaluation of rapid urban growth for sustainable development in the Pearl River Delta, China, *International Planning Studies*, 2: 193–210.

Yeh, A. and Li, X., 1998, Sustainable land development model for rapid growth areas using GIS, *International Journal of Geographical Information Science*, 2: 169–189.

Zhang, H.B., Sato, N., Izumi, T., Hanaki, K., and Aramaki, T., 2008, Modified RAMS-Urban Canopy Model for Heat Island Simulation in Chongqing, China, *Journal of Applied Meteorology and Climatology*, 47: 509–524.

Zhou, L., Dickinson, R., Tian, Y., Fang, J., Li, Q., Kaufmann, R., Tucker, C., and Myneni, R., 2004, Evidence for a significant urbanization effect on climate in China, *PNAS*, 26: 9540–9544.

20 Geospatial Information for Sustainable Development: A Case Study in Coastal East Africa

Yeqiao Wang, James Tobey, Amani Ngusaru, Vedast Makota, Gregory Bonynge, and Jarunee Nugranad

CONTENTS

20.1 INTRODUCTION

Sustainable development, as defined in "Our Common Future: The Report of the World Commission on Environment and Development [1]," is "development that meets the needs of the present without compromising the ability of future generations to meet their own needs." In 1999, the U.S. National Research Council added a specific environmental dimension defining sustainable development as "the reconciliation of society's developmental goals with its environmental limits over the long term [2]." Therefore sustainable development is a long-term goal whose implementation is reflected in a variety of action programs, of which Agenda 21—the Rio

Declaration on Environment and Development, adopted by world leaders in the 1992 U.N. Conference on Environment and Development held in Rio de Janeiro, Brazil, is the most prominent. The themes of sustainable development under consideration include biodiversity, deforestation, urbanization, environmental pollution, food production, freshwater supply, health and disease, coastal development, rural development, and climate change, among others. Chapter 40 of the Agenda 21 stresses the need for more and different types of data to be collected to track the status and trends of the Earth's ecosystems, natural resources, pollution, and socioeconomic variables for informed decision-making. The implementations of Agenda 21 was reaffirmed at the World Summit on Sustainable Development (WSSD), held in 2002 in Johannesburg, South Africa. The WSSD conference supported practical, result-oriented programs to alleviate poverty, environmental degradation, food insufficiency, and natural resource mismanagement.

While many of the Agenda 21 issues remain severe, the past decade has seen a significant improvement in the Earth observation data and information management systems. Improved geospatial technologies can help overcome a number of implementation challenges and offer a starting point on the path to sustainable development [3]. A key challenge for the international community is to make geographic information more accessible and useful to decision-makers working on sustainable development problems.

Spatial data and information, such as geographic location and elevation, are considered as framework foundation data that are central to most sustainable development applications. The data are readily accessible through Global Positioning Systems (GPS), digital elevation models and Shuttle Radar Topography Mission data. Other framework thematic data, such as land-use and land-cover maps, are essential to the planning, implementation, and monitoring of many of the action items identified by Agenda 21. As a contribution to the WSSD, the Geographic Information for Sustainable Development (GISD) initiative selected four case study regions, including the Upper Niger, East African Great Lakes, Kenya–Tanzania Coast, and Limpopo/Zambezi Basin, as efforts at application of Earth observation data, geographic information system (GIS) technologies, and field knowledge to address issues of natural resource management and sustainable development in Africa [4]. This chapter summarizes the Kenya–Tanzania coast study.

The coastal region of Kenya and Tanzania in East Africa has been undergoing notable changes. Primary coastal issues important to sustainable development include intensification of agriculture and mariculture, declining resource base, destruction of critical habitats such as mangrove forest, rapid expansion of coastal cities, increasing population, and overfishing. Selected field photographs illustrate examples of affected landscapes (Figure 20.1a–g). Extraction of living coral for conversion into lime is widespread along the entire coast. Production of lime by burning pieces of coral harvested from reefs was once a common source of coastal livelihood that is particularly destructive to both mangroves and coral reefs. Locals extract live and dead corals from reefs using pick axes, crowbars, and other implements and then bring corals ashore where they are piled into kilns and burned to produce lime for local construction and trade (Figure 20.1h). This activity is very fuel intensive and much of the fuel wood is from mangroves.

FIGURE 20.1 GIS-generated map illustrates the coastal districts of Tanzania mainland. The field photographs illustrate issues related to sustainable development in coastal regions such as intensification of agriculture by slash-and-burn practices (a–d), affected mangroves (e), increasing population and overfishing threaten important coral reefs and fish-breeding grounds (f). The corals are piled to be burned to produce lime for local building and trade (g).

The Tanzania coastal zone is recognized at the highest levels of government as an asset of virtually unparalleled socioeconomic and ecological importance. There are over 200 coastal villages that are homes to a quarter of the nation's population. The coastal region accounts for only 15% of the total land area, but about 75% of the nation's industry is located in the coastal zone. Coastal areas of Tanzania are extremely important for efforts to reduce poverty and promote food security and human development. In addition to these coastal issues, there is a lack of coastal maps for analysis and planning with geographically referenced data on coastal resources and land use; no comparable trend data; no sustained strategies for application of spatial information in coastal decision-making; and minimal in-country capacity for geospatial data preparation.

In Tanzania there are 11 marine protected areas and coastal management programs, in addition to a national Integrated Coastal Management program. Each of these efforts requires reliable knowledge and understanding in order to guide sustainable use of coastal resources, resolve human-induced problems, and improve governance systems [5]. The need for reliable information and understanding is becoming more evident as the complexity of the relationships among the environment, resources, and the economic and social well-being of resident coastal people is fully recognized. Decision-makers need data and tools to monitor and assess natural resource inventories, environmental change, and social change. Therefore, the objectives of this study were to analyze the change of coastal natural resources and land-use and land-cover types; to identify priority locations for coastal conservation, aquaculture, and land-use zoning; and to build capacity and provide a boost to a long-term sustained effort in the use of science and technology for coastal management.

Remote sensing is well documented as an effective tool for mapping and characterizing cultural and natural resources. The multispectral capabilities of remote sensing allow observation and measurement of biophysical characteristics, whereas the multitemporal and multisensor capabilities allow tracking of changes in these characteristics over time. A common use of satellite data is production of land-cover maps that indicate landscape patterns and human development processes [6,7]. Remote sensing change detection can be used to discern and simulate areas that have been altered by natural or anthropogenic processes [8,9]. Change detection methods have often been discussed in the literature [10,11] and issues in their application have been reviewed extensively [12,13]. Remote sensing has been identified as an effective tool to study otherwise difficult-to-reach and difficult-to-penetrate mangroves along coastal areas [14]. Landsat and SPOT imageries have been applied to mangrove studies through visual interpretation and classifications [15–20]. SIR-C radar data were applied to mangrove identification [21] and Shuttle Radar Topography Mission data were used for mapping the height and biomass of mangroves [22]. High spatial resolution satellite remote sensing data have been used in mangrove mapping [23–25].

Satellite remote sensing has been applied in Tanzania for general vegetation and land-cover mapping since the 1980s. A 1984 map of Tanzania vegetation types was compiled from the interpretation of 1974 Landsat Multispectral Scanner images at a scale of 1:2,000,000. Information on coastal natural resources, such as mangrove forests, was not available in this early map. Two recent efforts at application of satellite remote sensing for general land-cover mapping in Tanzania included the United Nations Africover Project and a natural resource mapping project executed by Hunting Technical Services (HTS) of the United Kingdom.

The HTS land-cover map was published at a 1:250,000 scale, which corresponds with existing topographic maps. The thematic maps were generated from interpretation of scale-controlled images. The mosaics of Landsat TM and SPOT satellite images used by HTS were acquired between May 1994 and July 1996. The HTS classification scheme used 34 land-cover categories. The United Nations Food and Agricultural Organization (FAO) hosted the Africover project. A goal of the Africover project is to establish a digital georeferenced database of land cover for African countries. Africover's product for Tanzania was produced by visual interpretation of Landsat and SPOT images acquired between February 1995 and June 1998. Land-cover categories were complied using an internationally adopted Land Cover Classification System [26]. The final map product was at a 1:200,000 scale. The above two map products provided good reference maps but were not suitable for the purpose of change study.

In this study, we developed land-use and land-cover geospatial data at a 1:50,000 scale for the coastal districts of Tanzania over two time periods using Landsat remote sensing data. We then used the land-use and land-cover datasets to analyze rates of resource and land-use change over approximately 10 years between 1990 and 2000. The study helped answer key management questions; identify priority locations for coastal action planning and conservation; and develop local-level coastal management action plans in support development decisions.

20.2 METHODS

20.2.1 Study Areas and Data Sources

The Tanzania coastline extends over 800 km from the border with Kenya in the north at about 4°37′ S, 39°06′ E to the border River of Ruvuma with Mozambique in the south at about 10°30′ S, 40°23′ E. We focused on the 14 mainland coastal districts of Tanzania, which consist of Bagamoyo, Ilala, Kibaha, Kilwa, Kinondoni, Kisarawe, Lindi, Mkuranga, Mtwara, Muheza, Pangani, Rufiji, Tanga, and Temeke (Figure 20.1).

We used Landsat TM scenes acquired between 1988 and 1990 with path/row numbers of P165/R66-67 and P166/Row63-66 and we refer to these scenes as the 1990 time period images (Figure 20.2a). We used Landsat-7 ETM+ images acquired in 2000 and 2001 with the same path/row numbers and we refer to these scenes as the 2000 time period images (Figure 20.2b). The TM and ETM+ images possess 30 m spatial resolution with multispectral coverage ranging from visible light to the middle infrared portion of the electromagnetic spectrum. In particular, these data were very effective in identifying mangrove areas because of their unique spectral feature (Figure 20.2c). We did not include thermal band data in this study. All images were orthorectified and georeferenced to the Universal Transverse Mercator (UTM) map projection and coordinate system. We adopted a land-use and land-cover classification scheme that consisted of nine general categories with 21 imbedded subcategories (Table 20.1). Among the land-use and land-cover types, urban and settled areas, mangrove forests, woodland categories, and cultivated lands were the topics of primary interest.

20.2.2 Field Verification

We conducted a field verification exercise to support image interpretation and land-cover mapping. The field observations provided essential independent reference data for identifying land-use and land-cover types within the Landsat scenes as well as for accuracy assessment. Since we used the ground reference data for interpretation of the two time periods of Landsat images, we paid special attention to the representative locations where land use and land cover had changed over the past 10 years. We visited Regional and District management offices to seek advice and guidance. District natural resource officers reviewed laminated hardcopies of the satellite imagery with us and identified areas of interest.

We employed a Trimble® GeoExplorer® 3 GPS unit to record the locations of field transects and points of interest. We recorded general characteristics of the landscape

FIGURE 20.2 Remote sensing data included Landsat TM images acquired between 1986 and 1993 (a) and Landsat-7 ETM+ images acquired between 2000 and 2001 (b). An example of Landsat ETM+ image along the southern coast of Rufiji River delta illustrates mangrove forests in dark color (c). Field photographs (d, e) identified conditions of mangrove forests with recorded latitude and longitude coordinates by GPS and compass facing directions where the photographs were taken.

and associated information using a data dictionary uploaded to the GeoExplorer® 3. Items in the GPS data dictionary included site and transect identification information, date, time, and a description of the landscape by characterizing the land use, vegetation canopy and canopy density, understory vegetation, and other site-specific comments. We also referenced land-cover types along transects on the Africover and HTS maps. The field survey covered a total of 3500 km by road. We imported GPS locations and the recorded attributes into a GIS database to facilitate integration of the data with the Landsat TM/ETM+ images. Also as part of this effort, we recorded georeferenced site photographs along transects and the points of interest using a Kodak® DC265 Field Imaging System. The equipment consists of a 12-channel

TABLE 20.1
Classification Scheme of the Land-Use and Land-Cover Types

Land-Cover Types	Code	Subclasses
Urban/settled	U	1. Urban/settled areas
Cultivated land	Ca	2. Cultivated agriculture (maize, rice)
	Ct	3. Cultivated tree (sisal, coconut, spice tree)
Forest	Fn	4. Natural forest
	Fp	5. Forest plantation
	Fm	6. Mangrove forest
Woodland	Wc	7. Closed woodland
	Wo	8. Open woodland
	Wcr	9. Woodland with crops
Bushland	Bd	10. Dense bushland
	Bo	11. Open bushland
	Bcr	12. Bushland with crops
Grassland	G	13. Grassland
	Gtf	14. Grassland; temporarily flooded
	Gcr	15. Grassland with crops
Barren land	BS	16. Bare soil
	Sc	17. Salt crust
	Sb	18. Sandy beaches
Water	IW	19. Inland water
	O	20. Ocean
Wetland	Sm	21. Swamp/marsh

Garmin® GPS III unit connected to a digital camera. The field photographs recorded both geographic location by latitude/longitude coordinates and the general compass bearing at the time of photography (Figure 20.2c and d). We recorded over 1500 georeferenced digital photographs that, when augmented with GPS data from the GeoExplorer® 3, effectively identified locations and characteristics of coastal land-use and land-cover types on Landsat images.

20.2.3 Image Interpretation

We delineated land-use and land-cover types by visual interpretation and heads-up digitizing within a GIS environment. We chose manual interpretation instead of digital image classification for several reasons. First, spectral signatures for some land-use and land-cover types are not unique in the study areas. For example, agriculture in coastal Tanzania is always interspersed with woodland or grassland areas, resulting in a landscape that is difficult to achieve a reliable result on image classification. Rural settlements are typically constructed with natural materials and are therefore often difficult to separate from their surroundings. Cultivated trees could be confused with other open and closed woodland and forested areas. Mapping land-use and land-cover types in this developing country require a comprehensive understanding of the characteristics of the landscape. Visual image interpretation and subsequent manual delineation served this purpose well. Secondly, as the data infrastructure was not complete in Tanzania, conventional postclassification improvement using GIS data was impossible. Thirdly, the local management officers preferred hard copy maps generated from vector format GIS data, rather than maps produced from the raster data resulting from digital image classification. Finally, local professionals felt more confident using manual interpretation in order to produce future updates of the land-cover and land-use data. Given these considerations, we chose manual interpretation to produce the land-use and land-cover maps and to conduct the subsequent change analysis. To reduce inconsistency that might be introduced in the image interpretation process we crosschecked the images in both time periods for quality assurance.

A field referencing exercise with input from local experts helped correct and improve identification of the interpreted land-cover types (Figure 20.3a through c). Besides conventional accuracy assessment measures, we conducted a 3-day workshop following the completion of the interpretation of the Landsat images. We invited natural resource officers from each of the districts and national government agencies. Those professionals had knowledge of the landscape and changing history of their own districts. We asked all participants to bring helpful information to the workshop about land cover in their respective districts, including vegetation types, agriculture and forestry activities, urban development statistics, and locations of mangrove replanting. Participants worked on the assignment actively and marked on the hardcopy maps where errors or incorrect interpretations were found, and introduced their recommendations on corrections (Figure 20.3d–f). The results of the workshop were used to revise our land-cover interpretation. This verification gave additional scientific credibility to the land-use and land-cover maps produced from the data at the conclusion of this effort.

20.3 RESULTS

The interpretation and mapping process produced land-cover maps for the 1990 and 2000 time periods for each of the coastal districts. Upon completion of the image interpretation, we calculated the total areas for each of the land-use and land-cover types from the two time periods for the coastal region (Table 20.2). We then used the administrative boundaries of the coastal districts to determine areas of the land-cover types within each district (Table 20.3). Figure 20.4 illustrates examples of land-cover maps for the 1990 and 2000 time periods for the districts of Kisarawe and Mkulanka, Rufiji, and Mtwara. These maps provided visual presentation for the identification of their respective geographic locations and change patterns.

The data indicate that urban and settlement areas increased from 38,877 has in the 1990 time period to 75,043 ha in 2000, a 93% increase in about 10 years. Urban land cover mostly increased in surrounding major coastal cities, such as Dar es Salaam, Tanga, Kilwa, Lindi, and Mtwara. Comparison of the resulting land-cover maps shows that areas of mangrove forests declined by 3828 ha from 112,135 ha in the 1990 time period to 108,307 ha in 2000. Conversion from closed or dense

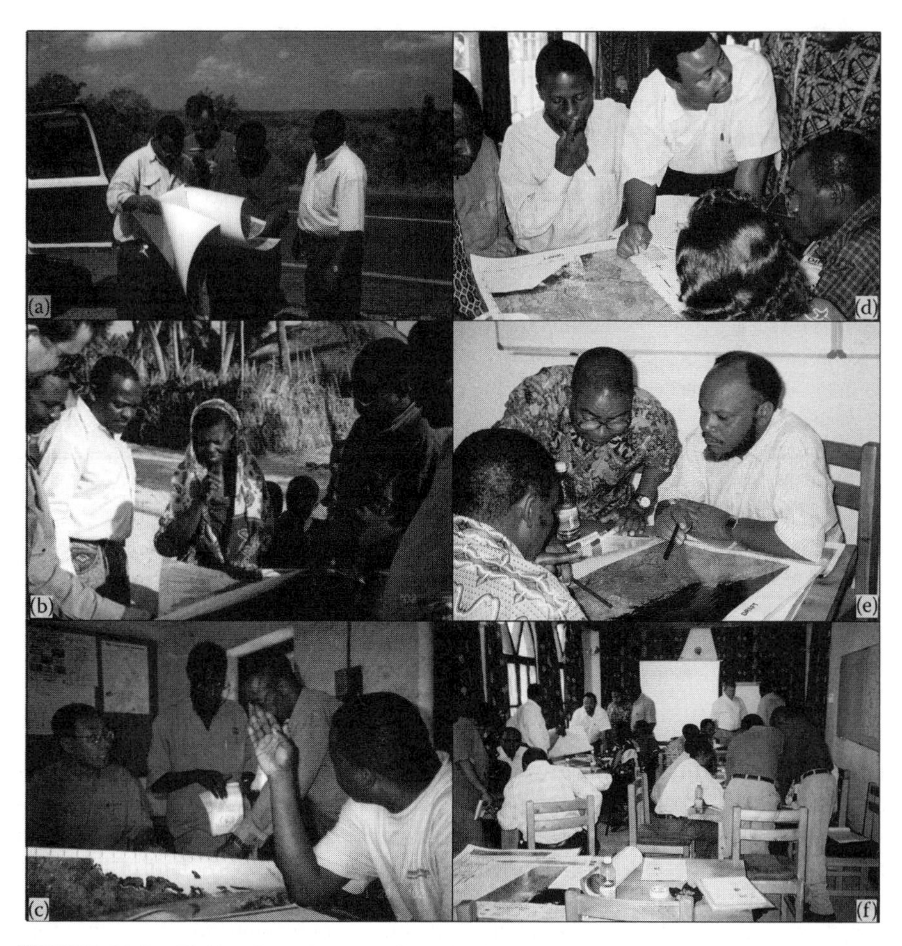

FIGURE 20.3 Field referencing exercise with input from local knowledge helped in the identification of land-cover types (a–c). Participants of the accuracy assessment workshop worked on the assignment and marked on the hardcopy maps where errors or incorrect interpretations were found, and introduced their recommendations on corrections (d–f).

woodland into open woodland and woodland with scattered agriculture was one of the major changes in the coastal areas over the time period. The area of closed woodland declined by 149,200 ha, while the area of woodland with cultivation increased by 78,285 ha and the area of open woodland increased by 8583 ha. Human impacts were the main reason for the change.

The increase in urban and settlement areas coincides with a parallel increase in population. Census data show that Dar es Salaam has gone from a city of 356,000 in 1967 to a city of 2.5 million in 2002. Between 1988 and 2002, the population of Dar es Salaam grew by 83%. Poverty and lack of livelihood opportunities were thought to be the key reasons for the rural-to-urban migration.

Mangrove areas declined in the districts of Rufiji, Kilwa, and Kisarawe and slightly increased in Tanga and Mtwara. However, an increase in area does not

TABLE 20.2
Land-Cover Types in Tanzania Coast Districts between 1990 and 2000 Time Periods

Land-Cover Types	Area (ha) 1990	Area (ha) 2000	Change in Area (ha)	Percent Change (%)
Bare soil	17,281	36,087	18,806	109
Urban/settled	38,877	75,043	36,166	93
Grassland	363,177	500,373	137,196	38
Grassland/flooded	108,189	136,647	28,458	26
Woodland with crops	571,163	649,448	78,285	14
Cultivated tree	419,491	425,767	6276	1
Open woodland	1,012,068	1,020,651	8583	1
Mangrove forest	112,135	108,307	−3828	−3
Open bushland	930,941	905,090	−25,851	−3
Wetland	13,688	12,179	−1509	−11
Cultivated agriculture	182,614	161,994	−20,620	−11
Closed woodland	1,189,373	1,040,173	149,200	−13
Grassland with crops	580,749	455,722	125,027	−22

necessarily mean that the mangrove ecosystem is in good condition. Further studies about mangrove ecosystem functions and density of the mangrove trees are necessary to answer biological questions in sustainable development. The Rufiji River delta is the largest estuarine mangrove forest that contains nearly half of the mangrove area in Tanzania. The mangrove environment in Rufiji is still quite intact. It supports a large number of other plants and animals indigenous to mangroves, thus presenting a unique ecological unit. The interpretation results indicate that 51,121 and 49,032 ha of mangroves existed in the district in 1990 and 2000, respectively.

Kilwa is situated in the south of the Rufiji delta. It is a sparsely populated district. The area has extensive shallow water with several islands, all of which are surrounded by coral, seagrass, seaweed, and mangroves. The mangrove areas were relatively stable within the district with approximately 22,454 and 21,777 ha of mangrove forest in 1990 and 2000 time periods, respectively.

Mtwara District also has extensive mangrove cover. The Mnazi Bay and Ruvuma River estuary are located in this region with extensive mangrove forests. This area has not been seriously exploited and its natural resources are in relatively good condition compared with those of other areas along the Tanzania coastline. Our data show that mangrove forests were 9486 ha in 1990 and 9643 ha in 2000. The increase could be associated with the active implementation of the mangrove management project (MMP) [27,28]. For example, newly planted mangroves were observed during the field investigation and verified on comparing the 1990 and 2000 mangrove maps. Natural regeneration of mangroves has been observed and reported in the Mtwara District as well.

TABLE 20.3
Land-Cover Changes of Coastal Districts between the 1990 and 2000 (ha)

	Agriculture		Bushland		Forest/Woodland		Mangrove		Urban	
District	**1990**	**2000**	**1990**	**2000**	**1990**	**2000**	**1900**	**2000**	**1900**	**2000**
Bagamoyo	76,914	73,050	187,218	198,205	474,896	437,830	5198	4603	2252	4012
Ilala	0	0	0	0	578	764	0	25	2868	5102
Kibaha	16,943	13,016	6536	5529	55,010	47,992	0	0	2365	7814
Kilwa	120,515	109,038	384,036	418,826	645,313	641,591	22,454	21,777	2070	4438
Kinondoni	0	0	0	2320	2318	1243	371	327	5843	9718
Kisarawe	35,150	19,915	128,739	95,558	256,233	273,389	0	0	1319	3054
Lindi	115,087	115,346	293,240	320,686	456,324	422,935	4419	4357	5213	9695
Mkuranga	34,607	43,494	15,051	4496	142,104	157,996	4940	4242	492	1436
Mtwara	48,200	51,563	87,299	83,669	209,850	206,397	9486	9643	6551	12,412
Muheza	48,886	63,236	120,984	124,005	232,842	226,019	7156	6976	2281	3713
Pangani	16,342	15,558	22,543	19,557	59,791	69,572	2335	2259	352	580
Rufiji	45,863	45,579	209,488	198,102	673,224	630,957	51,121	49,032	2054	5292
Tanga	8043	6457	11,567	11,217	3390	4912	2335	2853	2309	3185
Temeke	35,555	31,509	2188	3826	2153	419	2320	2213	2908	4592

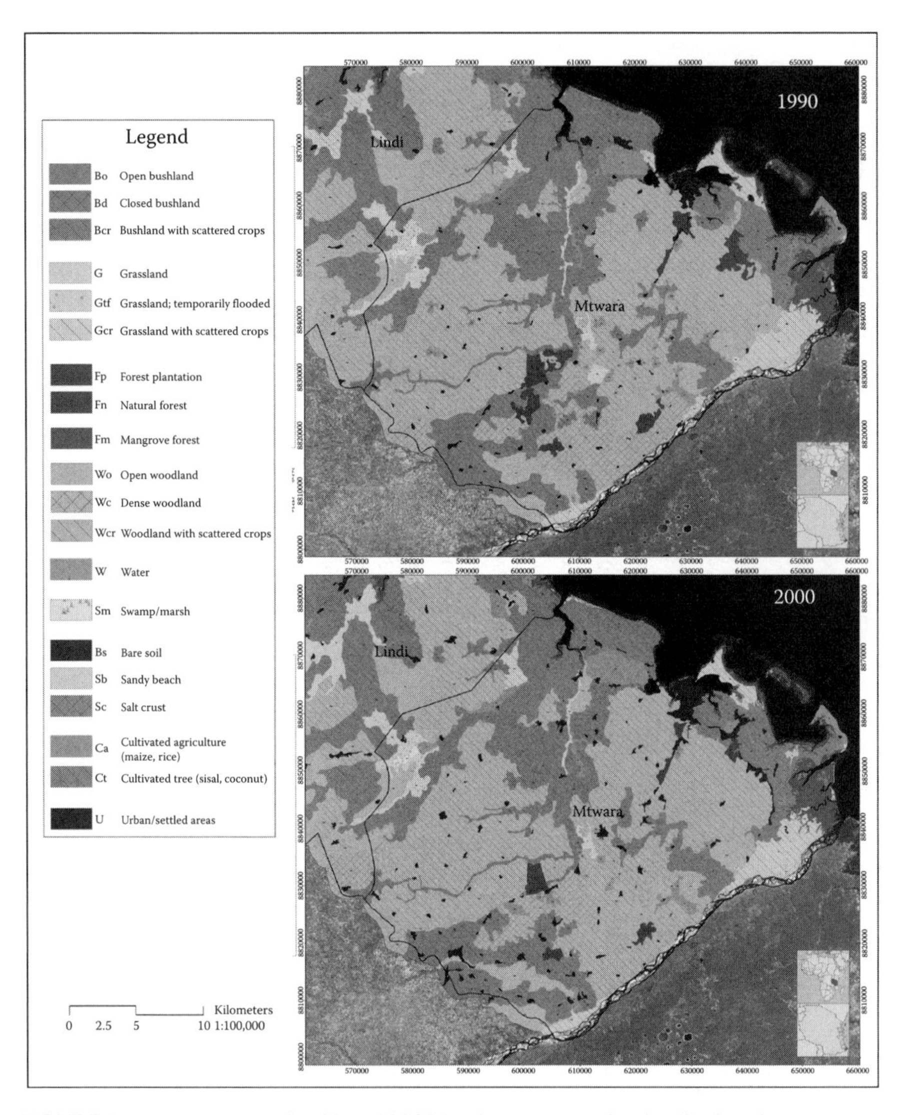

FIGURE 20.4 Examples of 1990 and 2000 land-cover maps for the district of Mtwara.

Pangani, Tanga, and Muheza districts comprise the northern zone of the MMP. Status of mangrove forests was reported in the northern zone [28]. The report stated that since the MMP started in 1994, there were no documented cases of clear cutting of mangroves for agriculture and salt farming, as well as no fuel wood cutting for salt boiling or lime burning. Nearly all the large degraded mangrove sites were replanted [29]. Illegal harvesting of mangrove products was reduced. This was attributed to increased mangrove protection efforts as well as an increase in people's awareness of the importance of the mangrove ecosystem.

The district around the urban center of Dar es Salaam has a high demand for timber products. Clearance of mangrove forests was identified in the past. Rapid rural population migration to the city has resulted in the conversion of large areas of mangrove forest to residential, business, and agricultural use. The growth rate of the city over the most recent decade was estimated at 4.8% [29]. Mangrove forests in suburban areas surrounding the city have been affected.

Conversion of closed woodland could be the result of increasing population in coastal areas and exploitation of wood resources. Without appropriate land-use management and sustainable agricultural practices, the loss of woodland will continue with associated consequences such as desertification, soil erosion, sedimentation, and the loss of coral and marine life resulting from sedimentation and turbidity in coastal waters. The implication of this study is that terrestrial land use, especially agriculture and forestry, is critical to sustainable coastal development and planning.

20.4 CONCLUSION AND DISCUSSION

Balancing natural resource management and the needs for sustainable development is a challenging task that affects virtually all of humanity. Geographic information and technologies are central to the transition from traditional environmental management to sustainable development. For African countries, some of the lessons learned on the path toward sustainable development include "needs-driven as opposed to prescriptive approaches with provision of information in appropriate and usable forms are most likely to result in effective application of geographic information [4]."

There is an increasing need for the use of science-based decisions for policy-making in Tanzania. Information on land-use and land-cover change is critically important. Coastal and natural resource managers have recognized the value of this type of information for resource management and sustainable development. This reflects the level of attention given to scientific and technological issues on the part of coastal managers.

The integrative quality of geographic information that links social, economic, and environmental data opens new opportunities for collaboration among natural scientists, social scientists, and decision-makers at all levels. The intersection of resource use, land-cover change, poverty, and environmental management, with their attendant social and economic consequences, is at the forefront of coastal and marine management in Tanzania. Successful implementation of national priorities addressing poverty alleviation and integrated coastal management requires objective scientific information to assist with developing policy priorities, understanding cause-and-effect linkages between human activities and ecosystem changes, formulating management strategies, and devising conservation measures. The results from this study demonstrate and encourage the use of geospatial technologies and information in environmental and natural resources monitoring in order to obtain information necessary to support decision-making.

The resources harvested from mangrove habitat and their resulting products are an important economic and ecological resource in Tanzania. It is estimated that over 150,000 people make their living directly from mangrove resources [29]. The importance of mangroves to Tanzania has been recognized and all mangrove areas are legally protected. It was also recognized that the area of mangrove forest does not give a full picture of the health condition of mangrove ecosystems. Field assessments confirmed that data on changed areas underrepresented the threats to mangrove forests in Tanzania and the potential for future loss. Less accessible areas such as Rufiji appear to be in good condition, but the health of mangrove forests closer to urban areas and growing settlements are far less robust. The key to their protection lies in the wise management and use of mangrove habitat, and in the enforcement of existing rules and regulations.

Major causes of the degradation of mangrove forests are the overharvesting of mangroves for firewood, charcoal-making, building poles, and boat-making and the clear-cutting of mangrove areas for agriculture, solar salt pans, road construction, urbanization, and commercial development. While mangrove forest is protected by law, controlled harvesting of mangrove poles is permitted where mangroves are ecologically stable and have sufficient regeneration potential. Most of the mangroves in the southern districts were affected by this type of commercial harvesting. An indicator of cutting pressure is the number of stumps of cut trees per area of measurement [30,31]. Legal mangrove pole and fuel wood harvesting have been monitored and recorded by mangrove management zonal offices. For example, data in Table 20.4 show that the dominant poles' harvest size is between 5 and 10 cm in diameter at buttress. Overharvesting may degrade the genetic pool of mangroves through the removal of trees with straight trunks, leaving behind only trees with badly formed stems. Illegal mangrove cutting also occurs. Overall, however, greater awareness of the value of mangrove resources among resource managers and communities has reduced the scale of degradation of mangrove vegetation.

TABLE 20.4
Recent Mangrove Poles and Fuel Wood Harvest in Tanzania

	Northern Zone	Central Zone	Southern Zone	Summary
Class 1 score		12	23	35
Class 2 score	462	95	81	638
Class 3 score	1321	1046	227	2594
Class 4 score		14,500	558	15,058
Class 5 score		967	152	1119
Fuel wood (m^3)	255	32	147	434

Notes: Mangrove poles are sold in scores of 20 poles each. One score is composed of 20 poles.

Class 1: Poles over 20 cm in diameter at buttress and not more than 30 cm in diameter at 1.3 m above ground.

Class 2: Poles over 15 cm in diameter and not more than 20 cm at buttress.

Class 3: Poles over 10 cm and not more than 15 cm in diameter at buttress.

Class 4: Poles over 5 cm and not more than 10 cm in diameter at buttress.

Class 5: Not more than 4 cm in diameter.

Economic well-being and the environment are intimately connected in coastal areas. A decline in coastal productivity of mangroves, seagrass, coral reefs, and associated nearshore fisheries has a direct and significant negative impact on coastal villages. The protection of environmental resources for people who depend on them for income generation and consumption is critical for the survival of coastal villages, for poverty reduction, and for slowing rural-to-urban migration.

Geospatial information helps us to obtain scientific consensus on the nature of conservation problems and increases the legitimacy and acceptability of policy options. Remote sensing and the derivatives of land-cover data from this study provided a foundation for integrated coastal resource planning, targeted field studies, and monitoring. With the base information developed, regular monitoring and proper management decision could be pursued for the sustainability of the mangrove resources in Tanzania. Recent applications of the data by the government of Tanzania include the development of a permitting and zoning scheme for earthen pond aquaculture in Mkuranga District, and analysis of land-cover change around Saadani National Park in Bagamoyo District.

Questions remain such as how could the resulting geospatial data be used to implement more effective decision-making? What is expected to change in terms of behavior or use of a resource in new management? Additional efforts are needed to expand commitment to resource management including shared knowledge and informed public awareness of conditions of the past and present, and understanding of the changes in resources that may generate a sense of urgency. The datasets helped connect visually the resource use and degradation, demonstrate environmental limits and impacts of resource use, and indicate benefits and successes of management policies. The enhanced capacity helped in the spread of technology and information and generate data and information that enable planning, policy, and management actions.

Geospatial data lie at the heart of many Agenda 21 issues. The values of geospatial data have been recognized and demands are increasing. Moving beyond the current capacity for application of geospatial data in Africa would require greater coordination among data providers, scientific communities, and end users. As identified by the Committee on the Geographic Foundation for Agenda 21, the earliest baseline against which future change can be compared often comes from historical legacy data. In many instances legacy data may be digitized, imported into a GIS, and analyzed in conjunction with more recent geographic data, such as satellite remote sensing data. The time scales over which change can be detected are extended through use of legacy data [4]. The datasets and change analysis out of this study provided the baseline information for inventory and monitoring of natural resource and environmental management toward sustainable development in coastal Tanzania. It is also a demonstration of how remote sensing and geospatial technology and local knowledge can be integrated into improved decision-making in coastal zone management.

ACKNOWLEDGMENTS

This study was made possible through the support provided by the Center for Economic Growth and Agricultural Development, Office of Agriculture and Food

Security, and the United States Agency for International Development (USAID) through its GISD program. The opinions expressed herein are those of the authors and do not necessarily reflect the views of USAID. We wish to express our appreciation to the professionals and officials associated with the offices of the Tanzania coastal districts for their assistance and contributions during the fieldwork and accuracy assessment workshop. The NASA Scientific Data Purchase (SDP) project provided the Landsat remote sensing data.

REFERENCES

1. WCED (World Commission on Environment and Development), 1987, *Our Common Future*, Oxford University Press, New York.
2. NRC (National Research Council), 1999, *Our Common Journey: A Transition Toward Sustainability*, National Academy Press, Washington, DC.
3. Brooner, W.C., 2002, Promoting sustainable development with advanced geospatial technologies, *Photogrammetric Engineering and Remote Sensing*, 68(3): 198–205.
4. Jensen, J.R., Botchwey, K., Brennan-Galvin, E., Johannsen, C.J., Juma, C., Mabogunje, A.L., Miller, R.B., Price, K.P., Reining, P.A.C., Skole, D.L., Stancioff, A., and Taylor, D.R.F., 2002, *Down To Earth: Geographic Information for Sustainable Development in Africa*, 155pp., National Academy of Sciences National Research Council, Washington, DC.
5. Wang, Y., Tobey, J., Bonynge, G., Nugranad, J., Makota, V., Ngusaru, A., and Traber, M., 2005, Involving geospatial information in the analysis of land cover change along Tanzania coast, *Coastal Management*, 33(1): 89–101.
6. Parmenter, A.P., Hansen, A., Kennedy, R., Cohen, W., Langner, U., Lawrence, R., Maxwell, B., Gallant, A., and Aspinall, R., 2003, Land use and land cover change in the Greater Yellowstone Ecosystem: 1975–95, *Ecological Applications*, 13: 687–703.
7. Wang, Y. and Moskovits, D.K., 2001, Tracking fragmentation of natural communities and changes in land cover: Applications of landsat data for conservation in an urban landscape (Chicago Wilderness), *Conservation Biology*, 15(4): 835–843.
8. Jantz, C.A., Goetz, S.J., and Shelley, M.A., 2003, Using the SLEUTH urban growth model to simulate the land use impacts of policy scenarios in the Baltimore–Washington metropolitan region, *Environment and Planning (B)*, 31(2): 251–271.
9. Wilkinson, D.W., Parker, R.C., and Evans, D.L., 2008, Change detection techniques for use in a statewide forest inventory program, *Photogrammetric Engineering and Remote Sensing*, 74(7): 893–901.
10. Rogan, J., Franklin, J., and Roberts, D.A., 2002, A comparison of methods for monitoring multitemporal vegetation change using thematic mapper imagery, *Remote Sensing of Environment*, 80: 143–156.
11. Hayes, D.J. and Sader, S.A., 2001, Comparison of change-detection techniques for monitoring tropical forest clearing and vegetation regrowth in a time series, *Photogrammetric Engineering and Remote Sensing*, 67(9): 1067–1075.
12. Coppin, P., Jonckheere, I., Nackaerts, K., Muys, B., and Lambin, E., 2004, Digital change detection methods in ecosystem monitoring: A review, *International Journal of Remote Sensing*, 25(9): 1565–1596.
13. Kennedy, R.E., Townsend, P.A., Gross, J., Warren, C., Bolstad, P., Wang, Y., and Adams, P., 2009, Remote sensing change detection and natural resource monitoring for managing natural landscapes, *Remote Sensing of Environment*, 113: 1382–1396.
14. Wang, Y., Ngusaru, A., Tobey, J., Makota, V., Bonynge, G., Nugranad, J., Traber, M., Hale, L., and Bowen, R., 2003, Remote sensing of mangrove change along the Tanzania coast, *Marine Geodesy*, 26(1–2): 1–14.

15. Gang, P.O. and Agatsiva, J.L., 1992. The current status of mangroves along the Kenyan coast, a case study of Mida Creek mangroves based on remote sensing, *Hydrobiologia*, 247, 29–36.
16. Dutrieux, E., Denis, J., and Populus, J., 1990, Application of SPOT data to a base-line ecological study the Mahakam Delta mangroves East Kalimantan, Indonesia, *Oceanologica Acta*, 13, 317–326.
17. Gao, J., 1998, A hybrid method toward accurate mapping of mangroves in a marginal habitat from SPOT multispectral data, *International Journal of Remote Sensing*, 19(10): 1887–1899.
18. Green, E.P., Mumby, P.J., Edwards, A.J., Clark, C.D., and Ellis, A.C., 1998, The assessment of mangrove areas using high resolution multispectral airborne imagery, *Journal of Coastal Research*, 14(2): 433–443.
19. Ramirez-Garcia, P., Lopez-Blanco, J., and Ocana, D., 1998, Mangrove vegetation assessment in the Santiago River mouth, Mexico, by means of supervised classification using Landsat TM imagery, *Forest Ecology and Management*, 105: 217–229.
20. Rasolofoharinoro, M., Blasco, F., Bellan, M.F., Aizpuru, M., Gauquelin, T., and Denis, J., 1998, A remote sensing based methodology for mangrove studies in Madagascar, *International Journal of Remote Sensing*, 19(10): 1873–1886.
21. Pasqualini, V., Iltis, J., Dessay, N., Lointier, M., Gurlorget, O., and Polidori, C., 1999, Mangrove mapping in North-Western Madagascar using SPOT-XS and SIR-C radar data, *Hydrobiologia*, 413: 127–133.
22. Simard, M., Zhang, K., Rivera-Monroy, V.H., Ross, M.S., Ruiz, P.L., Castañeda-Moya, E., Twilley, R.R., and Rodriguez, E., 2006, Mapping height and biomass of mangrove forests in Everglades National Park with SRTM elevation data, *Photogrammetric Engineering and Remote Sensing*, 72(3): 299–311.
23. Wang, L., Sousa, W.P., Gong, P., and Biging, G.S., 2004, Comparison of IKONOS and QuickBird images for mapping mangrove species on the Caribbean coast of Panama, *Remote Sensing of Environment*, 91: 432–440.
24. Mumby, P.J. and Edwards, A.J., 2002, Mapping marine environments with IKONOS imagery: Enhanced spatial resolution can deliver greater thematic accuracy, *Remote Sensing of Environment*, 2–3: 248–257.
25. Wang, L., Silvan, J., and Sousa. W., 2008, Neural network classification of mangrove species from multiseasonal IKONOS imagery, *Photogrammetric Engineering and Remote Sensing*, 74(7): 921–927.
26. Di Gregorio, A. and Jensen, L.J.M., 2000, *Land-cover Classification System (LCCS): Classification Concepts and User Manual*, 179pp., Food and Agriculture Organization of the United Nations, Rome.
27. Semesi, A.K., 1991, Management plan for the mangrove ecosystem of mainland Tanzania, Vols. 11 (Vols.1–10 joined), Ministry of Tourism, Natural Resources and Environment (MTNRE), Forestry and Beekeeping Division, Catchment Forestry Project, Dar es Salaam, 292pp.
28. Mbwambo, Z.D., 2001, *Status of the Mangrove Forest in the Northern Zone, Mangrove Management Project*, Northern Zone, Tanga, Tanzania.
29. TCMP (Tanzania Coastal Management Councial), 2001, *State of the Coast 2001: People and the Environment*, Tanzania Coastal Management Partnership, Dar es Salaam, Tanzania.
30. Wagner, G., 2003, State of mangrove forests in Tanzania, manuscript submitted to the Tanzania Coastal Management Partnership, University of Dar es Salaam, Department of Marine Biology, Dar es Salaam, Tanzania.
31. Njana, R., 2001, Mangrove Management Project Central Zone Project Status Report, Kibiti, Rufiji, Tanzania, 19pp.

Index

I

J

K

Q

R

T

U

V

W

Z